Exploring the Universe

Contributors:

Aristotle
Isaac Asimov
Francis Bacon
Roger Bacon
Lincoln Barnett
Lloyd V. Berkner
Hermann Bondi
Sir William Bragg
P. W. Bridgman
Jacob Bronowski
Herbert Butterfield
Norman Campbell
John Christianson
Arthur C. Clarke
I. Bernard Cohen
James B. Conant
Edward Uhler Condon
René Descartes
Lee Edson
Albert Einstein
Galileo Galilei
George Gamow
A. R. Hall
Fred Hoyle
David Hume
Aldous Huxley
Thomas H. Huxley
Christian Huygens

Sir James Jeans
Arthur Koestler
Pierre Simon de Laplace
Stephen Leacock
Giacomo Leopardi
A. C. B. Lovell
Henry Margenau
Sir Isaac Newton
José Ortega y Gasset
Thomas O. Paine
Blaise Pascal
Henri Poincaré
Eugene Rabinowitch
Eric M. Rogers
Bertrand Russell
Giorgio de Santillana
Erwin Schrödinger
Harlow Shapley
Bernard Shaw
J. W. N. Sullivan
Walter Sullivan
Rudolph Thiel
Arnold J. Toynbee
Warren Weaver
Alfred North Whitehead
Jerome B. Wiesner
W. P. D. Wightman
Abraham Wolf

EXPLORING
THE UNIVERSE

SECOND EDITION

Prepared by AMERICAN FOUNDATION FOR
CONTINUING EDUCATION

Edited by LOUISE B. YOUNG

New York   OXFORD UNIVERSITY PRESS   1971

PREFACE

The years of work we at the American Foundation for Continuing Education have devoted to *Exploring the Universe* have deepened our concern about the level of citizen understanding of science. We are convinced that the low level of understanding of science among otherwise well-informed and thoughtful laymen is one of the most critical problems confronting adult education in the United States today.

It is an interesting fact, for instance, that most well-educated men today know less about Newton's three laws of motion than their counterparts did two centuries ago; yet these principles form the foundation of our new space age. In the eighteenth century most educated men knew a good deal about the science of their day. Newtonian mechanics was discussed by poets, artists, and philosophers, as well as by scientists. Good books were written for the laymen by great masters, courses and demonstrations were offered, and interest in science ran high even among the ladies of the court. But today, discouraged by the very size and difficulty of the fare with which they are presented, many intelligent citizens have little interest in science. After struggling through a required high-school science course, they are happy to close the book and put the whole subject out of their minds.

Our concern is a concern for democracy itself because the lack of understanding of science raises in dramatic form a number of unsettling questions. It poses the issue of the proper relationship between laymen and experts; it compels us to consider the adequacy of the decision-making process in a free society; and it challenges us to justify the basic right of the people to govern themselves. In the past this right has been made operative by the conviction that well-informed and thoughtful men are competent to do so: that the citizen can in fact understand issues, evaluate conflicting opinions, and make intelligent decisions on matters of public policy. In our time, when public policy has become increasingly intermeshed with science and hurtled forward by the accelerating rate of scientific discovery, unless the lay citizen develops an adequate understanding of the nature of science and of the limits of scientific expertise, he becomes less able to understand a growing number of the most significant public policy issues and less competent to make sound judgments about them.

The first edition of *Exploring the Universe* was the outgrowth of three

v

years of research and experimentation by the American Foundation for Continuing Education to develop study-discussion programs in science for adults. Work was started in 1959 on a grant from the Fund for Adult Education. In 1960 the National Science Foundation made a grant to the AFCE to support the research and testing of these readings as the first unit of a long-range program designed to meet the needs of adult education in science. Published in 1963, *Exploring the Universe* has been used in adult educational programs and in many colleges and universities throughout the country. It also provided the framework for a series of eleven television programs. Hosted by Dave Garroway and featuring well-known scientists, this series was widely shown on educational television stations.

The seven years that have elapsed since that time have brought many exciting new scientific developments in the areas covered by *Exploring the Universe* and a thorough revision was recently deemed necessary. In editing this revised version we have been struck by the wealth of new information that these few years have produced: the discovery of quasars and pulsars, the rejection of the continuous creation theory, the new information about the moon and the other planets, as well as the brilliant technological achievements in space and the thrilling events of the Apollo 11 landing. These areas are all covered by new selections. Readings that had become out-of-date have been omitted or re-edited in order to bring the information back up to the frontiers of scientific thinking.

Two other units have also been published in the long-range program entitled Perspectives in Science. *The Mystery of Matter* (Oxford University Press, 1965) deals with the nature of matter, both living and non-living, and the implications of the atomic age. *Evolution of Man* (Oxford University Press, 1970) considers the evolutionary process as it applies to man and the pressing modern problems of man's control of his environment, his reproduction, and evolution.

Each of the books in this series is designed to illuminate one of the important areas in which science has had a major impact on modern life. They are designed to assist the thoughtful adult to develop a deeper understanding of the nature, methods, and limitations of science as the basis for sound judgments on public policy. Although each book can stand alone as a text or reference book, they are used to best advantage as a series. The material in each unit has been chosen to complement the readings in the other volumes. The texts have been cross-referenced and every effort has been made to avoid any unnecessary repetition of material. Together these books present a broad survey of the most important and exciting problems in science today.

The readings in this first volume begin with selections that provide a solid basis of understanding of the methods and nature of science. They then lead on into the discoveries of the great pioneers, Galileo, Kepler, and Newton. They trace the development of the principles on which the space exploration has been built, give a glimpse of the mystery and majesty of the universe that man has begun to explore, and suggest the spirit and significance of the search itself.

In several important ways this book and the two subsequent volumes represent an unusual approach to the problem of appreciating and understanding science. First, by showing historically how scientific ideas develop rather than by simply explaining the present state of scientific knowledge, these selections assist the reader to gain insight into how science works, and under what conditions important advances are made, and why scientific theories change from time to time. Second, from the original writings of many different scientists, the reader learns, in the scientists' own words, the ways in which they pursue their visions and the attitudes they have toward science. It is our hope that by following the scientist into his realm the reader will learn, not only something scientific, but something of science and will sense the qualities that give science its place in today's world. Third, unlike the usual textbook or exposition of science, there are many questions asked in these readings that are not answered. In each part a number of different points of view is presented with no editorial decision made about which is the right point of view. The reader is never told what to think. The aim of these books is to stimulate thought about some of the basic issues raised and to examine the different arguments about these issues. Confronted by differences of views, the reader will be encouraged to think through his own ideas and to deepen his understanding of the part science plays in his world.

The readings in this volume are intended to be read over a period of eleven weeks and are divided into eleven separate parts. Generally, the first articles in each part are descriptive and expository of a scientific concept. The articles toward the end of a part turn to an exploration of the human response or the religious, philosophical, and social aspects of the science material. As the articles in each part are selected to complement one another and are arranged to bring into focus a fundamental aspect of science, they should be read in the order in which they appear. Many of the selections have been excerpted. Titles have sometimes been changed to indicate better the nature of the selection's function in this book. In some cases footnotes and illustrations have been omitted because in this context they do not contribute to the central issue; editorial notes and illustrations have been added where they were deemed useful. Cross references to illustrations are enclosed either by parentheses or by brackets. Those enclosed in parentheses refer to figures based on illustrational concepts of the author of the article; those enclosed in brackets refer to illustrations that are either added by the editor for clarification or that appear in other articles but serve as well to illustrate points in the article where the bracketed reference appears. We have included an index of subjects and authors and a glossary of scientific terms appearing in the readings. This glossary contains definitions and equations, as well as a few simple derivations for those who prefer a mathematical explanation. At the end of each group of readings, a list of appropriate readings will help the layman to continue his own education in this very important field of human endeavor. You will see that a few of the books are mentioned frequently throughout the reading lists. You may wish to purchase these books, which will serve as valuable supplementary material. The titles listed are just a sample of the many fine science books available for the layman, and more are being published each month. Consult your local librarian and bookstore for other suggestions of good books on these subjects.

As with other adult education programs which the AFCE has prepared and published, we do not represent or support any particular viewpoint with respect to the public policy questions or the philosophical, religious, or social issues contained in these readings. The AFCE is a not-for-profit educational organization devoted to the development of study-discussion programs in the liberal disciplines for adults concerned with their own continuing education.

We are grateful to the many scientists and educators throughout the country who assisted us in developing the idea of a study-discussion program in science. Whereas final responsibility for the selections in this volume rests with the editor, we should especially like to thank Dr. Joseph Kaplan, Professor of Physics, University of California at Los Angeles; Dr. Victor F. Weisskopf, Professor of Physics, Massachusetts Institute of Technology and Director of Cern Laboratory, Geneva, Switzerland; and Dr. Paul B. Weisz, Professor of Biology, Brown University.

                                                                                   L.B.Y.

Winnetka, Illinois
February 1971

CONTENTS

Contents

Part 7   WHY DO THEORIES CHANGE?
Newton's Theory of Gravity

Einstein's Theory of Relativity

Theory and Common Sense

Part 8   HOW WAS THE UNIVERSE CREATED?
The Galaxies

Theories of the Origin of the Universe

Part 9   IS THERE OTHER LIFE IN THE UNIVERSE?
A Reconsideration of Man's Place in the Universe

The Origin of the Solar System and Theories of Stellar Evolution

Life in the Universe

Exploring the Universe

1

What Is the Nature of Science?

Generations of writers have sought a simple definition of science that incorporates the range of qualities implicit in the term. These qualities vary from connotations of hard-mindedness like *precise, factual, measurable, logical, calculable* to connotations of scope and free expression like *intuitive, creative, elegant, imaginative, exploratory*. Here are the efforts of a few well-known scientists to express in words the nature of science.

What Is the Nature of Science?

The study of nature that is a part of man's intellectual life is a delight to all scientists. In their hands, science is not just a business of collecting facts or stating laws or directing experiments. It is above all an *art* of sensing the best choice of view or the most fruitful line of investigation for a growing understanding of nature.

Eric M. Rogers
Physics for the Inquiring Mind (1960)

The Creative Aspects of Science *

by Jacob Bronowski (1956)

What is the insight with which the scientist tries to see into nature? Can it indeed be called either imaginative or creative? To the literary man the question may seem merely silly. He has been taught that science is a large collection of facts; and if this is true, then the only seeing which scientists need do is, he supposes, seeing the facts. He pictures them, the colorless professionals of science, going off to work in the morning into the universe in a neutral, unexposed state. They then expose themselves like a photographic plate. And then in the darkroom or laboratory they develop the image, so that suddenly and startlingly, it appears, printed in capital letters, as a new formula for atomic energy.

Men who have read Balzac and Zola are not deceived by the claims of these writers that they do no more than record the facts. The readers of Christopher Isherwood do not take him literally when he writes: "I am a camera." Yet the same readers solemnly carry with them from their school days this foolish picture of the scientist fixing by some mechanical process the facts of nature. I have had, of all people, a historian tell me that science is a collection of facts, and his voice had not even the irony of one filing cabinet reproving another.

It seems impossible that this historian had ever studied the beginnings of a scientific discovery. The Scientific Revolution can be held to begin in the year 1543 when there was brought to Copernicus, perhaps on his deathbed, the first printed copy of the book he had written about a dozen years earlier. The thesis of this book is that the earth moves around the sun. When did Copernicus go out and record this fact with

* From Jacob Bronowski, *Science and Human Values.* Julian Messner, A Division of Simon & Schuster, Inc., New York. Copyright © 1956, 1965 by J. Bronowski. Reprinted by permission of Julian Messner.

3

his camera? What appearance in nature prompted his outrageous guess? And in what odd sense is this guess to be called a neutral record of fact?

Copernicus found that the orbits of the planets would look simpler if they were looked at from the sun and not from the earth. But he did not in the first place find this by routine calculation. His first step was a leap of imagination—to lift himself from the earth, and put himself wildly, speculatively into the sun. "The earth conceives from the sun," he wrote; and "the sun rules the family of stars." We catch in his mind an image, the gesture of the virile man standing in the sun, with arms outstretched, overlooking the planets. Perhaps Copernicus took the picture from the drawings of the youth with outstretched arms which the Renaissance teachers put into their books on the proportions of the body. Perhaps he knew Leonardo's drawing of his loved pupil Salai. I do not know. To me, the gesture of Copernicus, the shining youth looking outward from the sun, is still vivid in a drawing which William Blake in 1780 based on all these: the drawing which is usually called *Glad Day* (Figure 1-2).

Kepler's mind, we know, was filled with just such fanciful analogies; and we know what they were. Kepler wanted to relate the speeds of the planets to the musical intervals. He tried to fit the five regular solids into their orbits. None of these likenesses worked, and they have been forgotten; yet they have been and they remain the stepping stones of every creative mind. Kepler felt for his laws by way of metaphors, he searched mystically for likenesses with what he knew in every strange corner of nature. And when among these guesses he hit upon his laws, he did not think of their numbers as the balancing of a cosmic bank account, but as a revelation of the unity in all nature. To us, the analogies by which Kepler listened for the movement of the planets in the music of the spheres are farfetched; but are they more so than the wild leap by which Rutherford and Bohr found a model for the atom in, of all places, the planetary system? . . .

The scientist looks for order in the appearances of nature by exploring such likenesses. For order does not display itself of itself; if it can be said to be there at all, it is not there for the mere looking. There is no way of pointing a finger or a camera at it; order must be discovered and, in a deep sense, it must be created. What we see, as we see it, is mere disorder.

This point has been put trenchantly in a fable by Professor Karl Popper. Suppose that someone wished to give his whole life to science. Suppose that he therefore sat down, pencil in hand, and for the next twenty,

Figure 1-1. *"The Ancient of Days Striking the First Circle of the Earth,"* *by William Blake.* (The New York Public Library)

Figure 1-2. *"Glad Day," by William Blake.* (The New York Public Library)

thirty, forty years recorded in notebook after notebook everything that he could observe. He may be supposed to leave out nothing: today's humidity, the racing results, the level of cosmic radiation and the stock market prices and the look of Mars, all would be there.

He would have compiled the most careful record of nature that has ever been made; and, dying in the calm certainty of a life well spent, he would of course leave his notebooks to the Royal Society. Would the Royal Society thank him for the treasure of a lifetime of observation? It would not. It would refuse to open his notebooks at all, because it would know without looking that they contain only a jumble of disorderly and meaningless items. . . .

The progress of science is the discovery at each step of a new order which gives unity to what had long seemed unlike. Faraday did this when he closed the link between electricity and magnetism. Clerk Maxwell did it when he linked both with light. Einstein linked time with space, mass with energy, and the path of light past the sun with the flight of a bullet; and spent his dying years in trying to add to these likenesses another, which would find a single imaginative order between the equations of Clerk Maxwell and his own geometry of gravitation.

When Coleridge tried to define beauty, he returned always to one deep thought: beauty, he said, is "unity in variety." Science is nothing else than the search to discover unity in the wild variety of nature—or more exactly, in the variety of our experience. Poetry, painting, the arts are the same search, in Coleridge's phrase, for unity in variety. Each in its own way looks for likenesses under the variety of human experience. . . .

The creative act is alike in art and in science; but it cannot be identical in the two; there must be a difference as well as a likeness. For example, the artist in his creation surely has open to him a dimension of freedom which is closed to the scientist. I have insisted that the scientist does not merely record the facts, but he must conform to the facts. The sanction of truth is an exact boundary which encloses him, in a way in which it does not constrain the poet or the painter. . . .

Science is the creation of concepts and their exploration in the facts. It has no other test of the concept than its empirical truth to fact. Truth is the drive at the center of science; it must have the habit of truth, not as a dogma but as a process.

The discoveries of science, the works of art are exploration—more, are explosions, of a hidden likeness. The discoverer or artist presents in them two aspects of nature and fuses them into one. This is the act of creation, in which an original thought is born, and it is the same act in original science and original art.

J. Bronowski
Science and Human Values (1956, 1965).

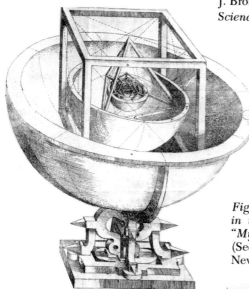

Figure 1-3. Planetary orbits embedded in the regular solids. From Kepler's "Mysterium Cosmographicum," 1596. (See "Kepler," in Glossary. Ed.) (The New York Public Library)

Figure 1-4. "Portrait of a Lady with an Ermine," by Leonardo da Vinci. "In the museum at Cracow there is a painting by Leonardo da Vinci called 'Portrait of a Lady with an Ermine': it shows a girl holding a stoat in her arms. . . . Leonardo has matched the stoat in the girl. In the skull under the long brow, in the lucid eyes, in the stately, brutal beautiful and stupid head of the girl, he has rediscovered the animal nature; and done so without malice, almost as a matter of fact" (J. Bronowski, "Science and Human Values," 1956). (Taurgo Slides)

Order must be discovered and, in a deep sense, it must be created.

J. Bronowski
Science and Human Values (1956, 1965).

Figure 1-5. *"Head" (about* 1935), *by Julio Gonzalez.* (Collection, The Museum of Modern Art, New York)

Two Definitions of Science [*]
by James B. Conant (1951)

There is the static view of science and the dynamic. The static places in the center of the stage the present interconnected set of principles, laws, and theories, together with the vast array of systematized information: in other words, science is a way of explaining the universe in which we live. The proponent of this view exclaims "How marvelous it is that our knowledge is so great!" If we consider science solely as a fabric of knowledge, the world would still have all the cultural and practical benefits of modern science, even if all the laboratories were closed tomorrow. This fabric would be incomplete, of course, but for those who are impressed with the significance of science as "explanations" it would be remarkably satisfactory. How long it would remain so, however, is a question.

The dynamic view in contrast to the static regards science as an activity; thus, the present state of knowledge is of importance chiefly as a basis for further operations. From this point of view science would disappear completely if all the laboratories were closed; the theories, principles, and laws embalmed in the texts would be dogmas; for if all the laboratories were closed, all further investigation stopped, there could be no re-examination of any proposition. I have purposely overdrawn the picture. No one except in a highly argumentative mood would defend either the extreme static or the extreme dynamic interpretation of the natural sciences. But the presentation of elementary science in school and college and the popular accounts almost necessarily take the dogmatic form, and therefore the American citizen is apt to be unconsciously drawn much too far in the one direction. The worker in the laboratory, however, clearly would not be there if he were not primarily concerned with science as an exploration. To understand him and his predecessors who have advanced the sciences since the sixteenth century one can hardly overemphasize the dynamic nature of science.

At all events, this is my own prejudice and I shall make no attempt to conceal it. My definition of science is, therefore, somewhat as follows: *science is an interconnected series of concepts and conceptual schemes that have developed as a result of experimentation and observation and are fruitful of further experimentation and observations.*[1] In this definition the emphasis is on the word "Fruitful." Science is a speculative enterprise. The validity of a new idea and the significance of a new experimental finding are to be measured by the consequences—consequences in terms of other ideas and other experiments. Thus conceived, science is not a quest for certainty; it is rather a quest which is successful only to the degree that it is continuous.

The theory that there is an ultimate truth, although very generally held by mankind, does not seem useful to science except in the sense of a horizon toward which we may proceed, rather than a point which may be reached.

Gilbert Lewis
The Anatomy of Science (1926)

---

1. Italics by editor.

Suggestions for Further Reading

° Bronowski, J.: *Science and Human Values,* Harper Torchbooks, Harper & Row, Publishers, New York, 1959. Chapters I and II, from which the quotations used in this part are taken, describe the creative and imaginative aspects of scientific discovery. The selection in Part 11 is taken from Chapter III. We strongly recommend reading the rest of this very fine little book.

Bulletin of the Atomic Scientists, February, 1959. A collection of very interesting essays on the relationship of science and art.

° Poincaré, Henri: *The Value of Science,* Dover Publications, Inc., New York; *Science and Method,* Dover Publications, Inc., New York. These two books are both beautifully written analyses of the philosophical significance and values of science.

"Innovation in Science," *Scientific American,* September, 1958. An issue devoted entirely to the creative aspects of scientific method.

° Waddington, C. H.: *The Scientific Attitude,* Penguin Books, Inc., Baltimore, 1948, chap. iv. The impact of science on art.

Whyte, Lancelot Law (ed.): *Aspects of Form,* Indiana University Press, Bloomington, Ind., 1951. An excellent collection of essays discussing the significance of form in art and science as well as in other fields.

° Paperback edition

2

Is There a Scientific Method?

Science as an activity is a continuing search whose nature is determined by its methods. Readings in this part examine the techniques by which scientists are constantly pushing back the frontiers of knowledge. Is there any special scientific method, or is science simply the synthesis of many different common-sense methods tailored to the problem at hand?

Is There a Scientific Method?

Introduction

I N PART 1 CONANT described two views of science, the static and the dynamic. The static view places small importance on the method by which knowledge is accumulated. Once discovered, the body of knowledge itself is the significant feature. The dynamic view, however, being essentially an activity, a way of increasing our understanding of the physical world, places the methods of science in the center of the picture. In fact, science becomes identified with its methods. Most scientists view their work in this dynamic sense. It is this aspect which lends it its exciting, imaginative, and creative character.

Scientists who are working at the frontiers of knowledge rarely have the time or inclination to step back and analyze the methods that they are using. If asked to explain their techniques, they would probably say that they follow no conscious method. It is just something they do by intuition, by feel, by "the seat of their pants." On the other hand, there have been a number of theorists and philosophers who have postulated a method which scientists do follow although perhaps unconsciously, a method which is unique to science and constitutes the key to its success. This view of scientific method was first formulated during the nineteenth century, an era in which knowledge in all areas of thought seemed very secure and settled, so that it is not surprising that the descriptions of scientific thinking that emerged from these studies were characteristically rigid and formalized. The scientific method was described typically as follows: "(1) a problem is recognized and an objective formulated; (2) all the relevant information is collected; (3) a working hypothesis is formulated; (4) deductions from the hypotheses are drawn; (5) the deductions are tested by actual trial; (6) depending on the outcome, the working hypothesis is accepted, modified, or discarded." [1] This process, if properly followed step by step in correct sequence with safeguards observed at every point of the way, would end in fruitful scientific results. It was almost a mechanical concept: feed in the proper ingredients at one end of the machine, turn the proper mental wheels, and scientific advances would emerge at the other end.

The modern view of science, as presented by Conant in these readings, is a much more flexible system. It recognizes that scientists, especially the most creative ones, do not always follow the method in its strict form. True, they use various features of it at different times in

1. See pp. 29–30. Ed.

their researches, but not necessarily in the prescribed sequence, often skipping steps or working backwards from hypotheses or intuitive hunches to the facts. In this sense, Conant says, there is not one scientific method but many scientific methods working together to produce "an interconnected series of concepts and conceptual schemes that have developed as a result of experimentation and observation and are fruitful of further experimentation and observations."

The cooperative nature of the scientific process is particularly important in understanding this more flexible interpretation of scientific method. By working together, the various scientists supplement one another, some contributing the careful experimental data, others using the data as the basis of a new theory, and still others recognizing ways in which the theory can be tested and used to predict new phenomena. Through scientific societies and publications the scientists cross-check one another, so that, taken together, the whole web of scientific activity does represent a method that has been enormously effective in pushing back the frontiers of knowledge.

In Part 1 the search for truth was presented as one of the principal aims of science. In these readings the question is raised about the nature of scientific truth. Some of the processes by which information is collected and conclusions drawn are examined. They are in many respects the same processes that are used to solve problems in everyday life, but they are performed with an extra degree of care and precision. The methods of deductive and inductive reasoning are analyzed from a logical standpoint, and the surprising fact is brought out that science cannot prove a theory or hypothesis to be definitely true. On the other hand, it can show that certain statements are incorrect; that is, they can be definitely disproven. This characteristic of scientific method, as we will see in later parts, is to a large extent responsible for the tentative, changing nature of scientific truth.

The examples of scientific methods presented in these readings are chosen from the seventeenth century because it was not until that time that the several different ways of studying nature came together to form modern science. As Conant says, "The 'new philosophy' or the 'experimental philosophy' was the result of the union of three main streams of thought and action. These may be designated as (1) speculative thinking (2) deductive reasoning (3) cut-and-try experimentation." Taken together, these three elements constitute the basic methods still used by scientists today.

Scientific Method

In questions of science the authority of a thousand is not worth the humble reasoning of a single individual.

Galileo Galilei (1564–1642)

We Are All Scientists [*]
by Thomas H. Huxley (1863)

Scientific investigation is not, as many people seem to suppose, some kind of modern black art. You might easily gather this impression from the manner in which many persons speak of scientific inquiry, or talk about inductive and deductive philosophy, or the principles of the "Baconian philosophy." I do protest that, of the vast number of cants in this world, there are none, to my mind, so contemptible as the pseudo-scientific cant which is talked about the "Baconian philosophy."

To hear people talk about the great Chancellor—and a very great man he certainly was,—you would think that it was he who had invented science, and that there was no such thing as sound reasoning before the time of Queen Elizabeth! Of course you say, that cannot possibly be true; you perceive, on a moment's reflection, that such an idea is absurdly wrong. . . .

The method of scientific investigation is nothing but the expression of the necessary mode of working of the human mind. It is simply the mode at which all phenomena are reasoned about, rendered precise and exact. There is no more difference, but there is just the same kind of difference, between the mental operations of a man of science and those of an ordinary person, as there is between the operations and methods of a baker or of a butcher weighing out his goods in common scales, and the operations of a chemist in performing a difficult and complex analysis by means of his balance and finely-graduated weights. It is not that the action of the scales in the one case, and the balance in the other, differ in the principles of their construction or manner of working; but the beam of one is set on an infinitely finer axis than

[*] From Thomas H. Huxley, *Darwiniana*, 1863. Appleton-Century-Crofts, Division of Meredith Publishing Company, New York.

the other, and of course turns by the addition of a much smaller weight.

You will understand this better, perhaps, if I give you some familiar example. You have all heard it repeated, I dare say, that men of science work by means of induction and deduction, and that by the help of these operations, they, in a sort of sense, wring from Nature certain other things, which are called natural laws, and causes, and that out of these, by some cunning skill of their own, they build up hypotheses and theories. And it is imagined by many, that the operations of the common mind can be by no means compared with these processes, and that they have to be acquired by a sort of special apprenticeship to the craft. To hear all these large words, you would think that the mind of a man of science must be constituted differently from that of his fellow men; but if you will not be frightened by terms, you will discover that you are quite wrong, and that all these terrible apparatus are being used by yourselves every day and every hour of your lives.

There is a well-known incident in one of Molière's plays, where the author makes the hero express unbounded delight on being told that he had been talking prose during the whole of his life. In the same way, I trust, that you will take comfort, and be delighted with yourselves, on the discovery that you have been acting on the principles of inductive and deductive philosophy during the same period. Probably there is not one who has not in the course of the day had occasion to set in motion a complex train of reasoning, of the very same kind, though differing of course in degree, as that which a scientific man goes through in tracing the causes of natural phenomena.

A very trivial circumstance will serve to exemplify this. Suppose you go into a fruiterer's shop, wanting an apple,—you take up one, and, on biting it, you find it is sour; you look at it, and see that it is hard and green. You take up another one, and that too is hard, green, and sour. The shopman offers you a third; but, before biting it, you examine it, and find that it is hard and green, and you immediately say that you will not have it, it must be sour, like those that you have already tried.

Nothing can be more simple than that, you think; but if you will take the trouble to analyse and trace out into its logical elements what has been done by the mind, you will be greatly surprised. In the first place, you have performed the operation of induction. You found that, in two experiences, hardness and greenness in apples went together with sourness. It was so in the first case, and it was confirmed by the second. True, it is a very small basis, but still it is enough to make an induction from; you generalise the facts, and you expect to find sourness in apples where you get hardness and greenness. You found upon that a general law, that all hard and green apples are sour; and that, so far as it goes, is a perfect induction. Well, having got your natural law in this way, when you are offered another apple which you find is hard and green, you say, "All hard and green apples are sour; this apple is hard and green, therefore this apple is sour." That train of reasoning is what logicians call a syllogism, and has all its various parts and terms—its major premise, its minor premise, and its conclusion. And, by the help of further reasoning, which, if drawn out, would have to be exhibited in two or three other syllogisms, you arrive at your final determination, "I will not have that apple." So that, you see, you have, in the first place, established a law by induction, and upon that you founded a deduction, and reasoned out the special conclusion of the particular case. Well now, suppose, having got your law, that at some time afterwards, you are discussing the qualities of apples with a friend: you will say to him "It is a very curious thing,—but I find that all hard and green apples are sour!" Your friend says to you, "But how do you know that?" You at once reply, "Oh, because I have tried them over and over again, and have always found them to be so." Well, if we were talking science instead of common sense, we should call that an experimental verification. And, if still opposed, you go further, and say, "I have heard from the people in Somersetshire and Devonshire, where a large number of apples are grown, that they have observed the same thing. It is also found to be the case in Normandy, and in North America. In short, I find it to be the universal experience of mankind wherever attention has been directed to the subject." Whereupon, your friend, unless he is a very unreasonable man, agrees with you, and is convinced that you are quite right in the conclusion you have drawn. He believes, although perhaps he does not know he believes it, that the more extensive verifications are,—that the more frequently experiments have been made, and results of the same kind arrived at,—that the more varied conditions under which the same results are attained, the more certain is the ultimate conclusion, and he disputes the question no further. He sees that the experiment has been tried under all sorts of conditions, as to time, place, and people, with the same result; and he says with you, therefore, that the law you have laid down must be a good one, and he must believe it.

In science we do the same thing;—the philosopher exercises precisely the same faculties, though in a much more delicate manner. In scientific inquiry it becomes a matter of duty to expose a supposed law to every possible kind of verification, and to take care, moreover, that this is done intentionally, and not left to a mere accident, as in the case of the apples. And in science, as in common life, our confidence in a law is in exact proportion to the absence of variation in the result of our experimental verifications. For instance, if you let go your grasp of an article you may have in your hand, it will immediately fall to the ground. That is a very common verification of one of the best established laws of nature—that of gravitation. The method by which men of science establish the existence of that law is exactly the same as that by which we have established the trivial proposition about the sourness of hard and green apples. But we believe it in such an extensive, thorough, and unhesitating manner because the universal experience of mankind verifies it, and we can verify it ourselves at any time; and that is the strongest possible foundation on which any natural law can rest.

So much, then, by way of proof that the method of establishing laws in science is exactly the same as that pursued in common life. Let us now turn to another matter (though really it is but another phase of the same question), and that is, the method by which, from the relations of certain phenomena, we prove that some stand in the position of causes towards the others.

I want to put the case clearly before you, and I will therefore show you what I mean by another familiar example. I will suppose that one of you, on coming down in the morning to the parlour of your house, finds that a tea-pot and some spoons which had been left in the room on the previous evening are gone,—the window is open, and you observe the mark of a dirty hand on the window-frame, and perhaps, in addition to that, you notice the impress of a hob-nailed shoe on the gravel outside [Figure 2-1]. All these phenomena have struck your attention instantly,

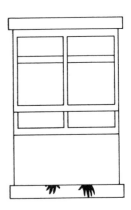

Figure 2-1. The mystery of the missing teapot and spoons. (Figure added by editor.)

and before two seconds have passed you say, "Oh, somebody has broken open the window, entered the room, and run off with the spoons and the tea-pot!" That speech is out of your mouth in a moment. And you will probably add, "I know there has; I am quite sure of it!" You mean to say exactly what you know; but in reality you are giving expression to what is, in all essential particulars, an hypothesis. You do not *know* it at all; it is nothing but an hypothesis rapidly framed in your own mind. And it is an hypothesis founded on a long train of inductions and deductions.

What are those inductions and deductions, and how have you got at this hypothesis? You have observed, in the first place, that the window is open; but by a train of reasoning involving many inductions and deductions, you have probably arrived long before at the general law— and a very good one it is—that windows do not open of themselves; and you therefore conclude that something has opened the window. A second general law that you have arrived at in the same way is, that tea-pots and spoons do not go out of a window spontaneously, and you are satisfied that, as they are not now where you left them, they have been removed. In the third place, you look at the marks on the window-sill, and the shoe-marks outside, and you say that in all previous experience the former kind of mark has never been produced by anything else but the hand of a human being; and the same experience shows that no other animal but man at present wears shoes with hob-nails in them such as would produce the marks in the gravel. I do not know, even if we could discover any of those "missing links" that are talked about, that they would help us to any other conclusion! At any rate the law which states our present experience is strong enough for my present purpose. You next reach the conclusion, that as these kinds of marks have not been left by any other animals than men, or are liable to be formed in any other way than by a man's hand and shoe, the marks in question have been formed by a man in that way. You have, further, a general law, founded on observation and experience, and that, too, is, I am sorry to say, a very universal and unimpeachable one,—that some men are thieves; and you assume at once from all these premises—and that is what constitutes your hypothesis—that the man who made the marks outside and on the window-sill, opened the window, got into the room, and stole your tea-pot and spoons. You have now arrived at a *vera causa;*—you have assumed a cause which, it is plain, is competent to produce all the phenomena you have observed. You can explain all these phenomena only by the hypothesis of a thief. But that is a hypothetical conclusion, of the justice of which you have no absolute proof at all; it is only rendered highly probable by a series of inductive and deductive reasonings.

I suppose your first action, assuming that you are a man of ordinary common sense, and that you have established this hypothesis to your own satisfaction, will very likely be to go for the police, and set them on the track of the burglar, with the view to the recovery of your property. But just as you are starting with this object, some person comes in, and on learning what you are about, says, "My good friend, you are going on a great deal too fast. How do you know that the man who really made the marks took the spoons? It might have been a monkey that took them, and the man may have merely looked in afterwards." You would probably reply, "Well, that is all very well, but you see it is contrary to all experience of the way tea-pots and spoons are abstracted; so that, at any rate, your hypothesis is less probable than mine." While you are talking the thing over in this way, another friend arrives. And he might say, "Oh, my dear sir, you are certainly going on a great deal too fast. You are most presumptuous. You admit that all these occurrences took place when you were fast asleep, at a time when you could not possibly have known anything about what was taking place. How do you know that the laws of Nature are not suspended during the night? It may be that there has been some kind of supernatural interference in this case." In point of fact, he declares that your hypothesis is one of which you cannot at all demonstrate the truth and that you are by no means sure that the laws of Nature are the same when you are asleep as when you are awake.

Well, now, you cannot at the moment answer that kind of reasoning. You feel that your worthy friend has you somewhat at a disadvantage. You will feel perfectly convinced in your own mind, however, that you are quite right, and you say to him, "My good friend, I can only be guided by the natural probabilities of the case, and if you will be kind enough to stand aside and permit me to pass, I will go and fetch the police." Well, we will suppose that your journey is successful, and that by good luck you meet with a policeman; that eventually the burglar is found with your property on his person, and the marks correspond to his hand and to his boots. Probably any jury would consider those facts a very good experimental verification of your hypothesis, touching the cause of the abnormal phenomena observed in your parlour, and would act accordingly.

Now, in this suppositious case, I have taken phenomena of a very common kind, in order that you might see what are the different steps in an ordinary process of reasoning, if you will only take the trouble to analyse it carefully. All the operations I have described, you will see, are involved in the mind of any man of sense in leading him to a conclusion as to the course he should take in order to make good a robbery and

punish the offender. I say that you are led, in that case, to your conclusion by exactly the same train of reasoning as that which a man of science pursues when he is endeavouring to discover the origin and laws of
the most occult phenomena. The process is, and always must be, the
same; and precisely the same mode of reasoning was employed by Newton and Laplace in their endeavours to discover and define the causes of
the movements of the heavenly bodies, as you, with your own common
sense, would employ to detect a burglar. The only difference is, that the
nature of the inquiry being more abstruse, every step has to be most
carefully watched, so that there may not be a single crack or flaw in
your hypothesis. A flaw or crack in many of the hypotheses of daily life
may be of little or no moment as affecting the general correctness of the
conclusions at which we may arrive; but, in a scientific inquiry, a fallacy, great or small, is always of importance, and is sure to be in the
long run constantly productive of mischievous, if not fatal results.

Do not allow yourselves to be misled by the common notion that an
hypothesis is untrustworthy simply because it is an hypothesis. It is
often urged, in respect to some scientific conclusion, that, after all, it is
only an hypothesis. But what more have we to guide us in nine-tenths of
the most important affairs of daily life than hypotheses, and often very
ill-based ones? So that in science, where the evidence of an hypothesis is
subjected to the most rigid examination, we may rightly pursue the same
course. You may have hypotheses and hypotheses. A man may say, if he
likes, that the moon is made of green cheese; that is an hypothesis. But
another man, who has devoted a great deal of time and attention to the
subject, and availed himself of the most powerful telescopes and the results of the observations of others, declares that in his opinion it is probably composed of materials very similar to those of which our own earth
is made up: and that is also only an hypothesis. But I need not tell you
that there is an enormous difference in the value of the two hypotheses.
That one which is based on sound scientific knowledge is sure to have a
corresponding value; and that which is a mere hasty random guess is
likely to have but little value. Every great step in our progress in discovering causes has been made in exactly the same way as that which I
have detailed to you. A person observing the occurrence of certain facts
and phenomena asks, naturally enough, what process, what kind of operation known to occur in Nature applied to the particular case, will un-
ravel and explain the mystery? Hence you have the scientific hypothesis;
and its value will be proportionate to the care and completeness with
which its basis has been tested and verified. It is in these matters as in
the commonest affairs of practical life: the guess of the fool will be folly,
while the guess of the wise man will contain wisdom. In all cases, you

see that the value of the result depends on the patience and faithfulness with which the investigator applies to his hypothesis every possible kind of verification.

Editor's Note on Scientific Reasoning and Terms

In describing scientific method, Thomas Huxley mentions *inductive* and *deductive* reasoning. Both of these methods of thinking contribute in different ways to the progress of science.

The first, which is the basis of all experimental science, proceeds from the particular to the general. Observations and experimental evidence are collected and systemized. Then a generalization is made based on these data. For example, after watching the moon's position for several years a scientist might "infer inductively" that the moon travels around the earth regularly, about 13 times a year and as long as there is no evidence to the contrary we assume that it will continue to do so. The more observations that are made which follow the rule, the more firmly established it becomes. We say that the *probability* of its being true increases but, rigorously speaking, the evidence of any number of cases is insufficient to prove it; while just one observation which goes against the rule is sufficient to disprove it. This one case may turn up at any time. However, for science a high degree of probability is good enough as a working basis. Scientists do trust and use the inductive method: Believing instinctively in the uniformity of nature, they assume that what has been found to be true up to the present time will continue to be true tomorrow.

Some rules or generalizations have been found to have such a broad application and such a high degree of probability that they have been dignified by the term, *Scientific Laws.* Many laws, especially those in physics, express a mathematical relationship between measurements of several physical quantities—as, for example, Newton's 2nd Law of Motion: force = mass $\times$ acceleration.

Deductive reasoning is the type most commonly used in mathematics and logic—the method is probably familiar to you from a study of Euclidean geometry in school. It starts with the assumption of general concepts—in science these assumptions are called postulates or hypotheses, in mathematics and logic they are called axioms or premises. From these generalizations logical conclusions are drawn. In science, the truth of the conclusion is tested by experiment or observation. For instance, Copernicus made the general assumption that all the planets revolve about the sun. Using this assumption, he calculated theoretically the positions of the other planets as seen from the earth at different times of the year. Finally, he compared these calculations with the observed positions of the planets and found a good correspondence between them. This verification of the original assumption gave it the standing of a scientific *theory*.

In logic, deductive reasoning takes the form of a *syllogism*, having two premises and a conclusion following logically from them. Take the classic example:

> All men are mortal. (major premise)
> Socrates is a man. (minor premise)
> Therefore Socrates is mortal. (conclusion)

Now one point that is interesting to note, especially in connection with scientific work, is that a false conclusion cannot follow logically from true premises. Therefore we know that *if* the premises are true the conclusion must also be true. However, a true conclusion can follow logically from false premises, as in the following example:

> All angels are mortal.
> Socrates is an angel.
> Therefore Socrates is mortal.

In general, if the conclusion of a deductive argument is shown to be true, it cannot be concluded that the premises are necessarily true. They may be, but they may also be false. This is an important point to remember in scientific procedures. As in the example of the Copernican Theory, an hypothesis (which is really just a good guess and whose truth is not yet established) is assumed. Then logical conclusions are drawn from it and these conclusions are tested by experiment. If they are found to check with observation, the hypothesis is accepted as a working basis or theory (although, as we now see, it has not been rigorously *proven* to be true). On the other hand, if the conclusion is shown to be false, the hypothesis is discarded. (We know that false conclusions cannot logically follow from true premises.) Again, we see that it is easier to prove something false than to prove it true.

In the last paragraph, we used the term *hypothesis* to mean a working hunch or single tentative guess, the truth of which is to be tested by experiment. This is the restricted scientific meaning of the word. However, you will find many writers using the term more loosely, and often interchangeably with the word *theory* which, strictly speaking, should refer to a broader, more definitely established conceptual scheme. The difference between theory and hypothesis is a matter of degree, both in breadth of application and extent of experimental verification. But since no sharp line can be drawn between the two terms, it is not surprising that there should be some difference of opinion about their use.

Sometimes it is necessary for a scientist to make a reasoned guess about physical quantities and relationships which have not been measured. For instance, suppose an astronomer is measuring the intensity of light from a star which he has observed to be increasing in brightness from night to night. He makes a measurement on Monday, Tuesday, and Wednesday and finds that each night there has been a constant increase in brightness. On Thursday night, there are clouds obscuring that part of the sky. On Friday he continues his observations and finds that there has been twice as great an increase in the 48-hour period since Wednesday night as there had been in the 24-hour periods previously measured. He might then *interpolate* and assume that half of this increase occurred between Wednesday night and Thursday night and half between Thursday night and Friday night, following the pattern he had found to be true earlier in the week of an approximately constant increase in each 24-hour period. *Interpolation* means making a reasoned guess within a known range. It is a fairly safe procedure if used with caution, although it is obvious that the assumption is not necessarily correct. In our example, the star could have stayed the same between Wednesday and Thursday and taken a sudden jump in brightness on Friday. Experience and scientific judgment enter into making a sound interpolation.

If the scientist makes a guess that goes beyond the known range, it is called *extrapolation*. For instance, if on the basis of the observations quoted, our astronomer concluded that on Saturday there would also be the same increase in brightness, he would be extrapolating. He might even go much farther and figure that in a month, at that same rate, there would be an increase of 30 times as much. As you can see, extrapolation requires much more experience and judgment than interpolation. In the example we have given, for instance, it is known that some stars increase in brightness for a certain length of time and then start decreasing. Extrapolation is risky but when used by experienced and imaginative scientists it can lead to some of the most fruitful speculations and theories.

Concerning the Alleged Scientific Method °

by James B. Conant (1951)

Those who favor the use of the word science to embrace all the activities of the learned world are inclined to believe in the existence of *a* sci-

° From James B. Conant, *Science and Common Sense.* Yale University Press, New Haven. Copyright © 1951 by Yale University Press. Reprinted by permission.

entific method. Indeed, a few go further and not only claim the existence of *a* method but believe in its applicability to a wide variety of practical affairs as well. For example, a distinguished American biologist declared not long ago that "Men and women effectively trained in science *and in the scientific method,* usually ask for evidence, almost automatically." He was referring not to scientific matters but to the vexing problems which confront us in everyday life—in factories, offices, and political gatherings.

One cannot help wondering where the author of such a categorical statement obtained his evidence. But this is perhaps making a debater's point. The significance of the statement is that it reflects a persistent belief in the correctness of the analysis of science presented by Pearson in *The Grammar of Science.* Throughout the volume Karl Pearson refers to science as the classification of facts, and in his summary of the first chapter he writes as follows: "The scientific method is marked by the following features: (a) careful and accurate classification of facts and observation of their correlation and sequence; (b) the discovery of scientific laws by aid of the creative imagination; (c) self-criticism and the final touchstone of equal validity for all normally constituted minds." With (b) and (c) one can have little quarrel since all condensed statements of this type are by necessity incomplete, but from (a) I dissent entirely. And it is the point of view expressed in this sentence that dominates Pearson's whole discussion. It seems to me, indeed, that one who had little or no direct experience with scientific investigations might be completely misled as to the nature of the methods of science by studying this famous book.

If science were as simple as this very readable account would have us believe, why did it take so long a period of fumbling before scientists were clear on some very familiar matters? Newton's famous work was complete by the close of the seventeenth century. The cultured gentlemen of France and England in the first decade of the eighteenth century talked in terms of a solar system almost identical with that taught in school today. The laws of motion and their application to mechanics were widely understood. This being the case it might be imagined that the common phenomenon of combustion would have been formulated in terms of comparable clarity once people put their minds on scientific problems. Yet it was not until the late 1770's that the role of oxygen in combustion was discovered. Another hotly debated problem, the spontaneous generation of life, was an open question as late as the 1870's. Darwin convinced himself and later the scientific world and later still the educated public of the correctness of the general idea of evolution because of his theory as to the mechanism by which evolution might have occurred. Today the basic idea of the evolutionary development of higher plants and animals stands almost without question, but Darwin's mechanism has been so greatly altered that we may say a modern theory has evolved. And we are no nearer a solution of the problem of how life originated on this planet than we were in Darwin's day.

The stumbling way in which even the ablest of the scientists in every generation have had to fight through thickets of erroneous observations, misleading generalizations, inadequate formulations, and unconscious prejudice is rarely appreciated by those who obtain their scientific knowledge from textbooks. It is largely neglected by those expounders of the alleged scientific method who are fascinated by the logical rather than the psychological aspects of experimental investigations. Science as I have defined the term[1] represents one segment of the much larger field of accumulative knowledge. The common characteristic of all the theoretical and practical investigations which fall within this framework —a sense of progress—gives no clue as to the *activities* of those who have advanced our knowledge. To attempt to formulate in one set of logical rules the way in which mathematicians, historians, archaeologists, philologists, biologists, and physical scientists have made progress would be to ignore all the vitality in these varied undertakings. Even within the narrow field of the development of "concepts and conceptual schemes from experiment" (experimental science) it is all too easy to be fascinated by oversimplified accounts of the methods used by the pioneers.

To be sure, it is relatively easy to deride any definition of scientific activity as being oversimplified, and it is relatively hard to find a better substitute. But on one point I believe almost all modern historians of the natural sciences would agree and be in opposition to Karl Pearson. There is no such thing as *the* scientific method. If there were, surely an examination of the history of physics, chemistry, and biology would reveal it. For as I have already pointed out, few would deny that it is the progress in physics, chemistry, and experimental biology which gives everyone confidence in the procedures of the scientist. Yet, a careful examination of these subjects fails to reveal any *one* method by means of which the masters in these fields broke new ground.

THE BIRTH OF EXPERIMENTAL SCIENCE IN THE SEVENTEENTH CENTURY

As I interpret the history of science, the sudden burst of activity in the seventeenth century which contemporaries called the "new philosophy" or the "experimental philosophy" was the result of the union of

1. See p. 9. Ed.

three streams of thought and action. These may be designated as (1) speculative thinking (2) deductive reasoning (3) cut-and-try or empirical experimentation. The first two are well illustrated by the writings of the learned men of the middle ages. The professor of law and theology as well as the teacher of mathematics and logic from the eleventh to the seventeenth century was concerned with a rational ordering of general ideas and the development of logical processes. In so doing, they extended to some degree the philosophical and mathematical ideas of the ancient Greeks and laid the foundations for the science of mechanics, the first of the branches of physics to take on modern dress.

A simple illustration of deductive reasoning is to be found in recalling one's experience in school with plane geometry. A set of postulates or axioms is given; then by logical processes of deduction many conclusions follow. Similarly, less formal and rigid general ideas—speculative ideas—can be manipulated by logical procedures which, however, frequently lack the rigor of mathematical reasoning. The discussion of general speculative ideas and the more detailed handling of mathematics, it should be noted, involve processes of thought which are believed to be sufficient unto themselves. No one feels impelled to appeal to observation in building a purely rational system of ideas.

The sudden burst of interest in the seventeenth century in the new experimental philosophy was to a considerable extent the result of a new curiosity on the part of thoughtful men. Practical matters ranging from agriculture and medicine to the art of pumping, the working of metals, and the ballistics of cannon balls began to attract the attention of learned professors or inquiring men of leisure. The early history of science is full of examples where the observation of a practical art by a scientist suggested a problem. *But the solution of a scientific problem is something quite different from the advances which had hitherto been made by the empirical experimentation of the agriculturist or the workman.* The new element which was introduced was the use of deductive reasoning. This was coupled with one or more generalizations often derived from speculative ideas of an earlier time. The focus of attention was shifted from an immediate task of improving a machine or a process to a curiosity about the phenomena in question. New ideas or concepts began to be as important as new inventions. The experimentation of the skilled artisans or the ingenious contriver of machines and processes became joined to the mathematical mode of reasoning of the learned profession. But it took many generations before deductive reasoning and experimentation could be successfully combined and applied to many areas of inquiry.

SPECULATIVE IDEAS, WORKING HYPOTHESES, AND CONCEPTUAL SCHEMES

Science we defined . . . as "an interconnected series of concepts and conceptual schemes that have developed as a result of experimentation and observation and are fruitful of further experimentation and observations." A conceptual scheme when first formulated may be considered *a working hypothesis on a grand scale.* From it one can deduce, however, *many* consequences, each of which can be the basis of chains of reasoning yielding deductions that can be tested by experiment. *If these tests confirm the deductions in a number of instances, evidence accumulates tending to confirm the working hypothesis on a grand scale, which soon becomes accepted as a new conceptual scheme.* Its subsequent life may be short or long, for from it new deductions are constantly being made which can be verified or not by careful experimentation.

In planning the experiments to test the deductions it became necessary, as science advanced, to make more precise and accurate many vague common sense ideas, notably those connected with measurement. Old ideas were clarified or new ones introduced. These are the new concepts which are often quite as important as the broad conceptual schemes. It is often much more difficult than at first sight appears to get a clear-cut yes or no answer to a simple experimental question. And the broader hypotheses must remain only speculative ideas until one can relate them to experiment.

An understanding of the relationship between broad speculative ideas and a wide conceptual scheme is of the utmost importance to an understanding of science. A good example is furnished by the history of the atomic theory. The notion that there were fundamental units—ultimate particles—of which matter was composed goes back to ancient times. But expressed merely in general terms this is a speculative idea and can hardly be considered an integral part of the fabric of science until it becomes the basis of a working hypothesis on a grand scale from which deductions capable of experimental test can be made. This particular speculative idea or working hypothesis on a grand scale became a new conceptual scheme only after Dalton had shown, about 1800, how fruitful it was in connection with the quantitative chemical experimentation that had been initiated by the chemical revolution. Here is an instance where we can see in some detail the origins of a working hypothesis, while in other instances we are uncertain how the idea came to the proponent's mind.

The great working hypotheses in the past have often originated in the minds of the pioneers as a result of mental processes which can best be described by such words as "inspired guess," "intuitive hunch," or "brilliant flash of imagination." Rarely if ever do they seem to have been the product of a careful examination of all the facts and a logical analysis of various ways of formulating a new principle. Pearson and other nineteenth-century writers about the methods of science largely overlooked this phenomenon. They were so impressed by the classification of facts and the drawing of generalizations from facts that they tended to regard this activity as all there was to science. Nowadays the pendulum has swung to the other extreme and some writers seem to concentrate attention on the development of new ideas and their manipulation, that is on theoretical science. Both points of view minimize the significance of the experiment. To my mind this distorts the history of science and, what is worse, confuses the layman who is interested in the scientific activity which is going on all about him. For these reasons and because of the author's own predilection, the present discussion of science and common sense emphasizes and re-emphasizes the interrelation of experiments and theory.

EXPERIMENTATION

The three elements in modern science already mentioned are: (1) speculative general ideas, (2) deductive reasoning, and (3) experimentation. We have discussed in a general way the manner in which new working hypotheses on a grand scale arise and how from them one may deduce certain consequences that can be tested by experiment. It has been implied that the art of experimentation long antedates the rise of science in the seventeenth century. If so, this is one way in which science and common sense are connected. A rather detailed analysis of experimentation in everyday life may serve a useful purpose at this point. For, as I hope to show, there is a continuous gradation between the simplest rational acts of an individual and the most refined scientific experiment. Not that the two extremes of this spectrum are identical. Quite the contrary: to understand science one must have an appreciation of just how common-sense trials differ from experiments in science.

Since the reader may well have been exposed at some time in his or her life to various statements about the alleged scientific method, it may be permissible to set up a few straw men and knock them down. I have read statements about the scientific method which describe fairly accurately the activity of an experimental scientist on many occasions (but not all). They run about as follows: (1) a problem is recognized and an objective formulated; (2) all the relevant information is collected (many a hidden pitfall lies in the word "relevant"!); (3) a working hypothesis is formulated; (4) deductions from the hypothesis are drawn; (5) the deductions are tested by actual trial; (6) depending on the outcome, the working hypothesis is accepted, modified, or discarded.

If this were all there was to science, one might say, in the words of a contemporary believer in *the* scientific method, that science as a method "consists of asking clear, answerable questions in order to direct one's observations which are made in a calm, unprejudiced manner, reported as accurately as possible and in such a way as to answer the questions that were asked to begin with; any assumptions that were held before the observations are now revised in the light of what has happened." But if one examines his own behavior whenever faced with a practical emergency (such as the failure of his car to start) he will recognize the preceding quotation [as] a description of what he himself has often done. Indeed, if one attempts to present the alleged scientific method in any such way to a group of discerning young people they may well come back with the statement that they have been scientists all their lives! . . .

TESTING DEDUCTIONS BY EXPERIMENT

Is there then no difference between science and common sense as far as method is concerned? Let us look at this question by considering in some detail first an everyday example of experimentation and then a scientific investigation. The type of activity by which the practical arts have developed over the ages is in essence a trial and error procedure. The same sort of activity is familiar in everyday life; we may call it experimentation. Let us take a very restricted and perhaps trivial example. Faced with a locked door and a bunch of keys lying on the floor, a curious man may wish to experiment with the purpose of opening the door in question. He tries first one key and then the other, each time essentially saying to himself, "If I place this key in the lock and turn it, then the result will either confirm or negate my hypothesis that this key fits the lock." This "if . . . then" type of statement is a recurring pattern in rational activity in everyday life. The hypothesis that is involved in such a specific trial is limited to the case at hand, the particular key in question. We may call it therefore a *limited working hypothesis.*

Let me turn now from an example of commonplace experimentation to a consideration of a scientific experiment. Let us examine the role of the limited working hypothesis in the testing of some scientific idea in the laboratory. For if you brought all the writers on scientific method with the most varied views together, I imagine that every one of them would agree that the testing of a deduction from a broad working hy-

pothesis (some would say a theory) was at least a part of science . . .
we are going to consider in some detail several instances of such procedures. Let us anticipate the story about atmospheric pressure only to the
extent of fixing our minds on one actual experiment. It makes little difference which one we choose, for we wish to center attention on the last
step, the actual experimental manipulation.

We will imagine that someone has related the broad hypothesis that
we live in a sea of air that exerts pressure to a particular experiment
with a particular piece of apparatus. Just before turning a certain stopcock, which we may assume to be the final step in the experiment, the
investigator may formulate his ideas in some such statement as "If all
my reasoning and plans are right, when I turn the stopcock, such and
such will happen." He turns the stopcock and makes the observation; he
can then say he has confirmed or failed to confirm his hypothesis. But,
strictly speaking, let it be noted, it is only a highly limited hypothesis
that has been tested by turning the stopcock and making the observation; this hypothesis may be thrown into some such form as "if I turn the
stopcock, then such and such will happen." The confirmation or not of
this extremely limited hypothesis is regarded as an experimental fact if
repetition yields the same result. The outcome of the experiment is
usually connected with the main question by a highly complex process
of thought and action which brings in many other concepts and conceptual schemes. . . . The point to be emphasized here is the existence of a
complicated chain of reasoning connecting the consequence deduced
from a broad hypothesis and the actual experimental manipulation; furthermore, . . . many assumptions, some conscious and some unconscious, are almost always involved in this chain of reasoning.

Now perhaps I may be permitted to jump from the scientist to the everyday experiments of a householder in his garage or a housewife in her
kitchen or an amateur with his radio set. If a car won't start, we certainly have a problem; we think over the various possibilities based on
our knowledge of automobiles in general and the particular car in question; we construct at least one working hypothesis (the gas tank is
empty!); we proceed to carry out a trial or test (an experiment) which
should prove the correctness of this particular working hypothesis; if we
are right then we believe we have found the trouble and proceed accordingly. (But how often have we been misled; perhaps there is more
than one trouble; perhaps the tank is empty and the battery run down
too!) Let us assume the simple hypothesis has led to a trial which consists of turning a particular switch or making a certain connection of
wires after various other manipulations have been performed. Then one
says to oneself, "Now at last if I turn the switch (or make the connec-

tion) the engine will turn over." The test is made and what is confirmed (or not) is a very limited working hypothesis hardly to be distinguished from the extremely limited working hypothesis of the scientists we have just been considering. Here science and common sense certainly seem to have come together. But notice carefully, they are joined only in the statement of the final operation. As we trace back the line of argument the differences become apparent. These are differences as to aims, as to auxiliary hypotheses and as to assumptions.

AIMS AND ASSUMPTIONS IN SCIENTIFIC EXPERIMENTATION

First, as regards aims, you want the car to start (or the radio set to work, to mention another example); you wish to reach a practical objective. The scientific experimenter on the other hand wants to test a deduction from a conceptual scheme (a theory), a very different matter. But we cannot leave the distinction between science and common sense resting on that point, important as it is. The conceptual scheme is not only being tested by the experimenter but it has given rise to the experiment. And that takes us back to our definition of science and our emphasis on the significance of the fruitfulness of a new conceptual scheme. The artisans who improved the practical arts over the centuries proceeded much as you or I do today when we are confronted with a practical problem. The aim of the artisan or the agriculturist was practical, the motivation was practical, though the objective was more general than just starting a given automobile. The workmen of the Middle Ages experimented and sometimes their results left a permanent residue because their contemporaries incorporated a new procedure into an evolving art. But the artisan rarely, if ever, bothered about the testing of the consequences of any general ideas. General ideas, logical thought, were the province of learned men. Drawing deductions from conceptual schemes was the type of activity known to the mathematicians and philosophers of the Middle Ages, not to the workmen. And with rare exceptions those who understood these recondite matters paid no attention to the workmen. . . . we shall consider examples of how the two activities —those of the logicians and the artisans—came together in the sixteenth and seventeenth centuries.

There is another important difference between the artisan and the scientist. The typical procedure of the artisan is very much like that of the housewife in the kitchen. Not only are the trials of new procedures for very practical and immediate ends, but the relevant information is largely unconnected with general ideas or theories. Until as late as the nineteenth century the practical man paid very little attention to the growing body of science. By and large the practical arts and science went their own ways during the seventeenth and eighteenth centuries. We may say that experimentation in the practical arts or in the kitchen is almost wholly empirical, meaning thereby to indicate the absence of a theoretical component. However, since the transition from common sense to science is gradual and continuous, one may well question if there is ever a total absence of a theoretical background. It can be argued that the concepts and conceptual schemes we take for granted in everyday life and share in common with our ancestors are not different in principle from the "well-established" ideas of science. This I believe to be true in the sense that infrared light is not different in principle from X rays; both forms of radiant energy are parts of a spectrum, but the two forms of light are certainly not interchangeable for most practical purposes. Quite the contrary. So too, common-sense ideas are distinct in many respects from the more abstract part of the scientific fabric. During the last two hundred years more and more of the material of science has become incorporated into our common-sense assumptions. But every age and cultural group has its own way of looking at the world. The pictures of the total universe which can be collected by anthropologists and students of cultural history, while having many assumptions in common, likewise show great divergences. Therefore, if the modern man in his garage "takes for granted" many things his grandfather would have believed impossible, this in no way invalidates the differentiation between common-sense ideas and scientific theories. (Though there is a wide, fuzzy, intermediary zone.)

THE DEGREE OF EMPIRICISM IN A SCIENCE OR A PRACTICAL ART

In analyzing the present relation of science to technology and medicine I have found it useful to use the term "degree of empiricism" to indicate the extent to which our knowledge can be expressed in terms of broad conceptual schemes. The same phrase is likewise useful, I am inclined to think, in connection with the history of both the sciences and the practical arts in the last three hundred years. The importance of the notion which lies back of the term, however, is that it may be of help to the layman who is confused about the relation of "pure" and "applied" science. In the last one hundred years science and technology have become so intertwined that even the practitioners in the field may be uncertain when they attempt to analyze the role of scientific theories. Yet anyone familiar with the physical sciences and modern industry would at once grant that there were wide differences in the extent to which scientists in different industries can apply scientific knowledge to the work at hand.

To illustrate what to me is a highly important point, let me contrast the business of making optical instruments with that of manufacturing rubber tires. The design of lenses and mirrors for telescopes, microscopes, and cameras is based on a theory of light which was developed 150 years ago and which can be expressed in simple mathematical terms. With the aid of this theory and a few measurements of the properties of the glasses used, it is possible to calculate with great accuracy the performance of optical equipment. Since theoretical knowledge is so complete in the field of optics, we may say that the degree of empiricism is low in this branch of physics. Because theory is so effective in the optical industry, the degree of empiricism is low. The manufacture of rubber tires is a totally different story. There is nothing comparable to the theory of light to provide a mathematical basis for calculation of what ingredients should be mixed with the rubber. The chemical change which is basic to the whole process is known as vulcanization but no one is very clear even today how to formulate it in theoretical terms. The action of the sulfur which was long thought to be an essential ingredient and of certain other chemicals known as "accelerators" is but little understood. The whole process has been built up by trial and error, by a procedure in which a vast number of experiments finally yielded knowledge in every way comparable to the knowledge of a first-rate chef. In this industry the degree of empiricism is high, and this reflects in turn the small extent to which the chemistry of rubber has been formulated in wide theoretical terms.

As in all comparative statements we must have some fixed points for standards. Therefore, without entering further into the philosophic analysis of common-sense knowledge (including the knowledge of an "art" such as cooking, or glass blowing, or metal making in the Middle Ages), we may take as an example of essentially empirical procedures those of the artisan before the advent of modern science and of the cook in a modern kitchen. Here one may arbitrarily say the degree of empiricism is practically 100 per cent. To find a case where the degree is so low that we can take it as zero on our scale, I call your attention to the work of the surveyor. The theoretical framework is here largely one branch of mathematics—geometry—and, except to a very slight extent in the building of the instruments and the efficiency of their handling, empirical procedures are conspicuous by their absence. Therefore, if the reader . . . whose acquaintance with science or technology is slight will envisage from time to time the surveyor with his transit and his measuring instruments on the one hand and the chef in the Grand Hotel on the other, he will have in mind a range of activities from zero to 100 as regards the degree of empiricism involved.

We shall return from time to time to the relation of scientific knowledge to the practical activities of the artisan, the agriculturist, and the medical man. We shall see that for an amazingly long time advances in science and progress in the practical arts ran parallel with few interconnecting ties. Thus if we take the birth of modern science as somewhere around 1600 (neglecting the long prenatal period which goes back to antiquity), one can say that it was two hundred years or more before the practical arts benefited much from science. Indeed, it would be my contention that it was not until the electrical and dyestuff industries were well started, about 1870, that science became of real significance to industry.

Let me conclude this discussion by pointing out that the degree of empiricism in a practical field today largely depends on the extent to which the corresponding scientific area can be formulated in terms of broad conceptual schemes. Therefore one may consider science as an attempt either to lower the degree of empiricism or to extend the range of theory. When scientific work is undertaken without regard for any practical application of the knowledge, it is convenient to speak of the activity as part of "pure" science. But there are certain overtones in the adjective which are unpleasant and seem to imply a hierarchy of values as between the scientists interested in theories and those interested in the practical arts. Therefore the phrase "basic science" is frequently employed. Almost all significant work of scientists today, I believe, comes under the heading of attempts to reduce the degree of empiricism; the distinction between one group and another is in the motivation. Those who are interested in the fabric of science as such are ready to follow any lead that gives promise of being fruitful in terms of extending theoretical knowledge. Others are primarily concerned with one of the ancient practical arts in modern dress; if it is some branch of industry, say metallurgy, they will be just as interested in widening the theoretical knowledge *in this field* as their colleagues in a university; they will be endeavoring to lower the degree of empiricism too, but in a limited area for a practical objective. The medical scientist is like the metallurgist except that his goal is not better metals but healthier people; both are working in applied science.

In the 1950's, therefore, we find a complex state of affairs. About three centuries ago the trial-and-error experimentation of the artisan was wedded to the deductive method of reasoning of the mathematician; the progeny of this union have returned after many generations to assist the "sooty empiric" in his labors. In so doing the applied scientist finds himself face to face with one of his distant ancestors, so to speak. For as he works in an industrial laboratory he will often find himself called on to carry out experiments for a practical purpose on nearly as empirical a basis as the artisan of the distant past. Particularly in those practical arts where the degree of empiricism is still high, men with the most advanced scientific training and using the latest equipment will often have to resort to wholly empirical procedures. On the one hand they will labor to reduce the degree of empiricism as best they can; on the other they must improve the art by using the knowledge and methods then at hand. In short, advances in science and progress in the practical arts today go hand in hand.

SCIENCE AS AN ORGANIZED ACTIVITY

The physical and biological sciences today consist of a closely interlocking set of principles and theories and a vast amount of classified information. They are also the product of a living organization. The theories, laws, data are to be found in libraries, herbariums, and museums; these are useful residues, deposits of the past, but essentially dead material. The activity we associate with the word "science" is the sum total of the potential findings of the workers in the laboratories; it is their plans, hopes, ambitions in the process of realization, week after week, year after year, that is the essence of modern science. Now this is a clear case, if there ever was one, of the whole being something quite different from the sum of the parts. For if the thousands of experimental scientists who are going to their laboratories tomorrow were not able to communicate with each other rapidly and easily there would be no modern science.

This is something far more complicated and far more important than the layman often realizes. Indeed, to a misunderstanding of the nature of science as an organized social activity today one can trace many a foolish statement and some practical blunders. It is amazing how much credence is given to self-deluded quacks or real charlatans, or how old wives' tales become accepted as scientific statements. The tendency to equate science with magic can be seen on almost every hand. This man tells you in all seriousness that he knows someone who knows how to stop an automobile engine at a distance of a mile by whistling the proper note; another believes that an untrained amateur has discovered a way of making real, natural rubber directly from garbage in one step; not to mention the whole host of bogus and pseudoscientific "cures" and remedies which still invade the field of medicine.

No one can be blamed for not detecting an absurd statement about alleged principles in physics, chemistry, or biology. All of us who have been engaged for many years in teaching these subjects or writing textbooks in these fields have had the experience of having to revise radically some of our basic statements to keep abreast of the advance of knowledge. But the first reaction of almost any scientist to a rumor about an alleged new step forward is one of incredulity. Maybe the step is a false one—he recalls at once the number of such instances in the field of his own experience. However, he is quite sure that in a short time the matter will be settled, unless some very revolutionary step is in the making. By the process of "publication" the new idea or new experimental finding will be made available to a host of scientists all over the world. Before long others will subject the report to critical examination if the matter is of real importance. No startling or even arresting alleged discovery will remain unnoticed.

It is not a question merely of repeating a series of calculations or checking experimental findings. There will be hundreds of implications for workers in the same or allied fields of inquiry and these will be followed up. If they fail to yield the expected result the original report may be dismissed by the scientific world as "another pipe dream." Eventually the author of the now discredited paper may discover his own error and publish a correction, or the whole matter may simply be allowed to drop. One could write a large volume on the erroneous experimental findings in physics, chemistry, and biochemistry which have found their way to print in the last hundred years; and another whole volume would be required to record the abortive ideas, self-contradictory theories and generalizations recorded in the same period.

The important fact which emerges from even a superficial study of the recent history of the experimental sciences (say, since 1850) is the existence of an organization of individuals in close communication with each other. Because of the existence of this organization new ideas spread rapidly, discoveries breed more discoveries, and erroneous observations or illogical notions are on the whole soon corrected. The deep significance of the existence of this organization is often completely missed by those who talk about science but have had no firsthand experience with it. Indeed, a failure to appreciate how scientists pool their information and by so doing start a process of cross fertilization in the realm of ideas has resulted in some strange proposals by politicians even in the United States. And in the Soviet Union we see what is apparently a deliberate attempt to alter drastically the nature of science as an organized social activity.

Science as a profession, we must remember, is a recent invention.

Some of the most important advances in the early development of physics and chemistry were made by amateurs. Indeed, in the examples given in the following chapters to illustrate the methods of science we shall encounter very few men who earned their living by scientific investigation or even by teaching science. As a rough generalization one may say that modern science started in the Italian universities in the sixteenth century; it flourished in the same environment until about the middle of the seventeenth century, and then the focus of activity is to be found in Paris and in London. The role of the universities does not become significant again until the nineteenth century. The seventeenth and eighteenth centuries were the period of the learned societies, particularly the Royal Society of London and the Academie des Sciences of Paris.

The significance of the Royal Society and the French academy of science lies in the fact that these *formal* institutions are the points of origin from which have grown the informal but highly complex organizations of modern science. The Royal Society was chartered by Charles II just after the Restoration but can trace its origin to the enthusiasm of a group of amateur scientists whom the accidents of party politics placed in Oxford in the Cromwellian period (1650–60). The French academy was the creation of Louis XIV in 1666 acting on the advice of Colbert. The intellectual parent of both is generally stated to be Francis Bacon, for in his unfinished fable, "The New Atlantis," published in 1626 shortly after his death, this ardent proponent of "the new experimental philosophy" (which he only partially understood and never practiced) described a House of Salomon which was a community of investigators and philosophers. An actual society founded in Rome in 1600, the Accademia dei Lincei, seems to have been a prototype of the organization Bacon envisaged. Galileo was a member of this academy which, as early as 1612, was stated to be a "gathering . . . which directs its labors diligently and seriously to studies as yet little cultivated." A generation later a group of Galileo's disciples in Florence established the Accademia del Cimento (1657) which flourished for ten years under the patronage of the two Medici brothers, Grand Duke Ferdinand II and Leopold, who had been pupils of Galileo. This Academy of Experiment was more nearly the forerunner of a twentieth-century research institute than of the eighteenth-century learned society, for the members were engaged in cooperative experimentation, of which more will be said in another chapter.

Both these Italian scientific societies were in the tradition of the literary clubs which flourished in the centers of Renaissance culture. The history of both the Royal Society and the French academy shows a certain ambiguity as between two goals; on the one hand, the members envisioned an active cooperating group of experimenters, and on the other merely a meeting place for reporting and discussing experimental findings, quaint observations and new ideas reported on a highly individualistic basis. The fact that the Royal Society never had any more support from the government than a royal blessing prevented any serious attempt to be much more than a focal point for discussion. The French monarchy, on the other hand, made grants to some of the members of the academy, and off and on during the course of a century thus supported scientists as a royal patron might support painters and men of letters.

Some of the expeditions organized and financed by these societies are of significance in the history of science. But the prime importance of the scientific societies lies in the fact that each undertook to publish a regular journal in which members and others could record their ideas and experimental results. Of the *Transactions* of the Royal Society whose publication was started in 1665 Huxley once said, "If all the books in the world except the Philosophical Transactions were destroyed, it is safe to say that the foundations of physical science would remain unshaken, and that the vast intellectual progress of the last two centuries would be largely, though incompletely, recorded." (One may doubt if those concerned today with the more descriptive sciences such as mineralogy and organic chemistry would quite agree with this remark made in the nineteenth century.)

Before the founding of the scientific societies and the establishment of regular quarterly or monthly magazines devoted to publishing the results of original work, news of scientific discoveries spread by letter. Then, from time to time, a scientific investigator would publish a small book giving his ideas and recounting his experiments. This practice of using separate books rather than communications to journals to announce scientific findings continued late into the nineteenth century. But the scientific journals became more and more important; books are now reserved for the purpose of summing up or amplifying the results published elsewhere. Today the scientific journals, not the scientific books, are the sources of information about what is going on among those who labor on the frontiers of knowledge.

To the uninitiated it would seem impossible for anyone to find his way through the mass of articles and reports which fill tens of thousands of pages every year. Actually the task, while time consuming, is far from hopeless for an investigator who makes a practice of following what the scientist calls "the current literature." In the first place, it must be remembered that by the beginning of the twentieth century the division and subdivision of the sciences had proceeded very far. The journals of the scientific societies still accepted communications covering a wide range of subjects, but specialized journals sprang up as early as the first half of the nineteenth century. Therefore, today one can keep abreast of the advance of science in a particular field of inquiry by reading a relatively small fraction of the monthly output of learned papers. In the second place, various elaborate schemes of indexing and abstracting have been devised. In some branches of science vast encyclopedias are published which summarize the results obtained under appropriate headings. A student bent on becoming an investigator soon learns how "to use the literature." Third, the custom has been established of referring to previous publications which bear on the subject under study. Finally, a great mass of irrelevant data and poorly digested reporting is eliminated by the activities of the editors of the journals. There is some danger in this process. More than one historic instance stands as evidence of originality taking such unusual forms as to cause a too conservative group of editors to regard the communication as erroneous or so fantastic as to warrant rejection. But there are so many different journals today that at the most there is likely to be but a short delay before publication. Informed criticism would say that far too much is published which might well be refused rather than that editors were too severe. . . .

Here I wish to emphasize that the methods of communicating scientific news on a reliable basis have evolved today to a point where there is little danger that any significant new discovery will remain unnoticed.

Examples of Scientific Method

The Torricellian Experiment [*]
by James B. Conant (1951)

For ages people have been aware of the fact that to drain a liquid out of a barrel there should be an opening near the bottom and one at the top; out of the lower orifice comes the wine, into the upper goes the air. Similarly, everyone knows that you can suck up liquid in a tube, close the top opening with your finger, and the liquid will not run out until the finger is removed; this is the principle of the pipette (Figure 2-2).

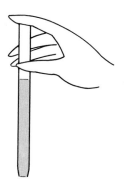

Figure 2-2. A pipette. If liquid is sucked up into a small tube and a finger pressed over the top end of the tube, the liquid will not run out.

Similar observations have been discussed since before the time of Aristotle. The standard explanation before the seventeenth century was couched in terms very similar to those which would be used by most of us today. "You have to have an opening in the top of the barrel to let the air in, otherwise the liquid won't run out." Of course, if pressed, the modern man (or woman) would have to admit to loose phraseology. He or she would probably say that the air coming in at the top was a consequence of the liquid running out. And sooner or later skillful and sympathetic cross-examination would elicit some information about the role

[*] From James B. Conant, *Science and Common Sense.*

of atmospheric pressure. Those who remembered their high-school physics might reply somewhat as follows: "The liquid in the pipette and in the barrel is prevented from running out by the pressure of the atmosphere; the purpose of taking your finger off the top of the pipette or opening a hole in the top of the barrel is to let the atmospheric pressure act on the top surface of the liquid. When that is done, the atmospheric pressure is essentially the same on the top and bottom of the liquid, which then runs out for the same reason a stone falls to earth."

The learned men of the Middle Ages, however, would have followed up the common-sense appraisal of the situation in very different terms. They would have taken quite literally the statement that the hole in the top of the barrel or the opening in the top of the pipette was necessary for the air to enter. For they would have started with the assumption that the universe was full. Therefore, they maintained, there could be a movement of the liquid out of the barrel only if some place were provided for the air that had to move to make room for the liquid; the place for this air was provided by the opening in which the air could come in.

The explanation in terms of a "full universe" satisfied generations of inquirers; it was part and parcel of the view of the world which comes from the writings of Aristotle as these were interpreted by the scholars of the Middle Ages. To do justice to this world view would require many chapters. One may caricature it by picking up one of the phrases used by the Aristotelians, namely that it was a universal principle that "nature abhors a vacuum." This principle was invoked to explain why the wine would not flow out of a single opening in a barrel; it was argued that if it did and nothing entered the barrel, a vacuum would result, and this was impossible. Which is, perhaps, only a picturesque way of stating that there must be an opening to let in the air, as many a person would say in an unreflective moment in the mid-twentieth century. The common-sense view of the world even today is more Aristotelian than we sometimes like to think.

The same principle that nature abhors a vacuum could be invoked to explain why one could suck up liquid into a tube or why a lift or suction pump would raise water. Consider the action of a modern version of this very ancient device (Figure 2-3), the old-fashioned pump, not long ago a feature of every kitchen. The operation of the handle raises the plunger and if the plunger is tight the water rises, is "sucked up" we often say. Why? Because otherwise there would be a vacuum, an Aristotelian would reply, and a vacuum is impossible. Again, some such explanation seems to have satisfied philosophers for generations. The first indication of a difficulty appears in the writings of Galileo. In his

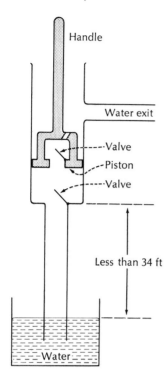

Handle

Water exit

----Valve

--Piston

----Valve

Less than 34 ft

Water

Figure 2-3. Cross-sectional diagram of a simple lift or suction pump. As the piston moves upward, water rises in the pipe.

Dialogues concerning Two New Sciences, published in 1638, he refers more or less in passing to the fact that a suction pump will not raise water more than a certain height. His explanation of this phenomenon need not detain us, for it was based on a poor analogy between the breaking of a column of water and the breaking of a long wire. But it is worth noting that this famous Florentine scientist missed the opportunity to make still another great contribution to the advance of science. Those who are inclined to think that the mere recognition of a problem by a great scientist will automatically produce the answer should ponder on this episode in the history of science.

A suction pump cannot raise water more than 34 feet. Galileo in discussing this phenomenon implies that it was first called to his attention by a workman. Suction pumps were no recent invention in Galileo's time. On the contrary, their history can be traced back for centuries; furthermore, practical men must have been long aware of the limitations on their use, for illustrations in Agricola's famous treatise on mining show suction pumps in tandem (Figure 2-4). It is strange that no discussion of the limit in the ability of a pump to raise water has been found

Figure 2-4. An illustration from Agricola's sixteenth-century book on mining, showing the use of pumps for removing water from mines. (The Bettmann Archive, Inc.)

earlier than that in Galileo's treatment. Perhaps any who thought about the subject attributed the failure to raise water more than a certain height to mechanical imperfections; indeed, the system of plungers and valves was very crude. But I am more inclined to think that the silence is merely a striking evidence of the vast gulf that separated artisans and learned men for centuries. There were those who operated mines, smelted ores, and operated pumps; they were improving the practical arts by endless cut-and-try experimentation. Then there were the professors and the learned men at princely courts who developed mathematics, deductive reasoning, and the embryo science of mechanics. Experimental science began when these two streams of human activity converged.

What Galileo missed, his pupil Torricelli found. In 1644, six years after the publication of Galileo's fruitless interpretation and two years after the master's death, Torricelli put down in writing some general but precise ideas about the atmosphere and atmospheric pressure. A broad hypothesis, we may call it, or a new conceptual scheme in the making. However you wish to characterize Torricelli's views, set forth in an exchange of letters with Cardinal Ricci, they represent a clean break with the Aristotelian notion of nature's abhorrence of a vacuum. In some way and at some date not recorded he saw that the limit of about 34 feet beyond which water would not rise in a suction pump might be a measure of the pressure of the atmosphere. He argued that if the earth were surrounded by a "sea of air" and if air had weight, there would be an air pressure on all objects submerged in this sea of air exactly as there is water pressure below the surface of the ocean.

Then follows a deduction from this hypothesis on a grand scale and the experimental confirmation of this deduction. If the column of water 34 feet high is sustained by the pressure of the atmosphere, the column of liquid nearly 14 times as heavy, namely liquid mercury, should be held up only 34/14 or 2$\frac{3}{7}$ feet. This is something that can be tested and it was. Again we are uncertain as to the time and place, but somewhere around 1640 and probably in Florence, Torricelli performed the classic experiment with which his name will be forever connected. . . .

This is one of the few revolutionary experiments which can be carried out with the simplest of equipment and understood with the minimum of sophistication in science or mathematics (Figure 2-5). Take a glass tube about a finger's width in diameter, 3 feet long, and sealed off at one end; fill it with liquid mercury (quicksilver); then placing the thumb or forefinger over the open end (carefully excluding any bubbles of air) invert the tube and submerge the end closed by your finger in an open vessel containing more mercury. Remove your finger and lo and behold, the mercury which had hitherto filled the tube now drops, leaving an empty space in the upper end of the tube. And the space is really empty, for you have just produced a vacuum as Torricelli did on the day when he first performed this experiment. Furthermore, you will have constructed a barometer. If you live close to sea level (and have carefully avoided air bubbles in filling the tube with the mercury) the

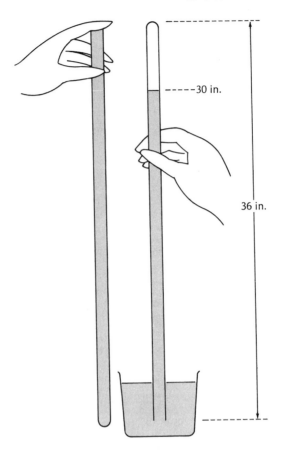

----30 in.

36 in.

Figure 2-5. In the Torricellian experiment, the tube is completely filled with mercury. The open end is sealed with a finger and the tube is inverted and placed with the "open" end in a dish of mercury. When the finger is removed, the mercury drops until the column is about 30 inches high.

height of the column will be approximately 30 inches. If you live at higher altitudes it will be less. Observed over a period of days, the level will fall and rise, as Torricelli soon discovered. This variation in atmospheric pressure is now a matter of daily news but it was more than a century before the relation between barometric changes and weather began to be noted.

At one stroke Galileo's disciple had invented a new instrument, tested the validity of one deduction from a working hypothesis on a grand scale, and produced a vacuum, which the Aristotelians maintained was impossible. The significance of the experiment for those studying the methods of science lies, however, in the fact that it constitutes a simple verification of *one* consequence of a very broad hypothesis or conceptual scheme. Historically we cannot be certain that the broad hypothesis preceded the experiment, for there is no record of how Torricelli came either to his idea or his experiment. Considering the recorded discussions of the phenomenon of the water pump by his great master, however, it seems extremely probable that the idea of atmospheric pressure preceded the Torricellian experiment.

One of the most baffling features of the advance of science is the unpredictable way in which revolutionary ideas have developed. Few if any pioneers have arrived at their important discoveries by a systematic process of logical thought. Rather, brilliant flashes of imagination or "hunches" have guided their steps—often at first fumbling steps. . . .

If we may assume the usual interpretation of Torricelli's mental processes to be correct, then one deduction from his new broad hypothesis was confirmed when he constructed the first barometer. A second important deduction was soon drawn by a French mathematician and tested by experiment. Here the pioneer was that extraordinary figure in the history of science and theology, Blaise Pascal. Hearing of the Torricellian experiment through the Parisian letter writer, Father Mersenne, he at once repeated the experiment in Rouen. He also set up a water barometer by arranging a series of tubes of water so that above a column of water 34 feet high there was a vacuum. (This, however, could be hardly considered an independent check of the Torricellian hypothesis.) The new consequence to which Pascal directed attention was this: if we live in a sea of air that exerts pressure, the situation is analogous to that which exists at the bottom of the ocean. Pascal and his contemporaries understood the phenomena connected with water pressure. The laws of hydrostatics had been formulated in the preceding century and were beautifully expounded in a volume by Pascal himself. The pressure below the surface of water in a reservoir, lake, or ocean depends on the depth. A submarine creature rising from the ocean floor toward the surface will be subjected to a regularly *diminishing* pressure. If we live in a sea of air, the same phenomenon should be observed in the atmosphere, Pascal argued. Torricelli had provided the necessary instrument to measure the pressure—his inverted tube with a column of mercury, the barometer.

Pascal, reasoning in some such fashion, arranged for his brother-in-law Perier to carry out a series of Torricellian experiments at different heights on a mountain in central France. He considered this test a crucial one. In a letter in 1647 he said that the experiment with the inverted tube of mercury gave one reason to believe that "it is not abhorrence of the vacuum that causes mercury to stand suspended, but really the pressure of the air which balances the weight of the mercury column." Nevertheless, he went on to say, the ancient principle of the abhorrence of a vacuum can also be invoked to explain the phenomenon. But he maintained that if on carrying out the Torricellian experiment at the top and at the foot of a mountain it should turn out that the height of the column were less in the first instance than in the second, then "it follows of necessity that the weight and pressure of the air are the sole cause of this suspension of the mercury, and not the abhorrence of a vacuum, for . . . one cannot well say that nature abhors a vacuum more at the foot of the mountain than at its summit."

Pascal's brother-in-law proved to be most obliging and in September, 1648, carried out his mission. The results were as expected. The height of the column in the Torricellian tube at the top of the mountain chosen (the Puy-de-Dome) was less than at the bottom by about three inches, and part way down the mountain the length of the column was a little more than at the top but definitely less than at the bottom. Perier further reported that the repetition of the experiment on the top gave the same result at five different points on the summit, some sheltered, some in the open, and one carried out when a cloud carrying rain drifted over the summit. An observer stationed at the bottom had been meanwhile watching one tube during the entire time; he had found that the level had remained unchanged (the atmospheric pressure during this period did not alter).

A second important deduction from the new hypothesis about a sea of air and atmospheric pressure had thus been verified.

The Story of the Great Experiment on the Weight of the Mass of Air *

by Blaise Pascal (1648)

It is no longer open to discussion that the air has weight. It is common knowledge that a balloon is heavier when inflated than when empty, which is proof enough. For if the air were light, the more the balloon was inflated the lighter the whole would be, since there would

* From Blaise Pascal, *The Physical Treatise of Pascal,* translated by I. H. B. and A. G. H. Spiers. Columbia University Press, New York, 1937. Reprinted by permission.

be more air in it. But since, on the contrary, when more air is put in, the whole becomes heavier, it follows that each part has a weight of its own, and consequently that the air has weight.

Whoever wishes for more elaborate proofs can find them in the writings of those who have devoted special treatises to the subject.

If it be objected that air is light when pure, but that the air that surrounds us is not pure, being mixed with vapor and impurities which alone give it weight, my answer is brief: I am not acquainted with "pure" air, and believe that it might be very difficult to find it. But throughout this treatise I am referring solely to the air such as we breathe, regardless of its component elements. Whether it be compound or simple, that is the body which I call the air, and which I declare to have weight. This cannot be denied, and I require nothing more for my further proof. . . .

If there were collected a great bulk of wool, say twenty or thirty fathoms high, this mass would be compressed by its own weight; the bottom layers would be far more compressed than the middle or top layers, because they are pressed by a greater quantity of wool. Similarly the mass of the air, which is a compressible and heavy body like wool, is compressed by its own weight, and the air at the bottom, in the lowlands, is far more compressed than the higher layers on the mountain tops, because it bears a greater load of air.

In the case of that bulk of wool, if a handful of it were taken from the bottom layer, compressed as it is, and lifted, in the same state of compression, to the middle of the mass, it would expand of its own accord; for it would then be nearer the top and subjected there to the pressure of a smaller quantity of wool. Similarly if a body of air, as found here below in its natural state of compression, were by some device transferred to a mountaintop, it would necessarily expand and come to the condition of the air around it on the mountain; for then it would bear a lesser weight of air than it did below. Hence if a balloon, only half inflated—not fully so, as they generally are—were carried up a mountain, it would necessarily be more inflated at the mountaintop, and would expand in the degree to which it was less burdened. The difference will be visible, provided the quantity of air along the mountain slope, from the pressure of which it is now relieved, has a weight great enough to cause a sensible effect.

THE STORY OF THE GREAT EXPERIMENT ON THE EQUILIBRIUM OF FLUIDS

At the time when I published my pamphlet entitled "New Experiments touching the Vacuum, et cetera," I used the phrase "Abhorrence of the Vacuum" because it was universally accepted and because I had not then any convincing evidences against it; but nevertheless sensed certain difficulties which made me doubt the truth of that conception. To clarify these doubts, I conceived at that very time the experiment here described, which I hoped would yield definite knowledge as a ground for my opinion. I have called it the Great Experiment on the Equilibrium of Fluids, because it is the most conclusive of all that can be made on this subject, inasmuch as it shows the equilibrium of air and quicksilver, which are, respectively, the lightest and the heaviest of all the fluids known in nature.

But since it was impossible to carry out this experiment here, in the city of Paris, and because there are very few places in France that are suitable for this purpose and the town of Clermont in Auvergne is one of the most convenient of these, I requested my brother-in-law, Monsieur Perier, counselor in the Court of Aids in Auvergne, to be so kind as to conduct it there. . . .

COPY OF THE LETTER SENT BY MONSIEUR PERIER
TO MONSIEUR PASCAL THE YOUNGER, SEPTEMBER 22, 1648

Monsieur,

At last I have carried out the experiment you have so long wished for. I would have given you this satisfaction before now, but have been prevented both by the duties I have had to perform in Bourbonnais, and by the fact that ever since my return the Puy-de-Dome, where the experiment is to be made, has been so wrapped in snow and fog that even in this season, which here is the finest of the year, there was hardly a day when one could see its summit, which is usually in the clouds and sometimes above them even while the weather is clear in the plains. I was unable to adjust my own convenience to a favorable state of the weather before the nineteenth of this month. But my good fortune in performing the experiment on that day has amply repaid me for the slight vexation caused by so many unavoidable delays.

I send you herewith a complete and faithful account of it, in which you will find evidence of the painstaking care I bestowed upon the undertaking, which I thought it proper to carry out in the presence of a few men who are as learned as they are irreproachably honest, so that the sincerity of their testimony should leave no doubt as to the certainty of the experiment.

The weather on Saturday last, the nineteenth of this month, was very unsettled. At about five o'clock in the morning, however, it seemed sufficiently clear; and since the summit of the Puy-de-Dome was then visible, I decided to go there to make the attempt. To that end I notified several people of standing in this town of Clermont, who had asked me to let them know when I would make the ascent. Of this company some were clerics, others laymen. Among the clerics was the Very Rev. Father Bannier, one of the Minim Fathers of this city, who has on several occasions been "corrector" (that is, father superior), and Monsieur Mosnier, canon of the Cathedral Church of this city; among the laymen were Messieurs la Ville and Begon, councilors to the Court of Aids, and Monsieur la Porte, a doctor of medicine, practicing here. All these men are very able, not only in the practice of their professions, but also in every field of intellectual interest. It was a delight to have them with me on this fine work.

On that day, therefore, at eight o'clock in the morning, we started off all together for the garden of the Minim Fathers, which is almost the lowest spot in the town, and there began the experiment in this manner.

First, I poured into a vessel six pounds of quicksilver which I had rectified during the three days preceding; and having taken glass tubes of the same size, each four feet long and hermetically sealed at one end but open at the other, I placed them in the same vessel and carried out with each of them the usual vacuum experiment [Figure 2–6]. Then, having set them up side by side without lifting them out of the vessel, I found that the quicksilver left in each of them stood at the same level, which was twenty-six inches and three and a half lines above the surface of the quicksilver in the vessel. I repeated this experiment twice at this same spot, in the same tubes, with the same quicksilver, and in the same vessel; and found in each case that the quicksilver in the two tubes stood at the same horizontal level, and at the same height as in the first trial.

That done, I fixed one of the tubes permanently in its vessel for continuous experiment. I marked on the glass the height of the quicksilver, and leaving that tube where it stood, I requested the Rev. Father Chastin, one of the brothers of the house, a man as pious as he is capable, and one who reasons very well upon these matters, to be so good as to observe from time to time all day any changes that might occur. With the other tube and a portion of the same quicksilver, I then proceeded

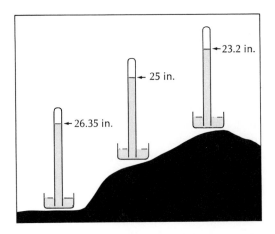

Figure 2-6. *Diagram of Monsieur Perier's experimental findings.* (Figure added by editor.)

with all these gentlemen to the top of the Puy-de-Dome, some five hundred fathoms above the convent. There, after I had made the same experiments in the same way that I had made them at the Minims', we found that there remained in the tube a height of only twenty-three inches and two lines of quicksilver; whereas in the same tube, at the Minims', we had found a height of twenty-six inches and three and a half lines. Thus between the heights of the quicksilver in the two experiments there proved to be a difference of three inches one line and a half. We were so carried away with wonder and delight, and our surprise was so great, that we wished, for our own satisfaction, to repeat the experiment. So I carried it out with the greatest care five times more at different points on the summit of the mountain, once in the shelter of the little chapel that stands there, once in the open, once shielded from the wind, once in the wind, once in fine weather, once in the rain and fog which visited us occasionally. Each time I most carefully rid the tube of air; and in all these experiments we invariably found the same height of quicksilver. This was twenty-three inches, and two lines, which yields the same discrepancy of three inches, one line and a half in comparison with the twenty-six inches, three lines and a half which had been found at the Minims'. This satisfied us fully.

Later, on the way down at a spot called Lafon de l'Arbre, far above the Minims' but much farther below the top of the mountain, I repeated the same experiment, still with the same tube, the same quicksilver, and the same vessel, and there found that the height of the quicksilver left in the tube was twenty-five inches. I repeated it a second time at the same spot; and Monsieur Mosnier, one of those previously mentioned, having the curiosity to perform it himself, then did so again, at the same spot. All these experiments yielded the same height of twenty-five inches, which is one inch, three lines and a half less than that which we had found at the Minims', and one inch and ten lines more than we had just found at the top of the Puy-de-Dome. It increased our satisfaction not a little to observe in this way that the height of the quicksilver diminished with the altitude of the site.

On my return to the Minims' I found that the [quicksilver in the] vessel I had left there in continuous operation was at the same height at which I had left it, that is, at twenty-six inches, three lines and a half; and the Rev. Father Chastin, who had remained there as observer, reported to us that no change had occurred during the whole day, although the weather had been very unsettled, now clear and still, now rainy, now very foggy, and now windy.

Here I repeated the experiment with the tube I had carried to the Puy-de-Dome, but in the vessel in which the tube used for the continuous experiment was standing. I found that the quicksilver was at the same level in both tubes and exactly at the height of twenty-six inches, three lines and a half, at which it had stood that morning in this same tube, and as it had stood all day in the tube used for the continuous experiment.

I repeated it again a last time, not only in the same tube I had used on the Puy-de-Dome, but also with the same quicksilver and in the same vessel that I had carried up the mountain; and again I found the quicksilver at the same height of twenty-six inches, three lines and a half which I had observed in the morning, and thus finally verified the certainty of our results. . . .

If you find any obscurities in this recital I shall be able in a few days to clear them up in conversation with you, since I am about to take a little trip to Paris, when I shall assure you that I am, Monsieur,

Your very humble and very affectionate servant,

Perier

This narrative cleared up all my difficulties and, I am free to say, afforded me great satisfaction. Seeing that a difference of twenty fathoms of altitude made a difference of two lines in the height of the quicksilver, and six or seven fathoms one of about half a line, facts which it was easy for me to verify in this city, I made the usual vacuum experiment on the top and at the base of the tower of St. Jacques de la Boucherie, which is some twenty-four or twenty-five fathoms high, and found a difference of more than two lines in the height of the quicksilver. I then repeated it in a private house ninety steps high, and found a clearly perceptible difference of half a line. These results are in perfect agreement with those given in M. Perier's narrative.

Any who care to do so, may, for themselves, confirm them at their pleasure.

CONSEQUENCES

From this experiment many inferences may be drawn, such as:

A method of ascertaining whether two places are at the same altitude, that is to say, equally distant from the center of the earth, or which of the two is the higher, however far apart they may be, even at antipodes—which would be impossible by any other means.

The unreliability of the thermometer in marking degrees of heat (which is not commonly recognized), as is shown by the fact that its liquid sometimes rises when the heat increases and sometimes, on the contrary, falls when the heat decreases, even though the thermometer be kept always in the same location.

Proof of the inequality of the pressure of the air, which, at the same degree of heat, is always greater in the lowest places.

On Pascal, Perier, von Guericke and Boyle °

by James B. Conant (1951)

In Pascal's mind, at least, the results were conclusive. Indeed, we might be tempted to say that after the Puy-de-Dome experiment, Torricelli's new idea had become a new conceptual scheme! Yet with such a conclusion as well as with Pascal's confidence, one may quarrel. The history of science since his time has shown the dangers of regarding the experimental verification of any *one* deduction as conclusive evidence as to the validity of a hypothesis.

It will be worth while to examine in some detail the experiment planned by Pascal and performed by Perier. First of all, it is worth noting that the new conceptual scheme could not be directly tested by experiment, though one can easily fall into the habit of speaking as though this were possible. In general, few if any hypotheses on a grand scale or conceptual schemes can be directly tested.

° From James B. Conant, *Science and Common Sense*.

A whole chain of reasoning connects a conceptual scheme with the experimental test. This may seem a trivial point, but it is not. More than one false step has resulted at this point; hidden assumptions as prickly as thorns abound; these may puncture a line of reasoning. What was observed in an experiment may not be actually related to the conceptual scheme in the way the experimenter had believed. . . . experimental errors can lead an experimenter astray. If we were to attempt to expound the difficult subject of twentieth-century physics, we should encounter the issue in a somewhat different form. We should then find that what appeared to be obvious common-sense assumptions involved in the line of reasoning connecting experiments with conceptual schemes had to be revised when either very high speeds or very small particles were involved.

Because of their significance to an understanding of the logic of experimentation, let us examine Perier's procedures with some care. Pascal's deduction from the new conceptual scheme may be stated somewhat as follows: "If the earth is surrounded by a sea of air and if air has weight, the pressure of the air will be less on the top of a mountain than at the foot." To translate this deduction into a specific experiment requires a long line of argument. If the height of the Torricellian column is a measure of the atmospheric pressure, then on performing the experiment on the top of the Puy-de-Dome and at the foot, the observed height will be less in the first instance than in the second, provided no other factors have influenced the pressure in the meantime or the height of the mercury column.

The "if" and the "provided" in this argument are of extreme importance. To guard against one obvious source of error, namely variations in the atmospheric pressure, Perier, as we have seen, had someone keep watch on a tube set up at the foot of the mountain during the entire time. To assure himself that in each case the height of the column was a true measure of the pressure, he says he carefully excluded all bubbles of air (for even a small one will on ascending into the vacuum cause the mercury to fall perceptibly). He does not state how he assured himself that he had been successful in his manipulations, and the accuracy with which his repeated observations agree with each other cannot but cause a skeptic to raise his eyebrows. Indeed, I am inclined to think he allowed his enthusiasm to affect the validity of his record. But there is no need to go into this interesting historical point; suffice it to say that standards of experimentation and recording had not been established by custom in 1648.

Perier performed certain manipulations at certain spots according to a careful set of rules. What he actually observed each time was a distance between two mercury levels; he probably used a wooden measuring stick graduated in inches and in lines (a twelfth of an inch). At the time of performing the experiment Perier was reasoning very much as the artisan or the housewife who makes a test of a new approach to a practical problem, for in effect he said, "If I perform Torricelli's experiment on this spot and there have been no errors in my manipulation and no unknown factors affecting the height of the mercury (a hypothesis limited to the case at hand), I shall observe a height of a mercury column less than is now being observed at the foot of the mountain." For all Pascal or Perier knew, a mercury column might be appreciably shorter at the top of a mountain than below for a variety of reasons: the relative densities of mercury and air might change (the ideas about gravity were then in process of formulation); the measuring stick might alter in length on being moved upward several thousand feet. Perier himself recognized that an open space and an enclosed building might make a difference, also a passing cloud. By repeating the experiment in and out of a chapel, when the air was clear and rainy, he tested the effect of certain *variables*, but the results he says were always the same.

The fact that so many variables exist in connection with testing a deduction by means of an actual experiment is the significant point. In this case, subsequent investigation has so far failed to reveal any variables that invalidate Perier's *qualitative* check on Pascal's deduction that the mercury column in a Torricellian tube is shorter on the top of a mountain than at the bottom.

YOUNG MEN AND AMATEURS: A DIGRESSION

At this point I propose to interrupt the story of seventeenth-century pneumatics to examine briefly an example of a recurring phenomenon in the history of science. I refer to the new waves of interest followed by important work that seem to sweep over the learned world from time to time. A new idea or new discoveries of new instruments open up a new area of inquiry: everyone rushes in and science advances in that field with astonishing speed, then progress lags and there may be a long period of marking time. The phenomenon to no small degree is connected, I believe, with the proverbial desire of young men to differ with their elders and to seek fields for new adventure. More than once the elderly scientists have said in essence, "Ah, that is a young man's game." The study of atmospheric pressure in the mid-seventeenth century was exactly that, for Torricelli was thirty-six when he wrote his famous letter to Cardinal Ricci; Pascal was twenty-four when he outlined the Puy-de-

Dome experiments; Boyle, whose work we are about to consider, was thirty-two when he started his line of investigation in this field.

These men, it must be remembered, were strictly amateurs; not for many years to come was experimental science to find a home in the universities; research laboratories and institutes were a full two centuries away. . . .

THE INVENTION OF THE VACUUM PUMP

One more amateur must be mentioned: the inventor of the vacuum pump, Otto von Guericke. He was a man of affairs, the mayor of Magdeburg who had played an active part in the Thirty Years' War and whose own city was sacked in 1631. His interest in the new experimental philosophy was probably not unconnected with the necessity of his being concerned with matters of military engineering. The story of how he developed his ideas about the atmosphere is obscure; he may have come quite independently to the same conclusions as did Torricelli. He certainly constructed a water barometer and built the first machine for pumping air out of a container. In retrospect his invention is an obvious adaptation of the suction part of a lift water pump. (The retrospect view of all discoveries and inventions must be used with care, lest one fall into the danger of being a "Monday morning quarterback.") Instead of sucking with a piston and cylinder on a column of water, as men had done for centuries in using suction pumps for raising water, von Guericke tried to suck water out of a full wooden cask using a brass pump. There were several models of this invention, and the usual partial success and failure of the pioneer attended his efforts. He attained his results only when he started to pump air as well as water from an enclosed container and finally ended by pumping out air alone. He also found that a spherical metal receptacle was necessary to stand the resulting atmospheric pressure. By 1654 he was able to carry out, before the Imperial Diet assembled at Ratisbon, the famous demonstration of the Magdeburg hemispheres (Figure 2-7). Two hollow bronze hemispheres were fitted carefully edge to edge and the air contained in the sphere thus formed removed by a pump. After evacuation the external atmospheric pressure held the two halves together so firmly that a team of eight horses could not pull them apart. Once air was admitted through a stopcock the hemispheres fell asunder.

This striking demonstration of von Guericke's may be regarded as the verification of another deduction from Torricelli's conceptual scheme. But for our purposes the use of von Guericke's pump by Robert Boyle is more instructive.

Figure 2-7. Otto von Guericke's Magdeburg hemispheres. (The Bettmann Archive, Inc.)

ROBERT BOYLE'S EXPERIMENTS

Boyle heard of the new pump through a book published in 1657 by a Jesuit professor at Würzburg (the communication of news of scientific discoveries was still on a most casual basis). Learning of this recent method of producing a vacuum, Boyle saw the possibility of testing still another deduction from the Torricellian conceptual scheme. His combination of logic and imagination represents a pattern repeated by many a successful investigator in the last three hundred years. More than one significant advance in science has come about because someone had the imagination to see that a new instrument made possible the testing of an important point. What Boyle proposed to do was to perform the equivalent of the Puy-de-Dome experiments in the laboratory. He had modified von Guericke's pump by providing an arrangement by which he could introduce the lower part of the Torricellian barometer into the vessel to be evacuated (Figures 2-8 to 2-10). Then, as he worked the pump and withdrew the air from above the mercury reservoir, the mercury column fell. . . . He was able to cause the column to fall almost but not quite to the level of the mercury in the reservoir. In other words he was able to evacuate the reservoir to a pressure something less than one-thirtieth of the original pressure but, as he suspected, his apparatus was not efficient enough to proceed further with the evacuation. When air was admitted into the receiver the mercury column rose to the usual height.

One would think that by the time Boyle published the account of his experiment the whole learned world would have accepted the new ideas. But the advance of science was slow in the mid-seventeenth century, in part because of the lack of scientific societies and scientific journals. In addition to an elaborate description of his pump, . . . the

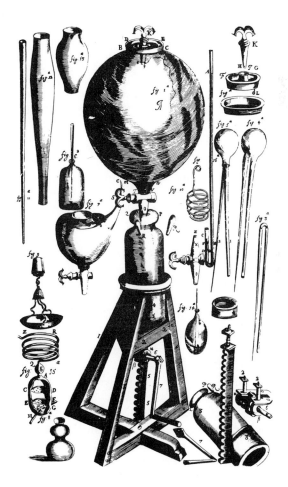

Figure 2-8. Boyle's first air pump. (University of Pennsylvania)

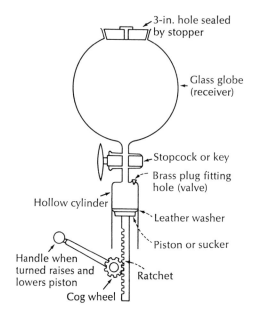

3-in. hole sealed
by stopper

Glass globe
(receiver)

Stopcock or key

Brass plug fitting
hole (valve)

Hollow cylinder

Leather washer

Piston or sucker

Handle when
turned raises and
lowers piston

Ratchet

Cog wheel

Figure 2-9. Simplified diagram of Boyle's pump. According to Boyle, "Upon the drawing down of the sucker (the valve being shut), the cylindrical space, deserted by the sucker, is left devoid of air; and therefore, upon the turning of the key, the air contained in the receiver rusheth into the emptied cylinder. . . . Upon shutting the receiver by turning the key, if you open the valve, and force up the sucker . . . you will drive out almost a whole cylinder full of air; but at the following exsuctions you will draw less and less of air out of the receiver into the cylinder because there will remain less and less air in the receiver itself. . . ."

author described many experiments which could be readily performed in a vacuum. . . . From these experiments . . . came quantitative measurements that resulted in the formulation of the famous Boyle's law relating to the volume and pressure of gas.[1]

The Role of Accident: Galvani's Discoveries *

by James B. Conant (1951)

Science is sometimes presented as though it were exclusively the work of high-powered mathematicians elaborating theories and sometimes as though it were all a matter of blind chance. As a consequence the reader is often confused in regard to the role of what appears to be an accidental observation. This is particularly true in connection with the development of new techniques and the evolution of new concepts from experiment. The case history which I recommend for a study of these subjects is the work of Galvani and Volta on the electric current. This case illustrates the fact that a chance observation may lead by a series of experiments (which must be well planned) to a new technique or a

° From James B. Conant, *Science and Common Sense.*
1. See "Boyle's law," in Glossary. Ed.

new concept or both. It also shows that in the exploration of a new phenomenon the experiments may be well planned without any "working
hypothesis" as to the nature of the phenomenon but that shortly an explanation is sure to arise. A new conceptual scheme will then be
evolved. This may be on a grand scale and have wide applicability or
may be strictly limited to the phenomenon in question. A test of the new
concept or group of concepts in either instance will probably lead to
new discoveries and the eventual establishment, modification, or overthrow of the conceptual scheme in question.

The story begins with certain observations made by Luigi Galvani, an
Italian physician and professor at Bologna, some time before 1786. This
investigator noted the twitching of a frog's leg when the crural nerves
were touched by a metallic scalpel in the neighborhood of an electrostatic machine from which sparks were drawn. *He followed up his observation:* this is the significant feature of the history of this episode. Time
and time again throughout the advance of science the consequences of
following up or not following up accidental discoveries have been very

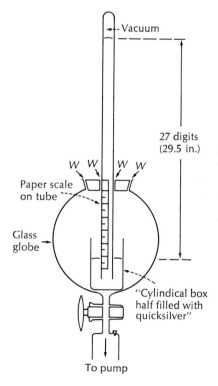

*Figure 2-10. Diagram of Boyle's ap-
paratus for removing the air above
the reservoir of a barometer. The
glass globe is the upper part of the
pump. W indicates where a sealing
wax was used. When the pump is operated, the mercury column decreases
in height.*

great. The analogy of a general's taking advantage of an enemy's error or a lucky break is clear. Pasteur once wrote that "chance favors only the prepared mind." This is excellently illustrated by the case at hand. The Dutch naturalist Swammerdam had previously discovered that if you lay bare the muscle of a frog in much the same way as Galvani did, grasp a tendon in one hand and touch the frog's nerve with a scalpel held in the other hand, a twitching will result. But Swammerdam never followed up his discovery. Galvani did. In his own words, "I had dissected and prepared a frog . . . and while I was attending to something else, I laid it on a table on which stood an electrical machine at some distance. . . . Now when one of the persons who were present touched accidentally and lightly the inner crural nerves of the frog with the point of a scalpel all the muscles of the legs seemed to contract again and again. . . . Another one who was there, who was helping us in electrical researches, thought that he had noticed that the action was excited when a spark was discharged from the conductor of the machine. Being astonished by this new phenomenon he called my attention to it, who at that time had something else in mind and was deep in thought. Whereupon I was inflamed with an incredible zeal and eagerness to test the same and to bring to light what was concealed in it."

Galvani did not succeed in bringing to light all that was concealed in the new phenomenon. But he proceeded far enough to make the subsequent discoveries inevitable. In a series of well-planned experiments he explored the obvious variables but without having a clear-cut over-all hypothesis. This is the usual situation when a new and totally unexpected phenomenon is encountered by a gifted experimenter. A series of working hypotheses spring to mind, are tested, and either discarded or incorporated into a conceptual scheme which gradually develops. For example, Galvani first determined whether or not sparks had to be drawn from the electrical machine in order to occasion twitching. He found "without fail there occurred lively contractions . . . at the same instant as that in which the spark jumped. . . ."

The nerves and muscle of the frog's leg constituted a sensitive detector of an electric charge. Galvani found that not only must a spark be passing from the electrostatic machine but the metallic blade of the scalpel must be in contact with the hand of the experimenter. In this way a small charge originating from the electrical disturbance, namely the spark, passed down the conducting human body through the scalpel to the nerve. So far the physician was on sound and fertile ground. There now occurred one of those coincidences which more than once have baffled an investigator initially but eventually have led to great advances. The frog's leg could under certain circumstances act not only as a sensitive electrical detector but as a source of electricity as well. When this happened the electricity self-generated, so to speak, actuated the detector. One can readily see that the superposition of these two effects could be most bewildering and misleading. This was particularly so since the conditions under which the frog's leg became a source of electricity were totally unconnected with any electrical phenomena then known. The variable was the nature of the metal, or rather metals, used. For Galvani discovered and duly recorded that the electrostatic machine could be dispensed with if the leg and nerve were connected by two *different* metals. Under these conditions the twitching occurred. (The experiment was usually performed as follows: a curved rod was made to touch simultaneously both a hook passing through the spinal cord of the frog and the "muscles of the leg or the feet.") "Thus, for example," wrote Galvani, "if the whole rod was iron or the hook was iron . . . the contractions either did not occur or were very small. But if one of them was iron and the other brass, or better if it was silver (silver seems to us the best of all the metals for conducting animal electricity) there occur repeated and much greater and more prolonged contractions."

Galvani had discovered the principle of the electric battery without knowing it. His two metals separated by the moist animal tissue were a battery, the frog's leg the detector. Every reader can perform the equivalent of Galvani's experiment himself. A copper coin and a silver one placed above and below the tongue when touched together produce in the tongue a peculiar "taste." A very small electric current flows, and our tongue records the fact through a series of interactions of electricity and nerves much in the same way as did Galvani's "prepared" frogs. Not having a suspicion of all this, however, Galvani developed a hypothesis on the grand scale to account for all the phenomena in terms of what was then known about electricity, which was derived entirely from experiment with electrostatic machines. Now that outside electrical disturbances had been found unnecessary (when he unwittingly used the *right* metallic combination!), the results, he says, "cause us to think that possibly the electricity was present in the animal itself." Galvani's following up of an accidental discovery by a series of controlled experiments led to a recording of the significant facts, but it was to be another Italian who developed the fruitful concept. It was Volta who in the late 1790's continuing the study of the production of electricity by the contact of two different metals, invented the electric battery as a source of what we now often call Galvanic electricity.

Examples of Unscientific Method

The male has more teeth than the female in mankind, and sheep, and goats, and swine. This has not been observed in other animals. Those persons which have the greatest number of teeth are the longest lived; those which have them widely separated, smaller, and more scattered, are generally more short lived.

> Aristotle
> *History of Animals* (4th century B.C.)

There are seven windows in the head, two nostrils, two eyes, two ears, and a mouth; so in the heavens there are two favorable stars, two unpropitious, two luminaries, and Mercury alone, undecided and indifferent. From which and many other similar phenomena of nature, such as the seven metals, etc., which it were tedious to enumerate, we gather that the number of planets is necessarily seven.

> Francesco Sizzi

In the space of one hundred and seventy-six years the Lower Mississippi has shortened itself two hundred and forty-two miles. That is an average of a trifle over one mile and a third per year. Therefore, any calm person, who is not blind or idiotic, can see that in the old Oolitic Silurian Period, just a million years ago next November, the Lower Mississippi River was upward of one million three hundred thousand miles long, and stuck out over the Gulf of Mexico like a fishing-rod. And by the same token any person can see that seven hundred and forty-two years from now the Lower Mississippi will be only a mile and three-quarters long, and Cairo and New Orleans will have joined their streets together, and be plodding comfortably along under a single mayor and a mutual board of aldermen. There is something fascinating about science. One gets such wholesale returns of conjecture out of such trifling investment of fact.

> Mark Twain
> *Life on the Mississippi* (1875)

Suggestions for Further Reading

° Beveridge, W. I. B.: *The Art of Scientific Investigation,* Vintage Books, Random House, Inc., New York, 1957. This book, which was written primarily as an aid to the young scientist, is also very interesting and instructive reading for the layman.

° Campbell, Norman: *What Is Science?* Dover Publications, Inc., New York, 1952. A well-expressed and informative discussion of scientific method.

Cohen, I. Bernard: *Science: The Servant of Man,* Little, Brown and Company, Boston, 1948. This book elucidates scientific method by describing case histories of a number of discoveries that have improved the well-being of mankind.

° Conant, James B.: *On Understanding Science,* Mentor Books, New American Library of World Literature, Inc., New York, 1951, and *Science and Common Sense,* Yale University Press, New Haven, Conn., 1951. These two books by Conant are very similar, the latter being essentially an amplification of the former. Conant uses case histories of important scientific concepts to illustrate the methods of science. In addition to the development of the concept of atmospheric pressure in the selection we use here, Conant also describes the development of a number of other fundamental concepts that the reader will find very helpful in gaining an understanding of scientific method.

° Popper, Karl: *Logic of Scientific Discovery,* Basic Books, Inc., Publishers, New York, or John Wiley & Sons, Inc., New York. This fine, authoritative book has become a classic exposition of scientific method. However, the reader will find the presentation more difficult than in the other books suggested here.

° Paperback edition

3

Is There an Order in Nature?

From early Greek times science has been based on the assumption that nature is orderly and that this order can be discovered and understood by man. The selections in this part examine the origin of this assumption, describe the different forms it took in ancient cosmology, and raise the question of its basic validity.

Is There an Order in Nature?

Introduction

THIS PART BEGINS the historical sequence of astronomy and cosmology that will be followed throughout the remaining parts. The selections take the reader back to ancient Egyptian and Greek times to the birth of some of the most basic scientific ideas. Historians of science agree that an important start toward understanding nature was made by the Greek natural philosophers. George Sarton, in his *Introduction to the History of Science,* says, "If science is more than an accumulation of facts; if it is not simply positive knowledge, but systematized knowledge; if it is not simply unguided analysis and haphazard empiricism, but synthesis; if it is not simply recording but constructive activity; then, undoubtedly, [ancient Greece] was its cradle." John Burnet goes further and says, ". . . it is an adequate description of science to say that it is 'thinking about the world in the Greek way.' That is why science has never existed except among people who came under the influence of Greece."

What was the Greek way of thinking about the world? The authors in this session suggest that the most important feature of this world picture was the hypothesis that "nature can be understood" and that the basis for understanding it is the discovery of an order that lies beneath the variety that we observe in the external world. Whitehead puts it this way: ". . . there can be no living science unless there is a widespread instinctive conviction in the existence of an *Order of Things,* and in particular of an *Order of Nature."* In fact, this assumption has become so basic to our whole world view that it may seem trivial to question its validity and to inquire about its origin. If any justification is needed for this inquiry, we have only to remember that there have been and still are civilizations in which magic or the acts of capricious gods are held responsible for the happenings of nature, and as we will see in Part 6, some of the findings of modern physics suggest that this instinctive belief in a natural order may not be as universally valid a principle as we had at first supposed.

The readings in this part, however, are designed to present the idea of an order in nature in its historical context. Benjamin Farrington, in his book on Greek Science, says, ". . . there is no human knowledge which cannot lose its scientific character when men forget the conditions under which it originated, the questions which it answered, and the functions it was created to serve. A great part of the mysticism and superstition

69

of educated men consists of knowledge which has broken loose from its
historical mooring." [1] It is important that we understand the ideas that
we have inherited from Greek thought, not only to use them more effectively, but also to be free of any mystical or superstitious attachment to
them.

The Greeks were the first to use effectively the ability to generalize
from perceptual material to conceptual schemes. Furthermore, they believed that this process of abstraction would tell them something significant about nature. The most outstanding product of this type of thinking
was mathematics and particularly the geometry that was developed by
the Pythagorean school and reached its culmination in the *Elements of
Euclid*. The materials throughout this book have been presented, as far
as possible, in a nonmathematical fashion in order to be more understandable to a general audience. However, it would be a grave misrep-
resentation to underestimate the importance of mathematics as a powerful aid to physical scientists.

Mathematics helps the scientist by providing a clear shorthand
language in which he can express himself without the vagueness or ambiguity of ordinary speech. For instance, the equation $a^2 + b^2 = c^2$, in
the figure below, expresses the Pythagorean theorem in the briefest and

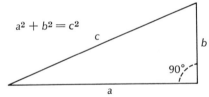

Figure 3-1. *The Pythagorean theorem.*

most concise manner. Try to put this theorem into words, and you will
see how much more complicated it is. For those who are familiar with
the language of mathematics the relationship is much easier to visualize
when it is expressed in mathematical form. Furthermore, by providing a
set of rules for manipulating the relationships, mathematics can suggest
new combinations and lead to broader generalizations. Descartes saw in
the Pythagorean theorem a method for describing the position of bodies
in three dimensional space and from this insight he invented the Cartesian coordinate system. Several centuries later, Einstein created a
broader synthesis using the Cartesian coordinates as a basis for the
frames of reference in his special theory of relativity. Mathematics gives
the scientists a method for recognizing and revealing relationships. In so
doing, it makes an important contribution to the search for an underlying form and order in nature.

1. Benjamin Farrington, *Greek Science*, Penguin Books, Inc., Baltimore, 1944.

The power of abstract reasoning is brilliantly demonstrated in the work of the Greek astronomers who, starting with a few simple observations of the positions of the stars and planets at different times of the year, devised vast and ingenious theoretical systems and even anticipated the heliocentric theory almost two thousand years before Copernicus. Two entirely different systems of astronomy were proposed to account for the set of known facts. Both of these systems represented a search for order in the apparent disorder of the planetary motions. Both of them accounted reasonably well for the observed facts and were able to predict the future positions within the limits of the accuracy possible at that time. In this search for order and unity in the heavens, the two schools of Greek astronomers made different simplifying assumptions. The Pythagorean school sought for harmonious numerical relationships in the spacing and the number of the planets. The school of Eudoxus and Ptolemy sought to explain the facts in terms of uniform circular motion. In retrospect, we believe today that the latter explanation was more artificial, that it represented an imposition of order on the diversity of nature, whereas the heliocentric theory, we believe, was the discovery of a real order in nature. But it could be argued that, if two different explanations account equally well for the facts, perhaps other valid explanations could also be found. Can one of these be said to be really true? In preferring the simpler explanation, have we discovered simplicity and order in nature itself or only a reflection of our own thoughts. As Arthur Eddington expressed it: "We have found a strange footprint on the shores of the unknown. We have devised profound theories, one after another, to account for its origin. At last, we have succeeded in reconstructing the creature that made the footprint. And lo! It is our own." [2]

2. As quoted by Tobias Dantzig in *Number: The Language of Science*, The Macmillan Company, 1944.

Ancient Cosmology

The Beginning of Astronomy [*]
by Eric M. Rogers (1960)

For centuries very primitive man must have watched the stars, perhaps wondering a little, taking the Sun for granted, yet using it as a guide, relying on moonlight for hunting, and even reckoning simple time by moons. Then slowly gathered knowledge was built into tradition with the help of speech. The Sun offered a rough clock by day and the stars by night.[1] For simple geography, sunrise marked a general easterly region, and sunset a westerly one; and the highest Sun (noon) marked an unchanging South all the year round. At night the pole-star marked a constant North.

As the year runs through its seasons the Sun's daily path changes from a low arc in winter to a high one in summer; and the exact direction of sunrise shifts round the horizon. Thus the Sun's path provided a calendar of seasons; and so did the midnight star-pattern, which shifts from night to night through the year.

With the age of cultivation, herding and agriculture made a calendar essential. It was necessary to foretell the seasons so that the ground could be prepared and wheat planted at the right time. Sheep, among the first animals domesticated, are seasonal breeders, so the early herdsmen also needed a calendar. Crude calendar-making seems easy enough to us today, but to simple men with no written records it was a difficult art to be practised by skilled priests. The priest calendar-makers were practical astronomers. They were so important that they were exempt from work with herds or crops and were paid a good share of food, just as in many savage tribes today.

When urban civilizations developed, clear knowledge of the apparent motions of Sun, Moon, and stars was gathered with growing accuracy and recorded. These regular changes were worth codifying for the pur-

[*] From Eric M. Rogers, *Physics for the Inquiring Mind: The Methods, Nature, and Philosophy of Physical Science*, Princeton University Press. © 1960 by Princeton University Press. Reprinted by permission.

1. An experienced camper can tell the time by the stars within a quarter hour.

72

pose of making predictions. In the great Nile valley, where one of the early civilizations grew, the river floods at definite seasons, and it was important for agriculture and for safety to predict these floods. And fishermen and other sailors at the mercy of ocean tides took careful note of tide regularities: a shifting schedule of two tides a day, with a cycle of big and little tides that follows the Moon's month. In cities, too, time was important: clocks and calendars were needed for commerce and travel.

Timekeeping promoted an intellectual development:[2] "In counting the shadow hours and learning to use the star clock, man had begun to use geometry. He had begun to find his bearings in cosmic and terrestrial space."

ASTRONOMY AND RELIGION

Apart from practical uses, why did early man attach such importance to astronomy and build myths and superstitions round the Sun, Moon, and planets?

The blazing Sun assumed an obvious importance as soon as man began to think about his surroundings. It gave light and warmth for man and crops. The Moon, too, gave light for hunters, lovers, travellers, warriors. These great lamps in the sky seemed closely tied to the life of man, so it is not surprising to find them watched and worshipped. The stars were a myriad lesser lights, also a source of wonder. Men imagined that gods or demons moved these lamps, and endowed them with powers of good and evil. We should not condemn such magic as stupid superstition. The Sun *does* bring welcome summer and the Moon *does* give useful light. The simple mind might well reason that Sun and Moon could be persuaded to bring other benefits. The very bright star, Sirius, rose just at dawn at the season of the Nile floods. If the Egyptians reasoned that Sirius caused the floods, it was a forgivable mistake—the confusion of *post hoc* and *proper hoc* that is often made today.

When a few bright stars were found to wander strangely among the rest, these "planets" (literally "wanderers") were watched with anxious interest. At a later time, early civilized man evolved a great neurotic superstition that man's fate and character are controlled by the Sun, Moon, and wandering planets. This superstition of astrology, built on earlier belief in magic, added drive—and profit—to astronomical observation.

Thus the growth of astronomy was interwoven with that of religion—

2. Lancelot Hogben, *Science for the Citizen*, George Allen & Unwin, Ltd., London, 1938.

and they still lie very close, since modern astronomy is bounded by the ultimate questions of the beginning of the world in the past and its fate in the future. The next two sections contain speculations on the early stages of that development.

SCIENCE, MAGIC, AND RELIGION

Science began in magic. Early man lived at the mercy of uncontrolled nature. Simple reasoning made him try to persuade and control nature as one would a powerful human neighbor. He tried simple imitative magic, such as jumping and croaking like a frog in the rain to encourage rainfall, or drawing animals on cave walls to promote success in hunting. He buried his dead near the hearth, with logical hope of restoring their warmth; and he gave them tools and provisions for future use. In a way, he was carrying out scientific experiments, with simple reasoning behind them. It was not his fault that he guessed wrong. The modern scientist disowns magicians because they refuse to learn from their results—that is the essential defect of superstition, a continuing blind faith. Primitive man, however, had neither the information nor the clear scientific reasoning to judge his magic.

As he carried out jumping ceremonies, or squatted before magic pictures, man could form the idea of presiding spirits: kindly deities who could help, malicious demons who brought disaster, powerful gods who controlled destiny. Like a child, man tried to please these gods so that they would grant good weather, health, plentiful game, fertility. The original reasons and purposes were then forgotten and the ceremonies continued by habit.

Speech was the essential vehicle of this development. Earliest man, just beginning speech, was forging his own foundations of thought, slowly and uncertainly. Other creatures communicate—bees dance well-coded news of honey, dogs bark with meaning—but man's speech opened up greater intellectual advances. In the course of a long development, it not only gave him a rich vocabulary for communicating information but it enabled him to store information in tales for later generations; and then it blossomed into a higher intellectual level with *words for abstract ideas*. Thus speech opened up a new field of *ideas and reasoning*.

Primitive religion wove together myths of gods, magic ceremonies, and tales of nature, in attempts to codify both the natural world and man's growing social system. Astronomy played an important part in this ceremonial religion. The priests—wise men of the village or tribe—were the calendar-makers, the first professional astronomers. Their successors were the powerful priesthood of the first urban civilizations. In early Babylon, for example, the priests were bankers, physicians, scientists, and rulers—they *were* the government. In their knowledge, and that of their craftsmen, there were the foundations of many sciences. The practical *information* was there, at first unrecognized, then held secret, then published in texts. Was it *science?*

SCIENCE, THE ART OF UNDERSTANDING NATURE

Curiosity and collecting knowledge go back before earliest man. Primitive man collected knowledge and used it: a beginning of applied science. Then, as a reasoner, he began to organize knowledge for use and thought. That is a difficult step, from individual examples to generalization—watch a child trying to do it. It is difficult to grasp the idea of a common behavior or a general law or an abstract quality. Yet that is the essential step in turning a "stamp collection" of facts into a piece of science. Science, as we think of it and use it now, never was just a pile of information. Scientists themselves, beginning perhaps with the early priesthood, are not just collectors. They dig under the facts for a more general understanding: they extract general ideas from observed events.

Scientists feel driven to *know*, know what happens, know how things happen; and, for ages, they have speculated why things happen. That drive to know was essential to the survival of man—a generation of children that did not want to find out, did not want to understand, would barely survive. That drive may have begun with necessity and fear; it may have been fostered by anxiety to replace capricious demons by a trustworthy rule. Yet there was also an element of wonder: an *intellectual delight* in nature, a delight in one's own sense of understanding, a delight in creating science. These delights may go back to primitive man's tales to his children, tales about the world and its nature, tales of the gods. We can read wonder and delight in stone-age man's drawings; he watched animals with intense appreciation and delighted in his art. And we meet wonder and delight in scientists of every age and make their science *an art of understanding nature.*

As scientists, we have travelled a long way from capricious gods to orderly rules; but all the way we have been driven by strong forces: a sense of urgent curiosity and a sense of delight.

Fear and anxiety, wonder and delight—these are two aspects of *awe*, mainspring of both science and religion. Lucretius held, 2000 years ago, that "Science liberates man from the terror of the gods."

FACTS

Before showing how astronomy was organized into great schemes of thought, we shall review the knowledge that you—or primitive man— could gain by watching the heavens. If you have lived in the country you will be familiar with most of this. If you have grown up in a town this will seem a confusing pile of information unless you go out and watch the sky—now is the time to observe.

THE SUN AS A MARKER

Each day the Sun rises from the eastern horizon, sweeps up in an arc to a highest point at noon, due South,[3] and down to set on the western horizon. It is too bright for accurate watching, but it casts a clear shadow of a vertical post. At noon, mid-day between sunrise and sunset, the shadow is shortest and points in the same direction, due North, every day of the year (Figure 3-2). The positions of the noon Sun in the sky

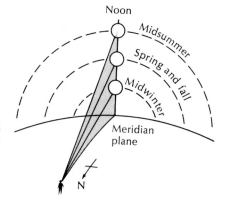

Figure 3-2. The meridian. Noon sun is due south (or due north). The meridian plane is a vertical plane passing through the sun's noon positions.

from day to day mark a vertical "meridian" (= mid-day) plane running N-S.

In winter the shadows are longer because the Sun sweeps in a low arc, rising south of East and setting south of West.[3] In summer the arc is much higher, shadows are shorter, daylight lasts longer. Half-way between these extremes are the "equinox" seasons, when day and night are equally long and the Sun rises due East, sets due West (Figure 3-3).

On the horizon—simple man's extension of the flat land he lived on —sunrise marked a general eastern region for him and the exact place

3. This description applies to the northern latitudes of the early civilizations.

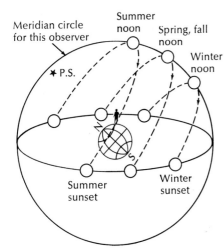

Meridian circle
for this observer

Summer
noon

Spring, fall
noon

Winter
noon

✶ P.S.

Figure 3-3. The path of the sun in the sky changes with the seasons.

Summer
sunset

Winter
sunset

of sunrise showed the season. The length of noonday shadows also provided a calendar. The shadow of a post made a rough clock. Although it told noon correctly, its other hours changed with the seasons—until some genius thought of tilting the post at the latitude-angle (parallel to the Earth's axis) to make a true sundial.

STARS

The stars at night present a constant pattern, in which early civilizations gave fanciful names to prominent groups (constellations). The whole pattern whirls steadily across the sky each night, as if carried by a rigid frame (Figure 3-4). One star, the pole-star, stays practically still while the rest of the pattern swings round it. Watch the stars for a few hours and you will see the pole-star, due North, staying still while the others move in circles around it. Or, point a camera at the sky with open lens, and let it take a picture of those circles (Figure 3-5). Night after night, year after year, the pattern rotates without noticeable change. These are the "fixed stars." [4] The pole-star is due North, in the N-S meridian plane of the noonday Sun. The star pattern revolves round it at an absolutely uniform rate. This motion of the stars gave early man a clock, and the pole-star was a clear North-pointing guide.[5]

4. However, if you could live for many centuries you would notice changes in the shape of some constellations. Stars are moving.
5. The procession of the equinoxes carries the Earth's spin-axis-direction slowly round in a cone among the stars, so only in some ages has it pointed to a bright star as pole-star. It is near to one now, our present pole-star, and was near to another

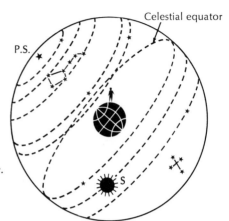

Figure 3-4. The star pattern revolves.

The simplest "explanation" or descriptive scheme for the stars is that they are shining lights embedded in a great spinning bowl, and we are inside at the center. That occurred to man long ago, and you would feel it true if you watched the sky for many nights. It was a clever thinker who extended the bowl to a complete sphere, of which we see only half at a time. This is the celestial sphere, with its axis running through the pole-star and a celestial equator that is an extension of the Earth's equator. The celestial sphere revolves steadily, carrying all the stars, once in 24 hours. The Sun is too dazzling for us to see the stars by day, so we only see the stars that are in the celestial hemisphere above us at night, when the Sun is in the other hemisphere below. The Sun's daily path across the sky is near the celestial equator; but it wobbles above and below in the course of the year, $23\frac{1}{2}°$ N in summer, $23\frac{1}{2}°$ S in winter (Figures 3-6 to 3-8).

Though the pattern of stars has unchanging shape, we do not see it in the same position night after night. As the seasons run, the part of the pattern overhead at midnight shifts westward and a new part takes its place, a whole cycle taking a year. Stars that set an hour after sunset are $1°$ lower in the West next time and set a few minutes earlier; and two weeks later they are level with the Sun and set at sunset. Thus, in 24 hours the celestial sphere makes a little more than one revolution: $360°$ + about $1°$. It is moving a little faster than the Sun, which makes one revolution, from due South to due South in the 24 hours from noon to

when the pyramids were built. In A.D. 1000 there was no bright pole-star—perhaps that lack delayed the growth of navigation.

noon. The celestial sphere of stars makes one extra complete revolution in a year.

SUN AND STARS

This difference between the Sun's daily motion and that of the stars (really due to the Earth's motion in its orbit round the Sun) was obvious, and suggested that the Sun is moved by a separate agent. The Sun-god became a central figure in many primitive religions, and his travels were carefully traced by shadows and recorded by alignments of great stones in ceremonial temples.

Instead of saying the star pattern "gains" (like a clock running fast) 1° a day, we take the very constant star motion as our standard and say the Sun lags behind it 1° a day. We may stick the Sun as well as the

Figure 3-5. A photograph of the sky near the Pole Star, taken in an eight-hour exposure. The Pole Star itself made the very heavy trail near the center. (Lick Observatory)

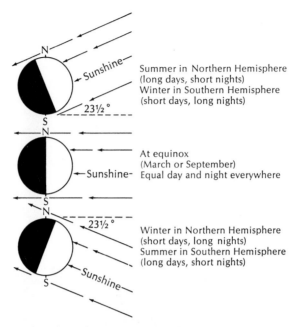

Summer in Northern Hemisphere
(long days, short nights)
Winter in Southern Hemisphere
(short days, long nights)

At equinox
(March or September)
Equal day and night everywhere

Winter in Northern Hemisphere
(short days, long nights)
Summer in Southern Hemisphere
(long days, short nights)

Figure 3-6. Earth and sunshine. Day and night at various seasons.

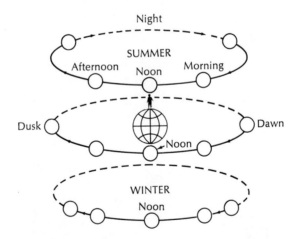

Figure 3-7. Path of the sun. Viewed from stationary earth, at various seasons.
Sun positions are labeled noon, afternoon, and so on, for observers in the
longitude of New York. If an observer could watch continuously, unob-
structed by the earth, he would see the sun perform the "spiral of circles"
sketched in Figure 3-8, during the course of six months from summer to win-
ter; then he would see the sun spiral upwards, revolving the same way, from
winter to summer.

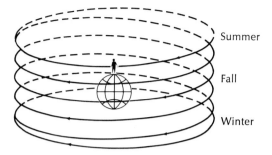

Figure 3-8. *The sun's spiral of circles, in the course of half a year of seasons.*

stars on the inside of the celestial sphere; but since the Sun lags behind the stars it does not stay in a fixed place in the starry sphere; it crawls slowly *backwards* round the inside of the sphere, making a complete circuit in a year. Thus we can picture the Sun's motion compounded of a *daily motion shared with the stars* of the celestial sphere and a *yearly motion backward through the star pattern.*

ECLIPTIC AND ZODIAC

This is a sophisticated idea, a piece of scientific analysis: to separate out the Sun's yearly motion from its daily motion across the sky with the star-pattern. Once this idea was clear it was easy to map the Sun's yearly track among the stars—not directly, because the Sun outglares the nearby stars by day, but by simple reference to the pattern of stars

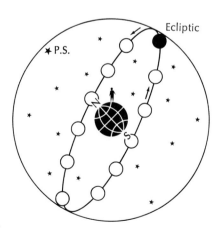

Figure 3-9. *The ecliptic, the sun's track through the star pattern in the course of a year. Here the daily motion is imagined "frozen."*

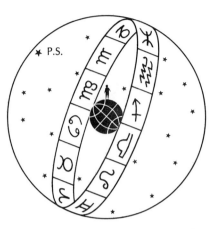

Figure 3-10. The zodiac was divided into twelve sections named after prominent star groups, or constellations. A belt of the celestial sphere, it is tilted 23½ degrees from the equator. The sun's yearly path (the ecliptic) runs along the middle line of this belt. The paths of moon and planets lie within this belt.

in the sky at midnight instead of noon. The Sun's yearly track is not the celestial equator but a tilted circle making 23½° with the equator. It is this tilt that makes the Sun's *daily* path across the sky change with the seasons. At the equinoxes the Sun's yearly track crosses the equator. In summer the tilted track has carried the daily path 23½° higher in the sky and in winter 23½° lower. This tilted yearly track is called the *ecliptic* (Figure 3-9).

As the Sun travels round the ecliptic in the course of the year, it

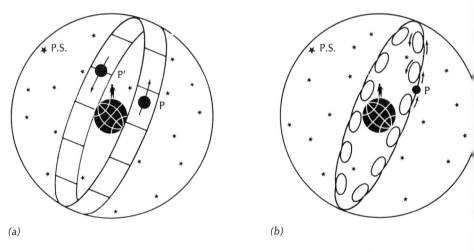

(a)   (b)

Figure 3-11. All the planets wander through the star pattern within the zodiac belt. Left, general region of a planet's path; right, in detail, a planet's path has loops—an epicycloid seen almost edge-on.

passes through the same constellations of stars at the same season year after year. This broad belt of constellations containing the ecliptic is called the *zodiac* (Figure 3-10). The constellations in it were given special names long ago by the astrologer priests, a named group for each month in the year.

A few bright "stars" do change their positions, and move so unevenly compared with Sun, Moon, and the rest that they are called *planets*, meaning wanderers. These planets, which look like very bright stars, with less twinkling, wander across the sky in tracks of their own *near the ecliptic* (Figure 3-11). They follow the general backward movement of the Sun and Moon through the constellations of the zodiac, but at different speeds and with occasional reverse motions. Primitive man must have observed the brighter planets but cannot have got any good use from his observations, unless like eclipses they were used to impress people.

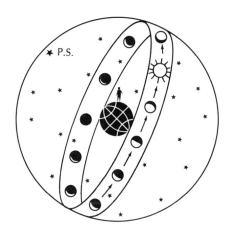

Figure 3-12. The zodiac belt with the positions of the moon, in various phases, in the course of a month. The daily motion of the celestial sphere is "frozen" here.

Thus the zodiac belt includes the Sun's yearly path, the Moon's monthly path (Figure 3-12), and the wandering paths of all the planets. In modern terms, the orbits of Earth, Moon, and planets all lie near to the same plane. Astrology assigned fate and character by the places in the zodiac occupied by Sun, Moon, and planets at the time of a man's birth.

THE PLANETS AND THEIR MOTIONS

Five wandering planets were known to the early civilizations, in addition to the Sun and Moon which were counted with them. These were:

Mercury and Venus, bright "stars" which never wander far from the Sun, but move to-and-fro in front of it or behind, so that they are seen only near dawn or sunset. Mercury is small and keeps very close to the Sun, so it is hard to locate. Venus is a great bright lamp in the evening or morning sky. It is called both the "evening star" and the "morning star"—the earliest astronomers did not realize that the two are the same.

Mars, a reddish "star" wandering in a looped track round the zodiac, taking about two years for a complete trip.

Jupiter, a very bright "star" wandering slowly round the ecliptic in a dozen years.

Saturn, a bright "star" wandering slowly round the ecliptic, in about thirty years.

Jupiter and Saturn make many loops in their track, one loop in each of our years.

When one of the *outer* planets, Mars, Jupiter or Saturn, makes a loop in its path, it crawls slower and slower eastward among the stars, comes to a stop, crawls in reverse direction westward for a while, comes to a stop, then crawls eastward again, like the Sun and Moon.

Figure 3-11 shows the looped tracks of planets through the star patterns. Once noticed, planets presented an exciting problem to early scientists: what gives them this extraordinary motion? . . . How can we explain (produce, predict) the strange motions of the planets, which excited so much wonder and superstition? We [will] study this [later] to show how scientific theory is made.

Early Concepts of Time °

by Rudolph Thiel (1957)

There is good reason why the human race has been searching the skies for so long. For astronomy has given us the ideas of space and time.

It is hard for us to sense what it must have been like for human beings to live upon an unknown Earth, knowing only that endless expanses opened out in all directions. For them there were landmarks only in the immediate neighborhood. The world into which each person was born seemed limited by the horizon, "as far as the eye could reach," but the horizon always moved along from one place to the next. The ultimate boundary discovered by the boldest travelers was the ocean, but it too stretched on into infinity. Real stability seemed to exist only in the firmament. The sky appeared to revolve constantly, but its lights again and again returned along the selfsame tracks. The Sun, Moon, and stars alone enabled men to find their way about their world; these alone made possible orientation upon Earth.

"Orientation" originally meant directing oneself toward the east, toward the rising Sun. Direction, in fact, exists only in relationship to the sky; on Earth, all directions were blurred. . . .

TIME WAS ONCE RELATIVE

Time was once relative not in a theoretical, Einsteinian sense, but perceptibly so in daily life. Each hour embraced a different span of time. At noon the hour was longer than in the evening or in the morning.

Nowadays, hours of different length are almost unimaginable. For us, time flows evenly. Yet human beings once thought otherwise. In fact, all earlier cultures had a sense of time entirely different from ours.

Men are not born with a concept of time. A child lives from moment to moment. The alteration of day and night sets up a rhythm of life to which everyone submits without thinking. In all probability the seasons, with their varying demands upon men, first produced a consciousness of regular, recurrent periods of time. The need to count days, to measure spans of time, coincides with a relatively developed form of life, such as farming. And astronomical observations were ready to hand when men, conscious of passing "time," began dividing up the day.

The first time designation was noon, the moment when the Sun's shadow is shortest. In order to determine it precisely, one of the oldest scientific instruments was invented, the vertical shadowstick. From this primitive device a logical but probably long and wearisome trail led to the sundial [Figures 3-13 to 3-15]. With the sundial, time was given concrete existence; it became, so to speak, something that human hands could make. The advance of the Sun's shadow across the graduated face of the sundial transformed an unconscious feeling more and more into a "conceptual" idea of a thing that moved, fled away, passed without man's intervention.

But to the Egyptian inventors of the sun clock, this something did not flow evenly; the speed of its passage varied. This may have come about by accident: it is noteworthy that the Egyptians marked out the dial of their sun clock in even divisions [Figure 3-13]. For man does have an instinct for symmetry; when he divides any space, he tends to divide it

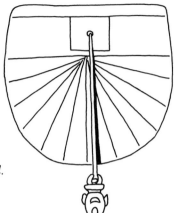

Figure 3-13. *An Egyptian sundial.*
(Figure added by editor.)

into segments of equal size. But the Sun passes overhead in an arc, and
its shadow therefore passed across the graduations of the dial in different periods of time. The Egyptians made each of their hours a different
length; moreover, they also adapted them to the season, the winter
hours being one seventh shorter than the hours of summer.

It is evident that the Egyptians had no feeling for time in our sense.
We cannot consider the idea of time an elementary component of our
minds, or even indispensable for the ordering of our environment.
"Time" is a fairly late abstraction, the end product of a long evolution in
which the clock played a crucial part. And since the first clock was a
sundial, in the beginning "time" moved at a varying, irregular pace.

So rooted did this idea become in the minds of the Egyptians that
they arranged their water clocks to register hours of varying duration.
Incredible though it seems, they went to great efforts to do so. In the
flow of water they had a direct image of steadily flowing time. But with
extraordinary skill and ingenuity they artificially produced irregularity

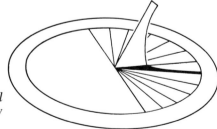

Figure 3-14. *Sundial with unequal
hourly divisions.* (Figure added by
editor.)

in a regular natural phenomenon, in order to make time flow in the only manner that seemed right to them: with the inconstancy of their sundials.

Centuries passed before they succeeded fully, as is evidenced from the inscription on the tomb of Prince Amenemhet, Keeper of the Seals of the New Kingdom, around 1500 B.C. Amenemhet is there praised for having at last succeeded in building a water clock that kept perfect time. His clock was able to dispense with the several jets of water formerly needed to measure the varying length of the hours. Its single stream could even be controlled to change the length of the hours with the seasons.

Figure 3-15. *A Chinese portable sundial with compass and level for setting the dial. The hour lines and numbers have been anglicized.* (Harvard College Observatory. Figure added by editor.)

The probable appearance of this ingenious clock can be deduced from a late Greek water clock (Figure 3-16)—in matters of technology the Greeks were very dependent upon the Egyptians. The inflow of water in this clock was regulated by a funnel-shaped valve controlled by movable weights, and the length of the hours could be altered from month to month with the aid of a scale mounted on a rotating drum. Every hour a slave had to shift the weights, and every month he changed the position of the drum.

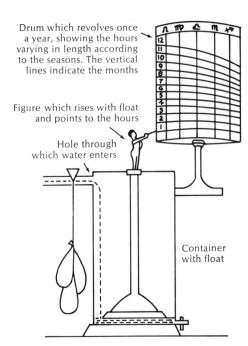

Drum which revolves once a year, showing the hours varying in length according to the seasons. The vertical lines indicate the months

Figure which rises with float and points to the hours

Hole through which water enters

Container with float

Figure 3-16. Ancient water clock. As water pours past the weighted control valves into the container, the float rises, carrying with it the figure that points to the hours on the drum. The drum revolves once a year, showing the hours varying in length accord- ing to the seasons, which are marked by the vertical lines indicating the months. Every hour a slave had to shift the weights, and every month he changed the position of the drum.

Egyptian time-measurement with its relative concept of time re- mained the model for all of antiquity. Athens, for example, was very proud of her Tower of the Winds, which in addition to a weather vane had no less than eight sundials, visible at a great distance from all direc- tions. They told the time with an error of no more than three or four minutes, we are informed—a type of accuracy which makes us smile. Of course the duration of every hour differed; but the Greeks undoubtedly felt the same confidence in their inconstant division of time as do we in our infallible chronometers whose hands circle the dial with military precision. Human beings can grow accustomed to anything—even to operating with two simultaneous and contradictory conceptions of time. The Babylonians and the Chinese did not fit their water clocks to their sundials; they let the water flow regularly. They therefore had two dif- ferent kinds of hours, the one changing and the other regular. As late as 1600, Father Ricci, the Jesuit who introduced pendulum clocks into China, adjusted some of them for absolute and some for relative time- measurement.

The Birth of Scientific Ideas °
by W. P. D. Wightman (1951)

Ideas like persons are born, have adventures, and die. But unlike most persons they do not disappear from this mortal stage; their ghosts walk, often to the confusion of new ideas. The birth of ideas, like that of persons, does not occur spontaneously, but is a culmination of travail following on, may be, centuries of gestation. Moreover their conception is often the result of a clash of cultures, if not of temperaments.

Until comparatively recent times it was believed that science, like almost every strand of Western civilisation, had its origin in the Eastern

Figure 3-17. Druidic ritual stones at Stonehenge, in southern England, were apparently used in part to mark the start of the seasons. Note how the sun rises directly over the "heel stone" when it is viewed in the center of the archway at summer solstice, June 20, 1964. (From CBS News Program, "Mystery of Stonehenge." Figure added by editor.)

outposts of the Greek confederation. It was admitted that the Greeks had borrowed a good deal of assorted information of a practical nature from the older cultures of the East, principally Egypt and what is loosely known as Babylonia. But all this information was of a purely "rule of thumb" nature and hardly merited the name of science. Our views on this matter have recently undergone a profound change. Due in part to the enlargement of our knowledge of the quantity and quality of the eastern achievements, due perhaps equally to the enlargement of our conception of what constitutes science, we are in no doubt that science was born with infinite pains and in many places, neither by the happy inspiration of one man, nor at any clearly defined epoch. The flint knappers who learnt by observation, imagination and trial how to extend the range of the appliances obtainable from the single material at their disposal were assisting at the birth of science.

It is significant that all the early historians agreed that science had its birth at the outposts of Greece, namely where the vigorous young civilisation was in intimate contact with a rich, if declining, culture. There was thus no doubt that the first speculative natural philosophers of whom we have any record were not creating something out of a cultural vacuum. To deny the title of men of science to those ingenious workers who created the technique of multiplication and division; who made an error of only one inch in the 755¾ feet base lines of the Great Pyramid; who discovered how to mark out the passing of the seasons by taking as a unit the lapse of time between two heliacal risings of the star Sirius— would be to narrow down the meaning of the term beyond what in this industrial age we should be willing to do. Yet, although the heritage of natural knowledge upon which the Greeks were able to draw is proving year by year to be far richer than was first supposed, nevertheless most contemporary historians would agree that although the Ionians did not create science they undoubtedly transformed it. The evidence for this is stated by Professor Gordon Childe, a writer who has done much to correct the former unjust estimate. In relation to arithmetic, the science in which Babylonia made the most notable progress, he says of the clay tablets which constitute the "documents": "It looks as if the research they disclose was really limited in its scope by consciously conceived possibilities of practical application. In any case no attempt was made to *generalise* the results" (italics mine). Herein lies the nature of the transformation effected by the Greeks. It was not merely that their problems were more "theoretical" in the sense of being more remote from the problems of daily activity—such a criterion for "science" would hardly be allowed in an age when the boundaries between "pure" and "applied" science are seen to be artificial and mere conveniences. What distinguishes the Greeks of the seventh century before our era from any previous thinkers of whom we have any record is that their concern is not with triangular fields but with triangles; their curiosity is not directed to the nature of the fire which is baking bricks but to the nature of fire itself. And they went further: they asked, if we can trust the testimony of Aristotle, writing three hundred years later, "What is the nature of Nature?" Lastly, with regard to mathematics, although in some respects they seem to have failed to garner the rich harvest of methods invented by their forerunners, they recognised, as the former did not, the necessity for deductive proof.

In a word, Greek thought differed from all that had gone before in respect both of generality and rigour. In general thinking, which opens up possibilities undreamed of by the severely "practical man," we pass from percepts—things to which we can point, like triangular fields—to concepts—creations of the process known as abstraction. If we ignore everything about a number of fields except the fact that they all happen to have three sides, we have arrived at the concept of triangularity. We cannot point to a triangle, because in order that triangularity may be manifested there must be something else—a hedge, a pencil line—and our abstraction has thereby become merged in the concrete instance. Moreover, since we cannot point to a triangle, we have to define it in words, that is, in the form of relations between other abstractions, such as straight lines and planes, none of which "exist" in nature. What makes this type of thought so immeasurably more powerful than thought (in so far as this is possible) about individual instances, is that we can worry out by deduction all the characteristics that must "go with" triangularity as such. Thence we return to the concrete individual, knowing that in so far as triangularity is manifested in it, so also is an infinite number of other possibilities which would never have suggested themselves by mere contemplation of a three-sided parcel of land. Science is, among other things, the generalisation of perceptual experience by means of adequate concepts; the concepts grow with the experience; history is the record of their growth; philosophy, among other things, the critique of their adequacy.

It must not be thought that the Greeks from the outset knew what they were doing; or at any rate could have expressed it in intelligible terms, because the terms themselves were created by their procedure. The conscious recognition came to birth painfully in the thought of Parmenides (ca. 500 B.C.) and especially of Plato.

In what follows we shall therefore take it as proved that so far as the existing evidence goes the birth of scientific IDEAS took place in Asia Minor in the seventh century B.C., on the ground that at that time and place the idea of nature as a whole first took shape, was explicitly stated, and systematically expounded. In our study of the growth of scientific ideas we shall therefore take as our starting point the same thinker as has traditionally been regarded as the founder of science, Thales of Miletos. . . .

Our aim therefore in Collingwood's phrase is to relive the ideas of the past in order that, by watching them grow, we may the better understand what scientific ideas really are. Incidentally we shall come to see, as President Conant has pointed out, how completely false is the belief in the triumphal march of science down the ages. Progress has resembled much more closely the infiltration of modern war, in which tactical objectives have been seized by bold leaps, and held precariously while the main body has been slowly consolidating the ground in the rear. This explains perhaps the frequent and disconcerting inconsistencies of great thinkers.

ALL NATURE FOR THEIR PROVINCE

There is no doubt that sometime in the seventh century before our era, probably between 630 and 620,[1] there was born at Miletos the man Thales, whom later Greeks counted as one of the Seven Sages. We can place the date of his birth fairly accurately because of the tradition reported by the historian Herodotus that he foretold an eclipse of the sun which put an end to the war between Lydians and the Medes. If this is true, it was a lucky shot; but even if it is no more than a fable, it gives us a good idea of the sort of problems people were interested in at that time. It also suggests that he was acquainted with Babylonian astronomy, in particular of the cycle of relationships between the sun and the moon as viewed from the earth, known as the Saros. Of his writings—if there were any—we know nothing; though Diogenes Laertius, a writer of the second century A.D., reports a belief that Thales wrote a book on astronomy for sailors—a significant suggestion even if unfounded. Of his reported sayings Aristotle mentions three: (a) All things are made of water on which the earth floats; (b) All things are full of "gods"; (c) The lodestone and amber have souls.

To anyone whose knowledge of science was restricted to the standpoint of the present day these sayings would appear as childish nonsense. It is only because we are aware of the gradual development and modification of these views by his contemporaries, and of the further fact that no one before Thales seems to have put forward views of a like kind, that we revere him as the founder of natural philosophy, and

1. All dates of pre-Socratic thinkers are to be regarded as approximate only.

through that, though at a considerable remove, of natural science. If we are to appreciate the greatness of Thales we must divest our minds at once of any idea that he was trying to found the science of chemistry on the orderly but ungeneralised observations of the copper smelters, glass-makers, and dyers of Egypt and Babylonia. Chemistry as we now know it is concerned with the composition and interaction of "pure" substance. . . . To take an example: pure salt is composed exclusively of the elements sodium and chlorine and contains no water; but *pure salt* is a product of modern laboratory technique and virtually did not exist in Thales' day. Thales on the other hand was dealing with things as they are, and not with things neatly sorted and cleaned up by chemists. His dictum then, though certainly not wholly true, was, at its face value, very far from being nonsense. The greater part of the earth's surface is water; water pervades every region of our atmosphere; life as we know it is impossible without water; water is the nearest approach to the alchemists' dream of a universal solvent; water disappears when fanned by the wind, and falls again from the clouds as rain; ice turns into water as does the snow that falls from the skies; and a whole country surrounded by a barren desert is fertile, rich, and populous because a huge mass of water sweeps through it annually. All these facts—the last particularly, for we know that he had spent much time in Egypt—gave Thales good grounds for thinking that *if there is any one thing at the basis of all nature,* that thing must be water. If there is any one thing! This supposition, that is to say, the asking as it were of this question, constitutes Thales' claim to immortality. The fact that he made a guess at the answer, and a pretty good guess at that, is of minor importance. If he had championed the cause of treacle as the sole "element" he would still have been rightly honoured as the father of speculative science. True, others before him (such as Homer and Hesiod) had sketched the origin of the world from one substance, but they were not content to deal with *verae causae,* that is with things whose existence can be verified by observation. To attempt to explain the origin and process of the world by having recourse to gods and spirits endowed with special powers, is merely to beg the question, since the existence of such beings can never be proved (nor of course disproved) by the means wherewith we know the world. In a word, it was Thales who first attempted to explain the variety of nature as the modifications of something in nature.

What of the other sayings of Thales? Here we are on more debatable ground. Most historians are agreed that they not only imply a theory as to the "go" of things, but also that the "go" is part of the stuff. . . . As usual Aristotle seems to be our safest guide, and he states that Thales "was led to this probably by the observation that the nutriment of all things was moist, and that heat itself is generated by moisture, and living beings live by it. . . ." Now if we take this in conjunction with the saying about the magnet we cannot doubt but that Thales regarded nature as a single immense organism composed of lesser organisms of which the earth was one, and also every tree and stone upon it.

At a superficial glance this attempt to explain the "go" of things is merely a bolder and yet more fantastic form of the myths in which the earlier poets, Homer and Hesiod, saw the "creation" of the earth out of the ocean, and filled every tree and brook with a god or nymph. But there is a subtle change in the emphasis; the gods and nymphs of mythology were human persons writ large—the products of the poetic fancy. The "gods" and "souls" of Thales were conceived, so it seems, as part of the nature of the natural object regarded as an organism. This is a vital step towards objectivity, that is towards the placing of man within the wider system of nature. It is not yet pure objectivity; that lies far ahead.

Why was this all-important step taken in such a way as to make the new world-view so fantastic? Here we have an example of the necessity to place ourselves as far as possible at the standpoint of the thinker we are considering. Divesting ourselves of all preoccupations with airborne microbes and the technique of sterilisation we see at once that the boundary between the living and the nonliving fades out. The cattle graze upon the grass; the grass feeds upon the earth; slime feeds upon the rocks; maggots grow out of slime. Everything grows out of something else. What is man that he should draw a line between the grass and the earth, between the slime and the rock? But everywhere for growth there must be water; and for the fishes of the sea this alone seems to suffice. So the whole world grows out of water.

Seen in this light, is the picture any more perverted than that in which the soil with its teeming population is regarded merely as an anchor and a sponge for the roots of the trees we plant?

We may agree then that Thales asked one of the most important questions in human history; that the answer was framed in terms intelligible to his contemporaries, and in a form perhaps fantastic but certainly not nonsensical to us. Criticism may however be levelled at it on two quite different points. First, that the answer, even if a real answer, was barren, in that no useful consequences could be drawn from it for the improvement of man's lot. Second, that in any case it was in fact no real answer. The first criticism would never have occurred to the Greeks of his day. Speculation on the nature of things was restricted to the rich and powerful for whom an adequate slave population provided those amenities which were likely to be desired by thinkers of such temper. The second point of weakness was recognised implicitly by his immediate successor, Anaximandros, namely that Thales had given no sort of clue as to *how* water could give rise to all the variety of natural beings unless it contained within itself something else with which it could react in some way or other; but in that case water as such could not be the primordial stuff. . . .

It is clear to us that a simple stuff can give rise to any form of variety only if it is either mixed with something else (in which case it would cease to be "the" material cause of nature) or capable of varying its relationships with respect to the space it occupies. The latter device was adopted by Anaximenes when he spoke of the formation of actual clouds, water, wind, earth as being formed from the ἀηρ by condensation and rarefaction. Though, as is inevitable with the pioneer, he is claiming for his principle of change more than it could achieve, yet when we think of the variation of form and texture which can be superinduced on substances held to be chemically (i.e. materially) identical (e.g. water, steam, and ice; diamond and graphite) by variation of the spatial relationships of identical units, we must concede to Anaximenes a permanent niche in the gallery of natural philosophers.

The first of the above alternatives, namely that there must be more than one primary substance (though the number may nevertheless be incomparably smaller than the huge variety of things which nature displays), was adopted by Empedokles of Akragas, who declared that four primary substances—air, fire, earth and water—are sufficient to account for all else.

Bless us, divine number, thou who generatest gods and men! O holy, holy tetraktys, thou that containest the root and the source of the eternally flowing creation! For the divine number begins with the profound, pure unity until it comes to the holy four; then it begets the mother of all, the all-comprising, the all-bounding, the first-born, the never-swerving, the never-tiring holy ten, the keyholder of all.

> Prayer of the Pythagoreans
> addressed to Tetraktys (the holy fourfoldness)
> (6th century B.C.)

The Music of the Spheres °
by Giorgio de Santillana (1947)

PYTHAGORAS

In the Island of Samos, about 590 B.C., was born the enigmatic character known to us as Pythagoras. When he was forty he removed to southern Italy, which was coming to be called Greater Greece because of the numerous Greek colonies on the coast, and there founded a brotherhood, with some monastic features, which lasted for about two hundred years and left a tradition which was still living in the early Christian centuries. Of his life we know little, and that for a significant reason. By his immediate followers, he was already recognized (like his near contemporaries, Zoroaster and Buddha) as one of those divine men who stand as mere names in history because their lives are at once transfigured into legend. He was, they said, the son of Apollo by a mortal woman. Historians recorded some of the miracles ascribed to his more than human powers. In fact, the later the story, the more wondrous. Here is one such transmission from late Antiquity; it is due to Hippolytus,[1] a Christian bishop.

[Pythagoras] . . . in his studies of nature, mingled astronomy and geometry and music and arithmetic. And thus he asserted that God is a monad, and examining the nature of number with especial care, he said that the universe produces melody and is put together with harmony, and he first proved the motion of the seven stars to be rhythm and melody. And in wonder at the structure of the universe, he decreed that at first his disciples should be silent, as it were mystae [adepts of the mysteries] who were coming into the order of the all; then when he thought they had sufficient education in the principles of truth, and had sought wisdom sufficiently in regard to stars and in regard to nature, he pronounced them pure and then bade them speak. He separated his disciples into two groups and called one esoteric [i.e., inner] and the other exoteric. . . . The early sect perished in a conflagration in Kroton in Italy. And it was the custom when one became a disciple for him to liquidate his property and to leave his money under a seal with Pythagoras, and he remained in silence sometimes three years sometimes five years, and studied. On being released from this, he mingled with the others and continued a disciple and made his home with them; otherwise he took his money and was sent off. . . .

    ° From Giorgio de Santillana, *The Origins of Scientific Thought,* The New American Library, Inc., New York. Copyright © 1961 by Giorgio de Santillana. Reprinted by permission of The New American Library, Inc.
    1. Quoted in Milton Charles Nahm (ed.), *Selections from Early Greek Philosophy,* Appleton-Century-Crofts, Inc., New York, 1947.

HARMONY AND NUMBER

Orpheus . . . had tamed animals and caused stones to rise into walls by the magic of his rhythms. Now, the myth of Orpheus contains a thought that took shape in the mind of Pythagoras. How can music possess this magical influence over the soul, and be the instrument of order already verified by the early legislators? There may be in the principle of life itself, and in the soul of man and of universal nature, chords that can answer to the touch of harmonious sound.

That is why Pythagoras taught that the soul is, or contains, a *harmonia*—the original word meaning not, as with us, a concord of several sounds, but the orderly adjustment of parts in a complex fabric, and, in particular, the tuning of a musical instrument. Now Pythagoras's one physical discovery, the original starting point of mathematical physics, was that the concordant intervals of the musical scale can be exactly expressed in terms of simple ratios. By changing the length of strings on a monochord stopped by a movable bridge, he found that the ratio of the octave is 1:2, of the fourth, 4:3, of the fifth, 3:2. These are the fixed intervals common to all Greek scales. The numbers which occur in these ratios are 1, 2, 3, 4, the sum of which is 10, the perfect number. So perfect and potent, indeed, that Pythagoras worshipped it as the Divine Ungenerated *Tetraktys*, "The source having the roots of ever-flowing nature"—a symbol of the Higher Unity wherein the One is unfolded.

Thus was the theory born: "all things are numbers"; which meant, in the old terms, "The nature of things is number." No use asking what it could mean precisely, in the moment of creative insight. There were in it centuries of thoughts and feelings unknown to us; the precise delicate twanging of the lyre bursting like stars upon the soul, turned into stars in the all-enveloping darkness of heaven; the One, the Monad of intelligent fire alone in the dark of Unlimit, populating that dark in mysterious array with units of fire like itself, as "the One begat the Two, the Two begat the Three, and the Three begat all things. . . .

"In approaching the moment of illumination," writes Cornford, the soul must have reached out with every power intent. . . . The final act of recognition must be overwhelming, because the truth, in such a moment of insight, is not presented as an intellectual formula, compact and comprehensible. It comes rather as an undefined mass of significance, fused in a glow of intense feeling. It may take years or generations for all the meaning and implications to be expressed in words. . . . When the feeling has passed, the thought is felt, an intellectual content distilled into the language of prose.

Of what happened in the early years of the sect little has reached us. Numbers seem to have been mainly figured as rows of monads. One such construction became fundamental. Take the row of integers as in Figure 3-18, and construct successive squares on them: the carpenter's squares or "gnomons" indicated by the lines "build" the successive squares of numbers: here is shown the virtue of the Odd, for the gnomons are the progression of odds.[2] The one alone is Odd-Even. The *Tetraktys* was a triangular number as in Figure 3-19, made upon the first

Figure 3-18

Figure 3-19

numbers: $1+2+3+4=10$, the perfect number, generated by the Monad and in turn conceived as generating all the other combinations of number and figure which made up what was called the *kosmos,* or "good array."

What could number come to mean in all this fermenting of ideas? There is no doubt that the Pythagoreans thought originally of physical patterns, such as are those engraved on the dice. Hence also the rather simple idea among some that shapes might be given by successive layers of gnomons or carpenter's squares, a naive kind of crystallography. And with it went a simple arithmetic of numbers made up of units, separated by gaps. But let us not forget that the "true" power of number, that which united knowledge with purification, had been discovered in the vibrating string. There it was that the *Tetraktys* rules, through the first four integers. Now, if the musical scale depends simply upon the imposition of definite proportions on the indefinite continuum of sound between high and low, might not the same principles, Limit and Unlimit, underlie the whole universe?

THE SYSTEM OF THE WORLD

It was the Pythagoreans who invented the astronomical system of the world as we understand it. This may seem unfair to the Babylonians, who knew the planets well and computed their positions with extreme

2. The series of square numbers is formed by the addition of successive odd numbers: $\boxed{1}+3=\boxed{4}+5=\boxed{9}+7=\boxed{16}+9=\boxed{25}$, etc. Ed.

accuracy. But for them it was really a matter of special positions, like heliacal risings and settings, whose recurrence when due was supposed to confirm the stable intentions of celestial powers. What those particular stars did in between was of little concern to them; it was simply represented by columns of figures on the tables. The Pythagoreans, true to their geometrical conceptions, thought of orbits on the celestial sphere, and imagined them, like the rest of the heavenly motions, as circular. They were proper motions, from west to east against the 24-hour rotation of the stars, not simply a "lag," as before imagined. The moving stars were then called "planets," or "wanderers." The different speeds suggested different distances, and it is thus that we have, with some occasional variants, the present order of the planets, with Mercury closest to the center and Saturn farthest off. The yearly period of the Sun put it in third place, between Venus and Mars. As for the nature of the Sun, which must surely be apart, there were many theories, as we shall see.

The Earth was for the first time declared to be a sphere, for reasons of symmetry, and also because its shadow on the moon during eclipses was observed to be circular [Figure 3-29]. The author of this theory is said by some to have been Pythagoras himself; by others, more reliably, Parmenides.

As if this were not momentous enough, there follows in one or two generations another no less momentous development, for the abstract geometrical way of thinking in the School gave them a prodigious freedom of imagination. Once it had become clear that sun, moon, and planets describe orbits from west to east, it probably was felt as strange that the whole heaven should rotate so much more rapidly in the opposite direction. Would it be possible to account for this by still another motion from west to east?

Philolaus, who seems to have been the chief theoretician of the period around 450 B.C., came up with the bold idea that it was really that the Earth was moving from west to east which caused the apparent rotation of the heavens the opposite way. This was a stroke of genius: the immensity of the heavens became at once more natural in repose. But Philolaus did not imagine, as we might think, the Earth turning around its center. He imagined it on the model of the moon, revolving closely around a center while keeping always the same face outward (Figure 3-20). And then he imagined another Earth on another circle just below it, keeping pace with us, and called it the "Counter-Earth." The only reference we have is a brief and unsympathetic account by Aristotle:

They also assume another earth, opposite to ours, which they call the "counter-earth," as they do not with regard to phenomena seek for their rea-

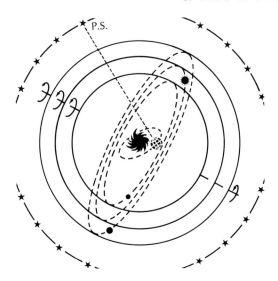

Figure 3-20. *Scheme of the Pythagorean, Philolaus, who pictured the earth swinging around a central fire once in twenty-four hours. This accounted for the daily motion of stars, sun, moon, and planets. Then spheres spinning slowly in the same direction carried the moon, sun, and planets.* (After Eric M. Rogers)

sons and causes, but forcibly make the phenomena fit their opinions and pre-conceived notions. . . . As ten is the perfect number they maintain that there must be ten bodies moving in the universe, and as only nine are visible they make the counter-earth the tenth.

This may have been for Philolaus a good reason, but certainly not the chief one. . . . there is at least one more good reason which is very evi-dent. The sphere of the universe was conceived of as defined and bounded by the outer envelope of fire, the Olympus.[3] It stood to reason that its center might be of the same nature, as suggested by pure geome-try. Philolaus postulated a Central Fire of the same Olympian nature, a "unitary harmony" around which the Earth and Counter-Earth revolve, with all the other celestial bodies, and he called it "Hearth of the all, watchtower of Zeus, altar, bond and measure of nature." In other words, the *archē* or principle must be at the strategic center. It is the only thing in the universe that does not move, while everything else does. The

3. This unconventional transformation of the abode of the gods may be prompted, as many other ideas of the sect, by relics of certain archaic images. A recurrent one is that of a mountain chain enclosing the course of the planets, called Elburz in Iran and Kwen Luen in China.

Central Fire remains then hidden, not only from us who always face "outward," but also from the antipodes, since it is screened by the Counter-Earth. But its light, some said, is visible on the moon as "ashen light."

As the Earth goes around the Central Fire in 24 hours, all the effects are produced that we expect of the Earth rotating on herself. A question which was sharply raised was that the daily motion of the Earth on her orbit would make the distance of the moon from the Earth vary considerably in the course of a day, with a consequent change in apparent diameter. It was answered that the orbit was small with respect to that of the moon—a few times the actual radius of the Earth, which would have to be considered anyway if the Earth stayed in the center. But the objection may have moved Hiketas of Syracuse, a later Pythagorean, and Heracleides of Pontus, who was a successor of Plato, to make the Earth rotate simply on herself, with the Central Fire placed inside it (Figure 3-21) and this was the beginning of other great developments.

Another even sharper question was asked of Philolaus: if the Earth swings around the Central Fire, should not the stars at the pole show a

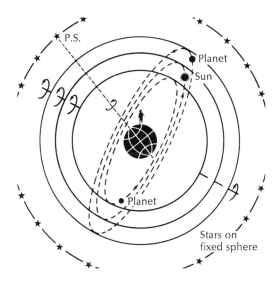

Figure 3-21. *A later Pythagorean view pictured the daily motion of stars, sun, moon, and planets as accounted for by the round earth spinning. Then spheres rotating slowly in the same direction carried the moon around once in a month, the sun around once in a year, and each planet around once in that planet's "year."* (After Eric M. Rogers)

parallax,[4] i.e., an apparent counter-swinging against the celestial background? The answer was that they are so distant as to make the effect unobservable. This was the correct answer and went on being so until Bessel's refined measurements in 1838.[5] With this the dimension of the heavens becomes really "astronomical" and immense. There were still at the time naturalists who made the sun be "pushed off" by the trade winds; a century or so earlier some of the "wise men" of Ionia had still thought the earth a pillar founded on the "deep," and surmounted by a brazen dome, which contained the "waters above." Now the Earth had become a "star" (a word which was still to arouse the ire of Inquisitors two thousand years later) moving like all other stars; the moon was "another Earth" and was supposed, as a matter of course, to have "trees and clouds and people on it. . . ."

RHYTHM AND MELODY

The idea behind these closed circling patterns of periods-within-periods is that of the "choral dance" led by heavenly bodies, or, as Hippolytus said, "rhythm and melody." The idea occurs consistently through the early works to link astronomy with music; it accounts charmingly for the "hesitating" curves described by the planets in heaven, with their stations and retrogressions which will give so much trouble to the later

4. In order to understand the meaning of parallax, hold a pencil up between your eyes and a window. Now close one eye and look at a distant object, lining it up with the edge of the pencil. Holding the pencil still, close that eye and open the other one. The pencil appears to have moved in relation to the distant object. Now try observing this same effect as you move the pencil farther away from you. The amount of parallactic displacement decreases as the angle A *decreases* (Figure 3-22). If you could move your eyes farther apart the displacement would increase. Prop the pencil up on a table and move a few feet to the left, then a few feet to the right, making an observation at both positions, and you will see that this is true. You have increased the so-called base and increased the angle A. You have therefore increased the parallactic displacement. When the optical base is known and the parallactic displacement can be measured, it is a simple geometrical problem to calculate the distance.

If you wanted to measure the displacement of very distant objects, you would, of course, use the widest possible optical base. In Figure 4-3 the different positions of the earth in its orbit (E_1, E_2, E_3, and E_4) provide the largest optical base for observing the displacement of the stars against farther stars S_1, S_2, S_3, and S_4. However the distance between S and E is very much greater than that shown in the figure, and the displacement is too small to be observed without powerful telescopes. Ed.

5. Galileo, once he had invented the telescope, affirmed that we would find a stellar parallax as soon as we had sufficiently precise instruments. The time came in 1838, when Bessel gave the parallax of 61 Cygni by alignment with more distant stars. In the same year Henderson measured Alpha Centauri, and Struve Alpha Lyrae. Even the nearest stars give a fraction of a second of arc. The modern unit for stellar distances is the parsec (parallax of one second).

mathematical astronomers. The precision of the motions, then, is that dictated by the rhythm of music.

But the over-all rotation of the sky, the obliquity of the ecliptic carrying all planetary motions in the opposite sense, seem to have suggested very soon, certainly close to the time of Parmenides (480 B.C.), the idea of separate concentric spheres as bearers of the individual motion. The "dance" is becoming geometrized, but the musical preconception remains deeply anchored. The cosmos was an ordered system which could be expressed in numerical ratios and which had partly revealed itself in the connection between the length of vibrating strings and their notes. The different radii of planetary spheres must then have harmonic ratios; they become comparable to the lengths of the string, and the angular velocities to the frequencies of vibration. Thus was born the idea of the "music of the spheres." The ones that revolve faster give a higher note than the slow ones, and all make up a harmony. "It seems to some thinkers," reports Aristotle,[6]

that bodies so great must inevitably produce a sound by their movements; even bodies on earth do so, although they are neither so great in bulk nor moving at so high a speed. . . . Taking this as their hypothesis, and also that the speeds of the stars, judged by their distances, are in the ratios of the musical consonances, they affirm that the sound of the stars as they revolve is concordant. To meet the difficulty that none of us is aware of this sound, they account for it by saying that the sound is with us right from birth and has thus no contrasting silence to show it up; for voice and silence are perceived by contrast with each other.

Two thousand years after Philolaus, Kepler tried his hand at reconstructing the "harmony of the world," but, not realizing the original

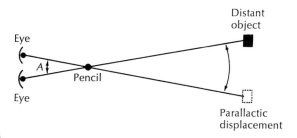

Figure 3-22

6. Aristotle, *Metaphysica*, translated by W. D. Ross, in W. D. Ross (ed.), Aristotle's *Works*, Oxford University Press, New York, 1908–1952, vol. 8.

meaning of the Greek word, he set up with infinite labor, out of the observed planetary velocities, a modern harmonic canon.

All this, including Kepler, may look very "archaic" compared to the
achievements of the hard-headed Mesopotamian computers who had
come before. It is. It turned out also to be more creative, reviving as it
did vast proto-historic schemes of thought, combining precision with
fantasy.

The way in which the Pythagoreans understood the cosmos corresponds to an almost lost state of consciousness—and we say lost because
although we have some information we have no real representation of
what ancient Greece meant by *mousikē*. It stands for an intrinsic union
of *logos*, melody, and motion. Our Western poetry is preceded by great
stretches of prose culture. Dante presupposes the Latin authors, the
Bible and Aquinas. But in Greece philosophical thought comes of a
background where technical prose was barely being born for the use of
craft books and chronicles, and all serious thought still couched in the
somewhat "poetic language" that Theophrastus notices in Anaximander.
Even in the time of Cicero, the old laws were still chanted. The poetry
of Homer or Pindar was essentially a "song," where the rhythm was dictated by the firm cadence and shape of the *logos* itself with its longs and
shorts. The choruses of tragedy, like the religious hymns, were slow choral dances whose circling was marked by a subtle alternation of forward
and back, a ripple and halting in the stance which brought out the free
"numbers" of the song. The rhythm was not superposed, as is the metro-
nome "beat" of modern music; the word and the act were as one body.
To circle in the dance meant to experience in one's being the timeless-
ness of first and last things, as expressed in the steady *logos* of the
words. It all became, as it were, the substrate of speech, a direct onto-
logical realization of the "divine" in immediate reality.

This is as much as to say that *mousikē* was not only an aesthetic experience as we mean it, but an activity closely tied to the poetic and the
ethical side of man's being: something which affects and transforms the
soul. An interchanging of tunes would make heroes out of cowards, as
the story of Tyrtaeus in Sparta is meant to show, and the legend of Orpheus tells how the hidden melody of things can hold man and nature
in its spell, once it is brought forth on the magic tones of the lyre.
Apollo, the god of Pythagoras, is the power both of light and music,
under whose strange song even stones had moved, "when Ilion like a
mist rose into towers."

Towards the fourth century B.C., the old *mousikē* tends to fall apart
into the more modern forms of music and poetry, and this may well be
one of the causes fo· the eclipse of the Pythagorean school which super-

venes at that time. Plato, in the *Republic* and then in the *Laws*, protests against the modern trend in the arts, which he calls a shameless mixing of styles and flouting of the laws, and advocates a system of education based on the pristine ways, right gymnastics and right music. The essence is, he adds, that harmony and rhythm should follow the "*logos* that is sung," and that the latter should be true, as the philosopher expects of the "*logos* that is not sung."

Against this background, the strict program of Pythagorean harmony stands out more clearly: silence, music, and mathematics. The soul must be kept clear to receive the true unspoken *logos*, "the secret language of the gods," in all its forms, from rhythm, number, and proportion to astronomy, and to grasp the connections between those forms which are inaccessible to discursive thought. The philosophical fragments of Philolaus and Archytas, in their fumbling, helpless style, make one think of men who would try to converse after a long sojourn in the region of silence. They lived among men, to be sure, they were versed in political counsel, they even taught theorems, but their own creative thinking they knew not how to talk about; it belonged to their search, their inner experience, and their unformulated intuitions.

THE HELIOCENTRIC THEORY

. . . Heracleides [of Pontus] suggested that the center of revolution for the two lower planets ought to be the sun itself (Figure 3-23); and as for the 24-hour revolution of the sky, he brought it back, reviving Philolaus's precedent, to a revolution of the earth itself. One perceives here the pursuit of some reasonably physical solution, and unmistakably in the Pythagorean vein, for it pointed the way to the sun's being the real Central Fire.

From the first the Pythagoreans had imagined a cosmos whose life came from the center outwards. . . .

The merit for having gone all the way belongs to Aristarchus of Samos, a great astronomer of c. 280 B.C. Like Heracleides and his own master Strato, he seems to have inclined to a corpuscular theory of reality, which remains the hallmark of the "physicalists," both Democritan and late Pythagorean. He probably started by generalizing Heracleides' construction and placing the center of all planetary orbits in the sun, with the sun still circling around the earth, according to what we call now the Tychonian system, because Tycho Brahe proposed it in 1577 (Figure 3-23). But he must have realized that this was only a way station towards a system making physical sense, for he came to the conclusion, as Archimedes reports, "that the fixed stars and the sun are immov-

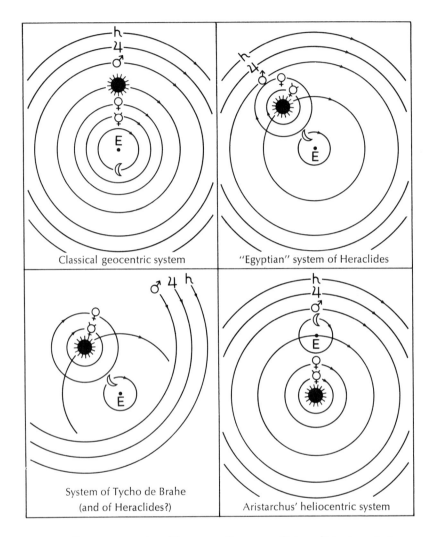

Classical geocentric system   "Egyptian" system of Heraclides

System of Tycho de Brahe   Aristarchus' heliocentric system
(and of Heraclides?)

☀ Sun   ☾ Moon   E Earth   ☿ Mercury   ♀ Venus   ♂ Mars   ♃ Jupiter   ♄ Saturn

Figure 3-23. Four early schemes for describing the solar system.

able, but that the earth is carried in a circle round the sun which is in
the middle of the course." The break-through had been achieved. This is
what we call the Copernican system, and it was invented within fifty
years of Aristotle's death.

Aristarchus was aware of the difficulties his system entailed, chief of which is the lack of an observable annual parallax of the fixed stars. Philolaus had already had to cope with that, but at least his own orbit of the earth was minuscule, and might not show up against the stars if they were kept at a reasonable distance. But now Aristarchus had placed the earth "in the third heaven," going the Great Orb in a year in the place of the sun (Figure 3-24). Could such a circle, too, be as nothing seen from the stars? It must be. We have for this, again, the reference of Archimedes himself in the *Sand Reckoner*:

He supposes . . . that the sphere of the fixed stars is of such a size that the circle, in which he supposes the earth to move, has the same ratio to the distance of the fixed stars as the center of the sphere has to the surface. But this is evidently impossible, for as the center of the sphere has no magnitude, it has no ratio to the surface.

The great Archimedes might have accepted this as only an image, instead of deploring his colleague's lack of precision. As an image, it can stand. It corresponds to saying there will be no parallax.

One wonders all the more, since such a passing slap at his colleague is really quite irrelevant to the subject Archimedes has undertaken to discuss, namely, is there a way to express numbers beyond the range of

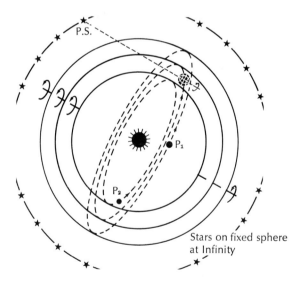

Figure 3-24. Aristarchus' scheme, showing only two specimen planets. P_1 might be Mars, Jupiter, or Saturn. P_2 might be Mercury or Venus. (After Eric M. Rogers)

the ordinary Greek notation, e.g., the number of grains of sand that
would be contained in the sphere of the cosmos. In fact, he has brought
in Aristarchus only to have the support of his authority in supposing the
cosmos much larger than commonly thought. He might have considered,
too, that the reasons for this poetic audacity on the part of Aristarchus
did not lie only in geometrical relativism. They were mainly physical,
and very novel at that. The only work of Aristarchus which has reached
us is a little treatise *On the Dimensions and Distances of the Sun and
Moon,* in which the first serious attempt is made at determining these
quantities of observation. He observed the angle [MES] between the
sun and moon at a time when the latter is half illuminated, that is, when
the angle [EMS] is a right angle (Figure 3-25): a method ingenious but
perforce miserably imprecise. He found 3° at most for the angle in the
sun (it is actually 10′). From this he deduced that the distance of the
sun must be at least eighteen times as great as the distance of the moon,
and hence also the sun at least three hundred times bigger than the
earth. A body of this size ought to be the center of gravity for the cosmos. It was all very well to say—as was being said—that celestial bodies move naturally and without effort in a circle, and that we should not
apply our terrestrial standards of weight and lightness to matters celestial, impassible, and uncorruptible. Philolaus had applied them, when
he "made of the earth a star," which implied conversely that the stars
are not without some earthiness either. Aristarchus had now lined up a
massive argument in favor of that point of view. It would be reasonable,
in fact, to assume that he made bold to suggest his Copernican hypothesis only after measuring the distance of the sun. He probably did no
more than sketch it out (Archimedes says he "outlines certain hypotheses"); and Archimedes may have thought that this was not a serious way of dealing with a well-worked-out problem, where masses of
quantitative data had to be accounted for. Still, it reveals on the part of

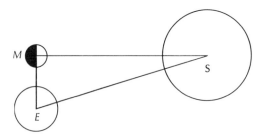

*Figure 3-25. Aristarchus' method for measuring the distance to the sun from
the earth, using the moon when it is half illuminated.*

the greatest mathematical physicist of antiquity, of the very man who had given the theory of centers of gravity, a strange insensibility to dynamic considerations. What Galileo was to feel intuitively, even before checking Copernicus's data, Archimedes did not feel at all. Was it the appalling esthetic disproportion involved in a scheme which left the solar system like a grain of sand in empty space? The fact is that he closed his mind to the new idea. This is almost the only clear fact which stands out in this strange episode of miscomprehension.

As for public opinion, it reacted almost not at all. This was too far removed from common sense to deserve comment. The philosophers reprimanded Aristarchus for speaking out of turn on a subject which they considered to be their proper domain. Cleanthes, the Stoic, scolded Aristarchus for impiety; others accused him of undermining the art of divination, which had recently been imported into Greece from the Near East. It was about the same general reaction which was to greet Copernicus. It was not strange. That was why the Pythagoreans had cloaked themselves in the magic and prophetic authority of the sect. A lone astronomer, speaking in physics only from scientific reasons, was alone indeed.

The astronomers themselves reacted coolly. They had by now become specialists, and had learned to mind their own business. They insisted that their business was to state the *how* and leave it at that. We know only of one who adopted the system a century later, Seleucus of Seleucia, an Oriental Greek from the Persian Gulf. For expert opinion, the system came probably too late to force a change of minds. Aristarchus had "saved the appearances"; he had outlined the proper symmetry of homocentric circles and velocities, but it was clear already that they could not be true sun-centered circles. . . . Then to what profit upset the visible order of the skies and put the system of the world on its head?

The Theory of Perfect Circles

Editor's Note

The heliocentric theory of Aristarchus represents the culmination of one school of Greek cosmology, which sprang from the teaching of the Pythagorean Brotherhood. In following the logical development of this line of thought up through the third century B.C. we have passed by several important philosophers who made significant contributions to Greek astronomy (see chronological table). Now we go back about 150 years, to pick up the other school of cosmology, which also developed out of the Pythagorean concept. This theory led to an entirely different solution to the problem of finding order in the cosmos, and it was this solution which was destined to be triumphant for almost 2,000 years.

Plato (428–348 B.C.) was deeply influenced by the Pythagorean idea of form and harmony as constituting the essential nature of things. He developed this idea into a philosophy of real forms, which are intelligible to the mind and of which the perceptible world is just a shadow or an imperfect copy.[1] Plato's contribution to astronomy represented a small and relatively unimportant part of his vast philosophical system. However, the cosmological theory set forth in *Timaeus*, his principal scientific dialogue, did strongly influence the philosophers who followed him. Plato believed like Pythagoras that mathematics provided the most direct approach to an understanding of the inner harmony of nature. He reasoned that since the sphere is the perfect three-dimensional shape, the universe must be a sphere and all motion of heavenly bodies must be composed of uniform circular motions.

And he gave the universe the figure which is proper and natural. . . . Wherefore he turned it, as in a lathe, round and spherical, with its extremities equidistant in all directions from the centre, the figure of all figures most perfect and most like to itself, for he deemed the like more beautiful than the unlike.

1. Plato's theory of pure forms, which we have not the space to present here, bears interesting analogies to the importance of form in modern atomic physics. To readers interested in understanding more about these aspects of Platonic theory we recommend Collingwood's *The Idea of Nature*, part I, chap. II, and de Santillana's *The Origins of Scientific Thought*, chap. 12. See Suggestions for Further Reading.

CHRONOLOGICAL TABLE

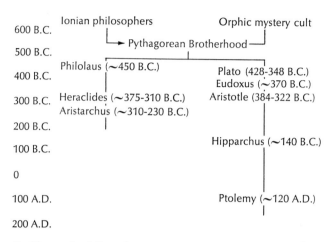

Figure 3-26. *The early philosophers.*

To the whole he gave, on the outside round about, a surface perfectly finished and smooth for many reasons. It had no need of eyes, for nothing visible was left outside it; nor of hearing, for there was nothing audible outside it; and there was no breath outside it requiring to be inhaled. . . . He allotted to it the motion of the seven motions which is most bound up with understanding and intelligence. Wherefore, turning it round in one and the same place upon itself, he made it move with circular rotation; all the other six motions (i.e., straight motion up and down, forward and back, right and left) he took away from it and made it exempt from their wanderings. And since for this revolution it had no need of feet, he created it without legs and without feet. . . . Smooth and even and everywhere equidistant from the centre, a body whole and perfect, made up of perfect bodies. . . . (*Timaeus*, Plato.)

The Triumph of a Theory *
by Eric M. Rogers (1960)

[The first successful attempt to account for the planetary motions using the principle of uniform circular motion was devised by Eudoxus about 370 B.C.] He studied geometry and philosophy under Plato, then travelled in Egypt, and returned to Greece to become a great mathemati-

---

* From Eric M. Rogers, *Physics for the Inquiring Mind: The Methods, Nature, and Philosophy of Physical Science.*

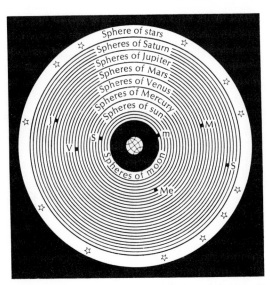

Figure 3-27. *Eudoxus' scheme of many concentric spheres. Each body, sun, moon, or planet, had several spheres spinning steadily around different axes. The combination of these motions succeeded in imitating the actual motions of sun, moon, and even planets across the star pattern.*

cian and the founder of scientific astronomy. Gathering Greek and Egyptian knowledge of astronomy and adding better observations from contemporary Babylon, he devised a scheme that would save the phenomena.

The system of a few spheres, one for each moving body, was obviously inadequate. A planet does not move steadily along a circle among the stars. It moves faster and slower, and even stops and moves backwards at intervals. The Sun and Moon move with varying speeds along their yearly and monthly paths.[1] Eudoxus elaborated that scheme into a vast family of concentric spheres, like the shells of an onion (Figure 3-27). Each *planet* was given several adjacent spheres spinning about different axes, one within the next: 3 each for Sun and Moon, 4 each for the planets: and the usual outermost sphere of all for the stars. Each sphere was carried on an axle that ran in a hole in the next sphere outside it, and the axes of spin had different directions from one sphere to the next. The combined motions, with suitably chosen spins, imitated

1. For example, the four seasons, from spring equinox to midsummer to autumn equinox, &c., are unequal. The Babylonians in their schemes for regulating the calendar by the new Moon had, essentially, time-graphs of the uneven motions of Moon and Sun.

the observed facts. Here was a system that was simple in form (spheres) with a simple principle (uniform spins), adjustable to fit the facts—by introducing more spheres if necessary. In fact, this was a good theory.

To make a good theory, we must have basic principles or assumptions that are simple; and we must be able to derive from them a scheme that fits the facts reasonably closely. Both the usefulness of a theory and our aesthetic delight in it depend on the simplicity of the principles as well as on the close fitting to facts. We also expect fruitfulness in making predictions, but that often comes with these two virtues of simplicity and accuracy. To the Greek mind, and to many a scientific mind today, a good theory is a simple one that can save all the phenomena with precision. Questions to ask, in judging a good theory, are, "Is it as simple as possible?" and "Does it save the phenomena as closely as possible?" If we also ask, "Is it *true?*", that is not quite the right requirement. We could give a remarkably *true* story of a planet's motion by just reciting its locations from day to day through the last 100 years; our account would be true, but so far from simple, so spineless, that we should call it just a list, not a theory. The earlier Greek pictures with real crystal spheres had been like myths or tales for children—simple teaching from wise men for simple people. But Eudoxus tried to devise a successful machine that would express the actual motions and predict their future. He probably considered his spheres geometrical constructions, not real globes, so he had no difficulty in imagining several dozen of them spinning smoothly within each other. He gave no mechanism for maintaining the spins—one might picture them as driven by gods or merely imagined by mathematicians.

Here is how Eudoxus accounted for the motion of a planet, with four spheres (Figure 3-28). The planet itself is carried by the innermost, embedded at some place on the equator. The outermost of the four spins round a North-South axle once in 24 hours, to account for the planet's daily motion in common with the stars. The next inner sphere spins with its axle pivoted in the outermost sphere and tilted $23\frac{1}{2}°$ from the N-S direction, so that *its* equator is the ecliptic path of the Sun and planets. This sphere revolves in the planet's own "year" (the time the planet takes to travel round the zodiac), so its motion accounts for the planet's general motion through the star pattern.[2] These two spheres are equivalent to two spheres of the simple system, the outermost sphere of stars that carried all the inner ones with it, and the planet's own sphere. The third and fourth spheres have equal and opposite spins about axes inclined at a small angle to each other. The third sphere has its axle pivoted in the zodiac of the second, and the fourth carries the

2. In terms of our view today, the spin of the outermost sphere corresponds to the Earth's daily rotation; the spin of the next sphere corresponds to the planet's own motion along its orbit round the Sun; the spins of the other two spheres combine to show the effect of viewing from the Earth which moves yearly around the Sun.

planet itself embedded in the equator. Their motions combine to add the irregular motion of stopping and backing to make the planet follow a looped path. The complete picture of this three dimensional motion is difficult to visualize.

With 27 spheres in all, Eudoxus had a system that imitated the observed motions quite well: he could save the phenomena. The basis of his scheme was simple: perfect spheres, all with the same center at the Earth, spinning with unchanging speeds. The mathematical work was far from simple: a masterpiece of geometry to work out the effect of four spinning motions for each planet and choose the speeds and axes so that the resultant motion fitted the facts. In a sense, Eudoxus used harmonic analysis—in a three-dimensional form!—two thousand years before Fourier. It was a good theory.

But not very good: Eudoxus knew there were discrepancies, and more accurate observations revealed further troubles. The obvious cure—add more spheres—was applied by his successors. One of his pupils, after consulting Aristotle, added 7 more spheres, greatly improving the agreement. For example, the changes in the Sun's motion that make the four seasons unequal were now predicted properly. Aristotle himself was worried because the complex motion made by one planet's quartet of

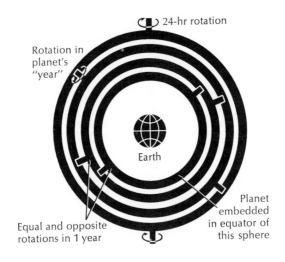

Figure 3-28. *Part of Eudoxus' scheme used four spheres to imitate the motion of a planet. The sketch shows the machinery for one planet. The outermost sphere spins once in twenty-four hours. The next inner sphere rotates once in the planet's "year." The two innermost spheres spin with equal and opposite motions, once in our year, to produce the planet's epicycloid loops.*

spheres would be handed on, unwanted, to the next planet's quartet. So he inserted extra spheres to "unroll" the motion between one planet and the next, making 55 spheres in all. The system seems to have stayed in use for a century or more till a simpler geometrical scheme was devised. (An Italian enthusiast attempted to revive it 2000 years later, with 77 spheres.)

Aristotle (340 B.C.), the great teacher, philosopher, and encyclopedic scientist, was the "last great speculative philosopher in ancient astronomy." He had a strong sense of religion and placed much of his belief in the existence of God on the glorious sight of the starry heavens. He delighted in astronomy and gave much thought to it. In supporting the scheme of concentric spinning spheres, he gave . . . [the same reason as Plato] *the sphere is the perfect solid shape.* . . . By the same token, the Sun, Moon, planets, stars must be spherical in form. The heavens, then, are the region of perfection, of unchangeable order and circular motions. The space between Earth and Moon is unsettled and changeable, with vertical fall the natural motion.

For ages Aristotle's writings were the only attempt to systematize the whole of nature. They were translated from language to language, carried from Greece to Rome to Arabia, and back to Europe centuries later, to be copied and printed and studied and quoted as the authority. Long after the crystal spheres were discredited and replaced by eccentrics, those circles were spoken of as spheres; and the medieval schoolmen returned to crystal spheres in their short-sighted arguments, and believed them real. The distinction between the perfect heavens and the corruptible Earth remained so strong that Galileo, 2000 years later, caused great annoyance by showing mountains on the Moon and claiming the Moon was earthy; and even he, with his understanding of motion, found it hard to extend the mechanics of downward earthly fall to the circular motion of the heavenly bodies.

Aristotle made a strong case for the Earth itself being round. He gave theoretical reasons:

(i) *Symmetry:* a sphere is symmetrical, perfect.

(ii) *Pressure:* the Earth's component pieces, falling *naturally* towards the center, would press into a round form.

and experimental reasons (Figure 3-29):

(iii) *Shadow:* in an eclipse of the Moon, the Earth's shadow is always circular: a flat disc could cast an oval shadow.

(iv) *Star heights:* even in short travels Northward or Southward, one sees a change in the position of the star pattern.

This mixture of dogmatic "reasons" and experimental common sense is typical of him, and he did much to set science on its feet. His whole teaching was a remarkable life work of vast range and enormous influence. At one extreme he catalogued scientific information and listed stimulating questions; at the other extreme he emphasized the basic problems of scientific philosophy, distinguishing between the *true physical causes* of things and *imaginary schemes to save the phenomena.*

The Shape of Heaven and Earth °

by Aristotle (4th century B.C.)

The shape of the heaven must be spherical. That is most suitable to its substance, and is the primary shape in nature. But let us discuss the question of what is the primary shape, both in plane surfaces and in solids. Every plane figure is bounded either by straight lines or by a circumference; the rectilinear is bounded by several lines, the circular by one only. Thus since in every genus the one is by nature prior to the many, and the simple to the composite, the circle must be the primary plane figure. Also, if the term "perfect" is applied, according to our previous definition, to that outside which no part of itself can be found, and addition to a straight line is always possible, to a circle never, the circumference of the circle must be a perfect line: granted therefore that the perfect is prior to the imperfect, this argument too demonstrates the priority of the circle to other figures. By the same reasoning the sphere is the primary solid, for it alone is bounded by a single surface, rectilinear solids by several. The place of the sphere among solids is the same as that of the circle among plane figures. . . .

Again, since it is an observed fact, and assumed in these arguments, that the whole revolves in a circle, and it has been shown that beyond the outermost circumference there is neither void nor place, this provides another reason why the heaven must be spherical. For if it is bounded by straight lines, that will involve the existence of place, body, and void. A rectilinear body revolving in a circle will never occupy the same space, but owing to the change in position of the corners there will at one time be no body where there was body before, and there will be body again where now there is none. It would be the same if it were of some other shape whose radii were unequal, that of a lentil or an egg for example. All will involve the existence of place and void

° From Aristotle, *On the Universe,* translated by W. K. C. Guthrie, Loeb Classical Library, Harvard University Press, Cambridge, 1939. Reprinted by permission of Harvard University Press.

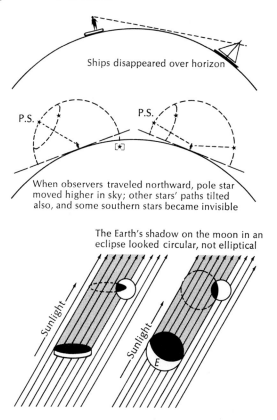

Ships disappeared over horizon

P.S.   P.S.

When observers traveled northward, pole star
moved higher in sky; other stars' paths tilted
also, and some southern stars became invisible

The Earth's shadow on the moon in an
eclipse looked circular, not elliptical

Sunlight

Sunlight

E

Figure 3-29. Evidences for a round earth.

outside the revolution, because the whole does not occupy the same
space throughout. . . .

For ourselves, let us first state whether the earth is in motion or at
rest. Some, as we have said, make it one of the stars, whereas others
put it at the centre but describe it as winding and moving about the
pole as its axis. But the impossibility of these explanations is clear if we
start from this, that if the earth moves, whether at the centre or at a distance from it, its movement must be enforced: it is not the motion of
the earth itself, for otherwise each of its parts would have the same
motion, but as it is their motion is invariably in a straight line towards
the centre. The motion therefore, being enforced and unnatural, could
not be eternal; but the order of the world is eternal.

Secondly, all the bodies which move with the circular movement are observed to lag behind and to move with more than one motion, with the exception of the primary sphere: the earth therefore must have a similar double motion, whether it move around the centre or as situated at it. But if this were so, there would have to be passings and turnings of the fixed stars. Yet these are not observed to take place: the same stars always rise and set at the same places on the earth.[1]

Thirdly, the natural motion of the earth as a whole, like that of its parts, is towards the centre of the Universe: that is the reason why it is now lying at the centre. It might be asked, since the centre of both is the same point, in which capacity the natural motion of heavy bodies, or parts of the earth, is directed towards it; whether as centre of the Universe or of the earth. But it must be towards the centre of the Universe that they move, seeing that light bodies like fire, whose motion is contrary to that of the heavy, move to the extremity of the region which surrounds the centre. It so happens that the earth and the Universe have the same centre, for the heavy bodies do move also towards the centre of the earth, yet only incidentally, because it has its centre at the centre of the Universe. As evidence that they move also towards the centre of the earth, we see that weights moving towards the earth do not move in parallel lines but always at the same angles to it:[2] therefore they are moving towards the same centre, namely that of the earth. It is now clear that the earth must be at the centre and immobile. To our previous reasons we may add that heavy objects, if thrown forcibly upwards in a straight line, come back to their starting-place, even if the force hurls them to an unlimited distance.

From these considerations it is clear that the earth does not move, neither does it lie anywhere but at the centre. In addition the reason for its immobility is clear from our discussions. If it is inherent in the nature of earth to move from all sides to the centre (as observation shows), and of fire to move away from the centre towards the extremity, it is impossible for any portion of earth to move from the centre except under constraint; for one body has one motion and a simple body a simple motion, not two opposite motions, and motion from the centre is the opposite of

1. The criticism depends on the analogy with the planets, following which A. assumes that if the earth moved with a motion of its own, as well as being carried round in the motion of the first heaven, its proper motion would be in the plane of the ecliptic and not of the equator. Were this so, the fixed stars would exhibit to our eyes the irregularities which he describes by the words παρόδους καὶ τροπάς; the pole-star would appear to describe a circle in the sky, and the stars would not rise and set as they do.

2. I.e. at right angles to a tangent. Stocks explains the Greek as meaning that the angles at each side of the line of fall of any one body are equal. But does it not more naturally mean that the angles made by one falling body with the earth are similar to those made by another (Figure 3-30)?

motion towards it. If then any particular portion is incapable of moving from the centre, it is clear that the earth itself as a whole is still more incapable, since it is natural for the whole to be in the place towards which the part has a natural motion. If then it cannot move except by the agency of a stronger force, it must remain at the centre. This belief finds further support in the assertions of mathematicians about astronomy: that is, the observed phenomena—the shifting of the figures by which the arrangement of the stars is defined—are consistent with the hypothesis that the earth lies at the centre. This may conclude our account of the situation and the rest or motion of the earth.

Its shape must be spherical. For every one of its parts has weight until it reaches the centre, and thus when a smaller part is pressed on by a larger, it cannot surge round it, but each is packed close to, and combines with, the other until they reach the centre. To grasp what is meant we must imagine the earth as in the process of generation in the manner which some of the natural philosophers describe (except that they make external compulsion responsible for the downward movement: let us rather substitute the true statement that this takes place because it is the nature of whatever has weight to move towards the centre). In these systems, when the mixture existed in a state of potentiality, the particles in process of separation were moving from every side alike towards the centre. Whether or not the portions were evenly distributed at the extremities, from which they converged towards the centre, the same result will be produced. It is plain, first, that if particles are moving from

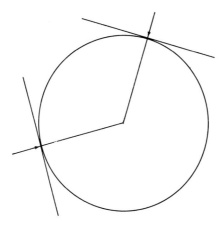

Figure 3-30

all sides alike towards one point, the centre, the resulting mass must be similar on all sides; for if an equal quantity is added all round, the extremity must be at a constant distance from the centre. Such a shape is a sphere. But it will make no difference to the argument even if the portions of the earth did not travel uniformly from all sides towards the centre. A greater mass must always drive on a smaller mass in front of it, if the inclination of both is to go as far as the centre, and the impulsion of the less heavy by the heavier persists to that point. . . .

If then the earth has come into being, this must have been the manner of its generation, and it must have grown in the form of a sphere: if on the other hand it is ungenerated and everlasting, it must be the same as it would have been had it developed as the result of a process. Besides this argument for the spherical shape of the earth, there is also the point that all heavy bodies fall at similar angles, not parallel to each other; this naturally means that their fall is towards a body whose nature is spherical. Either then it *is* spherical, or at least it is natural for it to be so, and we must describe each thing by that which is its natural goal or its permanent state, not by any enforced or unnatural characteristics.

Further proof is obtained from the evidence of the senses. (i) If the earth were not spherical, eclipses of the moon would not exhibit segments of the shape which they do. As it is, in its monthly phases the moon takes on all varieties of shape—straightedged, gibbous and concave—but in eclipses the boundary is always convex. Thus if the eclipses are due to the interposition of the earth, the shape must be caused by its circumference, and the earth must be spherical. (ii) Observation of the stars also shows not only that the earth is spherical but that it is of no great size, since a small change of position on our part southward or northward visibly alters the circle of the horizon, so that the stars above our heads change their position considerably, and we do not see the same stars as we move to the North or South. Certain stars are seen in Egypt and the neighbourhood of Cyprus, which are invisible in more northerly lands, and stars which are continuously visible in the northern countries are observed to set in the others. This proves both that the earth is spherical and that its periphery is not large, for otherwise such a small change of position could not have had such an immediate effect. For this reason those who imagine that the region around the Pillars of Heracles joins on to the regions of India, and that in this way the ocean is one, are not, it would seem, suggesting anything utterly incredible. They produce also in support of their contention the fact that elephants are a species found at the extremities of both lands, arguing that this phenomenon at the extremes is due to communication between the two. Mathematicians who try to calculate the circumference put it at 400,000 stades.[3]

From these arguments we must conclude not only that the earth's mass is spherical but also that it is not large in comparison with the size of the other stars.

The Triumph of a Theory ° (continued)
by Eric M. Rogers (1960)

NEW THEORIES: ECCENTRICS; EPICYCLES

In the school at Alexandria, the bold suggestion [of Aristarchus] of making the Earth spin and move round a central Sun did not find favor. A stationary central Earth remained the popular basis, but spinning

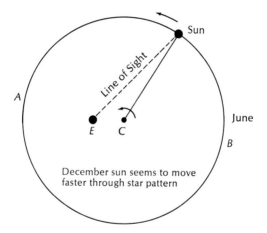

Figure 3-31. *The eccentric scheme for the sun pictures the sun carried around in a circular path by a radius that rotates at constant speed, as in the simplest system of spheres. The observer, on the earth, is off-center, so that he sees the sun move unevenly—as it does—faster in December, slower in June.*

° From Eric M. Rogers, *Physics for the Inquiring Mind: The Methods, Nature, and Philosophy of Physical Science.*

3. I.e. 9987 geographical miles. Prantl remarks that this is the oldest recorded calculation of the earth's circumference. . . . (The present-day figure in English miles is 24,902.)

concentric spheres made the model too difficult. Instead, the slightly uneven motion of the Sun around its "orbit" could be accounted for by a
single eccentric circle: the Sun was made to move steadily round a circle, with the Earth fixed a short distance off center. Then, as seen from
the Earth, the Sun would seem to move faster at some seasons (about
December, at A) and slower 6 months later (at B) (Figure 3-31). This
was still good theory. For good theory, the scheme should have an appealing simplicity and be based on simple assumptions. These needs
were met: a perfect circle of constant radius, and motion with *constant*
speed round it. Such constancies were necessary to the Greek mind—in
fact to any orderly scientific mind. Without them theory would degener-
ate into a pack of demons. Placing the Earth off-center was a regretta-
ble lapse from symmetry, but then the Sun's speed *does* behave
unsymmetrically—our summer is longer than our winter. A similar
scheme served for the Moon, but the planets needed more machinery
(Figures 3-32 and 3-33). Each planet was made to move steadily around
a circle, once around in its own "year," with the Earth fixed off center;
but then the whole circle, *planet's orbit and center*, was made to revolve
around the Earth once in 365 days. This added a small circular motion
(radius EC) to the large main one, producing the planet's epicycloid
track. The daily motion with the whole star pattern was superimposed
on this.

Another scheme to produce the same effect used a fixed main circle
(the deferent) with a radius arm rotating at constant rate (Figure 3-34).
The end of that arm carried a small sub-circle (the epicycle) (Figure 3-
35). A radius of that small circle carried the planet round at a steady
rate, once in 365 days. Though these schemes operated with circles, they

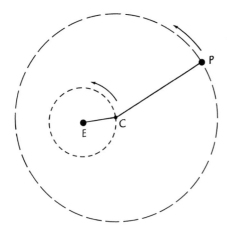

*Figure 3-32. The eccentric scheme
for a planet pictures each planet being
carried at the end of a radius that
rotates at constant speed; but this
whole circle—center, radius, and
planet—revolves once a year round the
eccentric earth. (To picture the motion, imagine the radius CP continued
out to be the handle of a frying pan.
Then imagine the frying pan given a
circular motion [of small radius EC]
around E, as center, by a housewife
who wants to melt a piece of butter
in it quickly. Then make the handle
CP revolve too—very slowly, for an
outer planet like Jupiter.)*

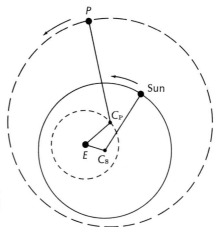

Figure 3-33. Sketch showing the machinery for motions of the sun and one planet.

were described more grandly in terms of spinning spheres and subspheres. For many centuries, astronomers thought in terms of such "motions of the heavenly spheres"—the spheres themselves growing more and more real as Greek delight in pure theory gave place to childish insistence on authoritarian truth.

Hipparchus (~ 140 B.C.), "one of the greatest mathematicians and astronomers of all time,"[1] made great advances. He was a careful observer, made new instruments and used them to measure star positions. A new star that blazed into prominence for a short time is said to have inspired him to make his great star catalogue, classifying stars by brightness and recording the positions of nearly a thousand by celestial latitude and longitude. He made the first recorded celestial globe. There were no telescopes,[2] nothing but human eyes looking through sighting holes or along wooden sticks. Simple instruments like dividers measured angles between one star (or planet) and another, or between star and plumb line. Yet Hipparchus tried to measure within ⅙°.

He practically invented spherical trigonometry for use in his studies of the Sun and Moon. He showed that eccentrics and epicycles are equivalent for representing heavenly motions. Adding his own observations to earlier Greek ones and Babylonian records, he worked out epicyclic systems for the Sun and Moon. The planets proved difficult for want of accurate data, so he embarked on new measurements.

From Greek observations dating back 150 years Hipparchus discov-

1. Sarton, *History of Science.*
2. That invention was seventeen centuries in the future. By magnifying a patch of sky it enables much finer measurements to be made.

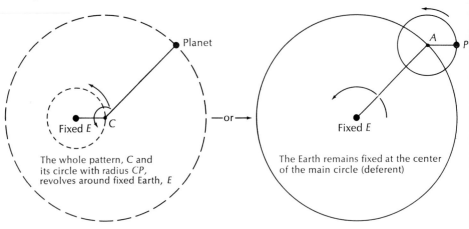

Figure 3-34. *The eccentric scheme and the epicycle scheme compared.*

ered a very small but very important astronomical creep: the "precession of the equinoxes" (Figure 3-36). At the spring equinox, midway between winter and summer, the Sun is at a definite place in the zodiac, and it returns there each year. But Hipparchus discovered that the Sun is not quite in the same patch of stars at the next equinox. In fact it gets around to the old patch of stars about 20 minutes too late, so that at the exact instant of equinox it is a little earlier on its zodiac path, about $\frac{1}{70}°$ after one year, nearly $1\frac{1}{2}°$ after a century. Hipparchus found this from changes in star longitudes between old records and new. Longitudes are reckoned along the zodiac, from the spring equinox where the equator cuts the ecliptic. Since all longitudes seemed to be changing by one or two degrees a century, Hipparchus saw that the zodiac girdle must be slipping round the celestial sphere at this rate, carrying all the stars with it, leaving the celestial equator fixed with the fixed Earth.[3]

3. The Sun's ecliptic path cuts the celestial equator in two points. When the Sun reaches either of these it is symmetrical with respect to the Earth's axis. Day and night are equal for all parts of the Earth: that is an equinox. Precession is a slow rotation of the whole celestial sphere, including the zodiac and the Sun, around the *axis of the ecliptic*, perpendicular to the ecliptic plane. From one century to the next, the creeping of precession brings a slightly different part of the zodiac belt to the equinox-points (*where ecliptic cuts equator*)—hence the name. The whole celestial sphere joins in this slow rotation round the ecliptic axis. This applies, for example, to the stars near the N-S axis, which is fixed with the Earth and $23\frac{1}{2}°$ from the ecliptic axis; so the motion carries the current pole-star away from the N-S axis and brings a new one in the course of time. Thus, in some centuries there is a bright star in the right position for pole-star, and in others there is no real star, only a blank in the pattern. In the 40-odd centuries between the building of the pyramids

This motion seems small—a whole cycle takes 26,000 years—yet it matters in astronomical measurements, and has always been allowed for since Hipparchus discovered it. The discovery itself was a masterpiece of careful observing and clever thinking.

Precession remained difficult to visualize until Copernicus, sixteen centuries later, simplified the story with a complete change of view. Even then it remained unexplained—unconnected with other celestial phenomena—until Newton gave a simple explanation. Discovered as a mysterious creep, precession is now a magnificent witness to gravitation.[4]

Hipparchus left a fine catalogue of stars, epicycle schemes, and good planetary observations—a magnificent memorial to a great astronomer. These had to wait two and a half centuries for the great mathematician Ptolemy to organize them into a successful theory.

Ptolemy (∼ A.D. 120) made a "critical reappraisal of the planetary records." He collected the work of Hipparchus and his predecessors, adding his own observations, evolving a first-class theory, and left a masterly exposition that dominated astronomy for the next fourteen centuries.

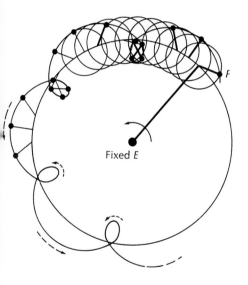

Figure 3-35. *The path of a planet as pictured by the epicycle scheme, showing how the two circular motions combine to produce the epicycloid pattern that is observed for a planet.*

Fixed E

F

and the present, this motion has accumulated a considerable effect. In fact, by examining the pyramid tunnels that were built to face the dog-star Sirius at midnight at the Spring equinox, we can guess roughly how long ago they were built.

4. See p. 158 for modern explanation. Ed.

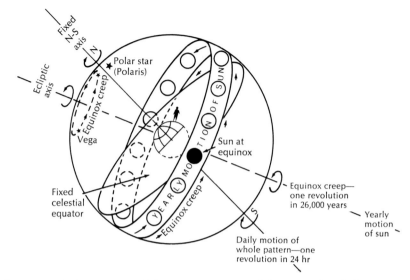

Figure 3-36. Precession of the equinoxes. In addition to the daily motion of the whole heavens around the N-S axis fixed in the fixed earth and the yearly motion of the sun around its ecliptic path in the zodiac band of stars, Hipparchus discovered a slow rotation of the whole pattern of stars around a different axis, the ecliptic axis, perpendicular to the zodiac belt.

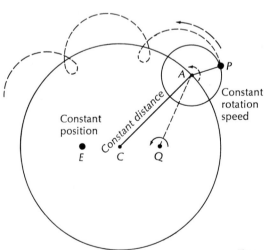

Figure 3-37. The Ptolemaic scheme imitated the motions of sun, moon, and planets very accurately.

The positions of the Sun, Moon, and planets, relative to the fixed stars, had been mapped with angles measured to a fraction of a degree. He could therefore elaborate the system of eccentric crystal spheres and epicycles and refine its machinery, so that it carried out past motions accurately and could grind out future predictions with success. He devised a brilliant *mathematical machine,* with simple rules but complex details, that could "save the phenomena" with century-long accuracy. In this, he neglected the crystal spheres as moving agents; he concentrated on the rotating spoke or radius that carried the planet around, and he provided sub-spokes and arranged eccentric distances. He expounded his whole system for Sun, Moon, and planets in a great book, the *Almagest.*

Ptolemy set forth this general picture: the heaven of the stars is a sphere turning steadily round a fixed axis in 24 hours; the Earth must remain at the center of the heavens—otherwise the star pattern would show parallax changes; the Earth is a sphere, and it must be at rest, for various reasons—e.g., objects thrown into the air would be left behind a moving Earth. The Sun moves round the Earth with the simple epicycle arrangement of Hipparchus, and the Moon has a more complicated epicyclic scheme.

In his study of the "five wandering stars," as he called the planets, Ptolemy found he could not "save the phenomena" with a simple epicyclic scheme. There were residual inequalities or discrepancies between theory and observation. He tried an epicycle scheme with the Earth eccentric, moved out from the center of the main circle. That was not sufficient, so he not only moved the Earth off center but also moved the center of uniform rotation out on the other side. He evolved the successful scheme shown in Figure 3-37. C is the center of the main circle; E is the eccentric Earth; Q is a point called the "equant," an equal distance the other side of C (QC=CE). An arm QA rotates with constant speed around Q, swinging through equal angles in equal times, carrying the center of the small epicycle, A, round the main circle. Then a radius, AP, of the epicycle rotates steadily, carrying the planet P. It was a desperate and successful attempt to maintain a scheme of circles, with constant rotations. The arm of the little circle carrying the planet rotated at constant rate. Ptolemy felt compelled to have an arm of the main circle also rotating at constant rate. To fit the facts, that arm could not be a radius from the center of the main circle, as in the simplest epicycle scheme. Nor could it be the arm from an eccentric Earth E. But he could save the phenomena with an arm from the equant point Q that did rotate at constant speed. Thus for each planet's main circle, there were three points, all quite close together, each with a characteristic constancy.

E	C	Q
the Earth in *fixed position*	the center of the main circle with arm CA of *constant length*	the equant point with arm QA rotating at *constant speed*

By choosing suitable radii, speeds of rotation, and eccentric distance EC (= CQ), Ptolemy could save the phenomena for all planets (though Mercury required a small additional circle). The main circle was given a different tilt for each planet, and the epicycle itself was tilted from the main circle.

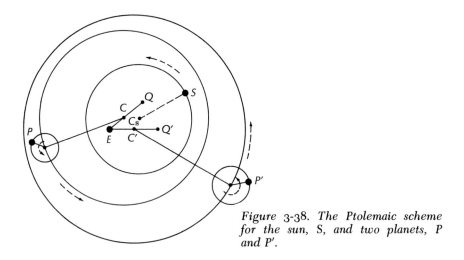

Figure 3-38. *The Ptolemaic scheme for the sun, S, and two planets, P and P'.*

Here was a gorgeously complicated system of main circles and subcircles, with different radii, speeds, tilts, and different amounts and directions of eccentricity. The system worked: like a set of mechanical gears, it ground out accurate predictions of planetary positions for year after year into the future, or back into the past. And, like a good set of gears, it was based on essentially simple principles: circles with *constant radii*, rotations with *constant speeds*, *symmetry* of equant (QC = CE), *constant* tilts of circles, and the Earth fixed in a constant position.[5]

5. If this insistence on circles seems artificial—a silly way of dealing with planetary orbits—remember:
   (i) that you have modern knowledge built into your own folklore,
   (ii) that though this now seems to you an unreal model, it is still fashionable as a method of analysis. Adding circle on circle in Greek astronomy corresponds to our use of a series of sines (projected circular motions) to analyze complex motions. Physicists today use such "Fourier analysis" in studying any repeating motion: analyzing

In the *Almagest*, Ptolemy described a detailed scheme for each planet and gave tables from which the motion of each heavenly body could be read off (Figure 3-38). The book was copied (by hand, of course), translated from Greek to Latin to Arabic and back to Latin as culture moved eastward and then back to Europe. There are modern printed versions with translations. It served for centuries as a guide to astronomers, and a handbook for navigators. It also provided basic information for that extraordinary elaboration of man's fears, hopes, and greed—astrology—which needed detailed records of planetary positions.

The Ptolemaic scheme was efficient and intellectually satisfying. We can say the same of our modern atomic and nuclear theory. If you asked whether it is true, both Greeks and moderns would question your word "true"; but if you offered an alternative that is simpler and more fruitful, they would welcome it.

musical sounds, predicting tides in a port, expressing atomic behavior. *Any* repeating motion however complex can be expressed as the resultant of simple harmonic components. Each circular motion in a Ptolemaic scheme provides two such components, one up-and-down, one to-and-fro. The concentric spheres of Eudoxus could be regarded as a similar analysis, but in a more complex form. Either scheme can succeed in expressing planetary motion to any accuracy desired, if it is allowed to use enough components.

Is Order Inherent in Nature or Imposed by Man?

The Instinctive Conviction of Order *

by Alfred North Whitehead (1925)

There can be no living science unless there is a widespread instinctive conviction in the existence of an *Order of Things,* and, in particular, of an *Order of Nature.* I have used the word *instinctive* advisedly. It does not matter what men say in words, so long as their activities are controlled by settled instincts. The words may ultimately destroy the instincts. But until this has occurred, words do not count. This remark is important in respect to the history of scientific thought. . . .

Of course we all share in this faith, and we therefore believe that the reason for the faith is our apprehension of its truth. But the formation of a general idea—such as the idea of the Order of Nature—and the grasp of its importance, and the observation of its exemplification in a variety of occasions are by no means the necessary consequences of the truth of the idea in question. Familiar things happen, and mankind does not bother about them. It requires a very unusual mind to undertake the analysis of the obvious. . . .

Obviously, the main recurrences of life are too insistent to escape the notice of the least rational of humans; and even before the dawn of rationality, they have impressed themselves upon the instincts of animals. It is unnecessary to labour the point, that in broad outline certain general states of nature recur, and that our very natures have adapted themselves to such repetitions.

But there is a complementary fact which is equally true and equally obvious:—nothing ever really recurs in exact detail. No two days are identical, no two winters. What has gone, has gone forever. Accordingly the practical philosophy of mankind has been to expect the broad recurrences, and to accept the details as emanating from the inscrutable womb of things beyond the ken of rationality. Men expected the sun to rise, but the wind bloweth where it listeth. . . .

* From Alfred North Whitehead, *Science and the Modern World,* The Macmillan Company, New York. Copyright 1925 by the Macmillan Company; copyright 1953 by Evelyn Whitehead. Reprinted by permission.

There have been great civilisations in which the peculiar balance of mind required for science has only fitfully appeared and has produced the feeblest result. For example, the more we know of Chinese art, of Chinese literature, and of the Chinese philosophy of life, the more we admire the heights to which that civilisation attained. For thousands of years, there have been in China acute and learned men patiently devoting their lives to study. Having regard to the span of time, and to the population concerned, China forms the largest volume of civilisation which the world has seen. There is no reason to doubt the intrinsic capacity of individual Chinamen for the pursuit of science. And yet Chinese science is practically negligible. There is no reason to believe that China if left to itself would have ever produced any progress in science. The same may be said of India. Furthermore, if the Persians had enslaved the Greeks, there is no definite ground for belief that science would have flourished in Europe. . . .

Faith in reason is the trust that the ultimate natures of things lie together in a harmony which excludes mere arbitrariness. It is the faith that at the base of things we shall not find mere arbitrary mystery. The faith in the order of nature which has made possible the growth of science is a particular example of a deeper faith. This faith cannot be justified by any inductive generalisation. It springs from direct inspection of the nature of things as disclosed in our own immediate present experience. There is no parting from your shadow. To experience this faith is to know that in being ourselves we are more than ourselves: to know that our experience, dim and fragmentary as it is, yet sounds the utmost depths of reality: to know that detached details merely in order to be themselves demand that they should find themselves in a system of things: to know that this system includes the harmony of logical rationality, and the harmony of aesthetic achievement: to know that, while the harmony of logic lies upon the universe as an iron necessity, the aesthetic harmony stands before it as a living ideal moulding the general flux in its broken progress towards finer, subtler issues.

Reality Is a Construction of Man °
by José Ortega y Gasset (1958)

For centuries and centuries the sidereal facts of this world were set clear before the eyes of humans; yet what those facts meant, what they

° From José Ortega y Gasset, *Man and Crisis*, translated from the Spanish by Mildred Adams, W. W. Norton & Company, Inc., New York. Copyright © 1958 by W. W. Norton & Company, Inc. Reprinted by permission.

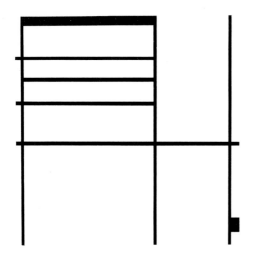

Figure 3-39. *The search for order.*
(Figures added by editor.)

"Composition in Black, White and Red" (1936) *by Piet Mondrian.* (Collection, The Museum of Modern Art, New York)

Snowflakes. (The Bettmann Archive, Inc.)

"Sculpture for the Blind" (1924) *in marble by Constantin Brancusi.* (Philadelphia Museum of Art)

Abacus, Oriental calculating device. (The Bettmann Archive, Inc.)

presented to man, what they made evident to him was by no means a reality, but quite the opposite—an enigma, a profound secret, a problem before which man trembled in terror.

Facts, then, come to be like the figures in hieroglyphic writing. Have you ever noted the paradoxical character of such figures? There they are, holding up their clean profiles to us so ostentatiously; but that very appearance of clarity is there for the purpose of presenting us with an enigma, of producing in us not clarity, but confusion. The hieroglyphic figure says to us, "You see me clearly? Good—now what you see of me is not my true being. I am here to warn you that I am not my own essential reality. My reality, my meaning, lies behind me and is hidden by me. In order to arrive at it, you must not fix your attention on me, nor take me for reality; on the contrary, you will have to interpret me, and this means that in order to arrive at the true and inward meaning of this hieroglyph, you must search for something very different from the aspect which its figures offer."

Science is the interpretation of facts. By themselves, facts do not give us reality; on the contrary, they hide it, which is to say that they present us with the problem of reality. If there were no facts, there would be no problem, there would be no enigma, there would be nothing hidden which it is necessary to de-hide, to dis-cover. The word which the Greeks used for truth is *aletheia,* which means discovery, to take away the veil that covers and hides a thing. Facts cover up reality; while we are in the midst of their innumerable swarmings we are in chaos and confusion. In order to discover reality we must for a moment lay aside the facts that surge about us, and remain alone with our minds. Then, on our own risk and account, we imagine a reality, or to put it another way, we construct an imaginary reality, a pure invention of our own; then, following in solitude the guidance of our own personal imagining, we find what aspect, what visible shapes, in short, what facts would produce that imaginary reality. It is then that we come out of our imaginative solitude, out from our pure and isolated mental state, and compare those facts which the imagined reality would produce with the actual facts which surround us. If they mate happily one with another, this means that we have deciphered the hieroglyph, that we have discovered the reality which the facts covered and kept secret.

That labor is what science is. As you see, it consists of two different operations: one purely imaginative and creative, which man produces out of his own most free substance; the other a confronting of that which is not man, but of that which surrounds him, the facts, the data. Reality is not a datum, not something given or bestowed, but a construction which man makes out of the given material.

The Harmony of Natural Law [*]
by Henri Poincaré (1913)

Is Mathematical analysis . . . only a vain play of the mind? It can
give to the physicist only a convenient language; is this not a mediocre
service, which, strictly speaking, could be done without; and even is it
not to be feared that this artificial language may be a veil interposed between reality and the eye of the physicist? Far from it; without this language most of the intimate analogies of things would have remained forever unknown to us; and we should forever have been ignorant of the
internal harmony of the world, which is, we shall see, the only true
objective reality.

The best expression of this harmony is law. Law is one of the most recent conquests of the human mind; there still are people who live in the
presence of a perpetual miracle and are not astonished at it. On the contrary, we it is who should be astonished at nature's regularity. Men demand of their gods to prove their existence by miracles; but the eternal
marvel is that there are not miracles without cease. The world is divine
because it is a harmony. If it were ruled by caprice, what would prove
to us it was not ruled by chance?

This conquest of law we owe to astronomy, and just this makes the
grandeur of the science rather than the material grandeur of the objects
it considers. It was altogether natural, then, that celestial mechanics
should be the first model of mathematical physics. . . .

Some people have exaggerated the role of convention in science; they
have even gone so far as to say that law, that scientific fact itself, was
created by the scientist. This is going much too far in the direction of
nominalism. No, scientific laws are not artificial creations; we have no
reason to regard them as accidental, though it be impossible to prove
they are not.

Does the harmony the human intelligence thinks it discovers in nature
exist outside of this intelligence? No, beyond doubt a reality completely
independent of the mind which conceives it, sees or feels it, is an impossibility. A world as exterior as that, even if it existed, would for us be
forever inaccessible. But what we call objective reality is, in the last
analysis, what is common to many thinking beings and could be com-

[*] From Henri Poincaré, *The Foundations of Science*, Dover Publications, Inc.,
New York, 1952.

mon to all, this common part, we shall see, can only be the harmony expressed by mathematical laws. It is this harmony then which is the sole objective reality, the only truth we can attain; and when I add that the universal harmony of the world is the source of all beauty, it will be understood what price we should attach to the slow and difficult progress which little by little enables us to know it better.

What we call thought (1) is itself an orderly thing, and (2) can only be applied to material, i.e. to perceptions or experiences, which have a certain degree of orderliness. This has two consequences. First, a physical organization, to be in close correspondence with thought (as my brain is with my thought) must be a very well-ordered organization, and that means that the events that happen within it must obey strict physical laws, at least to a very high degree of accuracy. Secondly, the physical impressions made upon that physically well-organized system by other bodies from outside obviously correspond to the perception and experience of the corresponding thought, forming its material, as I have called it. Therefore, the physical interactions between our system and others must, as a rule, themselves possess a certain degree of physical orderliness, that is to say, they too must obey strict physical laws to a certain degree of accuracy.

Erwin Schrödinger
What Is Life? (1944)

The man of science will act *as if* this world were an absolute whole controlled by laws independent of his own thoughts or acts; but whenever he discovers a law of striking simplicity or one of sweeping universality or one which points to a perfect harmony in the cosmos, he will be wise to wonder what role his mind has played in the discovery, and whether the beautiful image he sees in the pool of eternity reveals the nature of this eternity, or is but a reflection of his own mind.

Tobias Dantzig
Number: The Language of Science (1954)

Suggestions for Further Reading

° Collingwood, P. G.: *The Idea of Nature,* Galaxy Books, Oxford University Press, New York, 1960, part I. A philosophical analysis of the concepts of the Greek philosophers from Thales through Aristotle.

° De Santillana, Giorgio: *Origins of Scientific Thought*, Mentor Books, New American Library of World Literature, Inc., New York, 1961. In addition to the selection we have used, we recommend the rest of the book very highly for an authoritative treatment of various phases of scientific thought from 600 B.C. to A.D. 500.

° Farrington, Benjamin: *Greek Science*, Penguin Books, Inc., Baltimore, 1944. This account especially brings out "the connections of Greek science with practical life, with techniques, with the economic basis and productive activity of Greek society."

° Price, Derek J. de Solla: *Science Since Babylon*, Yale University Press, New Haven and London, 1961. The chapter "Celestial Clockwork in Greece and China" is interesting in connection with these readings.

° Whitehead, Alfred North: *Science and the Modern World*, Lowell Lectures, Mentor Books, New American Library of World Literature, Inc., New York, 1948, chaps. I and II. An examination of the philosophical significance of the Greek contributions to science and mathematics.

° Paperback edition

4

What Is Scientific Truth?

Truth has many shades of meaning, and confusion among them has led to bitter conflicts between science and society. In this part we find the questing, questioning spirit of science, as exemplified by Galileo, clashing with the prevailing conviction that organized religion and society need a firm and absolute truth as the basis for belief and security.

What Is Scientific Truth?

Introduction

SCIENTIFIC THEORIES never stand entirely alone on their intrinsic merits; their acceptance or rejection is often closely related to the climate of thought at the time of their conception. It has happened frequently in the history of science that an idea has been proposed simultaneously by different scientists working independently. The time was ripe for that particular idea. Conversely, we saw, in our study of Greek astronomy, that Aristarchus proposed the heliocentric theory in the third century B.C. but that other scientists did not adopt the idea for nearly 1,800 years. It is worthwhile to consider the social and human factors that enter into the acceptance of a scientific theory and to what extent it is justified to base religious or philosophical concepts on changing scientific theories.

The readings in Part 4 describe the revolution in thought that occurred in the Renaissance when the heliocentric theory began to challenge the geocentric theory that had dominated natural philosophy since the time of Aristotle. The new idea was more than a scientific theory; it involved a whole new world view, embracing religious, metaphysical, and even moral aspects. Scientific thought as represented by Aristotle had molded and dominated the entire view of the universe for so many years that it was unquestioningly accepted as representing an absolute form of truth. The independent, questioning spirit of the Renaissance was needed to break this hardened ground and prepare it for the beginning of a scientific revolution. The need to doubt, to question, and to disrupt all absolutes is an essential ingredient of the scientific spirit. As Gilbert Lewis [1] said, "The theory that there is an ultimate truth, although very generally held by mankind, does not seem useful to science except in the sense of a horizon toward which we may proceed."

Some scientific theories affect basic beliefs much more directly than do other theories. The central and really revolutionary idea embodied in the Copernican theory was that the earth moved. This idea was hard to accept from many different points of view. Rationally, it appeared to contradict common sense. If the earth moved, why were not men conscious of this motion? Scientifically, it posed problems of velocity that required the introduction of a whole new physics. As Cohen points out in his article, even today we have not fully assimilated the "new" ideas

1. See p. 9.

into our everyday thinking. The motion of the earth was difficult to reconcile with the scientific fact that no parallax had been observed in the position of the stars. Egocentrically, it was hard to accept because it displaced man from the center of the universe. Koestler suggests that it was not so much the representatives of the Roman Catholic Church itself as the ranks of entrenched professionals who opposed the new idea: "The inertia of the human mind and its resistance to innovation are most clearly demonstrated not, as one might expect, by the ignorant mass—which is easily swayed once its imagination is caught—but by professionals with a vested interest in tradition and in the monopoly of learning . . . they stretch, a solid and hostile phalanx of pedantic mediocrities, across the centuries." [2]

Certainly, the men whom Galileo was fighting, both the representatives of the church and the Aristotelians at the universities, were educated men, many of them interested themselves in advancing the progress of science. But both religiously and philosophically, the Copernican theory called for major changes in accepted doctrine, and these changes seemed hardly justified on the slim evidence in favor of the theory. Cardinal Bellarmine asked for proof that the earth moved. He said that, in the absence of such proof, the heliocentric theory could be allowed as an hypothesis but not accepted as absolute truth. If there were such proof, then the church would reinterpret the Scriptures ". . . rather than declare an opinion to be false which is proved to be true." [3] Galileo believed that he had such proof but he failed to convince many of his contemporaries. Actually, as we saw in Part 2, science cannot provide the kind of proof that Bellarmine was demanding. Science can disprove a theory, and Galileo went very far toward disproving Aristotle's model of the universe. It can also increase the probability that a certain conceptual scheme does represent a true picture of reality, but still it cannot prove it to be true in a positive and absolute sense. An understanding of this limitation of science could perhaps have helped to prevent the persecution and intolerance that were to follow, not only in this argument, but in later conflicts between science and society.

Shaw points up the basic dilemma when he says we must build a society that provides ". . . a large liberty to shock conventional people" and yet ". . . we must face the fact that society is founded on intolerance." [4] Can the need for a certain degree of stability in society be reconciled with the freedom required for creative science?

2. See pp. 191–92.
3. See p. 193.
4. See pp. 201, 202.

The Medieval Universe

All things whatsoever observe a mutual order; and this the form that maketh
    the universe like unto God.
Herein the exalted creatures trace the impress of the Eternal Worth, which is
    the goal whereto was made the norm now spoken of.
In the order of which I speak all things incline, by diverse lots, more near and
    less unto their principle;
Wherefore they move to diverse ports o'er the great sea of being, and each one
    with instinct given it to bear it on.
This beareth the fire toward the moon; this is the mover in the hearts of things
    that die; this doth draw the earth together and unite it.

Dante
Divine Comedy (c. 1307–21)

A View of the Medieval Cosmos °

by Herbert Butterfield (1957)

An introductory sketch of the medieval view of the cosmos must be
qualified first of all with the reservation that in this particular realm of
thought there were variations, uncertainties, controversies and developments which it would obviously be impossible to describe in detail. On
the whole, therefore, it would be well, perhaps, if we were to take
Dante's view of the universe as a pattern, because it will be easy to note
in parentheses some of the important variations that occurred, and at
the same time this policy will enable us to see in a single survey the
range of the multiple objections which it took the Copernican theory
something like a hundred and fifty years to surmount.

According to Dante, what one must have in mind is a series of
spheres, one inside another, and at the heart of the whole system lies the
motionless earth. The realm of what we should call ordinary matter is
confined to the earth and its neighbourhood—the region below the
moon; and this matter, the stuff that we can hold between our fingers
and which our modern physical sciences set out to study, is humble and

° From Herbert Butterfield, *The Origins of Modern Science,* The Macmillan Company, New York.© 1957 by G. Bell & Sons, Ltd., London. Reprinted by permission.

unstable, being subject to change and decay for reasons which we shall examine later. The skies and the heavenly bodies—the rotating spheres and the stars and planets that are attached to them—are made of a very tangible kind of matter too, though it is more subtle in quality and it is not subject to change and corruption. It is not subject to the physical laws that govern the more earthy kind of material which we have below the moon. From the point of view of what we should call purely physical science, the earth and the skies therefore were cut off from one another and were separate organisations for a medieval student, though in a wider system of thought they dovetailed together to form one coherent cosmos.

As to the ordinary matter of which the earth is composed, it is formed of four elements, and these are graded according to their virtue, their nobility. There is earth, which is the meanest stuff of all, then water, then air and, finally, fire, and this last comes highest in the hierarchy. We do not see these elements in their pure and undiluted form, however —the earthy stuff that we handle when we pick up a little soil is a base compound and the fire that we actually see is mixed with earthiness. Of the four elements, earth and water possess gravity; they have a tendency to fall; they can only be at rest at the centre of the universe. Fire and air do not have gravity, but possess the very reverse; they are characterised by levity, an actual tendency to rise, though the atmosphere clings down to the earth because it is loaded with base mundane impurities. For all the elements have their spheres, and aspire to reach their proper sphere, where they find stability and rest; and when even flame has soared to its own upper region it will be happy and contented, for here it can be still and can most endure. If the elements did not mix—if they were all at home in their proper spheres—we should have a solid sphere of earth at the heart of everything and every particle of it would be still. We should then have an ocean covering that whole globe, like a cap that fitted all round, then a sphere of air, which far above mountain-tops was supposed to swirl round from east to west in sympathy with the movement of the skies. Finally, there would come the region of enduring fire, fitting like a sphere over all the rest.

That, however, would be a dead universe. In fact, it was a corollary of this whole view of the world that ordinary motion up or down or in a straight line could only take place if there was something wrong— something displaced from its proper sphere. It mattered very much, therefore, that the various elements were not all in order but were mixed and out of place—for instance, some of the land had been drawn out above the waters, raised out of its proper sphere at the bottom, to provide habitable ground. On this land natural objects existed and, since they were mixtures, they might, for example, contain water, which as soon as it was released would tend to seek its way down to the sea. On the other hand, they might contain the element of fire, which would come out of them when they burned and would flutter and push its way upwards, aspiring to reach its true home. But the elements are not always able to follow their nature in this pure fashion—occasionally the fire may strike downwards, as in lightning, or the water may rise in the form of vapour to prepare a store of rain. On one point, however, the law was fixed: while the elements are out of their proper spheres they are bound to be unstable—there cannot possibly be restfulness and peace. Woven, as we find them, on the surface of the globe, they make a mixed and chancy world, a world that is subject to constant mutation, liable to dissolution and decay.

It is only in the northern hemisphere that land emerges, protruding above the waters that cover the rest of the globe. This land has been pulled up, out of its proper sphere, says Dante—drawn not by the moon or the planets or the ninth sky, but by an influence from the fixed stars, in his opinion. The land stretches from the Pillars of Hercules in the west to the Ganges in the east, from the Equator in the south to the Arctic Circle in the north. And in the centre of this whole habitable world is Jerusalem, the Holy City. Dante had heard stories of travellers who had found a great deal more of the continent of Africa, found actual land much farther south than he had been taught to consider possible. As a true rationalist he seems to have rejected "fables" that contradicted the natural science of his time, remembering that travellers were apt to be liars. The disproportionate amount of water in the world and the unbalanced distribution of the land led to some discussion of the whereabouts of the earth's real centre. The great discoveries, however, culminating in the unmistakable discovery of America, provoked certain changes in ideas, as well as a debate concerning the possibility of the existence of inhabited countries at the antipodes. There was a growing view that earth and water, instead of coming in two separate circles, the one above the other, really dovetailed into one another to form a single sphere.

All this concerns the sublunary region; but there is another realm of matter to be considered, and this, as we have already seen, comes under a different polity. The skies are not liable to change and decay, for they —with the sun, the stars and the planets—are formed of a fifth element, an incorruptible kind of matter, which is subject to a different set of what we should call physical laws. If earth tends to fall to the centre of the universe, and fire tends to rise to its proper sphere above the air itself, the incorruptible stuff that forms the heavens has no reason for discontent—it is fixed in its congenial place already. Only one motion is possible for it—namely, circular motion—it must turn while remaining in the same place. According to Dante there are ten skies, only the last of them, the Empyrean Heaven, the abode of God, being at rest. Each of the skies is a sphere that surrounds the globe of the earth, and though all these spheres are transparent they are sufficiently tangible and real to carry one or more of the heavenly bodies round on their backs as they rotate about the earth—the whole system forming a set of transparent spheres, one around the other, with the hard earth at the centre of all. The sphere nearest to the earth has the moon attached to it, the others carry the planets or the sun, until we reach the eighth, to which all the fixed stars are fastened. A ninth sphere has no planet or star attached to it, nothing to give visible signs of its existence; but it must be there, for it is the *primum mobile*—it turns not only itself but all the other spheres or skies as well, from east to west, so that once in twenty-four hours the whole celestial system wheels round the motionless earth. This ninth sphere moves more quickly than any of the others, for the spirits which move it have every reason to be ardent. They are next to the Empyrean Heaven.

In the system of Aristotle the spheres were supposed to be formed of a very subtle aethereal substance, moving more softly than liquids and without any friction; but with the passage of time the idea seems to have become coarsened and vulgarised. The successive heavens turned into glassy or crystalline globes, solid but still transparent, so that it became harder for men to keep in mind the fact that they were frictionless and free from weight, though the Aristotelian theory in regard to these points was still formally held.

The original beauty of this essentially Aristotelian system had been gravely compromised, however, by the improvements which had been made in astronomical observation since the time when it had been given its original shape; for even in the ancient world astronomy afforded a remarkable example of the progress which could be achieved in science by the sheer passage of time—the accumulating store of observations and the increasing precision in the recordings. Early in the Christian era, in the age of Ptolemy, the complications had become serious, and in the middle ages both the Arabs and the Christians produced additions to the intricacies of the system. The whole of the celestial machinery needed further elaboration to account for planets which, as viewed by the observer, now stopped in the sky, now turned back on their courses, now changed their distance from the earth, now altered their speed. However irregular the motion of the planets might seem to be, however curious the path that they traced, their behaviour must somehow or other be reduced to circular, even uniform circular motion—if necessary to a complicated series of circular motions each corrective of the other. Dante explains how Venus goes round with the sphere which forms the third of the skies, but as this does not quite correspond to the phenomena, another sphere which revolves independently is fixed to the sphere of the third sky, and the planet rides on the back of the smaller sphere (sitting like a jewel there, says Dante), reflecting the light of the sun. But writers varied on this point, and we meet the view that the planet was rather like a knot in a piece of wood, or represented a mere thickening of the material that formed the whole celestial sphere—a sort of swelling that caught the sunlight and shone with special brilliance as a result.

Writers differed also on the question whether the whole of the more elaborate machinery—the eccentrics or epicycles—as devised by Ptolemy and his successors, really existed in the actual architecture of the skies, though the theory of the crystalline spheres persisted until the seventeenth century. Since the whole complicated system demanded eighty spheres, some of which must apparently intersect one another as they turned round, some writers regarded the circles and epicycles as mere geometrical devices that formed a basis for calculation and prediction. And some men who believed that the nine skies were genuine crystalline spheres might regard the rest of the machinery as a mathematical way of representing those irregularities and anomalies which they knew they were unable properly to explain. In any case, it was realised long before the time of Copernicus that the Ptolemaic system, in spite of all its complications, did not exactly cover the phenomena as observed. In the sixteenth and seventeenth centuries we shall still find people who will admit that the Ptolemaic system is inadequate, and who will say that a new one must be discovered, though for understandable reasons they reject the solution offered by Copernicus. Copernicus himself, when he explained why his mind had turned to a possible new celestial system, mentioned amongst other things the divergent views that he found already in existence. The Ptolemaic system would be described as the Ptolemaic hypothesis, and we even find the Copernican theory described by one of its supporters as "the revision of the hypotheses." Many of us have gone too far perhaps in imagining a cast-iron Ptolemaic system, to the whole of which the predecessors of Copernicus were supposed to be blindly attached.

Finally, according to Dante, all the various spheres are moved by Intelligences or Spirits, which have their various grades corresponding to the degrees of nobility that exist in the physical world. Of these, the lowliest are the angels who move the sphere of the moon; for the moon

Figure 4-1. Views of the universe from ancient times to the Renaissance. (Figures added by editor.)

The Boat of the Sun traveling in the sky, an Egyptian concept. (The Bettmann Archive, Inc.)

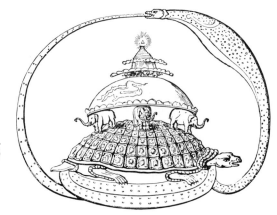

The Hindu earth. (The Bettmann Archive, Inc.)

The four elements: earth, water, fire, air.

A seventeenth-century engraving portraying the Ptolemaic system of the second century.

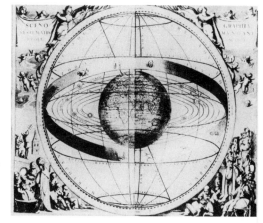

A view of the world in the Middle Ages.

The Globus Magnus of Tycho Brahe (1598).

is in the humblest of the heavens; she has dark spots that show her imperfections; she is associated with the servile and poor. (It is not the moon but the sun which affords the material for romantic poetry under this older system of ideas.) Through the various Intelligences operating by means of the celestial bodies, God has shaped the material world, only touching it, so to speak, through intermediaries. What He created was only inchoate matter, and this was later moulded into a world by celestial influences. Human souls, however, God creates with his own hands; and these, again, are of a special substance—they are incorruptible. . . .

In this whole picture of the universe there is more of Aristotle than of Christianity. It was the authority of Aristotle and his successors which was responsible even for those features of this teaching which might seem to us to carry something of an ecclesiastical flavour—the hierarchy of heavens, the revolving spheres, the Intelligences which moved the planets, the grading of the elements in the order of their nobility and the view that the celestial bodies were composed of an incorruptible fifth essence. Indeed, we may say it was Aristotle rather than Ptolemy who had to be overthrown in the sixteenth century, and it was Aristotle who provided the great obstruction to the Copernican theory.

Copernicus

The Copernican Revolution—1543 [*]

by Abraham Wolf (1950)

Niklas Koppernigk (whom we shall henceforward designate by the more familiar latinized form of his name, Nicolaus Copernicus) was born February 19, 1473, at Thorn, on the Vistula. His father was a merchant whose nationality, whether German or Polish, is still a matter of controversy. His mother was of German extraction. The father died in 1483, and Copernicus was brought up by an uncle who destined him for a career in the Church. After attending school at Thorn, Copernicus spent three years at the University of Cracow. Here his interest in mathematics and astronomy was awakened under the teaching of Albert Brudzewski, and he became accustomed to the use of astronomical instruments and to the observation of the heavens. After spending two years with his uncle, who was now Bishop of Ermland, Copernicus set out in 1496 for Italy, where, during the following ten years, he studied successively at the Universities of Bologna, Padua, and Ferrara. Law and medicine were his professed subjects of study during these years, but, although not much is known about his activities in Italy, there is good reason for supposing that he devoted much of his time there to the pursuit of astronomy, both theoretical and practical.

While at Bologna, Copernicus came into close personal touch with Comenico di Novara, the Professor of Astronomy there, a leader in that revival of Pythagorean ideals in natural philosophy which was awakening the Italian universities about that time. The two men observed together, and discussed, with a freedom unusual in the circles which Copernicus had hitherto frequented, the errors of Ptolemy's *Almagest*, and the possibility of improving upon the Ptolemaic system. There is little doubt that it was during his residence in Italy that Copernicus received the initial impulse towards that reform of astronomy which he achieved in his later, more secluded years.

[*] From Abraham Wolf, *A History of Science, Technology and Philosophy*, The Macmillan Company, New York, 1950. Reprinted by permission.

During his absence in Italy Copernicus had been appointed a Canon of Frauenburg Cathedral, in his uncle's diocese, but after his return home he continued to live with his uncle at his palace at Heilsberg until the Bishop's death in 1512. He then took up his duties at Frauenburg, where he lived, with occasional interruptions, for the remaining thirty years of his life. These thirty years were, to outward appearance, the most uneventful in the life of Copernicus. He shared in the business of his Chapter, did a little political work, and gave free medical advice to the poor of the district. But it was during these years that Copernicus thought out the details of his planetary system, marshalled the intricate array of calculations by which that system was eventually reduced from speculation to numerical precision, and slowly perfected the manuscript in which the fruits of all his labours were at last given to the world.

Copernicus from the first was well aware of the opposition, based on both learned and doctrinal grounds, which would be aroused by the publication of his novel views on the constitution of the solar system. Hence, while he kept revising his manuscript year after year, he hesitated to publish it. Some inkling of his actual opinions, however, leaked out, awakening discussion and curiosity; and about 1529 he circulated among his friends a manuscript *Commentariolus*. This little tract gives a descriptive account of his system in very nearly its final form, but with all the calculations omitted. . . . Again, about ten years later, Copernicus received a long visit from a young astronomer, Georg Joachim, better known by his Latin name Rheticus, who studied the still unpublished manuscript, and made its contents known to a wide circle by means of his printed *Narratio Prima* (1540).

It was to this Rheticus, three years later, that Copernicus, now old and ill, committed his manuscript when his friends at length prevailed upon him to publish it. The book was printed at Nürnberg and published in 1543; and the story goes that the first copy was given to Copernicus a few hours before his death on May 24, 1543.

The printed book bears the title *Nicolai Copernici Torimensis de revolutionibus orbium coelestium Libri VI*, and was dedicated to the reigning Pope, Paul III, whose interest and protection Copernicus claimed. The first edition, however, differs on almost every page from the original manuscript. The title itself is an addition, and there is reason to suppose that Copernicus would have preferred to call his work *De Revolutionibus* simply. The manuscript was lost for some two hundred years, but was rediscovered in time to serve as the basis for the Säkular-Ausgabe (Thorn, 1873), which is the authoritative edition of the text. (There is a German translation of the book by C. L. Menzzer, Thorn, 1879.)

For some years after the publication of Copernicus' book there was uncertainty as to whether his hypothesis was intended as a description of the actual motions of the Earth and planets, or merely as a computing device to facilitate the construction of planetary tables. This question was of all the more moment, as, in the state of religious opinion at the time, the acceptance or rejection of the teachings of Copernicus depended to a considerable degree upon the sense in which they were to be understood. The uncertainty arose chiefly out of the circumstances in which the book was published. Rheticus, who at first superintended the printing, was called away before it was completed, and he entrusted the supervision of the work to Andreas Osiander, a local Lutheran clergyman, who was a mathematician and a friend of Copernicus. Osiander was afraid that the doctrine of the Earth's motion would offend philosophers and strict Lutherans, and so he inserted a little preface of his own which stated that the whole was to be regarded as a mere computing device without prejudice to scriptural or physical truth.[1] The imposture was recognized by several friends of Copernicus from the first, and was finally exposed by Kepler. . . . It was probably well-intentioned; Osiander had previously advised Copernicus to insert such a deprecatory preface, but Copernicus had refused to do so. The preface seemed strangely at variance with the text, but could not be finally proclaimed to be an addition until the original manuscript was recovered. To Copernicus, imbued with Pythagorean ideas, the most elegant and harmonious mathematical representation of the planetary motions doubtless appeared to be the only true planetary theory.

COPERNICAN ASTRONOMY

In the dedicatory preface to his *De Revolutionibus* Copernicus immediately brings his readers face to face with the long-standing problem which it was his life's work to solve. *That problem was to ascertain what geometrical laws govern the motions of the planets in order to explain the apparent motions observed in the past, and to predict how the planets would move in the future. . . .*

Copernicus describes how, . . . casting about for fresh ideas, he turned to the classical writers to see what alternative theories they had to offer. He found that quite *a number of the early thinkers, such as Hicetas, Philolaus, and Heraclides of Pontus, had attributed some form of motion* (axial rotation or revolution in a closed orbit) *to the Earth;* and he quotes several classical writers to this effect. We cannot be certain whether Copernicus really derived his ideas, in the first instance, from

1. See pp. 160–61 for full text of Osiander's preface. Ed.

the writers whom he quotes, or whether he introduces their names merely for the sake of the impression which these would produce upon the readers of his day. We shall return to the question of the originality of Copernicus' conceptions later. In any case, *he uses these classic passages as an excuse for introducing his own system, in which the Earth rotates on its axis and revolves about the Sun as one of the planets.*

"Taking occasion thence," he writes, "I too began to reflect upon the Earth's capacity for motion. And though the idea appeared absurd, yet I knew that others before me had been allowed freedom to imagine what circles they pleased in order to represent the phenomena of the heavenly bodies. I therefore deemed that it would readily be granted to me also to try whether, by assuming the Earth to have a certain motion, demonstrations, more valid than those of others, could be found for the revolution of the heavenly spheres.

"And so, having assumed those motions which I attribute to the Earth further on in the book, I found at length, by much long-continued observation, that, if the motions of the remaining planets be referred to the revolution of the Earth, and calculated according to the period of each planet, then not only would the planetary phenomena follow as a consequence, but the order of succession and the dimensions of the planets and of all the spheres, and the heaven itself, would be so bound together

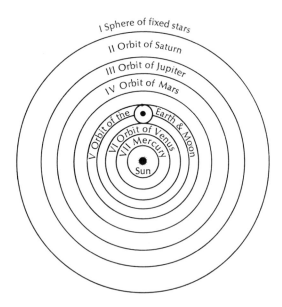

Figure 4-2. The universe according to Copernicus.

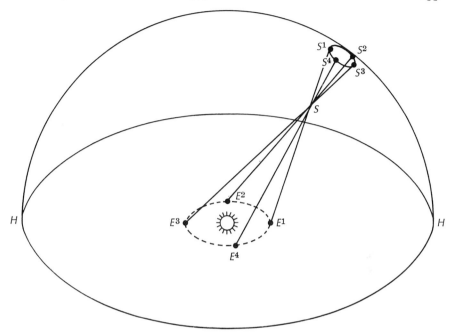

Figure 4-3. *Stellar parallax. To measure the displacement of very distant objects, the widest possible optical base is used. The different positions of the earth in its orbit* (E_1, E_2, E_3, *and* E_4) *provide the largest optical base for observing the displacement of the stars against farther stars* S_1, S_2, S_3, *and* S_4. *However the distance between S and E is very much greater than that shown in the figure, and the displacement is too small to be observed without powerful telescopes.*

that in no part could anything be transposed without the disordering of the other parts and of the entire universe" (*Preface*).

The general arrangement of the solar system as conceived by Copernicus, suppressing the refinements which he subsequently introduced, is shown in his well-known diagram, where Mercury, Venus, the Earth, Mars, Jupiter, and Saturn describe concentric orbits about the Sun in the centre (Figure 4-2).

"In the midst of all dwells the Sun. For who could set this luminary in another or better place in this most glorious temple than whence he could at one and the same time lighten the whole? . . . And so, as if seated upon a royal throne, the Sun rules the family of the planets as they circle around him. . . ." (I, 10).

Ever since the days of the Pythagoreans *objection had been made to all planetary hypotheses which involved any motion of the Earth, on the ground that any such motion would give rise to a corresponding apparent motion in the stars* (Figure 4-3). Such an apparent motion, though sought, was never observed. Copernicus anticipates this sort of criticism by supposing that the stars are at a distance from us incomparably greater than the radius of the Earth's orbit, so that our annual motion makes no difference to their apparent direction. This objection, however, became ever more serious as observation improved in accuracy yet failed to reveal any annual stellar parallax. It was finally removed within the last hundred years, when stellar parallax of a minute order was observed in certain stars.[2]

Copernicus was doubtless won over to the new point of view by its greater symmetry and coherence. These virtues would appeal to one imbued with Neo-Pythagorean ideas. *For the essence of Pythagoreanism was its insistence that the universe is to be described in terms of mathematical relations;* and that, of two geometrically equivalent planetary theories, the more harmonious and symmetrical was the more correct. But Copernicus had still to justify his point of view to the scholars of northern Europe, whose master was Aristotle rather than Pythagoras. He therefore devotes several of the early chapters of Book I to proving that the new system is just as much in accordance with Aristotelian physics as was the Ptolemaic system. *His problem is to confute the arguments by which Aristotle had asserted that the Earth is at rest at the centre of the universe, while at the same time preserving Aristotle's principles intact and using them as the basis of his own arguments.*

Copernicus reasons more soundly, however, from the principle that all motion is relative motion: "Every apparent change of position is due either to a motion of the object observed, or to a motion of the observer, or to unequal changes in the positions of both. . . . If then a certain motion be assigned to the Earth, it will appear as a similar but oppositely directed motion, affecting all things exterior to the Earth, as if we were passing them by" (I, 5). Copernicus applies this principle of the reciprocity of apparent motions in the first instance to account for the apparent diurnal rotation of the heavens: "If you will allow that the heavens have no part in this motion, but that the Earth turns from west to east, then, so far as pertains to the apparent rising and setting of the Sun, Moon and stars, you will find, if you think carefully, that these things occur in this way" (I, 5). Later he brings this same principle to bear upon the phenomena connected with the apparent annual circuit of the Sun: "If [this circuit] be transposed from being a solar to being a terrestrial [phenomenon], and it be granted that the Sun is at rest, then

2. See footnotes 4 and 5, p. 102. Ed.

the risings and settings of the constellations and the fixed stars, whereby they become morning and evening stars, will appear after the same manner [as before]" (I, 9).

The most convincing argument, however, in favour of the scientific superiority of the Copernican hypothesis was the simple explanation which it could give of certain peculiarities in the apparent motions of the planets. If one of these bodies (say a superior planet) is observed night after night, it is found, in general, to be moving slowly across the southern sky from west to east relatively to the background of stars. From time to time, however, this eastward motion is arrested and reversed, and the planet travels a short distance from east to west before resuming its normal eastward direction. The physical significance of these *stations* and *retrogressions* of the planets had always been an enigma to astronomers; but Copernicus was able to show that these inequalities arise as necessary consequences of the annual motion of the Earth. . . .

Ptolemy had represented this inequality in a planet's motion by supposing the planet to move upon an epicycle especially introduced for the purpose. This was equivalent to transferring the Earth's motion to the planet. But this had to be done for *each* planet, while Copernicus could explain this phenomenon in each planet by reference to a single motion of the Earth—a great gain in simplicity.

There are, however, further inequalities in the apparent paths of the planets arising, as we know now, from the ellipticity of their orbits. Moreover, the Sun's rate of apparent motion in the ecliptic is not quite uniform from day to day. In order to account for these phenomena Copernicus was obliged to refine upon the simple scheme of Figure 4–2, where the Earth and planets all describe concentric circles with the Sun in the centre. In constructing his detailed planetary orbits, a task which occupies most of the *De Revolutionibus,* Copernicus employs eccentrics and epicycles like those of the ancients, but, unlike Ptolemy, he is always careful to ensure that his circular motions are uniformly described relatively to the centres of the circles, and not merely to arbitrarily chosen points within the circles.

A whole book of the *De Revolutionibus* (the third) is devoted to the consideration of the various motions to be attributed to the Earth. Already, in the preliminary sketch of Book I, the phenomena of the seasons are shown to depend upon the Earth's revolving annually about the Sun while keeping the direction of its axis approximately invariable. In the more refined theory of Book III the Earth's orbit is assumed to be a circle having the Sun displaced a little way from its centre. Copernicus determines the direction of the apse-line of this orbit, and its eccentricity with regard to the Sun, after the manner of Hipparchus. His theory is complicated by an attempt to represent an (actual) progressive motion of the apse-line (suspected by the Arab al-Battani in the ninth century and confirmed by Copernicus), and to allow for certain (imaginary) fluctuations in this motion, and in the eccentricity of the orbit, suggested by some mediaeval observations of questionable accuracy, which he felt obliged to take into account.

An important feature of Copernicus' account of the Earth's motion is his explanation of the precession of the equinoxes. Hipparchus of Rhodes, who discovered this phenomenon about 150 B.C., attributed it to a slow rotation of the sphere of stars about the axis of the ecliptic. Copernicus introduces the modern explanation of precession as due to an alteration in the plane of the Earth's equator causing the Earth's axis to describe a cone in space. Here again he needlessly complicates his theory in order to bring it into conformity with certain ancient and mediaeval observations. Throughout his work he adopts an entirely uncritical attitude to traditional data of this kind, and makes no allowance for the possibility of serious errors of observation, fraud, or textual corruption. This involves his theories in needless intricacies, while at the same time revealing his remarkable skill in geometry. Occasionally he relies upon twenty-seven observations of his own, which fill only one page of a modern edition, and are, on his own admission, crude. "If only I can be correct to ten minutes of arc," he once said to Rheticus, "I shall be no less elated than Pythagoras is said to have been when he discovered the law of the right-angled triangle. . . ."

Following upon this investigation of the Earth's motion, we have a book devoted to lunar theory. The Moon's relation to the Earth was unaffected by the change of outlook which Copernicus initiated, and he made no additions to the lunar inequalities in longitude known to Ptolemy. But his methods of representing these inequalities are more satisfactory than those of the *Almagest*. According to Ptolemy's theory, the angular diameter of the Moon should be twice as great at some times as it is at others; Copernicus found a means of representing the Moon's motion in longitude as correctly as Ptolemy had done, but without such a gross exaggeration of the slight fluctuations in the apparent size of its disc. Copernicus, however, adopts with only a slight alteration Ptolemy's gross underestimate of the Sun's distance from the Earth as being only about 1,200 times the Earth's radius. This error clung to astronomy until the latter half of the seventeenth century, when the use of the telescope for precise celestial measurements made an accurate determination possible.

The last two books of the *De Revolutionibus* (V and VI) treat of the motions of the planets in longitude and in latitude respectively. . . .

Copernicus' treatment of the latitudes of the inferior planets is particularly complicated, and employs methods borrowed almost wholly from the *Almagest*.

It was an ambition of Copernicus to produce numerical planetary tables as accurate as any based on the geocentric hypothesis. From the theories set forth in the *De Revolutionibus* he constructed tables enabling the positions of the Sun, Moon, and planets at any given instant to be easily calculated. These tables, which form an essential feature of the book, were in fact an improvement upon those in current use, and this circumstance helped indirectly to make the new doctrine acceptable among astronomers. But the accuracy of the tables necessarily suffered through their being based upon a bare minimum of crude and often questionable observations embodied in a theory which was strained to conform to illusory physical laws. Such little improvement as could be made by a careful reconsideration of the data was effected a few years later by Copernicus' disciple Reinhold. Before the new cosmology could bear fruit in tables worthy of itself, however, there were needed the precise and systematic observations of Tycho Brahe, and the patient but adventurous genius of Kepler.

THE ORIGINALITY OF COPERNICUS

We may now face the problem of assessing the originality of Copernicus' contribution to astronomy. It cannot be denied that his debts to Ptolemy were considerable. From the *Almagest* he derived many of his observational data and geometrical devices, as well as the material for his star catalogue.

In one sense, however, Copernicus' debt to Ptolemy was insignificant, for the ideas with which he revolutionized European astronomy were entirely alien to the Alexandrians. The rudiments of these ideas are to be found, if anywhere, in the speculations of a few men who stood apart from the main current of opinion, and whose recorded teachings are scattered through classical and mediaeval literature. A study of these passages, so far as they were probably known to Copernicus, seems to make it clear that the basic ideas underlying his system did not originate with him. For instance, the complete heliocentric system had been anticipated, in its broad outlines, by Aristarchus of Samos [3] (c. 250 B.C.), who has been called on that account the "Copernicus of Antiq-

3. See p. 105. Ed.

uity," though Copernicus' relation to the ideas of Aristarchus is unfortunately obscure. From whatever source he derived his fundamental ideas, however, Copernicus' great and unquestionable contribution to astronomy must be held to lie in his elaboration of those ideas into a coherent planetary theory capable of furnishing tables of an accuracy not before attained. It is true that we can no longer regard the Sun, or the centre of the Earth's orbit, or any other origin of reference, as being at rest in space except as a matter of temporary convenience in treating some special problem. Nevertheless, the scientific utility of the heliocentric point of view, and its contacts with observed fact, have increased enormously since it was formulated by Copernicus. To him it merely represented the most symmetrical arrangement of the planets, and the simplest manner of accounting for their observed motions.

Entia non sunt multiplicanda praeter necessitatem.

Literally, beings are not to be multiplied without necessity; or, as translated freely in the *Encyclopedia Britannica*, 1911, "it is scientifically unsound to set up more than one hypothesis at once to explain a phenomenon." This principle is known as "Ockham's razor."
William of Ockham
             Commentarium in Libros IV Sententiarum Petri Lombardi (1318–23)

Preface to Copernicus' *De Revolutionibus*
by Andreas Osiander (1543)

TO THE READER CONCERNING THE HYPOTHESES OF THIS WORK

Since the novelty of the hypotheses of this work has already been widely reported, I have no doubt that some learned men have taken serious offence because the book declares that the earth moves, and that the sun is at rest in the centre of the universe; these men undoubtedly believe that the liberal arts, established long ago upon a correct basis, should not be thrown into confusion. But if they are willing to examine the matter closely, they will find that the author of this work has done nothing blameworthy. For it is the duty of an astronomer to compose the history of the celestial motions through careful and skillful observation. Then turning to the causes of these motions or hypotheses about them, he must conceive and devise, since he cannot in any way attain to the true causes, such hypotheses as, being assumed, enable the motions to be calculated correctly from the principles of geometry, for the future as well as for the past. The present author has performed both these duties excellently. For these hypotheses need not be true nor even probable; if they provide a calculus consistent with the observations, that alone is sufficient. Perhaps there is someone who is so ignorant of geometry and optics that he regards the epicycle of Venus as probable, or thinks that it is the reason why Venus sometimes precedes and sometimes follows the sun by forty degrees and even more. Is there anyone who is not aware that from this assumption it necessarily follows that the diameter of the planet in perigee should appear more than four times, and the body of the planet more than sixteen times, as great as in the apogee, a result contradicted by the experience of every age? In this study there are other no less important absurdities, which there is no need to set forth at the moment. For it is quite clear that the causes of the apparent unequal motions are completely and simply unknown to this art. And if any causes are devised by the imagination, as indeed very many are, they are not put forward to convince anyone that they are true, but merely to provide a correct basis for calculation. Now when from time to time there are offered for one and the same motion different hypotheses (as eccentricity and an epicycle for the sun's motion), the astronomer will accept above all others the one which is the easiest to grasp. The philosopher will perhaps rather seek the semblance of the truth. But neither of them will understand or state anything certain, unless it has been divinely revealed to him. Let us therefore permit these new hypotheses to become known together with the ancient hypotheses, which are no more probable; let us do so especially because the new hypotheses are admirable and also simple, and bring with them a huge treasure of very skillful observations. So far as hypotheses are concerned, let no one expect anything certain from astronomy, which cannot furnish it, lest he accept as the truth ideas conceived for another purpose, and depart from this study a greater fool than when he entered it. Farewell.

Letter to Copernicus from Cardinal Schoenberg Included by Copernicus in the Introduction to *De Revolutionibus*

Nicolaus Schoenberg, Cardinal of Capua, sends his greetings to Nicolaus Copernicus.

When several years ago I heard your diligence unanimously praised, I began to feel an increasing fondness for you and to deem our compatriots lucky on account of your fame. I have been informed that you not only have an exhaustive knowledge of the teachings of the ancient mathematicians, but that you have also created a new theory of the Universe according to which the Earth moves and the Sun occupies the basic and hence central position; that the eighth sphere [of the fixed stars] remains in an immobile and eternally fixed position and the Moon, together with the elements included in its sphere, placed between the spheres of Mars and Venus, revolves annually around the Sun; moreover, that you have written a treatise on this entirely new theory of astronomy, and also computed the movements of the planets and set them out in tables, to the greatest admiration of all. Therefore, learned man, without wishing to be inopportune, I beg you most emphatically to communicate your discovery to the learned world, and to send me as soon as possible your theories about the Universe, together with the tables and whatever else you have pertaining to the subject. I have instructed Dietrich von Rheden [another Frauenburg Canon] to make a fair copy of this at my expense and send it to me. If you will do me these favours, you will find that you are dealing with a man who has your interests at heart, and wishes to do full justice to your excellence. Farewell.

Rome, November 1, 1536

Galileo

Galileo's Discoveries of 1609 *
by I. Bernard Cohen (1960)

The gradual acceptance of Copernican ideas by one scholar here and another there was rudely interrupted in 1609, when a new scientific instrument changed the level and the tone of discussion of the Copernican and Ptolemaic systems to such a degree that the year overshadows 1543 in the development of modern astronomy.

It was in 1609 that man began to use the telescope to make systematic studies of the heavens. The revelations proved that Ptolemy made specific errors and important ones, that the Copernican system appeared to fit the new facts of observation, and that the moon and the planets in reality were very much like the earth in a variety of different ways and were patently unlike the stars. . . .

The scientist who was chiefly responsible for introducing the telescope as a scientific instrument, and who laid the foundations of the new physics, was Galileo Galilei. In 1609 he was a professor at the University of Padua, in the Venetian Republic, and was forty-five years old, which is considerably beyond the age when most men are likely to make profoundly significant scientific discoveries. The last great Italian, except for nobles and kings, to be known to posterity by his first name, Galileo was born in Pisa, Italy, in 1564, almost on the day of Michelangelo's death and within a year of Shakespeare's birth. His father sent him to the university of Pisa, where his sardonic combativeness quickly won him the nickname "wrangler." Although his first thought had been to study medicine—it was better paid than most professions—he soon found that it was not the career for him. He discovered the beauty of mathematics and thereafter devoted his life to it, physics and astronomy. We do not know exactly when or how he became a Copernican, but on his own testimony it happened earlier than 1597.

° From I. Bernard Cohen, *The Birth of a New Physics*, Doubleday & Company, Inc., Garden City, New York. Copyright © 1960 by Educational Services, Inc. Reprinted by permission.

Galileo made his first contribution to astronomy before he ever used a telescope. In 1604 a "nova" or new star suddenly appeared in the constellation Serpentarius.[1] Galileo showed this to be a "true" star, located out in the celestial spaces and not inside the sphere of the moon. That is, Galileo found that this new star had no measurable parallax and so was very far from the earth. Thus he delivered a nice blow at the Aristotelian system of physics because he proved that change could occur in the heavens despite Aristotle, who had held the heavens unchangeable and had limited the region where change may occur to the earth and its surroundings. Galileo's proof seemed to him all the more decisive in that it was the second nova to be observed and found to have no measurable parallax. . . .

Some time early in 1609 Galileo heard a report of the telescope, but without any specific information as to the way in which the instrument was constructed. He told us how:

A report reached my ears that a certain Fleming had constructed a spyglass by means of which visible objects, though very distant from the eye of the observer, were distinctly seen as if nearby. Of this truly remarkable effect several experiences were related, to which some persons gave credence while others denied them. A few days later the report was confirmed to me in a letter from a noble Frenchman at Paris, Jacques Badovere [a former pupil of Galileo], which caused me to apply myself wholeheartedly to inquire into the means by which I might arrive at the invention of a similar instrument. This I did shortly afterwards, my basis being the theory of refraction. First I prepared a tube of lead, at the ends of which I fitted two glass lenses, both plane on one side while on the other side one was spherically convex and the other concave. Then placing my eye near the concave lens I perceived objects satisfactorily large and near, for they appeared three times closer and nine times larger than when seen with the naked eye alone. Next I constructed another one, more accurate, which represented objects as enlarged more than sixty times. Finally, sparing neither labor nor expense, I succeeded in constructing for myself so excellent an instrument that objects seen by means of it appeared nearly one thousand times larger and over thirty times closer than when regarded with our natural vision.

Galileo was not the only observer to point the new instrument toward the heavens. It is even possible that two observers—Thomas Harriot in England and Simon Marius in Germany—may in some respects have been ahead of him. But there seems to be general agreement that the credit of first using the telescope for astronomical purposes may be given to Galileo because of "the persistent way in which he examined object after object, whenever there seemed any reasonable prospect of results following, by the energy and acuteness with which he followed up each clue, by the independence of mind with which he interpreted

1. The constellation known as *Serpentarius* in Kepler's day is now called *Ophiucus*. Ed.

his observations, and above all by the insight with which he realized their astronomical importance," as said Arthur Berry, British historian of astronomy. . . .

To see the way in which these events occurred, let us turn to Galileo's account of his discoveries, in a book which he called *Sidereus nuncius, The Starry Messenger* or *The Starry Message.* In its subtitle, the book is said to reveal "great, unusual, and remarkable spectacles, opening these to the consideration of every man, and especially of philosophers and astronomers." The newly observed phenomena, the title page of the book declared, were to be found "in the surface of the moon, in innumerable fixed stars, in nebulae, and above all in four planets swiftly revolving about Jupiter at differing distances and periods, and known to no one before the Author recently perceived them and decided that they should be named the Medicean Stars."

THE LANDSCAPE OF THE MOON

Immediately after describing the construction and use of the telescope, Galileo turned to results. He would "review the observations made during the past two months, once more inviting the attention of all who are eager for true philosophy to the first steps of such important contemplations."

The first celestial body to be studied was the moon, the most prominent object in the heavens (except for the sun), and the one nearest to us. The crude woodcuts accompanying Galileo's text cannot convey the sense of wonder and delight this new picture of the moon awoke in him. The lunar landscape, seen through the telescope (Figures 4-4 and 4-5), unfolds itself to us as a dead world—a world without color, and so far as the eye can tell, one without any life upon it. But the characteristic that stands out most clearly in photographs, and that so impressed Galileo in 1609, is the fact that the moon's surface looks like a ghostly *earthly* landscape. No one who looks at these photographs, and no one who looks through a telescope, can escape the feeling that the moon is a miniature earth, however dead it may appear, and that there are on it mountains and valleys, oceans, and seas with islands in them. To this day, we refer to those oceanlike regions as "maria" even though we know, as Galileo later discovered, that there is no water on the moon, and that these are not true seas at all.

The spots on the moon, whatever may have been said about them before 1609, were seen by Galileo in a coldly new and different light [Figure 4-6]. He found "that the surface of the moon is not smooth, uniform, and precisely spherical as a great number of philosophers believe

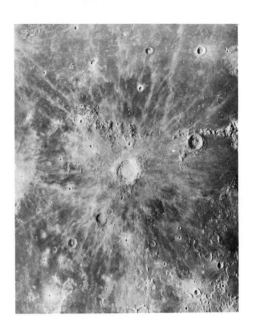

Figure 4-4. *A landscape similar to the earth's, but dead, was what impressed Galileo the first time he turned his telescope to the moon.* (Hale Observatories)

Figure 4-5. *Galileo was the first to see the craters on the moon. His observations killed the ancient belief that the moon was smooth and perfectly spherical.* (Hale Observatories)

it (and the other heavenly bodies) to be, but is uneven, rough, and full of cavities and prominences, being not unlike the face of the earth, relieved by chains of mountains and deep valleys." [See Figure 4-6.] As an example of Galileo's style in describing the earthlike quality of the moon, read the following:

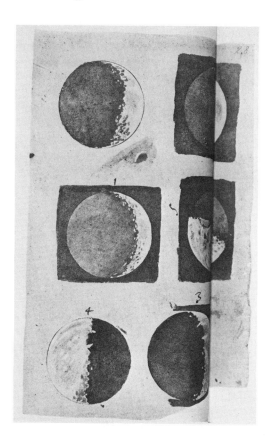

Figure 4-6. Galileo's own drawings of the moon. (Yerkes Observatory. Figure added by editor.)

Again, not only are the boundaries of shadow and light in the moon seen to be uneven and wavy, but still astonishingly many bright points appear within the darkened portion of the moon, completely divided and separated from the illuminated part and at a considerable distance from it. After a time these gradually increase in size and brightness, and an hour or two later they become joined with the rest of the lighted part which has now increased in size.

Meanwhile more and more peaks shoot up as if sprouting now here, now there, lighting up within the shadowed portion; these become larger, and finally they too are united with that same luminous surface which extends further. And on the earth, before the rising of the sun, are not the highest peaks of the mountains illuminated by the sun's rays while the plains remain in shadow? Does not the light go on spreading while the larger central parts of these mountains are becoming illuminated? And when the sun has finally risen, does not the illumination of plains and hills finally become one? But on the moon the variety of elevations and depressions appears to surpass in every way the roughness of the terrestrial surface, as we shall demonstrate further on.

Not only did Galileo describe the appearance of mountains on the moon; he also measured them. It is characteristic of Galileo as a scientist of the modern school that as soon as he found any kind of phenome-

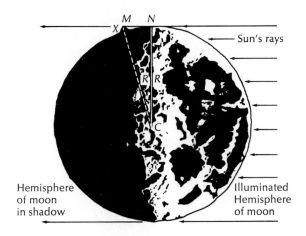

Figure 4-7. Galileo's measurement of the height of mountains on the moon was simple but convincing. The point N is the terminator (boundary) between the illuminated and nonilluminated portions of moon. The point M is a bright spot observed in the shadowed region; Galileo correctly surmised that the bright spot was a mountain peak whose base remained shadowed by the curvature of the moon. He could compute the moon's radius from the moon's known distance from the earth and could estimate the distance NM through his telescope. By the Pythagorean theorem, then, $CM^2 = MN^2 + CN^2$, or, since R is the radius and X the altitude of the peak,

$$(R+X)^2 \qquad = R^2 + MN^2$$

or

$$R^2 + 2RX + X^2 \quad = R^2 + MN^2$$

or

$$X^2 + 2RX - MN^2 = 0$$

which is easily solved for X, the altitude of the peak.

non he wanted to measure it. It is all very well to be told that the telescope discloses that there are mountains on the moon, just as there are mountains on the earth. But how much more extraordinary it is, and how much more convincing, to be told that there are mountains on the moon and that they are exactly four miles high! Galileo's determination of the height of the mountains on the moon has withstood the test of time, and today we agree with his estimate of their maximum height. . . . (For those who are interested, Galileo's method of computing the height of these mountains will be found in Figure 4-7.)

Galileo saw that the surface of the moon provided evidence that the earth is not unique. Since the moon resembles the earth, he had demonstrated that at least the nearest heavenly body does not enjoy that perfection attributed to all heavenly bodies by the classic authorities. Nor did Galileo make this a passing reference; he returned to the idea later in the book when he compared a portion of the moon to a specific region on earth: "In the center of the moon there is a cavity larger than all the rest, and perfectly round in shape. . . . As to light and shade, it offers the same appearance as would a region like Bohemia if that were enclosed on all sides by very lofty mountains arranged exactly in a circle."

EARTHSHINE

At this point Galileo introduces a still more startling discovery: earthshine. This phenomenon may be seen in the photograph reproduced in Figure 4-8. From the photograph it is plain, as may be seen when the moon is examined through a telescope, that there is what Galileo called a "secondary" illumination of the dark surface of the moon, which can be shown geometrically to accord perfectly with light from the sun reflected by the earth into the moon's darkened regions. It cannot be the moon's own light, or a contribution of starlight, since it would then be displayed during eclipses; it is not. Nor can it come from Venus or from any other planetary source. As for the moon's being illuminated by the earth, what, asked Galileo, is there so remarkable about this? "The earth, in fair and grateful exchange, pays back to the moon an illumination similar to that which it receives from her throughout nearly all the darkest gloom of night."

Here Galileo ends his description of the moon. The subject is one which, he told his readers, he would discuss more fully in his book on the *System of the World*. "In that book," he said, "by a multitude of arguments and experiences, the solar reflection from the earth will be shown to be quite real—against those who argue that the earth must be

Figure 4-8. *"In fair and grateful exchange," as Galileo put it, the earth contributes illumination to the moon. Earthshine shows on the portion of the moon
that otherwise would be in shadow.* (Yerkes Observatory)

excluded from the dancing whirl of stars (or heavenly bodies) for the
specific reason that it is devoid of motion and of light. We shall prove
the earth to be a wandering body surpassing the moon in splendor, and
not the sink of all dull refuse of the universe; this we shall support by an
infinitude of arguments drawn from nature." This was Galileo's first announcement that he was writing a book on the system of the world, a
work which was delayed for many years and which—when finally
published—resulted in Galileo's trial before the Roman Inquisition.

But observe what Galileo had proved thus far. He showed that the ancients were wrong in their descriptions of the moon; the moon is not the
perfect body they pictured, but resembles the earth, which therefore
cannot be said to be unique and consequently different from all the
heavenly objects. And if this was not enough, his studies of the moon
had shown that the earth shines. No longer was it valid to say that the
earth is not a shining object like the planets. And if the earth shines just
as the moon does, perhaps the planets may also shine in the very same
manner by reflecting light from the sun! Remember, in 1609 it was still
an undecided question whether the planets shine from internal light,

like the sun and the stars, or whether by reflected light, like the moon. As we shall see in a moment, it was one of Galileo's greatest discoveries to prove that the planets shine by reflected light, and that they encircle the sun in their orbits.

STARS GALORE

But before turning to that subject, let us state briefly some of Galileo's other discoveries. When Galileo looked at the fixed stars, he found that they . . . "appear not to be enlarged by the telescope in the same proportion as that in which it magnifies other objects, and even the moon itself." Furthermore, Galileo called attention to "the differences between the appearance of the planets and of the fixed stars" in the telescope. "The planets show their globes perfectly round and definitely bounded, looking like little moons, spherical and flooded all over with light; the fixed stars are never seen to be bounded by a circular periphery, but have rather the aspect of blazes whose rays vibrate about them and scintillate a great deal." Here was the basis of one of Galileo's great answers to the detractors of Copernicus. Plainly, the stars must be at enormous distances from the earth compared to the planets, if a telescope can magnify the planets to make them look like discs, but cannot do the same with the fixed stars.

Galileo related how he "was overwhelmed by the vast quantity of stars," so many that he found "more than five hundred new stars distributed among the old ones within limits of one or two degrees of arc." To three previously known stars in Orion's Belt and six in the Sword, he added "eighty adjacent stars." In several pictures he presented the results of his observations with a large number of newly discovered stars amongst the older ones. Although Galileo does not make the point explicitly, it is implied that one hardly needed to put one's faith in the ancients, since they had never seen most of the stars, and had spoken from woefully incomplete evidence. A weakness of naked-eye observation was exposed by Galileo in terms of "the nature and the material of the Milky Way." With the aid of the telescope, he wrote, the Milky Way has been "scrutinized so directly and with such ocular certainty that all the disputes which have vexed philosophers through so many ages have been resolved, and we are at last freed from wordy debates about it." Seen through the telescope, the Milky Way is "nothing but a congeries of innumerable stars grouped together in clusters. Upon whatever part of it the telescope is directed, a vast crowd of stars is immediately presented to view." And this was true not only of the Milky Way, but also of "the stars which have been called 'nebulous' by every astronomer up to this time," and which "turn out to be groups of very small stars arranged in a wonderful manner." . . .

Although Galileo's book ends with the description of the satellites of Jupiter, it will be wise, before we explore the implications of his research, to discuss three other astronomical discoveries made by Galileo with his telescope. The first was the discovery that Venus exhibits phases. For a number of reasons Galileo was overjoyed to discover that Venus exhibits phases. In the first place, it proved that Venus shines by reflected light, and not by a light of its own; this meant that Venus is like the moon in this regard, and also like the earth (which Galileo had previously shown to shine by reflected light of the sun). Here was another point of similarity between the planets and the earth, another weakening of the ancient philosophical barrier between earth and heavenly objects. Furthermore, as may be seen in Figure 4-9, top, if Venus moves in an orbit around the sun, not only will Venus go through a

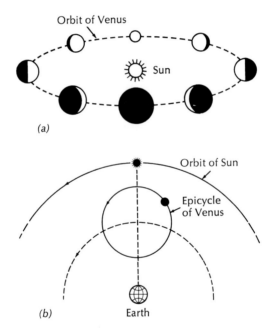

Figure 4-9. The phases of Venus, first observed by Galileo, were a powerful argument against the ancient astronomy. How the existence of phases accords with the system of Copernicus and how the change in the relative apparent diameter of Venus supports the concept of the planet having a solar orbit is shown at top. Why the phenomenon would be impossible in the Ptolemaic system is shown at bottom.

complete cycle of phases, but under constant magnification the different phases will appear to be of different sizes because of the change in the distance of Venus from the earth. . . . For instance, when Venus is at such a position as to enable us to see a complete circle or almost a complete circle, corresponding to a full moon, the planet is on the opposite side of its orbit around the sun from the earth, or is seen at its farthest distance from the earth. When Venus exhibits a half circle, corresponding to a quarter moon, the planet is not so far from the earth. Finally, when we barely see a faint crescent, Venus must be at its nearest point to the earth. Hence, we should expect that when Venus shows a faint crescent it would appear very large; when Venus shows the appearance of a quarter moon, it would be of moderate size; when we see the whole disc, Venus should be very small.

According to the Ptolemaic system, Venus (like Mercury) would never be seen far from the sun, and hence would be observed only as morning star or evening star near the place where the sun has either risen or set. The center of the orbit's epicycle would be permanently aligned between the center of the earth and the center of the sun and would move around the earth with a period of one year, just as the sun does. But it is perfectly plain, as may be seen in Figure 4-9, bottom, that in these circumstances we could never see the complete sequence of phases Galileo observed—and we can observe. For instance, the possibility of seeing Venus as a disc arises only if Venus is farther from the earth than the sun, and can never arise according to the principles of the Ptolemaic system. Here then was a most decisive blow against the Ptolemaic system.

We need not say much about two further telescopic discoveries of Galileo, because they had less force than the previous ones. The first was the discovery that sometimes Saturn appeared to have a pair of "ears," and that sometimes the "ears" changed their shape and even disappeared. Galileo never could explain this strange appearance, because his telescope could not resolve the rings of Saturn. But at least he had evidence to demonstrate how erroneous it was to speak of planets as perfect celestial objects, when they could have such queer shapes. One of his most interesting observations was of the spots on the sun, described in a book that bore the title *History and Demonstrations concerning Sunspots and their Phenomena* (1613). Not only did the appearance of these spots prove that even the sun was not the perfect celestial object described by the ancients, but Galileo was able to show that from observations of these spots one could prove the rotation of the sun, and even compute the speed with which the sun rotates upon its axis. Although the fact that the sun did rotate became extremely important in

Galileo's own mechanics, it did not imply, as he seems to have believed, that annual revolution of the earth around the sun must follow.

The Starry Messenger °
by Galileo Galilei (1610)

DISCOVERY OF JUPITER'S SATELLITES, JANUARY 7, 1610

I have now finished my brief account of the observations which I have thus far made with regard to the moon, the fixed stars, and the Galaxy. There remains the matter, which seems to me to deserve to be considered the most important in this work, namely, that I should disclose and publish to the world the occasion of discovering and observing four planets, never seen from the very beginning of the world up to our own times, their positions, and the observations made during the last two months about their movements and their changes of magnitude; and I summon all astronomers to apply themselves to examine and determine their periodic times, which it has not been permitted me to achieve up to this day, owing to the restriction of my time. I give them warning, however, again, so that they may not approach such an inquiry to no purpose, that they will want a very accurate telescope, and such as I have described in the beginning of this account.

On the seventh day of January in the present year, 1610, in the first hour of the following night, when I was viewing the constellations of the heavens through a telescope, the planet Jupiter presented itself to my view, and as I had prepared for myself a very excellent instrument, I noticed a circumstance which I had never been able to notice before, owing to want of power in my other telescope, namely, that three little stars, small but very bright, were near the planet; and although I believed them to belong to the number of the fixed stars, yet they made me somewhat wonder, because they seemed to be arranged exactly in a straight line, parallel to the ecliptic, and to be brighter than the rest of the stars equal to them in magnitude. The position of them with reference to one another and to Jupiter was as follows:

Figure 4-10   Ori.   ★   ★   ●   ★   Occ.

° From Galileo Galilei, *The Sidereal Messenger,* English translation by Edward Stafford Carlos, 1880, in Forest Ray Moulton and Justus J. Schifferes (eds.), *The Autobiography of Science,* Doubleday & Company, Inc., New York. Copyright 1945 by Justus J. Schifferes. Reprinted by permission.

On the east [Orient] side there were two stars, and a single one toward the west [Occident]. The star which was furthest toward the east, and the western star, appeared rather larger than the third.

I scarcely troubled at all about the distance between them and Jupiter, for, as I have already said, at first I believed them to be fixed stars; but when on January 8, led by some fatality, I turned again to look at the same part of the heavens, I found a very different state of things, for there were three little stars all west of Jupiter, and nearer together than on the previous night, and they were separated from one another by equal intervals, as the accompanying figure shows.

Figure 4-11   Ori.   ●   ★   ★   ★   Occ.

At this point, although I had not turned my thoughts at all upon the approximation of the stars to one another, yet my surprise began to be excited, how Jupiter could one day be found to the east of all the aforesaid fixed stars when the day before it had been west of two of them; and forthwith I became afraid lest the planet might have moved differently from the calculation of astronomers, and so had passed those stars by its own proper motion. I therefore waited for the next night with the most intense longing, but I was disappointed of my hope, for the sky was covered with clouds in every direction.

But on January 10 the stars appeared in the following position with regard to Jupiter,

Figure 4-12   Ori.   ★   ★   ●   Occ.

the third, as I thought, being hidden by the planet. They were situated just as before, exactly in the same straight line with Jupiter, and along the Zodiac.

When I had seen these phenomena, as I knew that corresponding changes of position could not by any means belong to Jupiter, and as, moreover, I perceived that the stars which I saw had always been the same, for there were no others either in front or behind, within a great distance, along the Zodiac—at length, changing from doubt into surprise, I discovered that the interchange of position which I saw belonged not to Jupiter but to the stars to which my attention had been drawn, and I thought therefore that they ought to be observed henceforward with more attention and precision.

Accordingly, on January 11 I saw an arrangement of the following kind, namely,

Figure 4-13   Ori.   ★   ★   ●   Occ

only two stars to the east of Jupiter, the nearer of which was distant from Jupiter three times as far as from the star further to the east; and the star furthest to the east was nearly twice as large as the other one; whereas on the previous night they had appeared nearly of equal magnitude. I therefore concluded, and decided unhesitatingly, that there are three stars in the heavens moving about Jupiter, as Venus and Mercury round the sun; which at length was established as clear as daylight by numerous other subsequent observations. These observations also established that there are not only three but four erratic sidereal bodies performing their revolutions round Jupiter, observations of whose changes of position made with more exactness on succeeding nights the following account will supply. I have measured also the intervals between them with the telescope in the manner already explained. Besides this, I have given the times of observation, especially when several were made in the same night, for the revolutions of these planets are so swift that an observer may generally get differences of position every hour.

January 12. At the first hour of the next night I saw these heavenly bodies arranged in this manner.

Figure 4-14   Ori.   ★ ★●   ★   Occ.

The satellite farthest to the east was greater than the satellite farthest to the west; but both were very conspicuous and bright; the distance of each one from Jupiter was two minutes. A third satellite, certainly not in view before, began to appear at the third hour: it nearly touched Jupiter on the east side, and was exceedingly small. They were all arranged in the same straight line, along the ecliptic.

January 13. For the first time four satellites were in view in the following position with regard to Jupiter.

Figure 4-15   Ori.   ★   ●★★★   Occ.

There were three to the west and one to the east; they made a straight line nearly, but the middle satellite of those to the west deviated a little from the straight line toward the north. The satellite farthest to the east was at a distance of 2′ from Jupiter; there were intervals of 1′ only between Jupiter and the nearest satellite, and between the satellites themselves, west of Jupiter. All the satellites appeared of the same size, and though small they were very brilliant and far outshone the fixed stars of the same magnitude.

January 14. The weather was cloudy. . . .

These are my observations upon the four Medicean planets, recently discovered for the first time by me; and although it is not yet permitted me to deduce by calculation from these observations the orbits of these bodies, yet I may be allowed to make some statements, based upon them, well worthy of attention.

DEDUCTIONS FROM THE PREVIOUS OBSERVATIONS CONCERNING
THE ORBITS AND PERIODS OF JUPITER'S SATELLITES

And, in the first place, since they are sometimes behind, sometimes before Jupiter, at like distances, and withdraw from this planet toward the east and toward the west only within very narrow limits of divergence, and since they accompany this planet alike when its motion is retrograde and direct, it can be a matter of doubt to no one that they perform their revolutions about this planet, while at the same time they all accomplish together orbits of twelve years' length about the center of the world. Moreover, they revolve in unequal circles, which is evidently the conclusion to be drawn from the fact that I have never been permitted to see two satellites in conjunction when their distance from Jupiter was great, whereas near Jupiter two, three, and sometimes all four have been found closely packed together. Moreover, it may be detected that the revolutions of the satellites which describe the smallest circles round Jupiter are the most rapid, for the satellites nearest to Jupiter are often to be seen in the east, when the day before they have appeared in the west, and contrariwise. Also the satellite moving in the greatest orbit seems to me, after carefully weighing the occasions of its returning to positions previously noticed, to have a periodic time of half a month.

Besides, we have a notable and splendid argument to remove the scruples of those who can tolerate the revolution of the planets round the sun in the Copernican system, yet are so disturbed by the motion of one moon about the earth, while both accomplish an orbit of a year's length about the sun, that they consider that this theory of the universe

must be upset as impossible: for now we have not one planet only revolving about another, while both traverse a vast orbit about the sun,
but our sense of sight presents to us four satellites circling about Jupiter,
like the moon about the earth, while the whole system travels over a
mighty orbit about the sun in the space of twelve years.

The Birth of a New Physics °
by I. Bernard Cohen (1960)

Prior to 1609 the Copernican system had seemed to men a mere mathematical speculation, a proposal made to "save the phenomenon." The
basic supposition that the earth was "merely another planet" had been
so contrary to all the dictates of experience and common sense that very
few men had faced up to the awesome consequences of the heliostatic
system. But after 1609, when men discovered through Galileo's eyes
what the universe was like, they had to accept the fact that the telescope showed the world to be non-Ptolemaic and non-Aristotelian, in
that the uniqueness attributed to the earth (and the physics based on
that supposed uniqueness) could not fit the facts. There were only two
possibilities open: One was to refuse to look through the telescope or to
refuse to accept what one saw when one did; the other was to reject the
physics of Aristotle and the astronomy of Ptolemy. . . .

Aristotelian physics, as we have seen, was based on two postulates
which could not stand the Copernican assault: One was the immobility
of the earth; the other was the distinction between the physics of the
earthly four elements and the physics of the fifth celestial element. So
we may understand that after 1610 it became increasingly clear that the
old physics had to be abandoned, and a new physics established—a
physics suitable for the moving earth required in the Copernican system.[1]

° From I. Bernard Cohen, *The Birth of a New Physics.*

1. Galileo's observations of the phases and relative sizes of Venus, and of the occasional gibbous phase of Mars, proved that Venus and presumably the other planets
move in orbits around the sun. There is no planetary observation by which we on
earth can prove that the earth is moving in an orbit around the sun. Thus all Galileo's discoveries with the telescope can be accommodated to the system invented by
Tycho Brahe just before Galileo began his observations of the heavens. In this Tychonic system, the planets Mercury, Venus, Mars, Jupiter and Saturn move in orbits
around the sun, while the sun moves in an orbit around the earth in a year. Furthermore, the daily rotation of the heavens is communicated to the sun and planets,
so that the earth itself neither rotates nor revolves in an orbit. The Tychonic system
appealed to those who sought to save the immobility of the earth while accepting
some of the Copernican innovations [Figure 3-23].

For most thinking men in the decades following Galileo's observations with the telescope the concern was not so much for the need of a new system of physics, as it was for a new system of the world. Gone forever was the concept that the earth had a fixed spot in the center of the universe, but it was now conceived to be in motion, never in the same place for any two immediately successive instants. Gone also was the comforting thought that the earth is unique, that it is an individual object without any likeness anywhere in the universe, that the uniqueness of man had given a uniqueness to his habitation. There were other problems that soon arose, of which one is the size of the universe. For the ancients the universe was finite, each of the celestial spheres, including that of the fixed stars, being a finite size and moving in its diurnal motion so that each part of it had a finite speed. If the stars were at an infinite distance, then they could not move in a daily circular motion around the earth with a finite speed, for the path of an object at an infinite distance must be infinitely long, and the time it takes to move an infinite distance is infinite. Hence in the geostatic system the fixed stars could not be infinitely far away. But in the Copernican system, when the fixed stars were not only fixed with regard to one another but were actually considered fixed in space, there was no limitation upon their distance.

Not all Copernicans considered the universe infinite, and Copernicus himself certainly thought of the universe as finite, as did Galileo. But others saw Galileo's discoveries as indicating the presence of innumerable stars at infinite distances, and the earth itself diminished to a speck. Nowhere can one see the disruption of "this little world of man," and what has been called "the realization of how slight a part that world plays in an enlarged and enlarging universe," better than in these lines of a sensitive clergyman and poet, John Donne:

> And new Philosophy calls all in doubt,
> The Element of fire is quite put out;
> The Sun is lost, and th' earth, and no mans wit
> Can well direct him where to looke for it.
> And freely men confesse that this world's spent,
> When in the Planets, and the Firmament
> They seeke so many new; then see that this
> Is crumbled out againe to his Atomies.
> 'Tis all in peeces, all cohaerence gone;
> All just supply, and all Relation.

THE PHYSICS OF A MOVING EARTH

Odd as it may seem, most people's views about motion are part of a system of physics that was proposed more than 2000 years ago and was experimentally shown to be inadequate at least 1400 years ago. It is a fact that presumably well-educated men and women tend even today to think about the physical world as if the earth were at rest, rather than in motion. By this I do not mean that such people really believe the earth is at rest; if questioned, they will reply that of course they "know" that the earth rotates once a day about its axis and at the same time moves in a great yearly orbit around the sun. Yet when it comes to explaining certain common physical events, these same people are not able to tell you how it is that these everyday phenomena can happen, as we see they do, on a moving earth. In particular, these misunderstandings of physics tend to center on the problem of falling objects, on the general concept of motion. Thus we may see exemplified the precept of the Renaissance genius Leonardo da Vinci, "To understand motion is to understand nature."

WHERE WILL IT FALL?

In his inability to deal with questions of motion in relation to a moving earth, the average person is in the same position as some of the greatest scientists of the past, which may be a source of considerable comfort to him. The major difference is, however, that for the scientist of the past the inability to resolve these questions was a sign of his time, whereas for the modern man such inability is, alas, a badge of ignorance. Characteristic of these problems is a woodcut of the sixteenth century showing a cannon pointing up in the air. [Consider] the question that is asked, "*Retombera-t-il?*" (Will it fall back down again?) If the earth is at rest, there is no doubt that the cannon ball fired straight up in the air would eventually come straight down again into the cannon. But will it on a moving earth? And if it will, why?

In a general way let us examine all the arguments. If supporters of the theory that the earth may move contended that the air must move along with the earth, and that an arrow shot up into the air would be carried along with the moving air and earth, the opponents had a ready reply: Even if the air could be supposed to move—a difficult supposition since there is no apparent cause for the air to move with the earth—would not the air move very much more slowly than the earth since it is so very different in substance and in quality? And even so would not, therefore, the arrow be left behind? And what of the high winds that a man in a tower should feel? . . .

The earth rotates upon its axis once in every 24 hours. At the equator the circumference of the earth is approximately 24,000 miles, and so the speed of rotation of any observer at the earth's equator is 1000 miles per hour. This is a linear speed of about 1500 feet per second. Conceive the following experiment. A rock is thrown straight up into the air. The time in which it rises is, let us say, two seconds, while a similar time is required for its descent. During four seconds the rotation of the earth will have moved the point from which the object was thrown a distance of some 6000 feet, a little over a mile. But the rock does not strike the earth one mile away; it lands very near the point from which it was thrown. We ask, How can this be possible? How can the earth be twirling around at this tremendous speed of 1000 miles per hour and yet we do not hear the wind whistling as the earth leaves the air behind it? Or, to take one of the other classical objections to the idea of a moving earth, consider a bird perched on the limb of a tree. The bird sees a worm on the ground and lets go of the tree. In the meanwhile the earth goes whirling by at this enormous rate, and the bird, though flapping its wings as hard as it can, will never achieve sufficient speed to find the worm—unless the worm is located to the west. But it is a fact of observation that birds do fly from trees to the earth and eat worms that lie to the east as well as to the west. Unless you can see your way clearly through these problems without a moment's thought, you do not really live modern physics to its fullest, and for you the statement that the earth rotates upon its axis once in 24 hours really has no meaning.

If the daily rotation presents a serious problem, think of the annual motion of the earth in its orbit. Let us compute the speed with which the earth moves in its orbit around the sun. There are 60 seconds in a minute and 60 minutes in an hour, or 3600 seconds in an hour. Multiply this number by 24 to get 86,400 seconds in a day. Multiply this by 365¼ days, and the result is a little more than 30 million seconds in a year. To find the speed at which the earth moves around the sun, we have to compute the size of the earth's orbit and divide it by the time it takes the earth to move through the orbit. This path is roughly a circle with a radius of about 93 million miles, and a circumference of about 580,-000,000 miles (the circumference of the circle is equal to the radius multiplied by 2π). This is equivalent to saying that the earth moves through about 3,000,000,000,000 feet in every year. The speed of the earth is thus

$$\frac{3,000,000,000,000 \text{ feet}}{30,000,000 \text{ seconds}} = 100,000 \text{ ft/sec}$$

Each of the questions raised about the rotating earth can be raised again in magnified form with regard to an earth moving in an orbit. This speed of 100,000 feet per second, or about 19 miles per second, shows us the great difficulty encountered at the beginning of the chapter. Let us ask this question: Is it possible for us to move at a speed of 19 miles per second and not be aware of it? Suppose we dropped an object from a height of 16 feet; it would take about one second to strike the ground. According to our calculation, while this object was falling the earth should have been rushing away underneath, and the object would strike the ground some 19 miles from the point where it was dropped! And as for the birds on the trees, if a bird hanging on a limb for dear life were to let go for an instant, it would be lost out in space forever. Yet the fact is that birds are not lost in space but continue to inhabit the earth and to fly about it singing gaily.

These examples show us how difficult it really is to face the consequences of an earth in motion. It is plain that our ordinary ideas are inadequate to explain the observed facts of daily experience on an earth that is either rotating or moving in its orbit. There should be no doubt, therefore, that the shift from the concept of a stationary earth to a moving earth necessarily involved the birth of a new physics.

UNIFORM LINEAR MOTION

Let us begin by a consideration of a limited problem: that of uniform linear motion. By this is meant motion proceeding in a straight line in such a way that if any two equal intervals of time are chosen, the distance covered in those two intervals will always be identical. This is the definition Galileo gave in his last and perhaps greatest book, *Discourses and Demonstrations Concerning Two New Sciences,* published in 1638, after his trial and condemnation by the Roman Inquisition. In this book Galileo presented his most mature views on dynamics, force in relation to motion. He emphasized particularly the fact that in defining uniform motion, it is important to make sure that the word "any" is included, for otherwise, he said, the definition would be meaningless. In this he was certainly criticizing some of his contemporaries and predecessors, and deservedly.

Suppose that there is such a motion in nature; we may ask with Galileo, what experiments could we imagine to demonstrate its nature? If we are in a ship or carriage moving uniformly in a straight line, what actually will happen to a weight allowed to fall freely? The answer, experiment will prove, is that in such circumstances the falling will be straight downward with regard to the frame of reference (say the cabin of a ship, or the interior of a carriage), and it will be so whether that frame of reference is standing still with regard to the outside environment or moving forward in a straight line at constant speed. Expressing it differently, we may state the general conclusion that no experiment can be performed within a sealed room moving in a straight line at constant speed that will tell you whether you are standing still or moving. In actual experience, we can often tell whether we are standing still or moving, because we can see from a window whether there is any relative motion between us and the earth. If the room is not closely sealed, we may feel the air rushing through and creating a wind. Or we may feel the vibration of motion or hear the wheels turning in a carriage, automobile, or railroad car. A form of relativity is involved here, and it was stated very clearly by Copernicus, because it was essential to his argument to establish that when two objects, such as the sun and earth, move relative to each other, it is impossible to tell which one is at rest and which one is in motion. Copernicus could point to the example of two ships at harbor, one pulling away from the other. A man on a ship asks which of the two, if either, is at anchor and which is moving out with the tide? The only way to tell is to observe the land, or a third ship at anchor. In present-day terms, we could use for this example two railroad trains on parallel tracks facing in opposite directions. We all have had the experience of watching a train on the adjacent track and thinking that we are in motion, only to find when the other train has left the station that we have been at rest all the time.

A LOCOMOTIVE'S SMOKESTACK AND A MOVING SHIP

But before we discuss this point further, an experiment is in order. This demonstration makes use of a toy train traveling along a straight track with what closely approximates uniform motion. The locomotive's smokestack contains a small cannon actuated by a spring, so constructed that it can fire a steel ball or marble vertically into the air. When the gun is loaded and the spring set, a release underneath the locomotive actuates a small trigger. In the first part of this experiment the train remains in place upon the track. The spring is set, the ball is placed in the small cannon, and the release mechanism triggered. In Figure 4-16, top, a scene of successive stroboscopic shots show us the position of the ball at equally separated intervals. Observe that the ball travels straight upward, reaches its maximum, then falls straight downward onto the locomotive, thus striking almost the very point from which it had been shot. In the second experiment the train is set into uniform motion, and the spring once again released. Figure 4-16, bottom, shows what happens. A

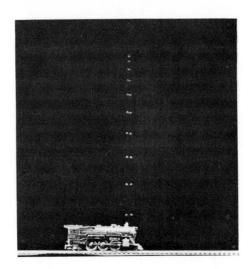

Figure 4-16. *A ball fired with a spring gun in the smokestack of a moving toy locomotive describes a parabola (bottom) and lands on the locomotive instead of going straight up and down as it does when the locomotive is standing still (top). These stroboscopic pictures, with exposures at intervals of one-thirtieth of a second, vividly illustrate one of Galileo's arguments on the behavior of falling bodies and settle the ancient debate about bodies dropped from the masts of moving ships. If the speed of the locomotive were absolutely uniform and if the ball met no air resistance, it would land in the smokestack. (In fact, even under the imperfect conditions of the experiment, the ball hits the smokestack more often than not.) Note that the ball attains the same height whether the locomotive is at rest or moving. Notice, too, that in the picture where the locomotive is standing still, the distances traveled by the ball in the intervals of exposure correspond almost exactly. On the ascent, gravity slows it down; on the descent, gravity speeds it up. (Berenice Abbott)*

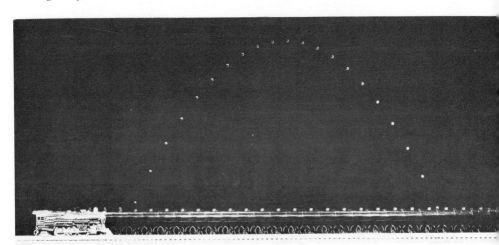

comparison of the two pictures will convince you, incidentally, that the upward and downward part of the motion is the same in both cases, and is independent of whether the locomotive is at rest or has a forward motion. . . . we are primarily concerned with the fact that the ball continued to move in a forward direction with the train, and that it fell onto the locomotive just as it did when the train was at rest. Plainly then, this particular experiment, at least to the extent of determining whether the ball returns to the cannon or not, will never tell us whether the train is standing still or moving in a straight line with a constant speed.

Even those who cannot explain this experiment can draw a most important conclusion. Galileo's inability to explain how Jupiter could move without losing its satellites did not destroy the phenomenon's effectiveness as an answer to those who asked how the earth could move and not lose its moon. Just as our train experiment—even if unexplainable— would be sufficient answer to the argument that the earth must be at rest because otherwise a dropped ball would not fall vertically downward to strike the ground at a point directly below, and a cannon ball shot vertically upward would never return to the cannon. . . .

Galileo would have nodded in approval at our experiment. In his day the experiment was discussed, but not often performed. The usual reference frame was a moving ship. This was a traditional problem, which Galileo introduced in his famous *Dialogue Concerning the Two Chief World Systems*, as a means of confuting the Aristotelian beliefs. In the course of this discussion, Galileo has Simplicio, the character in the dialogue who stands for the traditional Aristotelian, say that in his opinion an object dropped from the mast of a moving ship will strike the ship somewhere behind the mast along the deck. On first questioning, Simplicio admits that he has never performed the experiment, but he is persuaded to say that he assumes that Aristotle or one of the Aristotelians must have done this experiment or it would not have been reported. Ah no, says Galileo, this is certainly a false assumption, because it is plain that they have never performed this experiment. How can he be so sure? asks Simplicio, and he receives this reply: The proof that this experiment was never performed lies in the fact that the wrong answer was obtained. Galileo has given the right answer. The object will fall at the foot of the mast, and it will do so whether the ship is in motion or whether the ship is at rest. Incidentally, Galileo asserted elsewhere that he had performed such an experiment, although he did not say so in his treatise. Instead he said, "I, without observation, know that the result must be as I say, because it is necessary."

Why is it that an object falls to the same spot on the deck from the mast of a ship that is at rest and from the mast of a ship that is moving in a straight line with constant speed? For Galileo it was not enough that this should be so; it required some principle that would be basic to a system of dynamics that could account for the phenomena observed on a moving earth.

GALILEO'S DYNAMICS

Our toy train experiment . . . illustrates three major aspects of Galileo's work in dynamics. In the first place, there is the principle of inertia, toward which Galileo strove but which . . . awaited the genius of Isaac Newton for its modern definitive formulation. Secondly, the photographs of the distances of descent of the ball after successive equal intervals of time illustrate his principles of uniformly accelerated motion. Finally, in the fact that the rate of downward fall during the forward motion is the same as the rate of downward fall at rest, we may see an example of Galileo's famous principle of the composition of velocities. . . .

We shall examine these three topics by first considering Galileo's studies of accelerated motion in general, then his work dealing with inertia, and finally his analysis of complex motions.

In studying the problem of falling bodies, Galileo, we know, made experiments in which he dropped objects from heights, and—notably in the Pisan days of his youth—from a tower. Whether the tower was the famous Leaning Tower of Pisa or some other tower we cannot say; the records that he kept merely tell us that it was from some tower or other. Later on his biographer Viviani, who knew Galileo during his last years, told a fascinating story which has since taken root in the Galileo legend. According to Viviani, Galileo, desiring to confute Aristotle, ascended the Leaning Tower of Pisa, "in the presence of all other teachers and philosophers and of all students," and "by repeated experiments" proved "that the velocity of moving bodies of the same composition, unequal in weight, moving through the same medium, do not attain the proportion of their weight, as Aristotle assigned it to them, but rather that they move with equal velocity. . . ." Since there is no record of this public demonstration in any other source, scholars have tended to doubt that it happened, especially since in its usual telling and retelling it becomes fancier each time. Whether Viviani made it up, or whether Galileo told it to him in his old age, not really remembering what had happened many decades earlier, we do not know. . . .

Such an experiment, if performed, could only have the result of proving Aristotle wrong. In Galileo's day, it was hardly a great achievement to prove that Aristotle was wrong in only one respect. Pierre de la

Ramée (or Ramus) had some decades earlier made it known that everything in Aristotle's physics was unscientific. The inadequacies of the Aristotelian law of motion had been evident for at least four centuries, and during that time a considerable body of criticism had piled up. Although they struck another blow at Aristotle, experiments from the tower, whether the Tower of Pisa or any other, certainly did not disclose to Galileo a new and correct law of falling bodies. Yet formulation of the law was one of his greatest achievements.

To appreciate the full nature of Galileo's discoveries, we must understand the importance of abstract thinking, of its use by Galileo as a tool that in its ultimate polish was a much more revolutionary instrument than even the telescope. Galileo showed how abstraction may be related to the world of experience, how from thinking about "the nature of things," one may derive laws related to direct observation. . . .

Galileo demonstrated mathematically that a motion starting from rest in which the speed undergoes the same change in every equal interval of time (called *uniformly accelerated motion*) corresponds to traversing distances which are proportional to the squares of the elapsed times. Then Galileo showed by experiment that this ideal law is—within certain limits, we must interject—exemplified by motion on an inclined plane. From these two results, Galileo reasoned that in the absence of any air resistance, the motion of a freely falling body will always be accelerated according to this law. . . . When Robert Boyle, some thirty years later, was able to evacuate a cylinder, he showed that in such a vacuum all bodies fall with identical speeds no matter what their shapes. Thus proof was given of Galileo's assertion—an extrapolation from experience—that but for air resistance, all bodies fall at the same rate, with the same acceleration. Today we know the correct value of this acceleration (g) to be about 32 feet per second change of speed in each second. Hence, the speed of a falling body except for the almost negligible factor of air resistance, depends on the length of time during which it falls and not on its weight as Aristotle had supposed. . . .

In the *Discourses and Demonstrations Concerning Two New Sciences,* Galileo approached the problem of inertia chiefly in relation to his study of the path of a projectile, which he wanted to show was a parabola (Figure 4-17). Galileo considers a body sent out in a horizontal direction. It will then have two separate and independent motions. In the horizontal direction it will move with uniform velocity, except for the small slowing effect of air resistance. At the same time, its downward motion will be accelerated, just as a freely falling body is accelerated. It is the combination of these two motions that causes the trajectory to be parabolic. For his postulate that the downward component of the mo-

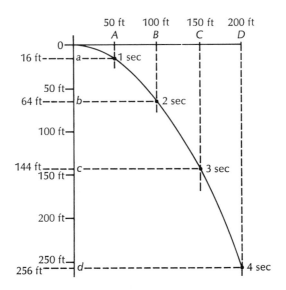

Figure 4-17. *To see how Galileo analyzed projectile motion, consider a shell fired horizontally from a cannon at the edge of a cliff at a speed of 50 feet per second. The points A, B, C, D show where the shell would be at the ends of successive seconds if there were no air resistance and no downward component; in this case there being a uniform horizontal motion, the shell going 50 feet in each second. In the downward direction, there is an accelerated motion. The points a, b, c, d show where the shell would be if it were to fall with no air resistance and no forward motion. Since the distance is computed by the law* $D = \frac{1}{2}AT^2$ *and the acceleration A is 32 feet per second, the distances corresponding to these times are:*

T	T²	½AT²	D
1 sec	1 sec²	16 ft/sec² × 1 sec²	16 ft
2 sec	4 sec²	16 ft/sec² × 4 sec²	64 ft
3 sec	9 sec²	16 ft/sec² × 9 sec²	144 ft
4 sec	16 sec²	16 ft/sec² × 16 sec²	256 ft

Since the shell actually has the two motions simultaneously, the net path is as shown by the curve.

For those who like a bit of algebra, let v be the constant horizontal speed and x the horizontal distance, so that $x = vt$. In the vertical direction let the distance be y, so that $y = \frac{1}{2}AT^2$.

Then, $x^2 = v^2t^2$, or $x^2/v^2 = t^2$, $2y/A = t^2$, and $x^2/v^2 = 2y/A$, or $y = (A/2v^2)x^2$, which is of the form $y = kx^2$, where k is a constant, and this is the classic equation of the parabola.

tion is the same as that of a freely falling body Galileo did not give an experimental proof, although he indicated the possibility of having one. He devised a little machine in which on an inclined plane a ball was projected horizontally, to move in a parabolic path.

Today we can easily demonstrate this conclusion by shooting one of a pair of balls horizontally, while the other is simultaneously allowed to fall freely from the same height. The result of such an experiment is shown in Figure 4-18, where a series of photographs taken stroboscopically at successive instants shows that although one of the balls is moving forward while the other is dropping vertically, the distances fallen in successive seconds are the same for both. This is the situation of a ball falling on a train moving at constant speed along a linear track. It falls *vertically* second after second just as it would if the train were at rest. Since it also moves horizontally at the same uniform speed as the train, its true path with respect to the earth is a parabola. Yet another modern example is that of an airplane flying horizontally at constant speed and releasing a bomb or torpedo. The downward fall is the same as if the bomb or torpedo had been dropped from the same height from an object

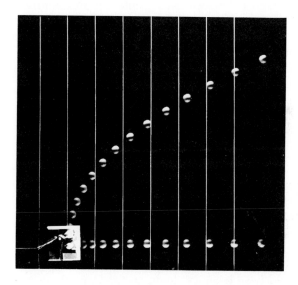

Figure 4-18. The independence of the vertical and horizontal components of projectile motion is illustrated in this stroboscopic photograph. In intervals of one-thirtieth of a second the projected ball falling along a parabolic path drops exactly the same distance as the ball allowed to fall vertically. (Berenice Abbott)

at rest, say a captive balloon on a calm day. As the bomb or torpedo falls from the airplane, it will continue to move forward with the horizontal uniform speed of the airplane and will, except for the effects of the air, remain directly under the plane. But to an observer at rest on the earth, the trajectory will be a parabola.

Finally consider a stone dropped from a tower. With respect to the earth (and for such a short fall the movement of the earth can be considered linear and uniform), falls straight downward. But with respect to the space determined by the fixed stars, it retains the motion shared with the earth at the moment of release, and its trajectory is therefore a parabola.

These analyses of parabolic trajectories are all based on the Galilean principle of separating a complex motion into two motions (or components) at right angles to each other. It is certainly a measure of his genius that he saw that a body could simultaneously have a uniform or nonaccelerated horizontal component of velocity and an accelerated vertical component—neither one in any way affecting the other. In every such case, the horizontal component exemplifies the tendency of a body that is moving at constant speed in a straight line to continue to do so, even though it loses physical contact with the original source of that uniform motion. This may also be described as a tendency of any body to resist any change in its state of motion, a property generally known since Newton's day as a body's *inertia*.

The Battle with Authority °
by Arthur Koestler (1959)

I shall take up the thread of Galileo's life at the point where his name suddenly burst into world fame through his discovery of the Jupiter planets. The *Star Messenger* was published in March 1610; in September, he took up his new post as "Chief Mathematician and Philosopher" to the Medicis in Florence; the following spring he spent in Rome.

The visit was a triumph. Cardinal del Monte wrote in a letter: "If we were still living under the ancient Republic of Rome, I verily believe that there would have been a column on the Capitol erected in Galileo's honor." The select Accademia dei Lincei (the lynx-eyed), presided by

° From Arthur Koestler, *The Sleepwalkers*, The Macmillan Company, New York. © 1959 by Arthur Koestler. Reprinted by permission of the Macmillan Company. Certain passages in the quoted material have been italicized for emphasis by Koestler.

Prince Federico Cesi, elected him a member and gave him a banquet; it was at this banquet that the word "telescope" was for the first time applied to the new invention. Pope Paul V received him in friendly audience, and the Jesuit Roman College honoured him with various ceremonies which lasted a whole day. The chief mathematician and astronomer of the College, the venerable Father Clavius, principal author of the Gregorian Calendar reform, who at first had laughed at the *Star Messenger,* was now entirely converted; so were the other astronomers at the College, Fathers Grienberger, van Maelcote and Lembo. They not only accepted Galileo's discoveries, but improved on his observations, particularly of Saturn and the phases of Venus. When the head of the College, the Lord Cardinal Bellarmine, asked for their official opinion on the new discoveries, they unanimously confirmed them.

This was of utmost importance. The phases of Venus, confirmed by the dozen of Jesuit astronomers, were incontrovertible proof that at least that planet revolved round the sun, that the Ptolemaic system had become untenable, and that the choice now lay between Copernicus and Brahe.[1] The Jesuit Order was the intellectual spearhead of the Catholic Church. Jesuit astronomers everywhere in Europe—particularly Scheiner in Ingoldstadt, Lanz in Munich, Kepler's friend Guldin in Vienna, and the Roman College in a body—began to support the Tychonic system as a half-way house to the Copernican. The Copernican system itself could be freely discussed and advocated as a working hypothesis, but it was unfavourably viewed to present it as established truth, because it seemed contrary to current interpretation of scripture—unless and until definite proof could be adduced in its favour. . . .

Within a brief period, Jesuit astronomers also confirmed the "earthly" nature of the moon, the existence of sunspots, and the fact that comets moved in outer space, beyond the moon. This meant the abandonment of the Aristotelian doctrine of the perfect and unchangeable nature of the celestial spheres. Thus the intellectually most influential order within the Catholic Church was at that time in full retreat from Aristotle and Ptolemy, and had taken up an intermediary position regarding Copernicus. They praised and feted Galileo, whom they knew to be a Copernican, and they kept Kepler, the foremost exponent of Copernicanism, under their protection throughout his life.

But there existed a powerful body of men whose hostility to Galileo never abated: the Aristotelians at the universities. The inertia of the human mind and its resistance to innovation are most clearly demonstrated not, as one might expect, by the ignorant mass—which is easily swayed once its imagination is caught—but by professionals with a

1. See footnote 1, p. 178, and Figure 3-23. Ed.

vested interest in tradition and in the monopoly of learning. Innovation is a twofold threat to academic mediocrities: it endangers their oracular authority, and it evokes the deeper fear that their whole, laboriously constructed intellectual edifice might collapse. The academic back-woodsmen have been the curse of genius from Aristarchus to Darwin and Freud; they stretch, a solid and hostile phalanx of pedantic medioc-rities, across the centuries. It was this threat—not Bishop Dantiscus or Pope Paul III—which had cowed Canon Koppernigk into lifelong si-lence. In Galileo's case, the phalanx resembled more a rearguard still firmly entrenched in academic chairs and preachers' pulpits.

❀   ❀   ❀

[During the next few years, Galileo's friends in Rome kept him in-formed about the attitude of the Church toward his work.

Letter to Galileo from Monsignor Ciampoli, 1615:]

Cardinal Barberini [the future Pope Urban VIII], who as you know from experience, has always admired your worth, told me only yesterday evening that with respect to these opinions he would like greater caution in *not going beyond the arguments used by Ptolemy and Copernicus*,[2] and finally in not ex-ceeding the limitations of physics and mathematics. For to explain the Scrip-tures is claimed by theologians as their field, and if new things are brought in, even by an admirable mind, not everyone has the dispassionate faculty of tak-ing them just as they are said. . . .

[Letter from Cardinal Piero Dini to Galileo a few days later:]

With Bellarmine I spoke at length of the things you had written. . . . And he said that as to Copernicus, there is no question of his book being prohib-ited; the worst that might happen, according to him, would be the addition of some material in the margins of that book to the effect that Copernicus had in-troduced his theory in order to save the appearances, or some such thing—just as others had introduced epicycles without thereafter believing in their exis-tence. And *with a similar precaution you may at any time deal with these mat-ters.* If things are fixed according to the Copernican system, [he said], it does not appear presently that they would have any greater obstacle in the Bible than the passage "(the sun) exults as a strong man to run his course," etc., which all expositors up to now have understood by attributing motion to the sun. And although I replied that this also could be explained as a concession to our ordinary forms of expression, I was told in answer that this was not a thing to be done in haste, just as the condemnation of any of these opinions was not to be passionately hurried. . . . I can only rejoice for you. . . .

The next utterance came from Bellarmine himself. It was a precise and authoritative statement of his attitude, and in view of his position as Consultor of the Holy Office, Master of Controversial Questions, etc., it amounted to an unofficial definition of the Church's attitude to Coperni-

2. I.e., that they are to be regarded as mathematical hypotheses only, in the sense of Osiander's preface.

cus. The statement was occasioned by Father Foscarini's book advocating the Copernican system, and couched in the form of a letter of acknowledgement; but it was clearly addressed to Galileo as well, whose name is expressly mentioned. The letter is dated 4 April, 1615:

My Very Reverend Father,

It has been a pleasure to me to read the Italian letter and the Latin paper you sent me. I thank you for both the one and the other, and I may tell you that I found them replete with skill and learning. As you ask for my opinion, I will give it as briefly as possible because, at the moment I have very little time for writing.

First, I say it seems to me that your Reverence and Signor Galileo act prudently when you content yourselves with speaking hypothetically and not absolutely, as I have always understood that Copernicus spoke. *For to say that the assumption that the Earth moves and the Sun stands still saves all the celestial appearances better than do eccentrics and epicycles, is to speak with excellent good sense and to run no risk whatever. Such a manner of speaking suffices for a mathematician.* But to want to affirm that the Sun, in very truth, is at the centre of the universe and only rotates on its axis without travelling from east to west, and that the Earth is situated in the third sphere and revolves very swiftly around the Sun, is a very dangerous attitude and one calculated not only to arouse all Scholastic philosophers and theologians but also to injure our holy faith by contradicting the Scriptures. . . .

Second, I say that, as you know, the council of Trent forbids the interpretation of the Scriptures in a way contrary to the common agreement of the holy Fathers. Now if your Reverence will read, not merely the Fathers, but modern commentators on Genesis, the Psalms, Ecclesiastes, and Joshua, you will discover that all agree in interpreting them literally as teaching that the Sun is in the heavens and revolves round the Earth with immense speed and that the Earth is very distant from the heavens, at the centre of the universe, and motionless. Consider, then, in your prudence, whether the Church can support that the Scriptures should be interpreted in a manner contrary to that of the holy Fathers and of all modern commentators, both Latin and Greek. . . .

Third, I say that, *if there were a real proof* that the Sun is the centre of the universe, that the Earth is in the third sphere, and that the Sun does not go round the Earth but the Earth round the Sun, then we should have to proceed with great circumspection in explaining passages of Scripture which appear to teach the contrary, and we should rather have to say that we did not understand them than declare an opinion to be false which is proven to be true. But I do not think there is any such proof *since none has been shown to me.* To demonstrate that the appearances are saved by assuming the sun at the centre and the earth in the heavens is not the same thing as to demonstrate that *in fact* the sun is in the centre and the earth in the heavens. I believe that the *first demonstration may exist, but I have very grave doubts about the second;* and in case of doubt one may not abandon the Holy Scriptures as expounded by the holy Fathers. . . .

[Galileo's] answer to Bellarmine was contained in a letter written at some date in May to Cardinal Dini:

To me, the surest and swiftest way to prove that the position of Copernicus is not contrary to Scripture would be to give a host of proofs that it is true and that the contrary cannot be maintained at all; thus, since no truths can contradict one another, this and the Bible must be perfectly harmonious. *But how can I do this, and not be merely wasting my time, when those Peripatetics who must be convinced show themselves incapable of following even the simplest and easiest of arguments?* . . .

Yet I should not despair of overcoming even this difficulty if I were in a place where I could use my tongue instead of my pen; and if I ever get well again so that I can come to Rome, I shall do so, in the hope of at least showing my affection for the holy Church. My urgent desire on this point is that no decision be made which is not entirely good. Such it would be to declare, under the prodding of an army of malign men who understand nothing of the subject, that Copernicus did not hold the motion of the earth to be a fact of nature, but as an astronomer merely took it to be a convenient hypothesis for explaining the appearances. . . .

Early in December [1615] Galileo arrived in Rome. . . . A Roman witness, Monsignor Querengo, describing Galileo in action:

We have here Signor Galileo who, in gatherings of men of curious mind, often bemuses many concerning the opinion of Copernicus, which he holds for true. . . . He discourses often amid fifteen or twenty guests who make hot assaults upon him, now in one house, now in another. But he is so well buttressed that he laughs them off; and although the novelty of his opinion leaves people unpersuaded, yet he convicts of vanity the greater part of the arguments with which his opponents try to overthrow him. Monday in particular, in the house of Federico Ghisileri, he achieved wonderful feats; and what I liked most was that, before answering the opposing reasons, he amplified them and fortified them himself with new grounds which appeared invincible, so that, in demolishing them subsequently, he made his opponents look all the more ridiculous. . . .

He is passionately involved in this quarrel, as if it were his own business, and he does not see and sense what it would comport; so that he will be snared in it, and will get himself into danger, together with anyone who seconds him. . . . For he is vehement and is all fixed and impassioned in this affair, so that it is impossible, if you have him around, to escape from his hands. And this is a business which is not a joke but may become of great consequence, and this man is here under our protection and responsibility. . . .

On March 5, 1616 the General Congregation of the Index issued . . . a decree:

And whereas it has also come to the knowledge of the said Congregation that the Pythagorean doctrine—which is false and altogether opposed to the Holy Scripture—of the motion of the Earth, and the immobility of the Sun, which is also taught by Nicolaus Copernicus in *De revolutionibus orbium coelestium,* and by Diego de Zuniga [in his book] on Job, is now being spread abroad and accepted by many—as may be seen from a certain letter of a Carmelite Father, entitled *Letter of the Rev. Father Paolo Antonio Foscarini, Carmelite, on the Opinion of the Pythagoreans and of Copernicus concerning the*

*Motion of the Earth, and the Stability of the Sun, and the New Pythagorean
System of the World, at Naples, Printed by Lazzaro Scoriggio, 1615:* wherein
the said Father attempts to show that the aforesaid doctrine of the immobility
of the sun in the centre of the world, and of the Earth's motion, is consonant
with truth and is not opposed to Holy Scripture. Therefore, in order that this
opinion may not insinuate itself any further to the prejudice of Catholic truth,
the Holy Congregation has decreed that the said Nicolaus Copernicus, *De revolutionibus orbium*, and Diego de Zuniga, *On Job*, be suspended until they
be corrected; but that the book of the Carmelite Father, Paolo Antonio Foscarini, be altogether prohibited and condemned, and that all other works likewise, in which the same is taught, be prohibited, as by this present decree it
prohibits, condemns, and suspends them all respectively. In witness whereof
the present decree has been signed and sealed with the hands and with the
seal of the most eminent and Reverend Lord Cardinal of St. Cecilia, Bishop of
Albano, on the fifth day of March, 1616.

Six days after the decree, Galileo was received by the Pope, in an audience which lasted three quarters of an hour. But while everything was
done to spare him public humiliation, he had been confidentially but
firmly enjoined to keep within the prescribed limits. This had happened
between the session of the Qualifiers on 23 February, and the publication of the decree. On Thursday, 25 February, there is the following
entry in the Inquisition file:

Thursday, 25 February 1616. The Lord Cardinal Mellini notified the Reverend Fathers, the Assessor, and the Commissary of the Holy Office that the
censure passed by the theologians upon the propositions of Galileo—to the effect that the Sun is the centre of the world and immovable from its place, and
that the Earth moves, and also with a diurnal motion—had been reported; and
His Holiness has directed the Lord Cardinal Bellarmine to summon before him
the said Galileo and admonish him to abandon the said opinion; and, *in case
of his refusal to obey*, that the Commissary is to enjoin on him, before a notary
and witnesses, a command to abstain altogether from teaching or defending
this opinion and doctrine and even from discussing it: and, if he do not ac-
quiesce therein, that he is to be imprisoned.

After the issue had been formally decided by the decree of 5 March
. . . the Duke ordered Galileo back to Florence. For the next seven
years he published nothing. . . .

Maffeo Barberini was elected to the papacy in 1623. . . . Barberini
had opposed the decree of Congregation and intervened in favour of
Galileo. . . . The [new] Pope showered favours on him—a pension for
Galileo's son, a precious painting, a gold and silver medal. He also provided him with a glowing testimonial, addressed to the new Grand
Duke, extolling the virtues and piety "of this great man, whose fame
shines in the heavens, and goes on earth far and wide. . . ."

Thus encouraged, and in the full sunshine of papal favour, Galileo, who was now past sixty, felt the road at last free to embark on his great apologia of Copernicus. . . . In February 1632, the first printed copies of the *Dialogue* [*on the Great World Systems*] came from the press. . . . By August, the book was confiscated, and in October Galileo was summoned to appear before the Inquisition in Rome.

The Sentence of the Inquisition °

(1633)

Whereas you, Galileo, son of the late Vincenzo Galilei, Florentine, aged seventy years, were in the year 1615 denounced to this Holy Office for holding as true the false doctrine taught by some that the Sun is the center of the world and immovable and that the Earth moves, and also with a diurnal motion; for having disciples to whom you taught the same doctrine; for holding correspondence with certain mathematicians of Germany concerning the same; for having printed certain letters, entitled "On the Sunspots," wherein you developed the same doctrine as true; and for replying to the objections from the Holy Scriptures, which from time to time were urged against it, by glossing the said Scriptures according to your own meaning: and whereas there was thereupon produced the copy of a document in the form of a letter, purporting to be written by you to one formerly your disciple, and in this divers propositions are set forth, following the position of Copernicus, which are contrary to the true sense and authority of Holy Scripture:

This Holy Tribunal being therefore of intention to proceed against the disorder and mischief thence resulting, which went on increasing to the prejudice of the Holy Faith, by command of His Holiness and of the Most Eminent Lords Cardinals of this supreme and universal Inquisition, the two propositions of the stability of the Sun and the motion of the Earth were by the theological Qualifiers qualified as follows:

The proposition that the Sun is the center of the world and does not move from its place is absurd and false philosophically and formally heretical, because it is expressly contrary to the Holy Scripture.

The proposition that the Earth is not the center of the world and immovable but that it moves, and also with a diurnal motion, is equally absurd and false philosophically and theologically considered at least erroneous in faith.

But whereas it was desired at that time to deal leniently with you, it was decreed at the Holy Congregation held before His Holiness on the twenty-fifth of February, 1616, that his Eminence the Lord Cardinal Bellarmine should order you to abandon altogether the said false doctrine and, in the event of your refusal, that an injunction should be imposed upon you by the Commissary of the Holy Office to give up the said doctrine and not to teach it to others, not to defend it, nor even discuss it; and failing your acquiescence in this injunction, that you should be imprisoned. And in execution of this decree, on the following day, at the Palace, and in the presence of his Eminence, the said Lord Cardinal Bellarmine, after being gently admonished by the said Lord Cardinal, the command was enjoined upon you by the Father Commissary of the Holy Office of that time, before a notary and witnesses, that you were altogether to abandon the said false opinion and not in future to hold or defend or teach it in any way whatsoever, neither verbally nor in writing; and, upon your promising to obey, you were dismissed.

And, in order that a doctrine so pernicious might be wholly rooted out and not insinuate itself further to the grave prejudice of Catholic truth, a decree was issued by the Holy Congregation of the Index prohibiting the books which treat of this doctrine and declaring the doctrine itself to be false and wholly contrary to the sacred and divine Scripture.

And whereas a book appeared here recently, printed last year at Florence, the title of which shows that you were the author, this title being: "Dialogue of Galileo Galilei on the Great World Systems"; and whereas the Holy Congregation was afterward informed that through the publication of the said book the false opinion of the motion of the Earth and the stability of the Sun was daily gaining ground, the said book was taken into careful consideration, and in it there was discovered a patent violation of the aforesaid injunction that had been imposed upon you, for in this book you have defended the said opinion previously condemned and to your face declared to be so, although in the said book you strive by various devices to produce the impression that you leave it undecided, and in express terms as probable: which, however, is a most grievous error, as an opinion can in no wise be probable which has been declared and defined to be contrary to divine Scripture.

Therefore by our order you were cited before this Holy Office, where, being examined upon your oath, you acknowledged the book to be written and published by you. You confessed that you began to write the said book about ten or twelve years ago, after the command had been imposed upon you as above; that you requested license to print it without, however, intimating to those who granted you this license that you had been commanded not to hold, defend, or teach the doctrine in question in any way whatever.

You likewise confessed that the writing of the said book is in many places drawn up in such a form that the reader might fancy that the arguments brought forward on the false side are calculated by their cogency to compel conviction rather than to be easy of refutation, excusing yourself for having fallen into an error, as you alleged, so foreign to your intention, by the fact that you had written in dialogue and by the natural complacency that every man feels in regard to his own subtleties and in showing himself more clever than the generality of men in devising, even on behalf of false propositions, ingenious and plausible arguments.

And, a suitable term having been assigned to you to prepare your defense, you produced a certificate in the handwriting of his Eminence the Lord Cardinal Bellarmine, procured by you, as you asserted, in order to defend yourself against the calumnies of your enemies, who charged that you had abjured and had been punished by the Holy Office, in which certificate it is declared that you had not abjured and had not been punished but only that the declaration made by His Holiness and published by the Holy Congregation of the Index had been announced to you, wherein it is declared that the doctrine of the motion of the Earth and the stability of the Sun is contrary to the Holy Scriptures and therefore cannot be defended or held. And, as in this certificate there is no mention of the two articles of the injunction, namely, the order not "to teach" and "in any way," you represented that we ought to believe that in the course of fourteen or sixteen years you had lost all memory of them and that this was why you said nothing of the injunction when you requested permission to print your book. And all this you urged not by way of excuse for your error but that it might be set down to a vainglorious ambition rather than to malice. But this certificate produced by you in your defense has only aggravated your delinquency, since, although it is there stated that said opinion is contrary to Holy Scripture, you have nevertheless dared to discuss and defend it and to argue its probability; nor does the license artfully and cunningly extorted by you avail you anything, since you did not notify the command imposed upon you.

And whereas it appeared to us that you had not stated the full truth with regard to your intention, we thought it necessary to subject you to a rigorous examination at which (without prejudice, however, to the matters confessed by you and set forth as above with regard to your said intention) you answered like a good Catholic. Therefore, having seen and maturely considered the merits of this your cause, together with your confessions and excuses above-mentioned, and all that ought justly to be seen and considered, we have arrived at the underwritten final sentence against you:

Invoking, therefore, the most holy name of our Lord Jesus Christ and of His most glorious Mother, ever Virgin Mary, by this our final sentence, which sitting in judgment, with the counsel and advice of the Reverend Masters of sacred theology and Doctors of both Laws, our assessors, we deliver in these writings, in the cause and causes at present before us between the Magnificent Carlo Sinceri, Doctor of both Laws, Proctor Fiscal of this Holy Office, of the one part, and you Galileo Galilei, the defendant, here present, examined, tried, and confessed as shown above, of the other part—

We say, pronounce, sentence, and declare that you, the said Galileo, by reason of the matters adduced in trial, and by you confessed as above, have rendered yourself in the judgment of this Holy Office vehemently suspected of heresy, namely, of having believed and held the doctrine—which is false and contrary to the sacred and divine Scriptures—that the Sun is the center of the world and does not move from east to west and that the Earth moves and is not the center of the world; and that an opinion may be held and defended as probable after it has been declared and defined to be contrary to the Holy Scripture; and that consequently you have incurred all the censures and penalties imposed and promulgated in the sacred canons and other constitutions, general and particular, against such delinquents. From which we are content that you be absolved, provided that, first, with a sincere heart and unfeigned faith, you abjure, curse, and detest before us the aforesaid errors and heresies and every other error and heresy contrary to the Catholic and Apostolic Roman Church in the form to be prescribed by us for you.

And, in order that this your grave and pernicious error and transgression may not remain altogether unpunished and that you may be more cautious in the future and an example to others that they may abstain from similar delinquencies, we ordain that the book of the "Dialogue of Galileo Galilei" be prohibited by public edict.

We condemn you to the formal prison of this Holy Office during our pleasure, and by way of salutary penance we enjoin that for three years to come you repeat once a week the seven penitential Psalms. Reserving to ourselves liberty to moderate, commute, or take off, in whole or in part, the aforesaid penalties and penance.

And so we say, pronounce, sentence, declare, ordain, and reserve in this and in any other better way and form which we can and may rightfully employ.

The Formula of Abjuration °
Galileo Galilei (1633)

I, Galileo, son of the late Vincenzo Galilei, Florentine, aged seventy years, arraigned personally before this tribunal and kneeling before you, Most Eminent and Reverend Lord Cardinals Inquisitors-General against heretical pravity throughout the entire Christian commonwealth, having before my eyes and touching with my hands the Holy Gospels, swear that I have always believed, do believe, and by God's help will in the future believe all that is held, preached, and taught by the Holy Catholic and Apostolic Church. But, whereas—after an injunction had been judicially intimated to me by this Holy Office to the effect that I must altogether abandon the false opinion that the Sun is the center of the world and immovable and that the Earth is not the center of the world and moves and that I must not hold, defend, or teach in any way whatsoever, verbally or in writing, the said false doctrine, and after it had been notified to me that the said doctrine was contrary to Holy Scripture—I wrote and printed a book in which I discuss this new doctrine already condemned and adduce arguments of great cogency in its favor without presenting any solution of these, I have been pronounced by the Holy Office to be vehemently suspected of heresy, that is to say, of having held and believed that the Sun is the center of the world and immovable and that the Earth is not the center and moves:

Therefore, desiring to remove from the minds of your Eminences, and of all faithful Christians, this vehement suspicion justly conceived against me, with sincere heart and unfeigned faith I abjure, curse, and detest the aforesaid errors and heresies and generally every other error, heresy, and sect whatsoever contrary to the Holy Church, and I swear that in future I will never again say or assert, verbally or in writing, anything that might furnish occasion for a similar suspicion regarding me; but, should I know any heretic or person suspected of heresy, I will denounce him to this Holy Office or to the Inquisitor or Ordinary of the place where I may be. Further, I swear and promise to fulfil and observe in their integrity all penances that have been, or that shall be, imposed upon me by this Holy Office. And, in the event of my contravening (which God forbid!) any of these my promises and oaths, I submit myself to all the pains and penalties imposed and promulgated in the sacred canons and other constitutions, general and particular, against

° From Giorgio de Santillana, *The Crime of Galileo.*

such delinquents. So help me God and these His Holy Gospels, which I touch with my hands.

Having recited, he signed the attestation:

I, the said Galileo Galilei, have abjured, sworn, promised, and bound myself as above; and in witness of the truth thereof I have with my own hand subscribed the present document of my abjuration and recited it word for word at Rome, in the convent of the Minerva, this twenty-second day of June, 1633.

I, Galileo Galilei, have abjured as above with my own hand.

It is not true that after reciting this abjuration, he muttered: *"Eppur si muove."* It was the world that said this—not Galileo.

Bertrand Russell
The Scientific Outlook (1931)

Toleration, Modern and Medieval °
by Bernard Shaw (1924)

We must face the fact that society is founded on intolerance. There are glaring cases of the abuse of intolerance; but they are quite as characteristic of our own age as of the Middle Ages. The typical modern example and contrast is compulsory inoculation replacing what was virtually compulsory baptism. But compulsion to inoculate is objected to as a crudely unscientific and mischievous anti-sanitary quackery, not in the least because we think it wrong to compel people to protect their children from disease. Its opponents would make it a crime, and will probably succeed in doing so; and that will be just as intolerant as making it compulsory. Neither the Pasteurians nor their opponents the Sanitarians would leave parents free to bring up their children naked, though that course also has some plausible advocates. We prate of toleration as we will; but society must always draw a line somewhere between allowable conduct and insanity or crime, in spite of the risk of mistaking sages for lunatics and saviors for blasphemers. We must persecute, even to the death; and all we can do to mitigate the danger of per-

° From Bernard Shaw, *Complete Plays with Prefaces*, Vol. II, Dodd, Mead and Company, Inc., New York, 1962. Reprinted by permission of the Society of Authors for the Bernard Shaw Estate.

secution is, first, to be very careful what we persecute, and second, to bear in mind that unless there is a large liberty to shock conventional people, and a well informed sense of the value of originality, individuality, and eccentricity, the result will be apparent stagnation covering a repression of evolutionary forces which will eventually explode with extravagant and probably destructive violence.

It has become a set piece in history to present Pope Urban VIII and his counselors as the bigoted oppressors of science. It would be possibly more accurate to say that they were the first bewildered victims of the scientific age. They had come into collision with a force of which they had not the faintest notion. In that sense, they are almost the polar opposite of the "progressive" rulers of the twentieth century, who are, one and all, bigoted believers in "scientism" while dealing with science in a no less highhanded manner. Still, the dramatic shape remains the same. (Giorgio de Santillana, *The Crime of Galileo*)

Suggestions for Further Reading

° Bronowski, Jacob, and Bruce Mazlish: *The Western Intellectual Tradition,* Harper Torchbook, Publishers, New York, 1960, chaps. 1–8. An excellent description of the intellectual and cultural history of the Renaissance from Leonardo through Galileo.

° Burtt, E. A.: *The Metaphysical Foundations of Modern Science,* Anchor Books, Doubleday & Company, Inc., Garden City, N.Y., 1954, chaps. I, II, and III. An analysis of the metaphysical and philosophical implications of the shift from the medieval to the modern view of man's place in the universe.

° Butterfield, Herbert: *The Origins of Modern Science,* The Macmillan Company, New York, 1959, chaps. I, II, and IV. A penetrating and interesting account of the scientific revolution.

° Cohen, I. Bernard: *The Birth of a New Physics,* Anchor Books, Doubleday & Company, Inc., Garden City, N.Y., 1960. A Science Study Series book that explores the emergence and development of new concepts in physics from Copernicus to Galileo. The Cohen selections in this part are taken from this book.

° De Santillana, Giorgio: *The Crime of Galileo,* Phoenix Books, The University of Chicago Press, Chicago, 1955. A scholarly and fascinating account of the conflicts of Galileo and the Roman Catholic Church.

° Hall, A. R.: *The Scientific Revolution,* Beacon Press, Boston, 1954, chaps. 1–5. A clear, well-written scientific history of this period.

° Koestler, Arthur: *The Sleepwalkers,* Grosset and Dunlap, New York, 1959, chaps. II–IV. An interesting account of the circumstances that led up to Galileo's trial by the Inquisition. Koestler comes to quite different conclusions concerning the treatment of Galileo than de Santillana does in his book (see above). The reader will find it very interesting to compare these two accounts of the conflict.

° Lodge, Oliver: *Pioneers of Science,* Dover Publications, Inc., New York, 1960, lectures I, IV, and V. Another historical account of the Copernican theory, Galileo's discoveries with the telescope, and his argument with the church.

° Russell, Bertrand: *The History of Western Philosophy,* Simon and Schuster, Inc., New York, 1945, book 3, part I, chap. VI. A fine analysis of the philosophical implication of the Copernican theory in its historical context.

° Paperback edition

5

What Is a Scientific Fact?

Science starts with facts and returns to them to prove
itself right or wrong. This part describes the dawning
recognition in the seventeenth century of the importance of facts as the touchstone of science and illustrates in the discovery of Kepler's laws how increasing
precision of measurement aids the advance of science.
What are scientific facts, and what are the limitations
involved in finding them?

What Is a Scientific Fact?

Introduction

SCIENCE, AS WE KNOW it today, was born in the Renaissance, the result of the union of logical thought with a new appreciation of the importance of experiment and observation. In the last two parts we followed the early history of abstract reasoning and saw how it laid the groundwork for scientific thinking. Now in this part we see how the inductive method became recognized as an essential part of science. The craftsmen and artisans of the Middle Ages had long used the cut-and-try empirical approach and had developed a store of practical knowledge, but, by and large, this information did not come to the attention of the scholastics who were theorizing about the physical world. Once the significance of empirical knowledge was recognized, many of the methods that had been developed by the artisans were taken over to help build scientific tools and perfect experimental techniques.

The importance of observation was mentioned as early as the thirteenth century in the writings of Roger Bacon, but these writings were not widely read at that time and did not have much influence until several hundred years later. The first important formulation of a philosophy of scientific method was made by Sir Francis Bacon. His precepts became the prototype of experimental or inductive method and still form the background for many of our beliefs about science today. The term *Baconian method* is sometimes loosely used as a synonym for *pure empiricism*. But Bacon believed that a proper balance should be struck between "empirics" and "dogmatics." He said, "The former, like ants, only heap up and use their store, the latter, like spiders, spin out their own webs. The bee, a mean between both, extracts matter from the flowers of the garden and the field; but works and fashions it by its own efforts." [1]

Fortunately, the recoil from rationality was not so violent that the power of abstract thought was entirely discredited. In Descartes, we have an example of one of the great men of the time who championed the power of reason. The recognition that these two methods must be combined was the idea of major importance for science that emerged from the Renaissance. Together they have forged a tool of unparalleled flexibility and power. As Whitehead puts it, the ". . . union of passionate interest in the detailed facts with equal devotion to abstract generalisation . . . forms the novelty in our present society." [2]

1. See p. 217.
2. See p. 210.

The belief in the validity of the experimental method rests on the assumption that significant information about the physical world is revealed through sensory experience. Bacon recognized the difficulty of this when he said ". . . by far the greatest hindrance and aberration of the human understanding proceeds from the dulness, incompetency, and deceptions of the senses." [3] Locke, Berkeley, and Hume, a century later, cast even graver doubts upon the validity of sensory data and the inductive process in general.

As scientists became more critical of their own methods and as better instruments became available to supplement the senses, the degree of accuracy used in scientific experiments grew increasingly important. The significance of this trend is illustrated by the work of Brahe and Kepler. Scientists began to discriminate between vague observations and scientific measurements, in which the accuracy was itself criticized and the range of error specified.

At first sight, it may seem paradoxical to talk about the range of error involved in facts. The confusion arises from a difference in the use of the word *fact* in everyday language and in scientific work. C. P. Snow makes the following comment on "scientific facts":

> Science starts with facts chosen from the external world. The relation between the choice, the chooser, the external world and the fact produced is a complicated one and brings us before questions of relativity and epistemology: but one gets through in the end, unless one is spinning a metaphysical veil for the sake of the craftsmanship, to an agreement upon "scientific facts." You can call them "pointer readings" as Eddington does, if you like. They are lines on a photographic plate, marks on a screen, all the "pointer readings" which are the end of the skill, precautions, inventions, of the laboratory. They are the end of the manual process, the beginning of the scientific. For from these "pointer readings," these scientific facts, the process of scientific reasoning begins: and it comes back to them to prove itself right or wrong. For the scientific process is nothing more nor less than a hiatus between "pointer readings"; one takes some pointer readings, makes a mental construction from them in order to predict some more.[4]

Fact in this sense does not have the same degree of certainty that it has in common usage. "Pointer readings" are true within the limits and conditions set by the experiment and may be proved false under better-controlled or more refined experimental conditions. Yet scientific facts do represent the closest approach to certain knowledge about some particular aspect of the physical world, and, as we will see in the story of *Kepler—Eight Minutes of Arc,* the knowledge of the degrees of accuracy of these facts is a significant factor in the advance of science.

3. See p. 216.
4. From C. P. Snow, *The Search,* Charles Scribner's Sons, New York, 1958.

The New Mentality

The Value of Experience °
by Roger Bacon (1267)

Without experience nothing can be sufficiently known. For there are two modes of acquiring knowledge, namely, by reasoning and experience. Reasoning draws a conclusion and makes us grant the conclusion, but does not make the conclusion certain, nor does it remove doubt so that the mind may rest on the intuition of truth, unless the mind discovers it by the path of experience; since many have the arguments relating to what can be known, but because they lack experience they neglect the arguments, and neither avoid what is harmful nor follow what is good. For if a man who has never seen a fire should prove by adequate reasoning that fire burns and injures things and destroys them, his mind would not be satisfied thereby, nor would he avoid fire, until he placed his hand or some combustible substance in the fire, so that he might prove by experience that which reasoning taught. But when he has had actual experience of combustion his mind is made certain and rests in the full light of truth. Therefore reasoning does not suffice, but experience does. . . .

He therefore who wishes to rejoice without doubt in regard to the truths underlying phenomena must know how to devote himself to experiment. For authors write many statements, and believe them through reasoning which they formulate without experience. Their reasoning is wholly false. For it is generally believed that the diamond cannot be broken except by goat's blood, and philosophers and theologians misuse this idea. But fracture by means of blood of this kind has never been verified, although the effort has been made and without that blood it can be broken easily. For I have seen this with my own eyes, and this is necessary, because gems cannot be carved except by fragments of this stone. . . .

Moreover, it is generally believed that hot water freezes more quickly

° From Roger Bacon, *Opus Majus,* translated by Robert B. Burke, University of Pennsylvania Press, Philadelphia. 1928. Reprinted by permission.

than cold water in vessels, and the argument in support of this is advanced that contrary is excited by contrary, just like enemies meeting each other. But it is certain that cold water freezes more quickly for anyone who makes the experiment. . . . If hot water and cold are placed in two vessels, the cold will freeze more quickly. Therefore all things must be verified by experiment.

The Scientific Revolution *

by Alfred North Whitehead (1925)

The new mentality is more important than the new science and the new technology. It has altered the metaphysical presuppositions and the imaginative contents of our minds; so that now the old stimuli provoke a new response. Perhaps my metaphor of a new colour is too strong. What I mean is just that slightest change of tone which yet makes all the difference. This is exactly illustrated by a sentence from a published letter of that adorable genius, William James. When he was finishing his great treatise on the *Principles of Psychology*, he wrote to his brother Henry James, 'I have to forge every sentence in the teeth of irreducible and stubborn facts.'

This new tinge to modern minds is a vehement and passionate interest in the relation of general principles to irreducible and stubborn facts. All the world over and at all times there have been practical men, absorbed in 'irreducible and stubborn facts': all the world over and at all times there have been men of philosophic temperament who have been absorbed in the weaving of general principles. It is this union of passionate interest in the detailed facts with equal devotion to abstract generalisation which forms the novelty in our present society. Previously it had appeared sporadically and as if by chance. This balance of mind has now become part of the tradition which infects cultivated thought. It is the salt which keeps life sweet. The main business of universities is to transmit this tradition as a widespread inheritance from generation to generation.

Galileo keeps harping on how things happen, whereas his adversaries had a complete theory as to why things happen. Unfortunately the two theories did not bring out the same results. Galileo insists upon 'irreducible and stubborn facts,' and Simplicius, his opponent, brings forward reasons, completely satisfactory, at least to himself. It is a great mistake to conceive this historical revolt as an appeal to reason. On the contrary, it was through and through an anti-intellectualist movement. It was the return to the contemplation of brute fact; and it was based on a recoil from the inflexible rationality of medieval thought.

Rules for Rightly Conducting the Reason *
by René Descartes (1637)

The *first* was never to accept anything for true which I did not clearly know to be such; that is to say, carefully to avoid precipitancy and prejudice, and to comprise nothing more in my judgment than what was presented to my mind so clearly and distinctly as to exclude all ground of doubt.

The *second*, to divide each of the difficulties under examination into as many parts as possible, and as might be necessary for its adequate solution.

The *third*, to conduct my thoughts in such order that, by commencing with objects the simplest and easiest to know, I might ascend by little and little, and, as it were, step by step, to the knowledge of the more complex; assigning in thought a certain order even to those objects which in their own nature do not stand in a relation of antecedence and sequence.

And the *last*, in every case to make enumerations so complete, and reviews so general, that I might be assured that nothing was omitted.

The Importance of Experiment †
by Francis Bacon (1620)

PREFACE

Those who have taken upon them to lay down the law of nature as a thing already searched out and understood, whether they have spoken

---

* From René Descartes, *Discourse on Method*, translated by Veitch, The Open Court Publishing Company, La Salle, Ill. Reprinted by permission.

† From Francis Bacon, *Novum Organum*, translated and edited by James Spedding et al., Houghton Mifflin Company, Boston, 1869–1872. Reprinted by permission.

in simple assurance or professional affectation, have therein done philosophy and the sciences great injury. For as they have been successful in inducing belief, so they have been effective in quenching and stopping inquiry; and have done more harm by spoiling and putting an end to other men's efforts than good by their own. Those on the other hand who have taken a contrary course, and asserted that absolutely nothing can be known,—whether it were from hatred of the ancient sophists, or from uncertainty and fluctuation of mind, or even from a kind of fulness of learning, that they fell upon this opinion,—have certainly advanced reasons for it that are not to be despised; but yet they have neither started from true principles nor rested in the just conclusion, zeal and affectation having carried them much too far. The more ancient of the Greeks (whose writings are lost) took up with better judgment a position between these two extremes,—between the presumption of pronouncing on everything, and the despair of comprehending anything; and though frequently and bitterly complaining of the difficulty of inquiry and the obscurity of things, and like impatient horses champing the bit, they did not the less follow up their object and engage with Nature; thinking (it seems) that this very question,—viz. whether or no anything can be known,—was to be settled not by arguing, but by trying. And yet they too, trusting entirely to the force of their understanding, applied no rule, but made everything turn upon hard thinking and perpetual working and exercise of the mind. . . .

Let there be therefore (and may it be for the benefit of both) two streams and two dispensations of knowledge; and in like manner two tribes or kindreds of students in philosophy—tribes not hostile or alien to each other, but bound together by mutual services;—let there in short be one method for the cultivation, another for the invention, of knowledge. . . .

And to make my meaning clearer and to familiarise the thing by giving it a name, I have chosen to call one of these methods or ways *Anticipation of the Mind*, the other *Interpretation of Nature*.

APHORISMS CONCERNING THE INTERPRETATION
OF NATURE AND THE KINGDOM OF MAN

1. Man, being the servant and interpreter of Nature, can do and understand so much and so much only as he has observed in fact or in thought of the course of nature: beyond this he neither knows anything nor can do anything.

2. Neither the naked hand nor the understanding left to itself can effect much. It is by instruments and helps that the work is done, which

are as much wanted for the understanding as for the hand. And as the instruments of the hand either give motion or guide it so the instruments of the mind supply either suggestions for the understanding or cautions.

3. Human knowledge and human power meet in one; for where the cause is not known the effect cannot be produced. Nature to be commanded must be obeyed; and that which in contemplation is as the cause is in operation as the rule.

9. The cause and root of nearly all evils in the sciences is this—that while we falsely admire and extol the powers of the human mind we neglect to seek for its true helps.

10. The subtlety of nature is greater many times over than the subtlety of the senses and understanding; so that all those specious meditations, speculations, and glosses in which men indulge are quite from the purpose,[1] only there is no one by to observe it. . . .

14. The syllogism consists of propositions, propositions consist of words, words are symbols of notions. Therefore if the notions themselves (which is the root of the matter) are confused and over-hastily abstracted from the facts, there can be no firmness in the superstructure. Our only hope therefore lies in a true induction.

19. There are and can be only two ways of searching into and discovering truth. The one flies from the senses and particulars to the most general axioms, and from these principles, the truth of which it takes for settled and immoveable, proceeds to judgment and to the discovery of middle axioms. And this way is now in fashion. The other derives axioms from the senses and particulars, rising by a gradual and unbroken ascent, so that it arrives at the most general axioms last of all. This is the true way, but as yet untried.

20. The understanding left to itself takes the same course (namely, the former) which it takes in accordance with logical order. For the mind longs to spring up to positions of higher generality, that it may find rest there; and so after a little while wearies of experiment. But this evil is increased by logic, because of the order and solemnity of its disputations.

21. The understanding left to itself, in a sober, patient, and grave mind, especially if it be not hindered by received doctrines, tries a little that other way, which is the right one, but with little progress; since the understanding, unless directed and assisted, is a thing unequal, and quite unfit to contend with the obscurity of things.

1. Literally, "are a thing insane." The meaning appears to be, that these speculations, being founded upon such an inadequate conception of the case, must necessarily be so wide of the truth that they would seem like mere madness if we could only compare them with it: like the aim of a man blindfolded to bystanders looking on.—J. Spedding.

22. Both ways set out from the senses and particulars, and rest in the highest generalities; but the difference between them is infinite. For the one just glances at experiment and particulars in passing, the other dwells duly and orderly among them. The one, again, begins at once by establishing certain abstract and useful generalities, the other rises by gradual steps to that which is prior and better known in the order of nature. . . .

24. It cannot be that axioms established by argumentation should avail for the discovery of new works; since the subtlety of nature is greater many times over than the subtlety of argument. But axioms duly and orderly formed from particulars easily discover the way to new particulars, and thus render sciences active.

25. The axioms now in use, having been suggested by a scanty and manipular experience and a few particulars of most general occurrence, are made for the most part just large enough to fit and take these in: and therefore it is no wonder if they do not lead to new particulars. And if some opposite instance, not observed or not known before, chance to come in the way, the axiom is rescued and preserved by some frivolous distinction; whereas, the truer course would be to correct the axiom itself. . . .

36. One method of delivery alone remains to us; which is simply this: we must lead men to the particulars themselves, and their series and order; while men on their side must force themselves for a while to lay their notions by and begin to familiarise themselves with facts.

37. The doctrine of those who have denied that certainty could be attained at all, has some agreement with my way of proceeding at the first setting out; but they end in being infinitely separated and opposed. For the holders of that doctrine assert simply that nothing can be known; I also assert that not much can be known in nature by the way which is now in use. But then they go on to destroy the authority of the senses and understanding; whereas I proceed to devise and supply helps for the same.

38. The idols and false notions which are now in possession of the human understanding, and have taken deep root therein, not only so beset men's minds that truth can hardly find entrance, but even after entrance obtained, they will again in the very instauration of the sciences meet and trouble us, unless men being forewarned of the danger fortify themselves as far as may be against their assaults. . . .

45. The human understanding is of its own nature prone to suppose the existence of more order and regularity in the world than it finds. And though there be many things in nature which are singular and unmatched, yet it devises for them parallels and conjugates and relatives which do not exist. Hence the fiction that all celestial bodies move in

perfect circles; spirals and dragons being (except in name) utterly rejected. Hence too the element of Fire with its orb is brought in, to make up the square with the other three which the sense perceives. Hence also the ratio of density of the so-called elements is arbitrarily fixed at ten to one. And so on of other dreams. And these fancies affect not dogmas only, but simple notions also.

46. The human understanding when it has once adopted an opinion (either as being the received opinion or as being agreeable to itself) draws all things else to support and agree with it. And though there be a greater number and weight of instances to be found on the other side, yet these it either neglects and despises, or else by some distinction sets aside and rejects; in order that by this great and pernicious predetermination[2] the authority of its former conclusions may remain inviolate. And therefore it was a good answer that was made by one who when they showed him hanging in a temple a picture of those who had paid their vows as having escaped shipwreck, and would have him say whether he did not now acknowledge the power of the gods,— "Aye," asked he again, "but where are they painted that were drowned after their vows?" And such is the way of all superstition, whether in astrology, dreams, omens, divine judgments, or the like; wherein men, having a delight in such vanities, mark the events where they are fulfilled, but where they fail, though this happen much oftener, neglect and pass them by. But with far more subtlety does this mischief insinuate itself into philosophy and the sciences; in which the first conclusion colours and brings into conformity with itself all that come after, though far sounder and better. Besides, independently of that delight and vanity which I have described, it is the peculiar and perpetual error of the human intellect to be more moved and excited by affirmatives than by negatives; whereas it ought properly to hold itself indifferently disposed towards both alike. Indeed in the establishment of any true axiom, the negative instance is the more forcible of the two.

47. The human understanding is moved by those things most which strike and enter the mind simultaneously and suddenly, and so fill the imagination; and then it feigns and supposes all other things to be somehow, though it cannot see how, similar to those few things by which it is surrounded. But for that going to and fro to remote and heterogeneous instances, by which axioms are tried as in the fire, the intellect is altogether slow and unfit, unless it be forced thereto by severe laws and overruling authority.

48. The human understanding is unquiet; it cannot stop or rest, and

2. Rather perhaps "prejudging the matter to a great and pernicious extent, in order that," &c. (*non sine magno et pernicioso praejudicio, quo, &c.*)—J. Spedding.

still presses onward, but in vain. Therefore it is that we cannot conceive of any end or limit to the world; but always as of necessity it occurs to us that there is something beyond. Neither again can it be conceived how eternity has flowed down to the present day; for that distinction which is commonly received of infinity in time past and in time to come can by no means hold; for it would thence follow that one infinity is greater than another, and that infinity is wasting away and tending to become finite. The like subtlety arises touching the infinite divisibility of lines, from the same inability of thought to stop. But this inability interferes more mischievously in the discovery of causes: for although the most general principles in nature ought to be held merely positive, as they are discovered, and cannot with truth be referred to a cause; nevertheless the human understanding being unable to rest still seeks something prior in the order of nature. And then it is that in struggling towards that which is further off it falls back upon that which is more nigh at hand; namely, on final causes: which have relation clearly to the nature of man rather than to the nature of the universe; and from this source have strangely defiled philosophy. But he is no less an unskilled and shallow philosopher who seeks causes of that which is most general, than he who in things subordinate and subaltern omits to do so.

49. The human understanding is no dry light, but receives an infusion from the will and affections; whence proceed sciences which may be called "sciences as one would." For what a man had rather were true he more readily believes. Therefore he rejects difficult things from impatience of research; sober things, because they narrow hope; the deeper things of nature, from superstition; the light of experience, from arrogance and pride, lest his mind should seem to be occupied with things mean and transitory; things not commonly believed, out of deference to the opinion of the vulgar. Numberless in short are the ways, and sometimes imperceptible, in which the affections colour and infect the understanding.

50. But by far the greatest hindrance and aberration of the human understanding proceeds from the dulness, incompetency, and deceptions of the senses; in that things which strike the sense outweigh things which do not immediately strike it, though they be more important. Hence it is that speculation commonly ceases where sight ceases; insomuch that of things invisible there is little or no observation. Hence all the working of the spirits inclosed in tangible bodies lies hid and unobserved of men. So also all the more subtle changes of form in the parts of coarser substances (which they commonly call alteration, though it is in truth local motion through exceedingly small spaces) is in like manner unobserved. And yet unless these two things just mentioned be searched out and

brought to light, nothing great can be achieved in nature, as far as the production of works is concerned. So again the essential nature of our common air, and of all bodies less dense than air (which are very many), is almost unknown. For the sense by itself is a thing infirm and erring; neither can instruments for enlarging or sharpening the senses do much; but all the truer kind of interpretation of nature is effected by instances and experiments fit and apposite; wherein the sense decides touching the experiment only, and the experiment touching the point in nature and the thing itself.

51. The human understanding is of its own nature prone to abstractions and gives a substance and reality to things which are fleeting. But to resolve nature into abstractions is less to our purpose than to dissect her into parts; as did the school of Democritus, which went further into nature than the rest. Matter rather than forms should be the object of our attention, its configurations and changes of configuration, and simple action, and law of action or motion; for forms are figments of the human mind, unless you will call those laws of action forms. . . .

59. . . . men believe that their reason governs words; but it is also true that words react on the understanding; and this it is that has rendered philosophy and the sciences sophistical and inactive. Now words, being commonly framed and applied according to the capacity of the vulgar, follow those lines of division which are most obvious to the vulgar understanding. And whenever an understanding of greater acuteness or a more diligent observation would alter those lines to suit the true divisions of nature, words stand in the way and resist the change. Whence it comes to pass that the high and formal discussions of learned men end oftentimes in disputes about words and names; with which (according to the use and wisdom of the mathematicians) it would be more prudent to begin, and so by means of definitions reduce them to order. Yet even definitions cannot cure this evil in dealing with natural and material things; since the definitions themselves consist of words, and those words beget others: so that it is necessary to recur to individual instances, and those in due series and order; as I shall say presently when I come to the method and scheme for the formation of notions and axioms. . . .

95. Those who have handled sciences have been either men of experiment or men of dogmas. The men of experiment are like the ant; they only collect and use: the reasoners resemble spiders, who make cobwebs out of their own substance. But the bee takes a middle course; it gathers its material from the flowers of the garden and of the field, but transforms and digests it by a power of its own. Not unlike this is the true business of philosophy; for it neither relies solely or chiefly on the pow-

ers of the mind, nor does it take the matter which it gathers from natural history and mechanical experiments and lay it up in the memory whole, as it finds it; but lays it up in the understanding altered and digested. Therefore from a closer and purer league between these two faculties the experimental and the rational, (such as has never yet been made) much may be hoped. . . .

98. Now for grounds of experience—since to experience we must come—we have as yet had either none or very weak ones; no search has been made to collect a store of particular observations sufficient either in number, or in kind, or in certainty, to inform the understanding, or in any way adequate. On the contrary, men of learning, but easy withal and idle, have taken for the construction or for the confirmation of their philosophy certain rumours and vague fames or airs of experience, and allowed to these the weight of lawful evidence. And just as if some kingdom or state were to direct its counsels and affairs, not by letters and reports from ambassadors and trustworthy messengers, but by the gossip of the streets; such exactly is the system of management introduced into philosophy with relation to experience. Nothing duly investigated, nothing verified, nothing counted, weighed, or measured, is to be found in natural history: and what in observation is loose and vague, is in information deceptive and treacherous. And if any one thinks that this is a strange thing to say, and something like an unjust complaint, seeing that Aristotle, himself so great a man, and supported by the wealth of so great a king, has composed so accurate a history of animals; and that others with greater diligence, though less pretence, have made many additions; while others, again, have compiled copious histories and descriptions of metals, plants, and fossils; it seems that he does not rightly apprehend what it is that we are now about. For a natural history which is composed for its own sake is not like one that is collected to supply the understanding with information for the building up of philosophy. They differ in many ways, but especially in this; that the former contains the variety of natural species only, and not experiments of the mechanical arts. For even as in the business of life a man's disposition and the secret workings of his mind and affections are better discovered when he is in trouble than at other times; so likewise the secrets of nature reveal themselves more readily under the vexations of art than when they go their own way. Good hopes may therefore be conceived of natural philosophy, when natural history, which is the basis and foundation of it, has been drawn up on a better plan; but not till then.

99. Again, even in the great plenty of mechanical experiments, there is yet a great scarcity of those which are of most use for the information of the understanding. For the mechanic, not troubling himself with the

investigation of truth, confines his attention to those things which bear upon his particular work, and will not either raise his mind or stretch out his hand for anything else. But then only will there be good ground of hope for the further advance of knowledge, when there shall be received and gathered together into natural history a variety of experiments, which are of no use in themselves, but simply serve to discover causes and axioms; which I call "*Experimenta lucifera*," experiments of *light*, to distinguish them from those which I call "*fructifera*," experiments of *fruit*.

Now experiments of this kind have one admirable property and condition; they never miss or fail. For since they are applied, not for the purpose of producing any particular effect, but only of discovering the natural cause of some effect, they answer the end equally well whichever way they turn out; for they settle the question.

100. But not only is a greater abundance of experiments to be sought for and procured, and that too of a different kind from those hitherto tried; an entirely different method, order, and process for carrying on and advancing experience must also be introduced. For experience, when it wanders in its own track, is, as I have already remarked, mere groping in the dark, and confounds men rather than instructs them. But when it shall proceed in accordance with a fixed law, in regular order, and without interruption, then may better things be hoped of knowledge. . . .

102. Moreover, since there is so great a number and army of particulars, and that army so scattered and dispersed as to distract and confound the understanding, little is to be hoped for from the skirmishings and slight attacks and desultory movements of the intellect, unless all the particulars which pertain to the subject of inquiry shall, by means of Tables of Discovery, apt, well arranged, and as it were animate, be drawn up and marshalled; and the mind be set to work upon the helps duly prepared and digested which these tables supply.

103. But after this store of particulars has been set out duly and in order before our eyes, we are not to pass at once to the investigation and discovery of new particulars or works; or at any rate if we do so we must not stop there. For although I do not deny that when all the experiments of all the arts shall have been collected and digested, and brought within one man's knowledge and judgment, the mere transferring of the experiments of one art to others may lead, by means of that experience which I term *literate*, to the discovery of many new things of service to the life and state of man, yet it is no great matter that can be hoped from that; but from the new light of axioms, which having been educed from those particulars by a certain method and rule, shall in

their turn point out the way again to new particulars, greater things may be looked for. For our road does not lie on a level, but ascends and descends; first ascending to axioms, then descending to works.

104. The understanding must not however be allowed to jump and fly from particulars to remote axioms and of almost the highest generality (such as the first principles, as they are called, of arts and things), and taking stand upon them as truths that cannot be shaken, proceed to prove and frame the middle axioms by reference to them; which has been the practice hitherto; the understanding being not only carried that way by a natural impulse, but also by the use of syllogistic demonstration trained and inured to it. But then, and then only, may we hope well of the sciences, when in a just scale of ascent, and by successive steps not interrupted or broken, we rise from particulars to lesser axioms; and then to middle axioms, one above the other; and last of all to the most general. For the lowest axioms differ but slightly from bare experience, while the highest and most general (which we now have) are notional and abstract and without solidity. But the middle are the true and solid and living axioms, on which depend the affairs and fortunes of men; and above them again, last of all, those which are indeed the most general; such I mean as are not abstract, but of which those intermediate axioms are really limitations.

The understanding must not therefore be supplied with wings, but rather hung with weights, to keep it from leaping and flying. Now this has never yet been done; when it is done, we may entertain better hopes of the sciences.

106. But in establishing axioms by this kind of induction, we must also examine and try whether the axiom so established be framed to the measure of those particulars only from which it is derived, or whether it be larger and wider. And if it be larger and wider, we must observe whether by indicating to us new particulars it confirm that wideness and largeness as by a collateral security; that we may not either stick fast in things already known, or loosely grasp at shadows and abstract forms; not at things solid and realised in matter. And when this process shall have come into use, then at last shall we see the dawn of a solid hope.

Irreducible and Stubborn Facts:
The Search for Precision

Kepler and Brahe [*]

by A. R. Hall (1954)

Kepler was the discoverer of the new descriptive laws of planetary motion, but he achieved more than this, for he made the first suggestions towards a physical theory of the universe adapted to the necessities of the new description. Although his life was given over to mathematical drudgery—in which he was aided by the much improved trigonometry of his day, and the invention of logarithms—Kepler was a man of vigorous and original scientific imagination. The mathematical operations he set himself could hardly have proved creative without it. In his first work [1] he sought for the divine canon in celestial architecture, and this pursuit ran through all his later computations. But Kepler was no empty theorist: the divine art of proportion, the harmony of the grand design of nature, was to be elucidated from the most precise mathematical observation of the universe. Hence the turning-point in his career, the necessary foundation for his work, was Kepler's encounter with the Danish astronomer, Tycho Brahe (1546–1601), who, after a quarrel with King Christian IV, and deserted his royally endowed observatory at Hveen to enter the service of the eccentric Emperor Rudolph II, patron of alchemists and astrologers, at Prague. Thither Kepler was drawn from Graz in Styria by the undigested mass of observations, of unparalleled accuracy, which Tycho had brought with him. . . .

In many ways the Danish astronomer's role in the early history of modern astronomy is [that of] . . . the first modern exponent of the art of disinterested observation and description. . . . And certainly Tycho was unique among early modern scientists in his insistence upon the

[*] From A. R. Hall, *The Scientific Revolution: 1500–1800*. Longmans, Green & Co., Inc., New York, 1954. Reprinted by permission of Longman Group Limited, Harlow, Essex.
1. Kepler's first work, *Mysterium Cosmographicum*, described the five planetary orbits as the areas circumscribed in the five regular polygons. See Figure 1-3; also "Kepler," in Glossary. Ed.

crucial importance of accurate quantitative measurement; always a desideratum in astronomy, certainly, but never previously handled with the analytical and inventive powers of Tycho, who first consciously studied methods of estimating and correcting errors of observation in order to determine their limits of accuracy. The most accurate predecessors of Tycho were not Europeans, but the astronomers who worked in the observatory founded at Samarkand by Ulugh Beigh, about 1420. Their results were correct to about ten minutes of arc (i.e. roughly twice as good as Hipparchus'); Tycho's observations were about twice as good again, falling systematically within about four minutes of modern values.[2] This result was achieved by patient attention to detail. The instruments at Hveen were fixed, of different types for the various kinds of angular measurement, and much larger than those commonly used in the past, so that their scales could be more finely divided. They were the work of the most skilful German craftsmen, whom Tycho encouraged by his patronage and direction. He devised a new form of sight, and a sort of diagonal scale [3] for reading fractions of a degree. In measuring either the longitude of a star, or its right ascension, it is most convenient to proceed by a way that requires an instrument measuring time accurately, and Tycho studied the improvement of clocks for this purpose: but he found that a new technique of his own by which observations were referred to the position of the sun was more trustworthy. He was the first astronomer in Europe to use the modern celestial coordinates reckoning star-positions with reference to the celestial equator,[4] not (as formerly) to the ecliptic. Another innovation in his practice was the observation of planetary positions not at a few isolated points in the orbit (especially when in opposition to the sun), but at frequent intervals so that the whole orbit could be plotted.

The techniques and standards of precision in astronomy, of which Tycho was the real founder, were evolved slowly over a period of thirty years to fulfil a very simple ambition. When he made his first observations with a home-made quadrant, he found that places given in star-catalogues were false, and that events such as eclipses occurred as much as two or three days from the predicted times. As the Copernican "Prutenic Tables" were computed from old observations they had brought in no significant improvement. The task Tycho set himself, therefore, was very simple: to plot afresh the positions of the brightest stars, and with

2. As was first pointed out by Robert Hooke, the unaided human eye is incapable of resolving points whose angular separation is less than about two minutes of arc, so that Tycho's work approximately attains the minimum theoretical limits of accuracy for instruments such as he used.

3. The so-called transversal points. See Figure 5-6. Ed.

4. See p. 78. Ed.

the fundamental map of the sky established to observe the motions of sun, moon and planets so that the elements of their orbits could be re-calculated without mistake. It does not seem that he undertook this with any violently partisan intent, but it is likely that he wished to show the falsity of the Copernicans' claim to have increased the accuracy of celestial mathematics, and to bolster up the geostatic doctrine by publishing unimpugnable tables and ephemerides calculated upon that assumption. To vindicate the "Tychonic" geostatic system [5] was the last ambition of his life, which he charged Kepler to fulfil. However, he was no slavish adherent to conventional ideas. He did not believe that apparent changes in the sky were due to meteors in the earth's atmosphere; he proved that comets were celestial bodies, and that the spheres could have no real existence as physical bodies since comets pass through them; and his description of the planetary motions is relativistically identical with that of Copernicus. As an astronomer, indeed, Tycho in no way belonged to the past: it was as a good Aristotelian natural-philosopher that he believed the earth incapable of movement.

Tycho's observations, including his catalogue of 1,000 star-places, have not proved of enduring value. The earliest observations that have an other than historical interest are those of the English astronomer, Flamsteed, early in the eighteenth century (error c. ten *seconds* of arc), for within about sixty years of Tycho's death the optically-unaided measuring instrument, still ardently defended by Hevelius of Dantzig, was beginning to pass out of use. Within a century Tycho's tables had been thoroughly revised by such astronomers as Halley, the Cassinis, Roemer and Flamsteed. In the interval, however, Kepler's discoveries based on Tycho's work had become recognized. To appreciate the relationship between Kepler and Tycho—the inventive mathematician and the patient observer—it must be realized that the balance of choice between Keplerian and Copernican astronomy is very narrow. Until measurements were available whose accuracy could be relied upon within a range of four minutes, or even less, there was no need to suppose that the planetary orbits were anything other than circles eccentric to the sun. Kepler, in plotting the orbit of Mars, from which he discovered the ellipticity of planetary orbits in general, was able to calculate the elements of a circular orbit which differed by less than ten minutes from the observations. It was only because he knew that Tycho's work was accurate within about half this range that he was dissatisfied and impelled to go further. *Kepler's famous "First Law" was thus the first instance in the history of science of a discovery being made as the result of a search for a theory, not merely to cover a given set of observations,*

5. See Figure 3-23. Ed.

but to interpret a group of refined measurements whose probable accuracy was a significant factor.[6] Discrimination between measurement in a somewhat casual sense, and scientific measurement, in which the quantitative result is itself criticized and its range of error determined, only developed slowly in other sciences during the course of the scientific revolution. . . .[7]

The invention of numerous scientific instruments during the seventeenth century, and their fertile use in many capacities, has long been associated with the revolution in scientific thought and method. The idea of science as a product of the laboratory (in the modern sense of the word) is indeed one of the creations of the scientific revolution. In no previous period had the study of natural philosophy or medicine been particularly linked with the use of specialized techniques or tools of inquiry, and though the surgeon or the astronomer had been equipped with a limited range of instruments, little attention was paid to their fitness for use or to the possibility of extending and perfecting their uses. More variants of the astrolabe, the most characteristic of all medieval scientific instruments, were designed during the last half-century of its use in Europe (c. 1575–1625) than in all its preceding history. The Greeks had known the magnifying power of a spherical vessel filled with water, but the lens was an invention of the eleventh century, the spectacle glass of the thirteenth, and the optical instrument of the seventeenth. Navigational instruments also were extremely crude before the later sixteenth century. It was not that ingenuity and craftsmanship were wholly lacking (for many examples of fine metalwork, for artistic and military purposes, prove the contrary); rather the will to refine and extend instrumental techniques was absent. On the other hand, it has justly been pointed out that the early strategic stages of the scientific revolution were accomplished without the aid of the new instruments. They were unknown to Copernicus, to Vesalius, to Harvey, Bacon and Gilbert. They leave no trace in the most important of Galileo's writings,

6. Italics by editor.

7. The range or margin of error has become significant as science has become more precise. Today workers in the exact sciences wherever possible specify the margin of error involved in their observations, as for instance:

Length AB = 6.03 in. ± 0.02 in.

or Length AB = 6.03 in. ± 0.33 per cent

These statements both mean that the length may be as much as 6.05 in. or as little as 6.01 in. Within that margin the exact figure is unknown, probably because of the limitations of the device used in making the measurement. The accuracy of an experimental result is limited by the least accurate element involved in the experiment. In the example above, even if the width is known much more accurately, say BC = 4.6285 in. ± 0.005 per cent, then a computation of the area using the measurement for length 6.03 in. ± 0.33 per cent will also have a margin of error of 0.33 per cent. Ed.

in which he reveals himself driven to measure small intervals of time by a device considerably less accurate than the ancient clepsydra. It is clear that, great as was the influence of the instrumental ingenuity of the seventeenth century upon the course of modern science, such ingenuity was not at all responsible for the original deflection of science into this new course.

Thus it would seem that the first factor limiting the introduction into scientific practice of higher standards of observation and measurement, or of more complex manipulations, lay in the nature of science itself. Only when the concept of scientific research had changed, as it had by the end of the first quarter of the seventeenth century, was it possible to pay attention to the attainment of these higher standards. A number of the new instruments of the seventeenth century were not the product of scientific invention, but were adopted for scientific purposes because the new attitude enabled their usefulness to be perceived. The balance was borrowed from chemical craftsmanship. The telescope was another craft invention, initially applied to military uses. The microscope was an amusing toy before it became a serious instrument of research. The air pump in the laboratory was an improved form of the common well pump—Otto von Guericke, its original inventor, had at the beginning exhausted vessels by pumping out water with which they had been filled. And inevitably the techniques used in the construction of the new scientific instruments were those already in existence; they were not summoned out of nothing by the unprecedented scientific demand. Some instruments were only practicable because methods of lathe-turning and screw-cutting had been gradually perfected during two or three centuries, partly owing to the more ready availability of steel tools, others, because it was possible to produce larger, stronger and smoother sheets or strips of metal. The techniques of glass grinding and blowing which provided lenses and tubes might have been turned to scientific use long before they were. The established skill of the astrolabe-maker could be devoted to the fabrication of other instruments requiring divided circles and engraved lines, that of the watchmaker to various computers and models involving exact wheel-work, and so on. Common craftsmanship held a considerable reservoir of ingenuity, when scientific imagination arrived to draw upon it.

The Celestial Palace of Tycho Brahe °
by John Christianson (1961)

[Tycho Brahe's] precise observations were made entirely without the aid of a telescope. It was not until 1609, eight years after Tycho's death, that Galileo first used the telescope for astronomical purposes. Tycho's instruments consisted of ingenious improvements on ancient devices, quadrants, armillary spheres and parallactic rulers. All were aimed at celestial objects by means of simple metal sights. . . .

In a tireless quest for accuracy Tycho filled his observatory with a succession of instruments, each more ambitious than the last. He recognized that by increasing the distance between the sights of his instruments, and by increasing the size of the calibrated arcs from which he read positions, he could gain accuracy. But he also came to realize that when he made his instruments too large, they bent and twisted of their own weight, introducing a new class of errors in place of those that size was supposed to eliminate. He recorded that this great observatory at Hven cost Frederick II "more than a tun of gold," the equivalent of about $1.5 million today. . . .

It was as instrument maker that Tycho excelled all astronomers who had lived before him. Some of the devices he built were of his own invention but most were refinements of traditional types. The basic instrument of astronomy in his day was the quadrant, a quarter-circle that was normally sighted from the angle to the arc. One of the quadrant's arms was made precisely horizontal and the other precisely vertical. The whole instrument was usually pivoted so that it could be rotated 360 degrees. When the sighting arm, or alidade, was pointed at a star or planet, the altitude was read off the 90-degree arc and the azimuth off a 360-degree circle within which the quadrant revolved.

Tycho designed at least two such instruments that had a radius of more than six feet. One was the *quadrans magnus chalibeus*, or great steel quadrant (Figure 5-1). A still more accurate instrument, rigidly fixed in a north-south plane—hence incapable of providing azimuth— was the *quadrans muralis sive Tichonicus*: the mural, or Tychonian, quadrant (Figure 5-5). This instrument did not survive, but Tycho declared its scale could be read directly to 10 seconds of arc and that by

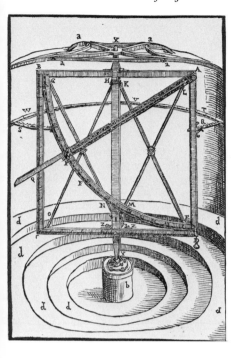

Figure 5-1. Great steel quadrant had as large an arc radius (76.5 inches) as the mural quadrant. Tycho said he depended chiefly on these two quadrants and the revolving azimuth quadrant. (The New York Public Library)

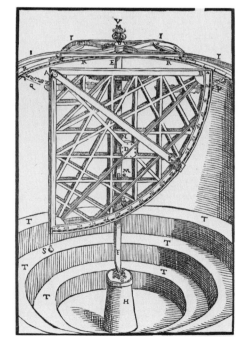

Figure 5-2. Revolving azimuth quadrant had an arc radius only 15 inches smaller than that of quadrant above. Tycho advised astronomers "to consider this construction as particularly commendable." (The New York Public Library)

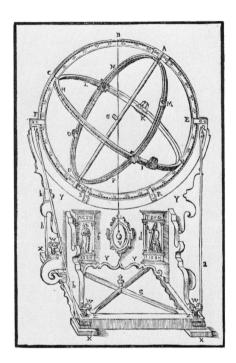

Figure 5-3. A variety of instruments illustrates Tycho Brahe's mechanical ingenuity. The equatorial armillary was a standard device for determining declination and right ascension of a celestial body. (The New York Public Library)

Figure 5-4. The great equatorial armillary was a typical Tychonic enlargement; the diameter of the revolving circle was 9 feet.

Figure 5-5. *Great Mural Quadrant, in which Tycho had "great faith," measured over 6 feet (76.5 inches) from the cylindrical front sight (in opening of the wall at upper left) to the rear sight, or pinnule, which could be moved along the arc. At lower right are two clocks of "the greatest possible accuracy." The quadrant was used to measure the altitude of celestial bodies as they crossed the north-south meridian. The full-size figure of Tycho was actually part of a wall painting depicting Uraniborg's multiple activities.* (The New York Public Library)

interpolation five seconds could be "read without difficulty." Even allowing that the 10-seconds calibrations were made by using so-called transversal points (Figure 5-6), Tycho's statement seems almost incredible.

Assuming, however, that he did indeed inscribe the arc of his mural quadrant to read to 10 seconds, he deceived himself in a way that was to become familiar in later science. He tried to extract from his observational method greater accuracy than its least accurate element permitted. In Tycho's method the fundamental source of inaccuracy was the human eye; unaided, it cannot resolve points that are separated by less than about two minutes of arc. Actually Tycho's stellar and planetary positions fall regularly within about four minutes of modern values. In other words, his mean error was less than half the distance that separates the twin stars Mizar and Alcor in the handle of the Big Dipper. (The ability to distinguish Mizar and Alcor was often used as a test of visual acuity in earlier times.)

Some indication of the way Tycho must have driven his instrument makers is revealed in the following passage: "Only this I wish to state here with regard [to these instruments], namely, that all of it has to be as nearly perfect as is possible in every respect and that, therefore, one should employ skillful craftsmen, who know how to carry out this sort of work artfully, or else can learn to do it. And even if they cannot perhaps do it all perfectly the first time, the constructor must not let himself be discouraged, but have the work repeated and improve the defects in every way until none is left."

But Tycho did more than simply put together elegant blown-up versions of old astronomical tools. He made technical innovations that

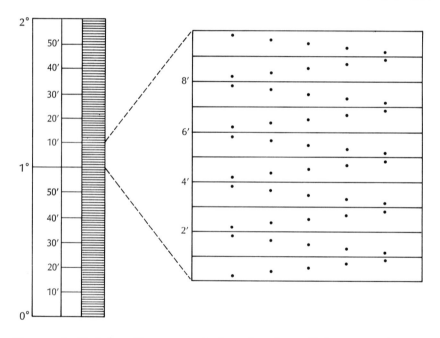

Figure 5-6. Tycho's calibration method employed so-called transversal points. The scale at left represents a full-size section of the arc of the mural quadrant. Tycho claimed that each minute was subdivided into six parts as shown enlarged at right. As the long sighting arm, the pinnule, moved over the face of the scale, the subdivision of each minute was indicated by the arm's position relative to the transversal points dividing the minute. Evidently Tycho did not realize that the unassisted eye limited his sighting accuracy to about 2 minutes of arc.

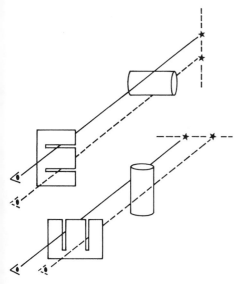

Figure 5-7. Tychonic sighting method used a cylinder as a front sight and two slits, with a spacing equal to the diameter of the cylinder, as a rear sight. In making a sighting (of altitude, at left; of azimuth, at right) the sights are aligned so that a celestial object appears equally bright on both sides of the cylinder when the eye is moved from one slit to the other.

greatly increased the quality of observations, particularly when these innovations were applied to his well-made giant instruments. Perhaps the most important were his instrument sights. The Tychonic sight had adjustable slits in the eyepiece in place of the customary peephole. The slits could be opened to admit the light of a faint star or closed to observe a star of greater magnitude. The front sight was a cylinder exactly as wide as the slits were far apart. The observer sighted through one slit and then the other until he saw the star just touching each side of the cylinder (Figure 5-7). Such a sighting device proved much more accurate than a simple pierced eyepiece.

In 1597 Frederick's successor, Christian IV, lost patience with Tycho's lordly ways and withdrew his fief at Hven. Tycho left his beloved Uraniborg and moved to Prague to serve the Holy Roman Emperor Rudolph II. It was there that Tycho met the young German mathematician Johannes Kepler and took him on as an assistant.

Tycho and Kepler turned out to be the most fortunate combination in the history of astronomy. Tycho's years of patient observation combined with Kepler's mystical, almost magical, skill at mathematics founded modern astronomy.

Kepler—Eight Minutes of Arc °
by Arthur Koestler (1959)

[Kepler's] *magnum opus,* published in 1609, bears the significant title:

A NEW ASTRONOMY Based on Causation
OF A PHYSICS OF THE SKY
derived from Investigations of the
MOTIONS OF THE STAR MARS
Founded on Observations of
THE NOBLE TYCHO BRAHE

Kepler worked on it, with interruptions, from his arrival at Benatek in 1600, to 1606. It contains the first two of Kepler's three planetary laws: (1) that the planets travel round the sun not in circles but in elliptical orbits, one focus of the ellipse being occupied by the sun; (2) that a planet moves in its orbit not at uniform speed but in such a manner that a line drawn from the planet to the sun always sweeps over equal areas in equal times. . . .

On the surface, Kepler's laws look as innocent as Einstein's $E = Mc^2$, which does not reveal, either, its atom-exploding potentialities. But the modern vision of the universe was shaped, more than by any other single discovery, by Newton's law of universal gravitation, which in turn was derived from Kepler's three laws. Although (owing to the peculiarities of our educational system) a person may never have heard of Kepler's laws, his thinking has been moulded by them without his knowledge; they are the invisible foundation of a whole edifice of thought.

Thus the promulgation of Kepler's laws is a landmark in history. They were the first "natural laws" in the modern sense: precise, verifiable statements about universal relations governing particular phenomena, expressed in mathematical terms. They divorced astronomy from theology, and married astronomy to physics. Lastly, they put an end to the nightmare that had haunted cosmology for the last two millennia: the obsession with spheres turning on spheres, and substituted a vision of material bodies not unlike the earth, freely floating in space, moved by physical forces acting on them.

The manner in which Kepler arrived at his new cosmology is fascinating; I shall attempt to re-trace the zig-zag course of his reasoning. For-

° From Arthur Koestler, *The Sleepwalkers,* The Macmillan Company, New York. © 1959 by Arthur Koestler. Reprinted by permission of The Macmillan Company.

tunately, he did not cover up his tracks, as Copernicus, Galileo and Newton did, who confront us with the result of their labours, and keep us guessing how they arrived at it. Kepler was incapable of exposing his ideas methodically, text-book fashion; he had to describe them in the order they came to him, including all the errors, detours, and the traps into which he had fallen. The *New Astronomy* is written in an unacademic, bubbling baroque style, personal, intimate, and often exasperating. But it is a unique revelation of the ways in which the creative mind works.

"What matters to me," Kepler explained in his Preface, "is not merely to impart to the reader what I have to say, but above all to convey to him the reasons, subterfuges, and lucky hazards which led me to my discoveries. When Christopher Colombus, Magelhaen and the Portuguese relate how they went astray on their journeys, we not only forgive them, but would regret to miss their narration because without it the whole, grand entertainment would be lost. Hence I shall not be blamed if, prompted by the same affection for the reader, I follow the same method."

Before embarking on the story, it will be prudent to add my own apology to Kepler's. Prompted by the same "affection for the reader" I have tried to simplify as far as possible a difficult subject: even so, the present chapter must of necessity be slightly more technical. . . . If some passages tax his patience, even if occasionally he fails to grasp a point or loses the thread, he will, I hope, nevertheless get a general idea of Kepler's odyssey of thought, which opened up the modern universe. . . .

[After] young Kepler's arrival at Benatek Castle, he was allotted the study of the motions of Mars which had defeated Tycho's senior assistant, Longomontanus, and Tycho himself.

"I believe it was an act of Divine Providence," he commented later on, "that I arrived just at the time when Longomontanus was occupied with Mars. For Mars alone enables us to penetrate the secrets of astronomy which otherwise would remain forever hidden from us."

The reason for this key position of Mars is that, among the outer planets, his orbit deviates more than the others' from the circle; it is the most pronouncedly elliptical. It was precisely for that reason that Mars had defied Tycho and his assistant: since they expected the planets to move in circles, it was impossible to reconcile theory with observation:

He [Mars] is the mighty victor over human inquisitiveness, who made a mockery of all the stratagems of astronomers, wrecked their tools, defeated their hosts; thus did he keep the secret of his rule safe throughout all past cen-

turies and pursued his course in unrestrained freedom; wherefore the most famous of Latins, the priest of nature Pliny, specially indicted him: MARS IS A
STAR WHO DEFIES OBSERVATION.

Thus Kepler, in his dedication of the *New Astronomy* to the Emperor
Rudolph II. The dedication is written in the form of an allegory of Kepler's war against Mars, begun under "Tycho's supreme command," patiently pursued in spite of the warning example of Rheticus who went
off his head over Mars, in spite of other dangers and terrible handicaps,
such as a lack of supplies owing to Rudolph's failure to pay Kepler's
salary—and so on to the triumphant end when the Imperial Mathemati-
cus, riding a chariot, leads the captive enemy to the Emperor's throne.

Thus Mars held the secret of all planetary motion, and young Kepler
was assigned the task of solving it. He first attacked the problem on traditional lines; when he failed, he began to throw out ballast and continued doing so until, by and by, he got rid of the whole load of ancient
beliefs on the nature of the universe, and replaced it by a new science.

As a preliminary, he made three revolutionary innovations to gain
elbow room, as it were, for tackling his problem. It will be remembered
that the centre of Copernicus' system was not the sun, but the centre of
the earth's orbit; [1] . . . already in the *Mysterium Cosmographicum* Kepler had objected to this assumption as physically absurd. Since the
force which moved the planets emanated from the sun, the whole system
should be centred on the body of the sun itself.

But in fact it was not. The sun occupies not the exact centre of the
orbit at C; it occupies one of the two foci of the ellipse at S (Figure 5-8).

Kepler did not know as yet that the orbit was an ellipse; he still regarded it as a circle. But even so, to get approximately correct results,
the centre of the circle had to be placed at C, and not in the sun. Accordingly, the question arose in his mind: if the force which moves the
planets comes from S, why do they insist on turning round C? Kepler
answered the question by the assumption that each planet was subject
to two conflicting influences: *the force of the sun, and a second force located in the planet itself.* This tug-of-war caused it now to approach the
sun, now to recede from him.

The two forces are, as we know, gravity and inertia. Kepler, as we
shall see, never arrived at formulating these concepts. But he prepared
the way for Newton by postulating two dynamic forces to explain the
eccentricity of the orbits. Before him, the need for a physical explanation was not felt; the phenomenon of eccentricity was merely "saved" by

1. See p. 157. Ed.

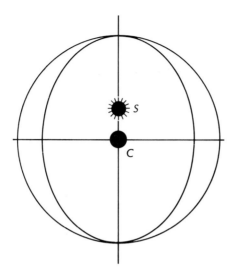

Figure 5-8

the introduction of an epicycle or eccenter, which made C turn round S. Kepler replaced the fictitious wheels by real forces.

For the same reason, he insisted on treating the sun as the centre of his system not only in the physical but in the geometrical sense, by making the distances and positions of the planets relative to the sun (and not relative to the earth or the centre C) the basis of his computations. This shift of emphasis, which was more instinctive than logical, became a major factor in his success.

His second innovation is simpler to explain. The orbits of all planets lie very nearly, but not entirely, in the same plane; they form very small angles with each other—rather like adjacent pages of a book which is nearly, but not entirely, closed. The planes of all planets pass, of course, through the sun—a fact which is self-evident to us, but not to pre-Keplerian astronomy. Copernicus, once again misled by his slavish devotion to Ptolemy, had postulated that the plane of the Martian orbit *oscillates in space;* and this oscillation he made to depend on the position of the earth—which, as Kepler remarks, "is no business of Mars." He called this Copernican idea "monstrous" (though it was merely due to Copernicus' complete indifference to physical reality) and set about to prove that the plane in which Mars moves passes through the sun, and does not oscillate, but forms a fixed angle with the plane of the earth's orbit. Here he met, for once, with immediate success. He proved, by sev-

eral independent methods, all based on the Tychonic observations, that
the angle between the planes of Mars and Earth remained always the
same, and that it amounted to 1°50'. He was delighted, and remarked
smugly that "the observations took the side of my preconceived ideas, as
they often did before."

The third innovation was the most radical. To gain more elbow room,
he had to get out of the strait-jacket of "uniform motion in perfect
circles"—the basic axiom of cosmology from Plato up to Copernicus and
Tycho. For the time being, he still let circular motion stand, but he
threw out uniform speed. Again he was guided mainly by physical considerations: if the sun ruled the motions, then his force must act more
powerfully on the planet when it is close to the source, less powerfully
when away from it; hence the planet will move faster or slower, in a
manner somehow related to its distance from the sun.

This idea was not only a challenge to antique tradition; it also reversed the original purpose of Copernicus. It will be remembered that
Copernicus' original motive for embarking on a reform of the Ptolemaic
system was his discontent with the fact that, according to Ptolemy, a
planet did not move at uniform speed around the centre of its orbit, but
only around a point at some distance from the centre. This point was
called the *punctum equans*—the point in space, from which the planet
gave the illusion of "equal motion." Canon Koppernigk regarded this arrangement as an evasion of the command of uniform motion, abolished
Ptolemy's equants, and added, instead, more epicycles to his system.
This did not make the planet's *real* motion either circular, or uniform,
but each wheel in the imaginary clockwork which was supposed to account for it, did turn uniformly—if only in the astronomer's mind.

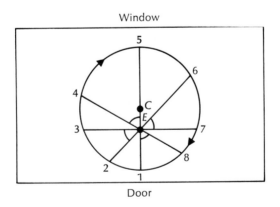

Figure 5-9   Door

When Kepler renounced the dogma of uniform motion, he was able to throw out the epicycles which Copernicus had introduced to save it. Instead, he reverted to the equant as an important calculating device (Figure 5-9).

Let the circle be the track of a toy train chugging round a room. When near the window it runs a little faster, near the door a little slower. Provided that these periodic changes of speed follow some simple, definite rule, then it is possible to find a *punctum equans,* "E," from which the train *seems* to move at uniform speed. The closer we are to a moving train, the faster it seems to move; hence the *punctum equans* will be somewhere between the centre C of the track and the door, so that the speed-surplus of the train when passing the window will be eliminated by distance, its speed-deficiency at the door compensated by closeness. The advantage gained by the introduction of this imaginary *punctum equans* is that, *seen from E,* the train seems to move uniformly, that is, it will cover equal angles at equal times—which makes it possible to compute its various positions *1, 2, 3,* etc., at any given moment.

By these three preliminary moves: (a) the shifting of the system's centre into the sun; (b) the proof that the orbital planes do not "oscillate" in space; and (c) the abolition of uniform motion, Kepler had cleared away a considerable amount of the rubbish that had obstructed progress since Ptolemy, and made the Copernican system so clumsy and unconvincing. In that system Mars ran on five circles; after the clean-up, a single eccentric circle must be sufficient—if the orbit was really a circle. He felt confident that victory was just around the corner, and before the final attack wrote a kind of obituary notice for classical cosmology:

Oh, for a supply of tears that I may weep over the pathetic diligence of Apianus [author of a very popular textbook] who, relying on Ptolemy, wasted his valuable time and ingenuity on the construction of spirals, loops, helixes, vortices and a whole labyrinth of convolutions, in order to represent that which exists only in the mind, and which Nature entirely refuses to accept as her likeness. And yet that man has shown us that, with his penetrating intelligence, he would have been capable of mastering Nature.

THE FIRST ASSAULT

Kepler's first attack on the problem is described in great detail in the sixteenth chapter of the *New Astronomy.*

The task before him was to define the orbit of Mars by determining the radius of the circle, the direction (relative to the fixed stars) of the axis connecting the two positions where Mars is nearest and farthest from the sun (perihelion and aphelion) (Figure 5-10), and the positions

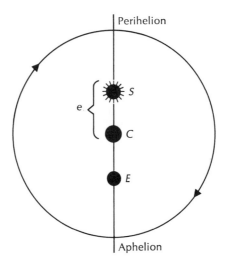

Perihelion

Aphelion

Figure 5-10

of the sun (S), orbital centre (C), and *punctum equans* (E), which all lie
on that axis. Ptolemy had assumed that the distance between E and C
was the same as between C and S, but Kepler made no such assumption,
which complicated his task even more.

   He chose out of Tycho's treasure four observed positions of Mars at
the convenient dates when the planet was in opposition to the sun. The
geometrical problem which he had to solve was, as we saw, to determine, out of these four positions, the radius of the orbit, the direction of
the axis, and the position of the three central points on it. It was a problem which could not be solved by rigorous mathematics, only by approximation, that is, by a kind of trial-and-error procedure which has to
be continued until all the pieces in the jig-saw puzzle fit together tolera-
bly well. The incredible labour that this involved may be gathered from
the fact that Kepler's draft calculations (preserved in manuscript) cover
nine hundred folio pages in small handwriting.

   At times he was despairing; he felt, like Rheticus, that a demon was
knocking his head against the ceiling, with the shout: "These are the
motions of Mars." At other times, he appealed for help to Maestlin (who
turned a deaf ear), to the Italian astronomer, Magini (who did the
same), and thought of sending an SOS to François Vieta, the father of
modern algebra: "Come, oh Gallic Appollonius, bring your cylinders and
spheres and what other geometer's houseware you have. . . ." But in the

end he had to slog it out alone, and to invent his mathematical tools as he went along.

Half-way through that dramatic sixteenth chapter, he burst out:

> If thou [dear reader] art bored with this wearisome method of calculation, take pity on me who had to go through with at least seventy repetitions of it, at a very great loss of time; nor wilst thou be surprised that by now the fifth year is nearly past since I took on Mars. . . .

Now, at the very beginning of the hair-raising computations in chapter sixteen, Kepler absentmindedly put three erroneous figures for three vital longitudes of Mars, and happily went on from there, never noticing his error. The French historian of astronomy, Delambre, later repeated the whole computation, but, surprisingly, his correct results differ very little from Kepler's faulty ones. The reason is, that toward the end of the chapter Kepler committed several mistakes in simple arithmetic—errors in division which would bring bad marks to any schoolboy—and these errors happen very nearly to cancel out his earlier mistakes. We shall see, in a moment, that, at the most crucial point of the process of discovering his Second Law, Kepler again committed mathematical sins which mutually cancelled out, and "as if by miracle" (in his own words), led to the correct result.

At the end of that breathtaking chapter, Kepler seems to have triumphantly achieved his aim. As a result of his seventy-odd trials, he arrived at values for the radius of the orbit and for the three central points which gave, with a permissible error of less than 2', the correct positions of Mars for all the ten oppositions recorded by Tycho. The unconquerable Mars seemed at last to have been conquered. He proclaimed his victory with the sober words:

> Thou seest now, diligent reader, that the hypothesis based on this method not only satisfies the four positions on which it was based, but also correctly represents, within two minutes, all the other observations. . . .

There follow three pages of tables to prove the correctness of his claim; and then, without further transition, the next chapter starts with the following words:

> Who would have thought it possible? This hypothesis, which so closely agrees with the observed oppositions, is nevertheless false. . . .

THE EIGHT MINUTES ARC

In the two following chapters Kepler explains, with great thoroughness and an almost masochistic delight, how he discovered that the hypothesis is false and why it must be rejected. In order to prove it by a further test, he had selected two specially rare pieces from Tycho's treasury of observations, and lo! they did not fit; and when he tried to adjust his model to them, this made things even worse, for now the observed positions of Mars differed from those which his theory demanded by magnitudes up to eight minutes of arc.

This was a catastrophe. Ptolemy, and even Copernicus, could afford to neglect a difference of eight minutes, because their observations were only accurate within a margin of ten minutes anyway. "But," the nineteenth chapter concludes, "but for us, who, by divine kindness were given an accurate observer such as Tycho Brahe, for us it is fitting that we should acknowledge this divine gift and put it to use. . . . Henceforth I shall lead the way toward that goal according to my own ideas. For, if I had believed that we could ignore these eight minutes, I would have patched up my hypothesis accordingly. But since it was not permissible to ignore them, those eight minutes [2] point the road to a complete reformation of astronomy: they have become the building material for a large part of this work. . . ."

It was the final capitulation of an adventurous mind before the "irreducible, obstinate facts." Earlier on, if a minor detail did not fit into a major hypothesis, it was cheated away or shrugged away. Now this time-hallowed indulgence had ceased to be permissible. A new era had begun in the history of thought: an era of austerity and rigour. As Whitehead has put it:

> All the world over and at all times there have been practical men, absorbed in "irreducible and stubborn facts": all the world over and at all times there have been men of philosophic temperament who have been absorbed in the weaving of general principles. It is this union of passionate interest in the detailed facts with equal devotion to abstract generalization which forms the novelty in our present society.

This new departure determined the climate of European thought in the last three centuries, it set modern Europe apart from all other civilizations in the past and present, and enabled it to transform its natural and social environment as completely as if a new species had arisen on this planet.

---

2. In order that the reader may visualize how small an angle is represented by 8 minutes of arc, we have drawn here to scale an angle which is ten times larger, that is, 80 minutes, or 1⅓ degrees. Try dividing the angle into ten equal parts and you will see how small a measurement this is. Ed.

Figure 5-11

The turning point is dramatically expressed in Kepler's work. In the *Mysterium Cosmographicum* the facts are coerced to fit the theory. In the *Astronomia Nova*, a theory, built on years of labour and torment, was instantly thrown away because of a discord of eight miserable minutes of arc. Instead of cursing those eight minutes as a stumbling block, he transformed them into the cornerstone of a new science.

What caused this change of heart in him? . . . some of the general causes which contributed to the emergence of the new attitude [are] the need of navigators, and engineers, for greater precision in tools and theories; the stimulating effects on science of expanding commerce and industry. But what turned Kepler into the first law-maker of nature was something different and more specific. It was *his introduction of physical causality into the formal geometry of the skies* which made it impossible for him to ignore the eight minutes arc. So long as cosmology was guided by purely geometrical rules of the game, regardless of physical causes, discrepancies between theory and fact could be overcome by inserting another wheel into the system. In a universe moved by real, physical forces, this was no longer possible. The revolution which freed thought from the stranglehold of ancient dogma, immediately created its own, rigorous discipline.

The Second Book of the *New Astronomy* closes with the words:

And thus the edifice which we erected on the foundation of Tycho's observations, we have now again destroyed. . . . This was our punishment for having followed some plausible, but in reality false, axioms of the great men of the past.

THE WRONG LAW

The next act of the drama opens with Book Three. As the curtain rises, we see Kepler preparing himself to throw out more ballast. The axiom of *uniform motion* has already gone overboard; Kepler feels, and hints, that the even more sacred one of *circular* motion must follow. The impossibility of constructing a circular orbit which would satisfy all existing observations, suggests to him that the circle must be replaced by some other geometrical curve.

But before he can do that, he must make an immense detour. For if the orbit of Mars is not a circle, its true shape can only be discovered by defining a sufficient number of points on the unknown curve. A circle is defined by three points on its circumference; every other curve needs more. The task before Kepler was to construct Mars' orbit without any preconceived ideas regarding its shape; to start from scratch, as it were.

To do that, it was first of all necessary to re-examine the motion of the earth itself. For, after all, the earth is our observatory; and if there is some misconception regarding its motion, all conclusions about the motions of other bodies will be distorted. Copernicus had assumed that the earth moves at uniform speed—not, as the other planets, only "quasi-uniformly" relative to some equant or epicycle, but *really* so. And since observation contradicted the dogma, the inequality of the earth's motion was explained away by the suggestion that the orbit periodically expanded and contracted, like a kind of pulsating jellyfish. It was typical of those improvisations which astronomers could afford so long as they felt free to manipulate the universe as they pleased on their drawing boards. It was equally typical that Kepler rejected it as "fantastic," again on the grounds that no physical cause existed for such a pulsation.

Hence his next task was to determine, more precisely than Copernicus had done, the earth's motion round the sun. For that purpose he designed a highly original method of his own. It was relatively simple, but it so happened that nobody had thought of it before. It consisted, essentially, in the trick of transferring the observer's position from earth to Mars, and to compute the motions of the earth exactly as an astronomer on Mars would do it.

The result was just as he had expected: the earth, like the other planets, did not revolve with uniform speed, but faster or slower according to its distance from the sun. Moreover, at the two extreme points of the orbit, the aphelion and perihelion (Figure 5-10) the earth's velocity proved to be, simply and beautifully, inversely proportional to distance.

At this decisive point, Kepler flies off the tangent and becomes airborne, as it were. Up to here he was preparing, with painstaking patience, his second assault on the orbit of Mars. Now he turns to a quite different subject. "Ye physicists, prick your ears," he warns, "for now we are going to invade your territory." The next six chapters are a report on that invasion into celestial physics, which had been out of bounds for astronomy since Plato.

A phrase seems to have been humming in his ear like a tune one cannot get rid of; it crops up in his writings over and again: there is a force in the sun which moves the planet, there is a force in the sun, there is a force in the sun. And since there is a force in the sun, there must exist some beautifully simple relation between the planet's distance from the sun and its speed. A light shines the brighter the nearer we are to its source, and the same must apply to the force of the sun: the closer the planet to it, the quicker it will move. This is his instinctive conviction, already expressed in the *Mysterium Cosmographicum;* but now, at last, he has succeeded in proving it.

In fact he has not. He has proved the inverse ratio of speed to distance only for the *two extreme points* of the orbit; and his extension of this "Law" to the *entire* orbit was a patently incorrect generalization. Moreover, Kepler knew this, and admitted it at the end of the thirty-second chapter, before he became airborne; but immediately afterwards, he conveniently forgot it. This is the first of the critical mistakes which "as if by a miracle" cancelled out, and led Kepler to the discovery of his Second Law. It looks as if his conscious, critical faculties were anaesthetized by the creative impulse, by his impatience to get to grips with the physical forces in the solar system.

Since he had no notion of the *momentum* which makes the planet persist in its motion, and only a vague intuition of *gravity* which bends that motion into a closed orbit, he had to find, or invent, a force which, like a broom, sweeps the planet around its path. And since the sun causes all motion, he let the sun handle the broom. This required that the sun rotate round its own axis—a guess which was only confirmed much later; the force which it emitted rotated with it, like the spokes of a wheel, and swept the planets along. But if that were the only force acting on them, the planets would all have the same angular velocity, they would all complete their revolutions in the same period—which they do not. The reason, Kepler thought, was the laziness or "inertia" of the planets, who desire to remain in the same place, and resist the sweeping force. The "spokes" of that force are not rigid; they allow the planet to lag behind; it works rather like a vortex or whirlpool. The power of the whirlpool diminishes with distance, so that the farther away the planet, the less power the sun has to overcome its laziness, and the slower its motion will be.

It still remained to be explained, however, why the planets moved in eccentric orbits instead of always keeping the same distance from the centre of the vortex. Kepler first assumed that apart from being lazy, they performed an epicyclic motion in the opposite direction under their own steam, as it were, apparently out of sheer cussedness. But he was dissatisfied with this, and at a later stage assumed that the planets were "huge round magnets" whose magnetic axis pointed always in the same direction, like the axis of a top; hence the planet will periodically be drawn closer to, and be repelled by the sun, according to which of its magnetic poles faces the sun.

Thus, in Kepler's physics of the universe, the roles played by gravity and inertia are reversed. Moreover, he assumed that the sun's power diminishes in direct ratio to distance. He sensed that there was something wrong here, since he knew that the intensity of light diminishes with the *square* of distance; but he had to stick to it, to satisfy his theorem of the ratio of speed to distance, which was equally false.

THE SECOND LAW

Refreshed by this excursion into the *Himmelsphysik,* our hero returned to the more immediate task in hand. Since the earth no longer moved at uniform speed, how could one predict its position at a given time? (The method based on the *punctum equans* had proved, after all, a disappointment.) Since he believed to have proved that its speed depended directly on its distance from the sun, the time it needed to cover a small fraction of the orbit was always proportionate to that distance. Hence he divided the orbit (which, forgetting his previous resolve, he still regarded as a circle) into 360 parts, and computed the distance of each bit of arc from the sun. The sum of all distances between, say, 0° and 85° was a measure of the time the planet needed to get there.

But this procedure was, as he remarked with unusual modesty, "mechanical and tiresome." So he searched for something simpler:

> Since I was aware that there exists an infinite number of points on the orbit and accordingly an infinite number of distances [from the sun] the idea occurred to me that the sum of these distances is contained in the *area* of the orbit. For I remembered that in the same manner Archimedes too divided the area of a circle into an infinite number of triangles.

Accordingly, he concluded, the area swept over by the line connecting planet and sun AS-BS (Figure 5-12) is a measure of the time required by the planet to get from A to B; *hence the line will sweep out*

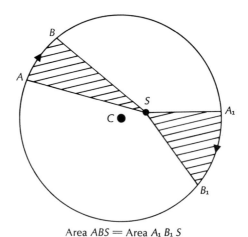

Figure 5-12   Area ABS = Area $A_1 B_1 S$

equal areas in equal times. This is Kepler's immortal Second Law (which he discovered before the First)—a law of amazing simplicity at the end of a dreadfully confusing labyrinth.

Yet the last step which had got him out of the labyrinth had once again been a faulty step. For it is not permissible to equate an area with the sum of an infinite number of neighbouring lines, as Kepler did. Moreover, he knew this well, and explained at length why it was not permissible. He added that he had also committed a second error, by assuming the orbit to be circular. And he concluded: "But these two errors —it is like a miracle—cancel out in the most precise manner, as I shall prove further down."

The correct result is even more miraculous than Kepler realized, for his explanation of the reasons why his errors cancel out was once again mistaken, and he got, in fact, so hopelessly confused that the argument is practically impossible to follow—as he himself admitted. And yet, by three incorrect steps and their even more incorrect defence, Kepler stumbled on the correct law. It is perhaps the most amazing sleepwalking performance in the history of science—except for the manner in which he found his First Law, to which we now turn.

THE FIRST LAW

The Second Law determined the variations of the planet's speed along its orbit, but it did not determine the shape of the orbit itself.

At the end of the Second Book, Kepler had acknowledged defeat in his attempts to define the Martian orbit—a defeat caused by a discrepancy of eight minutes arc. He had then embarked on an enormous detour, starting with the revision of the earth's motion, followed by physical speculations, and terminating in the discovery of the Second Law. In the Fourth Book he resumed his investigation of the Martian orbit where he had left off. By this time, four years after his first, frustrated attempts, he had become even more sceptical of orthodox dogma, and gained an unparalleled skill in geometry by the invention of methods all his own.

The final assault took nearly two years; it occupies chapters 41 to 60 of the *New Astronomy*. In the first four (41–44), Kepler tried for the last time, with savage thoroughness, to attribute a circular orbit to Mars and failed: this section ends with the words:

The conclusion is quite simply that the planet's path is not a circle—it curves inward on both sides and outward again at opposite ends. Such a curve is called an oval. The orbit is not a circle, but an oval figure.

But now a dreadful thing happened, and the next six chapters (45–50) are a nightmare journey through another labyrinth. This oval orbit is a wild, frightening new departure for him. To be fed up with cycles and epicycles, to mock the slavish imitators of Aristotle is one thing; to assign an entirely new, lopsided, implausible path for the heavenly bodies is quite another.

Why indeed an oval? There is something in the perfect symmetry of spheres and circles which has a deep, reassuring appeal to the unconscious—otherwise it could not have survived two millennia. The oval lacks all such archetypal appeal. It has an arbitrary form. It distorts that eternal dream of the harmony of the spheres, which lay at the origin of the whole quest. Who art thou, Johann Kepler, to destroy divine symmetry? All he has to say in his own defence is, that having cleared the stable of astronomy of cycles and spirals, he left behind him "only a single cart-ful of dung": his oval.

At this point, the sleepwalker's intuition failed him, he seems to be overcome by dizziness, and clutches at the first prop that he can find. He must find a physical cause, a cosmic *raison d'être* for his oval in the sky—and he falls back on the old quack remedy which he has just abjured, the conjuring up of an epicycle! To be sure, it is an epicycle with a difference: it has a physical cause. We have heard earlier on that while the sun's force sweeps the planet round in a circle, a second, antagonistic force, "seated in the planet itself," makes it turn in a small epicycle in the opposite direction. It all seems to him "wonderfully plausible," for the result of the combined movement is indeed an oval. But a very special oval: it has the shape of an egg, with the pointed end at the perihelion, the broad end at the aphelion.

No philosopher had laid such a monstrous egg before. Or, in Kepler's own words of wistful hindsight:

> What happened to me confirms the old proverb: a bitch in a hurry produces blind pups. . . . But I simply could not think of any other means of imposing an oval path on the planets. When these ideas fell upon me, I had already celebrated my new triumph over Mars without being disturbed by the question. . . . whether the figures tally or not. . . . Thus I got myself into a new labyrinth. . . . The reader must show tolerance to my gullibility.

The battle with the egg goes on for six chapters, and took a full year of Kepler's life. It was a difficult year; he had no money, and was down with "a fever from the gall." A threatening new star, the *nova* of 1604, had appeared in the sky; Frau Barbara [his wife] was also ill, and gave birth to a son—which provided Kepler with an opportunity for one of his dreadful jokes: "Just when I was busy squaring my oval, an unwelcome guest entered my house through a secret doorway to disturb me."

To find the area of his egg, he again computed a series of one hundred and eighty Sun-Mars distances and added them together; and this whole operation he repeated no less than forty times. To make the worthless hypothesis work, he temporarily repudiated his own, immortal Second Law—to no avail. Finally, a kind of snowblindness seemed to descend on him: he held the solution in his hand without seeing it. On 4 July, 1603, he wrote to a friend that he was unable to solve the geometrical problems of his egg; but "if only the shape were a perfect ellipse all the answers could be found in Archimedes' and Apollonius' work." A full eighteen months later, he again wrote to the same correspondent that the truth must lie somewhere halfway between egg-shape and circular-shape "just as if the Martian orbit were a perfect ellipse. But regarding that I have so far explored nothing." What is even more astonishing, he constantly used ellipses in his calculations—but merely as an *auxiliary* device to determine, by approximation, the area of his egg-curve— which by now had become a veritable fixation.[3] . . .

And yet, these years of wandering through the wilderness were not entirely wasted. The otherwise sterile chapters in the *New Astronomy* devoted to the egg hypothesis, represent an important further step towards the invention of the infinitesimal calculus. Besides, Kepler's mind had by now become so saturated with the numerical data of the Martian orbit, that when the crucial hazard presented itself, it responded at once like a charged cloud to a spark.

The hazard is perhaps the most unlikely incident in this unlikely story. It presented itself in the shape of a number which had stuck in Kepler's brain. The number was 0.00429.

When he at last realized that his egg had "gone up in smoke" and that Mars, whom he had believed a conquered prisoner "securely chained by my equations, immured in my tables," had again broken loose, Kepler decided to start once again from scratch.

He computed very thoroughly a score of Mars-Sun distances at various points of the orbit. They showed again that the orbit was some kind of oval, looking like a circle flattened inward at two opposite sides, so that there were two narrow sickles or "lunulae" left between the circle and the Martian orbit. The width of the sickle, where it is thickest, amounted to 0.00429 of the radius (Figure 5-13).

At this point Kepler, for no particular reason, became interested in the angle at M—the angle formed between the sun and the centre of the orbit, as seen from Mars. This angle was called the "optical equation." It

3. Copernicus, too, stumbled on the ellipse and kicked it aside; but Copernicus, who firmly believed in circles, had much less reason to pay attention to it than Kepler, who had progressed to the oval.

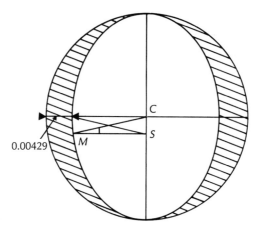

Figure 5-13

varies, of course, as Mars moves along its orbit; its maximum value is 5° 18′. This is what happened next, in Kepler's own words:

> I was wondering why and how a sickle of just this thickness (0.00429) came into being. While this thought was driving me around, while I was considering again and again that . . . my apparent triumph over Mars had been in vain, I stumbled entirely by chance on the secant [4] of the angle 5° 18′, which is the measure of the greatest optical equation. When I realized that this secant equals 0.00429, I felt as if I had been awakened from a sleep. . . .

It had been a true sleepwalker's performance. At the first moment, the reappearance of the number 0.00429 in this unexpected context must have appeared as a miracle to Kepler. But he realized in a flash that the apparent miracle must be due to a fixed relation between the angle at M and the distance to S, a relation which must hold true for any point of the orbit; only the manner in which he had stumbled on that relation was due to chance. "The roads that lead man to knowledge are as wondrous as that knowledge itself."

4. Mathematically, the secant of an acute angle of a right triangle is the ratio of the hypotenuse to the side adjacent to the angle in question. In the illustration below, the secant of the angle at M is the ratio MC:MS. These letters correspond to those in Figure 5-13. Ed.

Figure 5-14

Now at last, at long last, after six years of incredible labour, he held the secret of the Martian orbit. He was able to express the manner in which the planet's distance from the sun varied with its position, in a simple formula, a mathematical Law of Nature. *But he still did not realize that this formula specifically defined the orbit as an ellipse.*[5] Nowadays, a student with a little knowledge of analytical geometry would realize this at a glance; but analytical geometry came after Kepler. He had discovered his magic equation empirically, but he could no more identify it as the shorthand sign for an ellipse than the average reader of this book can; it was nearly as meaningless to him. He had reached his goal, but he did not realize that he had reached it.

The result was that he went off on one more, last, wild goose chase. He tried to construct the orbit which would correspond to his newly discovered equation; but he did not know how, made a mistake in geometry, and arrived at a curve which was too bulgy; the orbit was a *via buccosa*, chubby-faced, as he noted with disgust.

What next? We have reached the climax of the comedy. In his despair, Kepler threw out his formula (which denoted an elliptic orbit) because he wanted to try out an entirely new hypothesis: to wit, an elliptic orbit. It was as if the tourist had told the waiter, after studying the menu: "I don't want *côtellette d'agneau,* whatever that is; I want a lamb chop."

By now he was convinced that the orbit must be an ellipse, because countless observed positions of Mars, which he knew almost by heart, irresistibly pointed to that curve; but he still did not realize that his equation, which he had found by chance-plus-intuition, *was* an ellipse. So he discarded that equation, and constructed an ellipse by a different geometrical method. And then, at long last, he realized that the two methods produced the same result.

With his usual disarming frankness, he confessed what had happened:

Why should I mince my words? The truth of Nature, which I had rejected and chased away, returned by stealth through the backdoor, disguising itself to be accepted. That is to say, I laid [the original equation] aside, and fell back on ellipses, believing that this was a quite different hypothesis, whereas the two, as I shall prove in the next chapter, are one and the same . . . I thought and searched, until I went nearly mad, for a reason why the planet preferred an elliptical orbit [to mine]. . . . Ah, what a foolish bird I have been!

But in the List of Contents, in which he gives a brief outline of the whole work, Kepler sums up the matter in a single sentence:

5. In modern denotation, the formula is: $R = 1 + e \cos \beta$ where R is the distance from the sun, β the longitude referred to the centre of the orbit, and e the eccentricity. [See "ellipse," in Glossary. Ed.]

I show [in this chapter] how I unconsciously repair my error.

The remainder of the book is in the nature of a mopping-up operation after the final victory.

THE THIRD LAW

The *Harmonice Mundi* [published in 1619] contains . . . Kepler's Third Law of planetary motion. It says, put into modern terms, that the square of the periods of revolution of any two planets are as the cubes of their mean distances from the sun. Here is an illustration of it. Let the earth's distance from the sun be our unit measure, then Saturn's distance from the sun will be a little over nine units. The square root of 1 is 1; the square root of 9=3. The cube of 1 is 1, the cube of 3 is 27. Thus a Saturn year will be a little over twenty-seven earth years; in fact it is thirty years. Apologies for the coarse example—it is Kepler's own. Unlike his First and Second Laws . . . the Third Law was the fruit of nothing but patient, dogged trying. When after endless trials, he had at last hit on the square-to-cube ratio, he of course promptly found a reason why it should be just that and none other; I have said before that Kepler's *a priori* proofs were often invented *a posteriori*.

The exact circumstances of the discovery of the Third Law were again faithfully recorded by Kepler:

On March 8 of this present year 1618, if precise dates are wanted, (the solution) turned up in my head. But I had an unlucky hand and when I tested it by computations I rejected it as false. In the end it came back again to me on May 15, and in a new attack conquered the darkness of my mind; it agreed so perfectly with the data which my seventeen years of labour on Tycho's observations had yielded, that I thought at first I was dreaming. . . .

He had been searching for this Third Law, that is to say, for a correlation between a planet's period and its distance, since his youth. Without such a correlation, the universe would make no sense to him; it would be an arbitrary structure. If the sun had the power to govern the planets' motions, then that motion must somehow depend on their distance from the sun; but how? Kepler was the first who saw the problem —quite apart from the fact that he found the answer to it, after twenty-two years of labour. The reason why nobody before him had asked the question is that nobody had thought of cosmological problems in terms of actual physical forces. So long as cosmology remained divorced from physical causation in the mind, *the right question could not occur in that mind.* . . .

I have tried to re-trace the tortuous progress of his thought. Perhaps the most astonishing thing about it is the mixture of cleanness and uncleanness in his method. On the one hand, he throws away a cherished theory, the result of years of labour, because of those wretched eight minutes of arc. On the other hand he makes impermissible generalizations, knows that they are impermissible, yet does not care. . . .

We saw him plod, with infinite patience, along dreary stretches of trial-and-error procedure, then suddenly become airborne when a lucky guess or hazard presented him with an opportunity. What enabled him to recognize instantly his chance when the number 0.00429 turned up in an unexpected context was the fact that not only his waking mind, but his sleep-walking unconscious self was saturated with every conceivable aspect of his problem, not only with the numerical data and ratios, but also with an intuitive "feel" of the physical forces, and of the Gestalt-configurations which it involved. A locksmith who opens a complicated lock with a crude piece of bent wire is not guided by logic, but by the unconscious residue of countless past experiences with locks, which lend his touch a wisdom that his reason does not possess.

Limitations of the Inductive Method

Concerning Human Understanding [*]

by David Hume (1748)

As to past *Experience*, it can be allowed to give *direct* and *certain* information of those precise objects only, and that precise period of time, which fell under its cognizance: but why this experience should be extended to future times, and to other objects, which for aught we know, may be only in appearance similar; this is the main question on which I would insist. The bread, which I formerly eat, nourished me; that is, a body of such sensible qualities was, at that time, endued with such secret powers: but does it follow, that other bread must also nourish me at another time, and that like sensible qualities must always be attended with like secret powers? The consequence seems nowise necessary. At least, it must be acknowledged that there is here a consequence drawn by the mind; that there is a certain step taken; a process of thought, and an inference, which wants to be explained. These two propositions are far from being the same, *I have found that such an object has always been attended with such an effect,* and *I foresee, that other objects, which are, in appearance, similar, will be attended with similar effects.* I shall allow, if you please, that the one proposition may justly be inferred from the other: I know, in fact, that it always is inferred. But if you insist that the inference is made by a chain of reasoning, I desire you to produce that reasoning. The connexion between these propositions is not intuitive. There is required a medium, which may enable the mind to draw such an inference, if indeed it be drawn by reasoning and argument. What that medium is, I must confess, passes my comprehension; and it is incumbent on those to produce it, who assert that it really exists, and is the origin of all our conclusions concerning matter of fact.

[*] From David Hume, *An Enquiry Concerning Human Understanding,* 1748, Section IV, Part II.

Universal Assent °
by Norman Campbell (1921)

The community of sensations is our chief and final test that experiences are true sensations such as give information about the external world; if we apply other tests, it is only because this chief test is not available, and any other tests we may apply are based on the results which we are accustomed to obtain with this chief test. . . .

Other people agree with me much more closely about sensations than they do about any other kind of thought. The fact, expressed in that manner, is extremely familiar. If I am in a room when the electric light bulb bursts, not only I, but everyone else in the room (unless some of them are blind or deaf) hears the explosion and experiences the change from light to darkness. On the other hand, apart from sensations, we may all have been thinking about different things, remembering different things, following different trains of reasoning, and experiencing different desires. This community of sensations, contrasted with the particularity of other kinds of thoughts, leads naturally to the view that the sensations are determined by something that is not me or you or anybody else in the room, but is something external to us all; while the other thoughts, which we do not share, are parts of the particular person, experiencing them. This simple experience is probably the main reason why we have come to believe so firmly that there is an external world and that our perceptions received by our senses give us information about it. . . .

Science is the study of those judgments concerning which universal agreement can be obtained.

False facts are highly injurious to the progress of science, for they often endure long; but false views, if supported by some evidence, do little harm, for every one takes a salutary pleasure in proving their falseness.

Charles Darwin
The Descent of Man (1871)

° From Norman Campbell, *What Is Science?*, Dover Publications, Inc., New York, 1952. Reprinted by permission of the publisher.

The Scientist Imposes Order on His Data °
by Henry Margenau (1961)

Every observation, every measurement, indeed every perception, introduces errors. A measurement without error is an absurdity. Let the measurement (and measurement is after all only refined perception) be of a star's position in the sky, of the length of this table, of an automobile's speed, or of an electrical current; its outcome is never to be believed exactly. This is apparent in the circumstance that the same number will not ordinarily result when the measurement is repeated, regardless of the care taken in performing it. Characteristically, only the careless experimenter and the ignorant observer believe raw nature to be unambiguous. To be sure, the different numbers found by the astronomer for the latitude (and longitude) of a star in successive observations lie within a reasonable interval, and this convinces him that he is somewhere near a "true" value of the quantity he seeks, that nature is not fooling him with hallucinations. Yet in a very fundamental sense he is witnessing a behavior not unlike the lawless emissions of the microcosmic firefly; we thus see that even the macrocosm is not wholly without its vagaries, but that it confines them sufficiently so that the observer with some credulity can feign their absence: he can blame himself for nature's equivocality and call the departures from a true value "errors."

But what is the *true* value? Let us look into scientific practice. If ten measurements of a physical quantity yield ten slightly different values, not one of them is necessarily regarded as true. Their *arithmetic mean* is singled out for this distinction, even though it may not have occurred among the measurements. The justice of this choice is not provided by the ten measurements, nor by any finite number of observations; it comes from a belief in, or rather the postulation of, a certain uniformity of nature. Thus the very determination of a true value, and in the end the selection of whatever is believed to be true perceptory fact, involves a reference to law and order not immediately presented in the sensory structure of that fact. Here we find the clearest expression of the attitude which had led to the development of deductive science: it relies upon rational elements to straighten out erratic data. It does not ignore the latter's presence, nor does it accept them unrefined. It distills from them

° From Henry Margenau, *Open Vistas*, pp. 179–180. Yale University Press, New Haven. Copyright © 1961 Yale University Press. Reprinted by permission.

an essence and *calls* it *true*. But the nature of the essence is partly determined by the process of distillation.

A hypothesis or theory is clear, decisive, and positive, but it is believed by no
one but the man who created it. Experimental findings, on the other hand, are
messy, inexact things which are believed by everyone except the man who did
that work.

Harlow Shapley
Harvard Crimson (1935)

Suggestions for Further Reading

Bronowski, Jacob, and Bruce Mazlish: *The Western Intellectual Tradition*,
Harper & Row, Publishers, New York, 1960, chaps. 10, 11, and 12. A history
of the formation of the Royal Society and a discussion of the philosophies of
Hobbes, Locke, and Descartes.
° Burtt, E. A.: *The Metaphysical Foundations of Modern Science*, Anchor
Books, Doubleday & Company, Inc., Garden City, N.Y., 1954, chaps. IV, V,
and VI. A detailed and interesting discussion of Descartes as well as the philo-
sophical problems of knowledge, causality, sense perception, and the reality of
secondary qualities.
° Butterfield, Herbert: *The Origins of Modern Science*, The Macmillan
Company, New York, 1959, chaps. V, VI, and VII. A very fine analysis of the
use of the experimental method, the influence of Bacon and Descartes.
° Cohen, I. Bernard: *The Birth of a New Physics*, Anchor Books, Doubleday
& Company, Inc., Garden City, N.Y., 1960, chap. VI. A good discussion of
Kepler's laws.
° Conant, James B.: *Science and Common Sense*, Yale University Press,
New Haven, 1951, chap. 6. An explanation of the relationship between ab-
stract reasoning and quantitative experimentation.
° Lodge, Oliver: *Pioneers of Science*, Dover Publications, Inc., New York,
1960, lectures II and III. The story of Tycho Brahe's life and work and Kep-
ler's discovery of the laws of planetary motion.
° Russell, Bertrand: *A History of Western Philosophy*, Simon and Schuster,
Inc., New York, 1945, book 3, part I, chaps. VII, IX, XIII, XVI, and XVII.
An analysis of the philosophies of Francis Bacon, Descartes, Locke, Berkeley,
and Hume.

° Paperback edition

6

Is There a Limit to Man's Understanding of Nature?

A study of the development of two contradictory theories of the nature of light brings into focus the philosophical crisis arising from the discoveries of modern physics. Can a model or picture of reality be formulated that fits the known facts and is consistent with the assumption that there is an order in nature that man can understand?

Is There a Limit to Man's Understanding of Nature?

Introduction

LIGHT IS THE PRINCIPAL source of our knowledge of the universe. Without it, our world would be small indeed, confined to what we could touch, taste, hear, or smell. The light from the sun and stars was first used to estimate their relative sizes and distances from us. Later it was discovered that the messages we receive contain much more information about the bodies that are radiating or reflecting this light. The complicated code of these messages was gradually deciphered as scientists investigated the nature of light itself. It was found to contain data about the temperature, composition, size, and age of the stars as well as their velocity in relation to the earth. Although we still do not understand exactly why light behaves as it does, we have learned a great deal about how it behaves and have used this knowledge very effectively to build useful devices and learn more about other parts of the universe.

The readings in this part trace the search for a satisfactory theory of light from Newton and Huygens to the modern quantum theory and bring out the fact that the nature of light is still one of the greatest mysteries of science. In modern physics, theories of light merge with theories of the nature of matter and become an integral part of the broader theory known as quantum mechanics. At this point we have made an effort to include selections that bear directly on the nature of light itself, in order to keep the material within limits that can reasonably be handled in one part and have left the study of the nature of matter to another time when it can be handled in much greater detail. The modern concept of an electron is mentioned in these readings only insofar as the idea is necessary to an understanding of the uncertainty principle, which has had such important repercussions on philosophical thought.

As a result of quantum theory and the uncertainty principle, three of the most basic beliefs of science have been brought into question. The first of these is the uniformity of nature, which we discussed in Part 3. That "instinctive belief in the order of nature" that appeared to be essential to the growth of the scientific spirit has been, if not actually destroyed, at least seriously shaken. This situation is illustrated by a joke current among physicists that light behaves as waves on Mondays, Wednesdays, and Fridays and as quanta on the other days. It may seem surprising to the layman that scientists can continue to work effectively in this state of suspended judgment, when even the most basic assumption of scientific activity has been shaken. But actually, the state of mind represented here is a necessary facet of the scientific spirit. To some extent, science must always reserve final judgment, as we will see in Part 7, and must stand ready to change basic concepts in the face of new evidence.

The second tenet that has been shaken by modern physics is the ability of science to remain objective. In studying the very smallest units of mass and energy, the differentiation between subject and object becomes blurred, and this changed relationship poses the problem of how to preserve the objective scientific attitude as man attempts to probe further into the nature of light and matter.

The third fundamental concept that is questioned is the nature of causality. The causal relationship that has been extensively used in scientific reasoning was shaken by the discovery that we cannot form a continuous description of elementary particles. There are gaps in continuity due to the quantization of energy by which we perceive and measure events in space and time. Heisenberg's uncertainty principle is a mathematical statement of this difficulty. If it is theoretically impossible ever to formulate a complete description of the situation at one instant in time, then it is also impossible to predict what will happen as a result of this state. The picture presented by Laplace in 1814 that "we ought . . . to regard the present state of the universe as the effect of its anterior state and as the cause of the one which is to follow"[1] becomes forever inaccessible.

Taken all together, the questioning of these three basic tenets leads to an even more challenging problem. Have we, as Bridgman suggests, reached the limit of vision because the structure of nature does not correspond sufficiently with our thought processes? Or is the barrier we have reached only another hurdle to be surmounted in the search for order in the universe? Some noted scientists such as Einstein and Schrödinger take this view. "In this physical world," says Harlow Shapley, "there is no real chaos; all is in fact orderly; all is ordered by the physical principles. Chaos is but unperceived order; it is a word indicating the limitations of the human mind and the paucity of observational facts. The words 'chaos,' 'accidental,' 'chance,' 'unpredictable,' are conveniences behind which we hide our ignorance."[2]

1. See p. 305.
2. Harlow Shapley, *Of Stars and Men,* Beacon Press, 1958, 1964.

The Nature of Light

In the beginning God created the heaven and the earth. And the earth was without form, and void; and darkness *was* upon the face of the deep. And the Spirit of God moved upon the face of the waters.
And God said, Let there be light: and there was light.
And God saw the light, that it was good: and God divided the light from the darkness.
And God called the light Day, and the darkness he called night. And the evening and the morning were the first day.

Genesis, 1:1–5
King James Bible (1604–11)

Dispersion of Light [*]
by Sir Isaac Newton (1672)

In the year 1666, (at which time I applied myself to the grinding of optick glasses of other figures than spherical) I procured me a triangular glass prism to try therewith the celebrated phaenomena of colours. And in order thereto, having darkened my chamber, and made a small hole in my window-shuts, to let in a convenient quantity of the sun's light, I placed my prism at its entrance, that it might be thereby refracted to the opposite wall. It was at first a very pleasing divertisement, to view the vivid and intense colours produced thereby; but after a while applying myself to consider them more circumspectly, I became surprised, to see them in an oblong form; which, according to the received laws of refraction, I expected should have been circular. They were terminated at the sides with straight lines, but at the ends, the decay of light was so gradual that it was difficult to determine justly, what was their figure; yet they seemed semicircular.

Comparing the length of this colour'd Spectrum with its breadth, I found it about five times greater, a disproportion so extravagant, that it excited me to a more than ordinary curiosity to examining from whence

[*] From Sir Isaac Newton, *Principia Mathematica,* edited by Florian Cajori, from the translation by Andrew Motte. University of California Press, Berkeley, 1934. Reprinted by permission.

it might proceed. I could scarce think, that the various thicknesses of the glass, or the termination with shadow or darkness, could have any influence on light to produce such an effect; yet I thought it not amiss, first to examine those circumstances, and so try'd what would happen by transmitting light through parts of the glass of diverse thicknesses, or through holes in the window of diverse bignesses, or by setting the prism without, so that the light might pass through it, and be refracted, before it was terminated by the hole: But I found none of those circumstances material. The fashion of the colours was in all these cases the same.

Then I suspected, whether by any unevenness in the glass or other contingent irregularity, these colours might be thus dilated. And to try this, I took another prism like the former, and so placed it, that the light passing through them both might be refracted contrary ways, and so by the latter returned into that course from which the former had diverted it. For by this means I thought the regular effects of the first prism would be destroyed by the second prism, but the irregular ones more augumented, by the multiplicity of refractions. The event was, that the light, which by the first prism was diffused into an oblong form was by the second reduced into an orbicular one, with as much regularity as when it did not at all pass through them.

Then I began to suspect, whether the rays, after their trajection through the prism, did not move in curve lines, and according to their more or less curvity tend to divers parts of the wall. And it increased my suspicion, when I remembered that I had often seen a tennis ball struck with an oblique racket, describe such a curve line. For, a circular as well as a progressive motion being communicated to it by that stroke, its parts on that side where the motions conspire, must press and beat the contiguous air more violently than on the other, and there excite a reluctancy and reaction of the air proportionably greater. And for the same reason, if the rays of light should possibly be globular bodies, and by their oblique passage out of one medium into another, acquire a circulating motion, they ought to feel the greater resistance from the ambient ether, on that side, where the motions conspire, and thence be continually bowed to the other. But notwithstanding this plausible ground of suspicion, when I came to examine it, I could observe no such curvity in them. And besides (which was enough for my purpose) I observed, that the difference 'twixt the length of the image, and the diameter of the hole, through which the light was transmitted, was proportionable to their distance.

The gradual removal of these suspicions at length led me to the *Experimentum Crucis*, which was this: I took two boards, and placed one of them close behind the prism at the window, so that the light might pass through a small hole, made in it for the purpose, and fall on the other board, which I placed at about 12 feet distance, having first made a small hole in it also, for some of the incident light to pass through. Then I placed another prism behind this second board, so that the light trajected through both the boards might pass through that also, and be again refracted before it arrived at the wall. This done, I took the first prism in my hand, and turned it to and fro slowly about its axis, so much as to make the several parts of the image cast, on the second board, successively pass through the hole in it, that I might observe to what places on the wall the second prism would refract them. And I saw by the variation of those places, that the light, tending to that end of the image, towards which the refraction of the first prism was made, did in the second prism suffer a refraction considerably greater than the light tending to the other end. And so the true cause of the length of that image was detected to be no other, than the light is not similar or homo-genial, but consists of *Difform Rays, some of which are more Refrangi-ble than others;* so that without any difference in their incidence on the same medium, some shall be more Refracted than others; and therefore that, according to their *particular Degrees of Refrangibility,* they were transmitted through the prism to divers parts of the opposite wall.

Now I shall proceed to acquaint you with another more notable *Dif-formity* in its rays, wherein the origin of colours is unfolded: concerning which I shall lay down the doctrine first, and then for its examination give you an instance or two of the experiments, as a specimen of the rest.

The doctrine you will find comprehended and illustrated in the fol-lowing propositions:

1. As the rays of light differ in degrees of refrangibility so they also differ in their disposition to exhibit, this or that particular colour. Col-ours are not qualifications of light, derived from refractions, or reflec-tions of natural bodies (as 'tis generally believed) but original and con-nate properties, which in divers rays are divers. Some rays are disposed to exhibit a red colour and no other; some a yellow and no other, some a green and no other, and so of the rest. Nor are there only rays proper and particular to the more eminent colours, but even to all their inter-mediate gradations.

2. To the same degree of refrangibility ever belongs the same colour, and to the same colour ever belongs the same degree of refrangibility. The least refrangible rays are all disposed to exhibit a red colour, and contrarily those rays which are disposed to exhibit a red colour, are all the least refrangible: so the most refrangible rays are all disposed to ex-

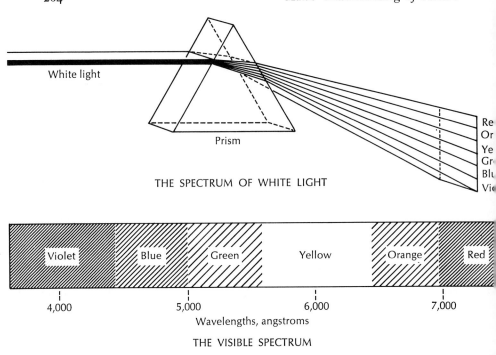

White light

Prism

THE SPECTRUM OF WHITE LIGHT

Re
Or
Ye
Gr
Blu
Vi

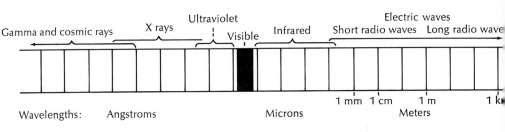

| Violet | Blue | Green | Yellow | Orange | Red |

4,000   5,000   6,000   7,000

Wavelengths, angstroms

THE VISIBLE SPECTRUM

Gamma and cosmic rays   X rays   Ultraviolet   Infrared   Electric waves
                                      Visible   Short radio waves   Long radio wave

1 mm 1 cm   1 m   1 k

Wavelengths: Angstroms   Microns   Meters

THE COMPLETE ELECTROMAGNETIC SPECTRUM

Figure 6-1. *The place of the visible spectrum in the electromagnetic spectrum.* (Figure added by editor.)

hibit a deep violet colour, and contrarily those which are apt to exhibit such a violet colour are all the most refrangible. And so to all the intermediate colours in a continued series belong intermediate degrees of refrangibility. And this Analogy 'twixt colours and refrangibility is very precise and strict; the rays always either exactly agreeing in both, or proportionally disagreeing in both [Figure 6-1].

3. The species of colour, and degree of refrangibility proper to any particular sort of rays, is not mutable by refraction, nor by reflection from natural bodies, nor by any other cause that I could yet observe. When any one sort of rays hath been well parted from those of other kinds, it hath afterwards obstinately retained its colour, notwithstanding my utmost endeavours to change it. I have refracted it with prisms, and reflected it with bodies, which in daylight were of other colours; I have intercepted it with the coloured film of air, interceded two compressed plates of glass; transmitted it through coloured mediums, and through mediums irradiated with other sorts of rays, and diversely terminated it; and yet could never produce any new colour out of it. It would by contracting or dilating become more brisk, or faint, and by the loss of many rays, in some cases very obscure and dark; but I could never see it changed in specie.

4. Yet seeming transmutations of colours may be made, where there is any mixture of divers sorts of rays. For in such mixtures, the component colours appear not, but, by their mutual allaying each other, constitute a middling colour. And therefore, if by refraction, or any other of the aforesaid causes, the difform rays, latent in such a mixture, be separated, there shall emerge colours different from the colour of the composition. Which colours are not new generated, but only made apparent by being parted; for if they be again entirely mixt and blended together, they will again compose that colour, which they did before separation. And for the same reason, transmutations made by the convening of diverse colours are not real; for when the difform rays are again severed, they will exhibit the very same colours which they did before they entered the composition; as you see blue and yellow powders, when finely mixed, appear to the naked eye, green, and yet the colours of the component corpuscles are not thereby really transmuted, but only blended. For when viewed with a good microscope they still appear blue and yellow interspersedly.

5. There are therefore two sorts of colours. The one original and simple, and the other compounded of these. The original or primary colours are red, yellow, green, blue, and a violet-purple, together with orange, indico, and an indefinite variety of intermediate gradations.

6. The same colours in species with these primary ones may be also produced by composition. For a mixture of yellow and blue makes green; of red and yellow makes orange; of orange and yellowish green makes yellow. And in general, if any two colours be mixed, which in the series of those generated by the prism are not too far distant one from another, they by their mutual alloy compound that color, which in the said series appeareth in the midway between them. But those which are

situated at too great a distance, do not so. Orange and indico produce not the intermediate green, nor scarlet and green the intermediate yellow.

7. But the most surprising, and wonderful composition was that of whiteness. There is no one sort of rays which alone can exhibit this. 'Tis ever compounded, and to its composition, are requisite all the aforesaid primary colours, mixed in a due proportion. I have often with admiration beheld that all the colours of the prism being made to converge, and thereby to be again mixed, as they were in the light before it was incident upon the prism, reproduced light, entirely and perfectly white, and not at all sensibly differing from a direct light of the sun, unless when the glasses, I used, were not sufficiently clear; for then they would a little incline it to their colour.

8. Hence therefore it comes to pass, that whiteness is the usual colour of light; for light is a confused aggregate of rays indued with all sorts of colours, as they were promiscuously darted from the various parts of luminous bodies. And of such a confused aggregate, as I said, is generated whiteness, if there be a due proportion of the ingredients; but if any one predominate, the light must incline to that colour; as it happens in the blue flame of brimstone; the yellow flame of a candle; and the various colours of the fixed stars.

9. These things considered, the manner how colours are produced by a prism is evident. For, of the rays, constituting the incident light, since those which differ in colour proportionally differ in refrangibility, they by their unequal refractions must be severed and dispersed into an oblong form in an orderly succession, from the least refracted scarlet, to the most refracted violet. And for the same reason it is, that objects when looked upon through a prism, appear coloured. For the difform rays, by their unequal refractions, are made to diverge towards several parts of the Retina, and these express the images of things coloured, as in the former case they did the sun's image upon a wall. And by this inequality of refractions, they become not only coloured, but also very confused and indistinct.

10. Why the colours of the rainbow appear in falling drops of rain, is also from hence evident. For those drops which refract the rays, disposed to appear purple, in greatest quantity to the spectator's eye, refract the rays of other sorts so much less, as to make them pass beside it; and such are the drops on the inside of the primary bow, and on the outside of the secondary or exterior one. So those drops, which refract in greatest plenty the rays, apt to appear red, toward the spectator's eye, refract those of other sorts so much more, as to make them pass beside

it; and such are the drops on the exterior part of the primary, and interior part of the secondary bow.

11. The odd phaenomena of an infusion of *Lignum Nephriticum,* leaf-gold, fragments of coloured glass, and some other transparently coloured bodies, appearing in one position of one colour, and of another in another, are on these grounds no longer riddles. For those are substances apt to reflect one sort of light, and transmit another; as may be seen in a dark room, by illuminating them with familiar or uncompounded light. For then they appear of that colour only, with which they are illuminated, but yet in one position more vivid and luminous than in another, accordingly as they are disposed more or less to reflect or transmit the incident colour.

12. From hence also is manifest the reason of an unexpected experiment which *Mr. Hook,* somewhere in his *Micrography* relates to have made with two wedge-like transparent vessels, filled the one with a red, the other with a blue liquor: namely, that though they were severally transparent enough, yet both together became opake; for if one transmitted only red, and the other only blue, no rays could pass through both.

13. I might add more instances of this nature, but I shall conclude with this general one. That the colours of all natural bodies have no other origin than this, that they are variously qualified, to reflect one sort of light in greater plenty than another. And this I have experimented in a dark room, by illuminating those bodies with uncompounded light of divers colours. For by that means any body may be made to appear of any colour. They have there no appropriate colour, but ever appear of the colour of the light cast upon them, but yet with this difference, that they are most brisk and vivid in the light of their own daylight colour. Minimum appeareth there of any colour indifferently, with which it is illustrated, but yet most luminous in red, and so bise appeareth indifferently of any colour, but yet most luminous in blue. And therefore minimum reflecteth rays of any colour, but most copiously those endowed with red, and consequently when illustrated with daylight; that is, with all sorts of rays promiscuously blended, those qualified with red shall abound most in the reflected light, and by their prevalence cause it to appear of that colour. And for the same reason bise, reflecting blue most copiously, shall appear blue by the excess of those rays in its reflected light; and the like of other bodies. And that this is the entire and adequate cause of their colours, is manifest, because they have no power to change or alter the colours of any sort of rays incident apart, but put on all colours indifferently, with which they are enlightened.

These things being so, it can be no longer disputed, whether there be colours in the dark, or whether they be the qualities of the objects we see, no nor perhaps, whether light be a body. For, since colours are the qualities of light, having its rays for their entire and immediate subject, how can we think those rays qualities also, unless one quality may be the subject of, and sustain another; which in effect is to call it substance. We should not know bodies for substances; were it not for their sensible qualities, and the principal of those being now found due to something else, we have as good reason to believe that to be a substance also.

Besides, who ever thought any quality to be a heterogeneous aggregate, such as light is discovered to be? But to determine more absolutely what light is, after what manner refracted, and by what modes or actions it produceth in our minds the phantasms of colours, is not so easie; and I shall not mingle conjectures with certainties.

Treatise on Light *
by Christian Huygens (1678)

It is inconceivable to doubt that light consists in the motion of some sort of matter. For whether one considers its production, one sees that here upon the Earth it is chiefly engendered by fire and flame which contain without doubt bodies that are in rapid motion, since they dissolve and melt many other bodies, even the most solid; or whether one considers its effects, one sees that when light is collected, as by concave mirrors, it has the property of burning as a fire does, that is to say it disunites the particles of bodies. This is assuredly the mark of motion, at least in the true Philosophy, in which one conceives the causes of all natural effects in terms of mechanical motions. This, in my opinion, we must necessarily do, or else renounce all hopes of ever comprehending anything in Physics.

And as, according to this Philosophy, one holds as certain that the sensation of sight is excited only by the impression of some movement of a kind of matter which acts on the nerves at the back of our eyes, there is here yet one reason more for believing that light consists in a movement of the matter which exists between us and the luminous body.

Further, when one considers the extreme speed with which light spreads on every side, and how, when it comes from different regions,

* From Christian Huygens, "Treatise on Light" in George Schwartz and Philip W. Bishop (eds.), *Moments of Discovery*, Basic Books, Inc., Publishers, New York. Copyright © 1958 by Basic Books, Inc., Publishers, New York. Reprinted by permission.

even from those directly opposite, the rays traverse one another without hindrance, one may well understand that when we see a luminous object, it cannot be by any transport of matter coming to us from this object, in the way in which a shot or an arrow traverses the air; for assuredly that would too greatly impugn these two properties of light, especially the second of them. It is then in some other way that light spreads; and that which can lead us to comprehend it is the knowledge which we have of the spreading of Sound in the air.

We know that by means of the air, which is an invisible and impalpable body, Sound spreads around the spot where it has been produced, by a movement which is passed on successively from one part of the air to another; and that the spreading of this movement, taking place equally rapidly on all sides, ought to form spherical surfaces ever enlarging and which strike our ears. Now there is no doubt at all that light also comes from the luminous body to our eyes by some movement impressed on the matter which is between the two; since, as we have already seen, it cannot be by the transport of a body which passes from one to the other. If, in addition, light takes time for its passage—which we are now going to examine—it will follow that this movement, impressed on the intervening matter, is successive; and consequently it spreads, as Sound does, by spherical surfaces and waves: for I call them waves from their resemblance to those which are seen to be formed in water when a stone is thrown into it, and which present a successive spreading as circles, though these arise from another cause, and are only in a flat surface.

To see then whether the spreading of light takes time, let us consider first whether there are any facts of experience which can convince us to the contrary. As to those which can be made here on the Earth, by striking lights at great distances, although they prove that light takes no sensible time to pass over these distances, one may say with good reason that they are too small, and that the only conclusion to be drawn from them is that the passage of light is extremely rapid.

It is true that we are here supposing a strange velocity that would be a hundred thousand times greater than that of Sound. For Sound, according to what I have observed, travels about 180 Toises [1] in the time of one Second, or in about one beat of the pulse. But this supposition ought not to seem to be an impossibility; since it is not a question of the transport of a body with so great a speed, but of a successive movement which is passed on from some bodies to others. I have then made no difficulty, in meditating on these things, in supposing that the emanation of light is accomplished with time, seeing that in this way all its phenom-

1. About 1,115 feet. Ed.

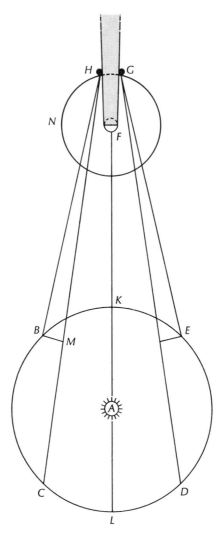

Figure 6-2. (Figure added by editor.)

ena can be explained, and that in following the contrary opinion everything is incomprehensible. For it has always seemed to me that even Mr.
Des Cartes, whose aim has been to treat all the subjects of Physics
intelligibly, and who assuredly has succeeded in this better than any
one before him, has said nothing that is not full of difficulties, or even
inconceivable in dealing with Light and its properties.

But that which I employed only as a hypothesis, has recently received great seemingness as an established truth by the ingenious proof of Mr. Römer which I am going here to relate, expecting him himself to give all that is needed for its confirmation. It is founded as is the preceding argument upon celestial observations, and proves not only that Light takes time for its passage, but also demonstrates how much time it takes, and that its velocity is even at least six times greater than that which I have just stated.

For this he makes use of the Eclipses suffered by the little planets which revolve around Jupiter, and which often enter his shadow: and see what is his reasoning [Figure 6-2]. Let A be the Sun, BCDE the annual orbit of the Earth, F Jupiter, GN the orbit of the nearest of his Satellites, for it is this one which is more apt for this investigation than any of the other three, because of the quickness of its revolution. Let G be this Satellite entering into the shadow of Jupiter, H the same Satellite emerging from the shadow.

Let it be then supposed, the Earth being at B some time before the last quadrature, that one has seen the said Satellite emerge from the shadow; it must needs be, if the Earth remains at the same place, that, after 42½ hours, one would again see a similar emergence, because that is the time in which it makes the round of its orbit, and when it would come again into opposition to the Sun. And if the Earth, for instance, were to remain always at B during 30 revolutions of this Satellite, one would see it again emerge from the shadow after 30 times 42½ hours. But the Earth having been carried along during this time to C, increasing thus its distance from Jupiter, it follows that if Light requires time for its passage the illumination of the little planet will be perceived later at C than it would have been at B, and that there must be added to this time of 30 times 42½ hours that which the Light has required to traverse the space MC, the difference of the spaces CH, BH. Similarly at the other quadrature when the Earth has come to E from D while approaching toward Jupiter, the immersions of the Satellite ought to be observed at E earlier than they would have been seen if the Earth had remained at D.

Now in quantities of observations of these Eclipses, made during ten consecutive years, these differences have been found to be very considerable, such as ten minutes and more; and from them it has been concluded that in order to traverse the whole diameter of the annual orbit KL, which is double the distance from here to the sun, Light requires about 22 minutes of time.

The movement of Jupiter in his orbit while the Earth passed from B to C, or from D to E, is included in this calculation; and this makes it evident that one cannot attribute the retardation of these illuminations or the anticipation of the eclipses, either to any irregularity occurring in the movement of the little planet or to its eccentricity.

If one considers the vast size of the diameter KL, which according to me is some 24 thousand diameters of the Earth, one will acknowledge the extreme velocity of Light. For, supposing that KL is no more than 22 thousand of these diameters, it appears that being traversed in 22 minutes this makes the speed a thousand diameters in one minute, that is 16⅔ diameters in one second or in one beat of the pulse, which makes more than 11 hundred times a hundred thousand Toises; since the diameter of the Earth contains 2,865 leagues, reckoned at 25 to the degree, and each league is 2,282 Toises, according to the exact measurement which Mr. Picard made by order of the King in 1669. But Sound, as I have said above, only travels 180 Toises in the same time of one second: hence the velocity of Light is more than six hundred thousand times greater than that of Sound. This, however, is quite another thing from being instantaneous, since there is all the difference between a finite thing and an infinite. Now the successive movement of Light being confirmed in this way, it follows, as I have said, that it spreads by spherical waves, like the movement of Sound.

But if the one resembles the other in this respect, they differ in many other things; to wit, in the first production of the movement which causes them; in the matter in which the movement spreads; and in the manner in which it is propagated. As to that which occurs in the production of Sound, one knows that it is occasioned by the agitation undergone by an entire body, or by a considerable part of one, which shakes all the contiguous air. But the movement of the Light must originate as from each point of the luminous object, else we should not be able to perceive all the different parts of that object, as will be more evident in that which follows. And I do not believe that this movement can be better explained than by supposing that all those of the luminous bodies which are liquid, such as flames, and apparently the sun and the stars, are composed of particles which float in a much more subtle medium which agitates them with great rapidity, and makes them strike against the particles of the ether which surrounds them, and which are much smaller than they. But I hold also that in luminous solids such as charcoal or metal made red hot in the fire, this same movement is caused by the violent agitation of the particles of the metal or of the wood; those of them which are on the surface striking similarly against the ethereal matter. The agitation, moreover, of the particles which engender the light ought to be much more prompt and more rapid than is that of the bodies which cause sound, since we do not see that the tremors of a body which is giving out a sound are capable of giving rise to Light, even as the movement of the hand in the air is not capable of producing sound.

Now if one examines what this matter may be in which the movement coming from the luminous body is propagated, which I call Ethereal matter, one will see that it is not the same that serves for the propagation of Sound. For one finds that the latter is really that which we feel and which we breathe, and which being removed from any place still leaves there the other kind of matter that serves to convey Light. This may be proved by shutting up a sounding body in a glass vessel from which the air is withdrawn by the machine which Mr. Boyle has given us, and with which he has performed so many beautiful experiments [Figure 6-3]. But in doing this of which I speak, care must be taken to

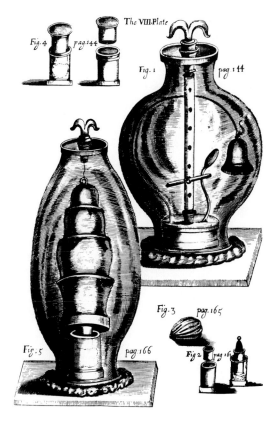

Figure 6-3. *At upper right a wood engraving of Boyle's apparatus for striking a bell in a vacuum.* (University of Pennsylvania. Figure added by editor.)

place the sounding body on cotton or on feathers, in such a way that it cannot communicate its tremors either to the glass vessel which encloses it, or to the machine; a precaution which has hitherto been neglected. For then after having exhausted all the air one hears no Sound from the metal, though it is struck.

One sees here not only that our air, which does not penetrate through glass, is the matter by which Sound spreads; but also that it is not the same air but another kind of matter in which Light spreads; since if the air is removed from the vessel the Light does not cease to traverse it as before.

And this last point is demonstrated even more clearly by the celebrated experiment of Torricelli, in which the tube of glass from which the quicksilver has withdrawn itself, remaining void of air, transmits Light just the same as when air is in it. For this proves that a matter different from air exists in this tube, and that this matter must have penetrated the glass or the quicksilver, either one or the other, though they are both impenetrable to the air. And when, in the same experiment, one makes the vacuum after putting a little water above the quicksilver, one concludes that the said matter passes through glass or water, or through both.

As regards the different modes in which I have said the movements of Sound and of Light are communicated, one may sufficiently comprehend how this occurs in the case of Sound if one considers that the air is of such a nature that it can be compressed and reduced to a much smaller space than that which it ordinarily occupies. And in proportion as it is compressed the more does it exert an effort to regain its volume; for this property along with its penetrability, which remains notwithstanding its compression, seems to prove that it is made up of small bodies which float about and which are agitated very rapidly in the ethereal matter composed of much smaller parts. So that the cause of the spreading of Sound is the effort which these little bodies make in collisions with one another, to regain freedom when they are a little more squeezed together in the circuit of these waves than elsewhere.

But the extreme velocity of Light, and other properties which it has, cannot admit of such a propagation of motion, and I am about to show here the way in which I conceive it must occur. For this, it is needful to explain the property which hard bodies must possess to transmit movement from one to another.

When one takes a number of spheres of equal size, made of some very hard substance, and arranges them in a straight line, so that they touch one another, one finds, on striking with a similar sphere against the first of these spheres, that the motion passes as in an instant to the last of them, which separates itself from the row, without one's being able to perceive that the others have been stirred [Figure 6-4]. And even that one which was used to strike remains motionless with them. Whence one sees that the movement passes with an extreme velocity which is the greater, the greater the hardness of the substance of the spheres.

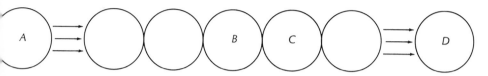

Figure 6-4. (Figure added by editor.)

But it is still certain that this progression of motion is not instantaneous, but successive, and therefore must take time. For if the movement, or the disposition to movement, if you will have it so, did not pass successively through all these spheres, they would all acquire the movement at the same time, and hence would all advance together; which does not happen. For the last one leaves the whole row and acquires the speed of the one which was pushed. Moreover there are experiments which demonstrate that all the bodies which we reckon of the hardest kind, such as quenched steel, glass, and agate, act as springs and bend somehow, not only when extended as rods but also when they are in the form of spheres or of other shapes. That is to say they yield a little in themselves at the place where they are struck, and immediately regain their former figure. For I have found that on striking with a ball of glass or of agate against a large and quite thick piece of the same substance which had a flat surface, slightly soiled with breath or in some other way, there remained round marks, of smaller or larger size according as the blow had been weak or strong. This makes it evident that these substances yield where they meet, and spring back: and for this time must be required.

I have then shown in what manner one may conceive Light to spread successively, by spherical waves, and how it is possible that this spreading is accomplished with as great a velocity as that which experiments and celestial observations demand. Whence it may be further remarked that although the particles are supposed to be in continual movement (for there are many reasons for this) the successive propagation of the waves cannot be hindered by this; because the propagation consists nowise in the transport of those particles but merely in a small agitation which cannot help communicating to those surrounding, notwithstanding any movement which may act on them causing them to be changing positions amongst themselves.

But we must consider still more particularly the origin of these waves, and the manner in which they spread. And first, it follows from what has been said on the production of Light, that each little region of a luminous body, such as the Sun, a candle, or a burning coal, generates its own waves of which that region is the centre. Thus in the flame of a candle, having distinguished the points A, B, C, concentric circles described about each of these points represent the waves which come from them. And one must imagine the same about every point of the surface and of the part within the flame.

But as the percussions at the centres of these waves possess no regular succession, it must not be supposed that the waves themselves follow

Figure 6-5. (Figure added by editor.)

one another at equal distance: and if the distances marked in the figure [Figure 6-5] appear to be such, it is rather to mark the progression of one and the same wave at equal intervals of time than to represent several of them issuing from one and the same centre.

After all, this prodigious quantity of waves which traverse one another without confusion and without effacing one another must not be deemed inconceivable; it being certain that one and the same particle of matter can serve for many waves coming from different sides or even from contrary directions, not only if it is struck by blows which follow one another closely but even for those which act on it at the same instant. It can do so because the spreading of the movement is successive. This may be proved by the row of equal spheres of hard matter, spoken of above. If against this row there are pushed from two opposite sides at the same time two similar spheres A and D, one will see each of them rebound with the same velocity which it had in striking, yet the whole row will remain in its place, although the movement has passed along its whole length twice over. And if these contrary movements happen to meet one another at the middle sphere, B, or at some other such as C, that sphere will yield and act as a spring at both sides, and so will serve at the same instant to transmit these two movements.

But what may at first appear full strange and even incredible is that the undulations produced by such small movements and corpuscles, should spread to such immense distances; as for example from the Sun or from the Stars to us. For the force of these waves must grow feeble in proportion as they move away from their origin, so that the action of each one in particular will without doubt become incapable of making itself felt to our sight. But one will cease to be astonished by considering how at a great distance from the luminous body an infinitude of waves, though they have issued from different points of this body, unite together in such a way that they sensibly compose one single wave only, which, consequently, ought to have enough force to make itself felt. Thus this infinite number of waves which originate at the same instant from all points of a fixed star, big it may be as the Sun, make practically only one single wave which may well have force enough to produce an impression on our eyes. Moreover from each luminous point there may come many thousands of waves in the smallest imaginable time, by the frequent percussion of the corpuscles which strike the Ether at these points: which further contributes to rendering their action more sensible.

The Two Theories of Light *
by Edward Uhler Condon (1955)

Three centuries ago, when men speculated about the nature of the in-fluence we call light, that which proceeds from a candle to the eye, some believed that the candle projects outward a stream of minute cor-puscles of light which fly through space and even through dense trans-parent materials such as glass. This is the corpuscular theory of light.

Others thought that the influence called light was a wave motion. These waves were supposed to travel in a space-filling material called the luminiferous ether, which was later identified with the space-filling ether presumed to be the means of propagating electric and magnetic ef-fects. This was the wave theory of light.

Newton favored the corpuscular theory, and Christian Huygens, a great Dutch contemporary of his, favored the wave theory.

The shadow of an object in light from a small source is quite sharp (Figure 6-6). There is an abrupt break between the surrounding bright-ness and the cast shadow. This is the pattern to be expected if light con-sists of a stream of particles moving out from the source in straight lines.

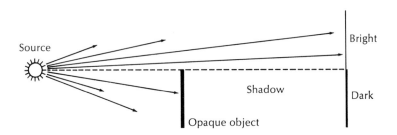

Figure 6-6. *In the corpuscular view, a shadow cast on a screen by an opaque object has a sharp edge if the light source is small or very distant.*

How waves behave when they encounter an obstacle is familiar to us from observing the ripples on a pond. Water waves are not cut sharply by an obstacle; instead, they bend partially around it and then continue, so that the wave motion can be seen on the side of the obstacle away

---

* From Edward Uhler Condon, "Physics," in James R. Newman (ed.), *What Is Science?*, Simon and Schuster, Inc., New York. Copyright © 1955 by James R. New-man. Reprinted by permission of Simon and Schuster, Inc.

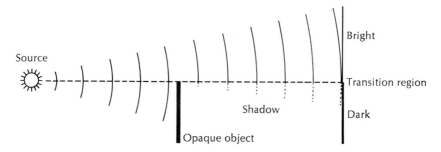

Figure 6-7. *In the wave view, the shadow does not have a sharp edge, but some light penetrates into the transition region.*

from the wave source. Thus if light were propagated by wave motion, we should expect the shadows cast by objects to have soft, diffused edges. The observed sharpness of shadow edges seems, therefore, to support the theory that light consists of a stream of particles. This course of reasoning was decisive in leading Newton to reject the wave theory (Figure 6-7).

But the problem is not so simple. Experiments performed many years after Newton revealed that the bending of waves into the region behind an obstacle is only noticeable when the obstacle is small compared with the length (distance from crest to crest measured in the direction of travel) of the waves; in the opposite case, where the wave length is small compared to the obstacle, the wave motion is inappreciable and apt to escape detection. In other words if the wave length is small compared to the object casting the shadow—the most common occurrence —only the most delicate tests will help clarify the issue between the corpuscular and the wave theory of light.

When experimental techniques and instruments had been sufficiently refined, a higher level of understanding was attained. Physicists were able to show that light really does bend somewhat into the shadowed region. The phenomenon is called diffraction. To be sure, a person who was determined to hold onto his belief in the corpuscular theory might start speculating that the material of the opaque screen exerts attractive forces on the light corpuscles that fly past near the edge of the screen and bends their trajectories into the region that otherwise would be the dark shadow. But exact and detailed studies of how much the light is bent into the shadow when passing objects of various shapes and sizes gave results that could not be accounted for in this way. On the other hand, they could be beautifully described according to the theory of

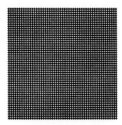
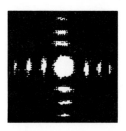

Figure 6-8. *The drawing at left shows the evenly spaced threads of the weave of silk cloth. At right is the pattern of light of a distant small source, like a street light, observed through such a closely woven cloth. The central image is round and white. The others are elongated and show rainbow colors with the red end away from the central image.*

wave propagation on the supposition that light of a particular color consists of waves of a particular wave length. Blue light in this view has a wave length of close to 4×10^{-5} cm; red light, at the other end of the visible spectrum, moves in waves about twice as long. The invisible ultra-violet radiations have a wave length shorter than the violet and the invisible infrared radiations have wave lengths longer than the red [Figure 6-1].

Figure 6-9. *The central part of the drawing shows how a very distant and small source of light appears when viewed through a small rectangular opening held at arm's length from the eyes, the rectangular opening being the shape shown in the lower right-hand corner. Note that there is a whole series of images extending in each direction and that they are more closely spaced in the direction corresponding to the long dimension of the rectangle. If the picture were executed in color, it would be seen that the more or less overlapping images of the separate blobs of light away from the central image are colored with the rainbow colors, the red end being farthest away from the central image. Such extra images and dozens of other similar experiments with light going through openings of other shapes are interpreted as indicating that light is propagated as a wave motion.*

It is the smallness of these wave lengths compared with the million times greater wave lengths for the ripples on a pond that made the diffraction effects more difficult to observe and resulted in their being hidden from the eyes of man so long. Nevertheless, now that these matters are more thoroughly understood, optical effects which directly exhibit the wave nature of light may be easily observed.

Look at a distant street light through a silk umbrella or a tightly stretched silk handkerchief held several feet in front of the eyes. Instead of seeing only the ordinary direct image of the street light, perhaps a little blurred by scattering in the fabric, you will see two additional series of images extending away from the central direct image in two perpendicular directions. The directions in which these images extend will be easily found to be along the directions of the threads of the fabric (Figures 6-8 and 6-9).

These additional images are quantitatively explainable on the wave theory of light. The essential thing here is that a wave motion spreads out from the distant street light and impinges on different parts of the fabric so that parts of the wave motion get through to the other side by going through the different regularly spaced openings on the fabric. The ideas involved may be more simply considered if we suppose that the light from the distant source goes through a screen having two very narrow and closely spaced slitlike openings in it as indicated in Figure 6-10. If one wants to go to the trouble he can make such an opaque screen

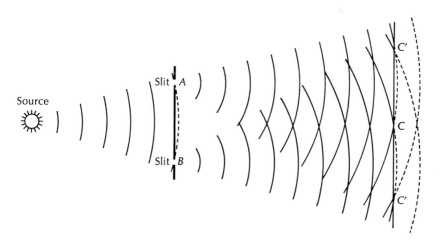

Figure 6-10. The passage of light through a screen having two slit-shaped openings.

from a densely exposed photographic plate, using a razor blade to rule two very narrow slits on it quite close together. The razor blade cuts away a narrow strip of the darkened film and allows the light to get through. When such a plate is used to view a distant light source it will be found that the direct image is also accompanied on each side by several weaker side images in directions at right angles to the direction of the rulings on the plate. Careful observation of the side images either as seen through the silk umbrella or as seen through the two-slit plate will show that the central image is white, but that the side images are colored like the rainbow, with the red end of the spectrum at the end farthest away from the central image.

Now we can return to the consideration of Figure 6-10. Suppose the screen is set up in such a way that slit A and slit B are at the same distance from source S. This is not essential but simplifies the discussion. Then the crests of waves from S will reach A and B at the same time. On the other side of the screen two new sets of waves will spread out from A and B as new centers and these will be synchronized because they are the result of feeding the new sources A and B with light coming from source S. Next we have to recognize that what affects any light-detecting device such as the eye is the resultant amplitude of all the different waves which reach it. Suppose C is a point equidistant from A and B. Then crests of waves from A will arrive at C at the same time as crests from B and will reinforce each other to make bigger waves. Thus we may expect the brightest illumination at C even though it lies within the shadow of the screen between A and B and therefore on the corpuscular view should be dark. Consider now what will be observed at a point C', a little above or below C. In moving from C to C' the distance from B is increased and the distance from A decreased. Suppose, in particular, that C' is so located that the difference between AC' and BC' is exactly half a wave length. Then at C' the crests of waves from A will fall on troughs of waves from B, and the waves will just cancel each other; in other words, assuming light is a wave motion, C' will be dark. Now move C' further up so that the difference between the distances from A and from B is equal to a whole wave length. In this case, again, the crest of each wave from A falls on the crest from B, intensifying the light. As we continue moving up C', it will alternately be light and dark as the differences between the paths is an integral multiple and an integral-and-a-half multiple of a wave length. This special property of waves of reinforcing each other or destroying each other according as crest falls on crest or crest on trough is called interference. It accounts quantitatively for the light images seen through the silk umbrella. For example the reason the red end of the spectrum is farther out than the blue is because the red wave length is greater than the blue; so it is necessary to go farther out to get a half wave length path difference with red light than it is with blue light.

During the nineteenth century a great many special cases of the behavior of light in going through obstacles having one or several holes of different shapes were analyzed carefully, both theoretically and experimentally. The results were found to be quantitatively describable in terms of the mathematical theory of wave propagation. Thus the wave theory of light became firmly established: students were taught that light is a wave motion and received good marks on examination papers for repeating this back to their professors.

However, new phenomena began to be studied which pointed once more toward the corpuscular theory. Effects . . . pointed strongly to the conclusion that a beam of light consists of a stream of particles of definite size, possessing energy and momentum content. Physicists were thus confronted with two independent sets of evidence. One set gave convincing "proof" that light is a wave motion and the other gave equally convincing "proof" that a beam of light is a stream of corpuscles. Light is emitted and absorbed as if it were a stream of particles but it moves from place to place past obstacles as if it were a wave motion. Clearly there is truth in both models, and a synthesis of the two views into a unified picture is urgently needed. This wave-particle duality is an unsolved dilemma of modern physics.

The Universe of Light °
by Sir William Bragg (1933)

Stars appear to us as bright points which move together across the sky. Astronomers have shown us long ago that in so doing they are telling us of the rotation of the earth rather than of their own movements. Very patient and skilled observation shows that there are relative motions amongst them, yet so small that the constellations have retained their form since man was first able to describe them. In the last hundred years the power of interpreting the messages brought by light has increased greatly. We are in touch with the stars. They are no longer a mere medley of bright points irregularly spread over the sky; nor even of suns that emit light like ours and are scattered through the depths of space. They have become to us parts of an active universe of which our own solar system is also a part. We can measure their distance, their weights, their luminosities, their compositions, their movements, and even estimate their historic past and their future. Our world of knowledge and perception has suddenly increased, and we begin to have understanding of its greater laws.

The very ancient science of astronomy has in times past relied almost entirely upon observations of the positions of the heavenly bodies. Light has been used as a means of measuring angles and so of plotting the apparent motions of sun, moon and stars upon a celestial sphere. The very great extension of astronomical knowledge which has occurred in the last hundred years had its beginnings in the examination of light itself and not only of the direction from which it came. The quality of the light from the sun and its intensity are characteristic of the nature and physical conditions of the source: their study provides new means of enquiry and, at the same time, gives new possibilities to the means that have been so long in use. . . .

THE ANALYSIS OF THE LIGHT FROM THE STARS

The first step in the acquirement of this knowledge is made when we use a prism or a grating to analyse the light.[1] In every case we find the same spectrum colours, running from red to violet, for our eyes perceive these alone of the complex which each star emits. But the emphasis in different parts of the spectrum is not always the same: it varies from star to star. A blue star is always hotter than a red star. The temperature of the star can in fact be determined from the position of greatest emphasis. This would not be possible were it not for a very remarkable natural law, namely, that the quality of the light emitted by a substance depends only, except in special cases, on the temperature and not upon the nature of the substance. If, for example, we gaze into a hollow in a glowing coal fire, a red hot cave with as small an opening as possible, where everything is at the same high temperature, we do not see the outlines of the pieces of coal inside, nor of pieces of metal or china if they are well inside the hole and have acquired the fire temperature. Being at the same temperature, coal, metal and china all send out radiation of the same quality and so they cannot be distinguished from one another. In a furnace where the temperature is raised by a blast the quality of the light is not the same as that of the light from a coal fire. It is whiter because the emphasis has been shifted towards the blue end of the spectrum: more blue rays are present. If a temperature of 6000° Cen-

1. The modern instrument used for this analysis is called a spectroscope. See Glossary. Ed.

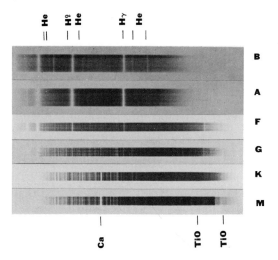

Figure 6-11. *A few star spectra showing the shift of the maximum intensities from hot white stars of class F to cooler red stars of class M.* (Yerkes Observatory)

tigrade could be reached the light would have the quality of daylight because that is the temperature approximately of the sun. Some of the stars are hotter still and the blue rays increase still more in relative intensity. Thus the temperature of a star is told by the quality of its spectrum, that is to say, by the relative distribution of energy among the various wave-lengths. . . .

Examples of star spectra are given in Figure 6-11.

A very simple experiment illustrates this connection between quality and temperature. We throw upon the screen the spectrum of the light from an electric arc, and then cut off the current. As the glow of the carbon fades and its temperature declines, we see that the spectrum disappears but the blue goes first and the red last.

SPECTRUM ANALYSIS APPLIED TO THE STARS

By a still closer observation of the spectrum of a star we can find its nature. When the light is analysed by the prism, it is found that the spectrum is, allowing for the effects of temperature, like that of any other glowing mass, as, for example, the carbon in the arc lamp, with an important exception. A number of clearly defined ranges of wave-length, generally exceedingly narrow ranges, are missing. . . . Illustrations of this peculiarity may be seen in the reproductions of spectra in Figure 6-11.

The explanation is ready to hand: We can base it on the analogies . . . [with] tuning forks, and the wireless sets. . . .

Our first point is simple enough. A single tuning fork gives out a very pure note: in other words, the sound which it emits consists of air waves of a definite frequency. This sound wave spreads out in all directions. If in one direction it passes by another tuning fork of exactly the same pitch, it will set this second fork going and of course it will spend energy in doing so. Thus the sound travelling in this special direction is weakened: there has been absorption. Nor is this loss made up by the sounding of the second fork, because the latter spreads its sound in all directions and cannot make good the loss in the particular direction. This is the basis of the absorption and production of colour. . . . It will be observed that the emitter and receiver are of the same pitch. Just so a radio station emits waves which are sharply adjusted within a certain range of frequency: if care were not taken in this respect broadcasting would be impossible. And if the receiver can be sharply tuned to the same frequency, it will absorb some of the energy distributed from the central station, which energy is diminished thereby. A wireless set can be made to radiate at least feebly, and if it does it emits waves of exactly the same pitch as that of the sending station to which it responds. Some wireless sets as we know can be tuned very sharply: if the radio station is sending out waves which adhere closely to their proper frequency such sets will only respond if the tuning is carefully done, and then the response will be strong.

In the same way an atom may be set into vibration by heating the body or gas of which it is a part. If the radiation which it emits meets a number of atoms capable of vibrating with exactly the same frequency —or it may be frequencies—its energy is absorbed in part and passes on diminished. And here again the matter of the exactness of tuning is important. An atom by itself, uninfluenced by neighbours, is like a good radio transmitter, its emitted vibrations lie within a very narrow range. We should say ranges rather than range, because an atom should be compared rather with a bell or a violin string than a tuning fork. The fork is exceedingly simple, emitting little more than one note; whereas bell and string and atom each emit a number of notes at one and the same time. Each note, however, is very sharply defined and can be tuned to with great exactness. This sharpness of tuning is characteristic of the atoms in a gas, for in this condition the atoms are independent of one another for most of the time: collisions occur, but the times during which the atoms are so close to one another as to influence each other's

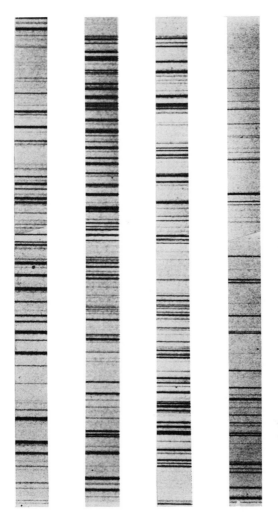

Figure 6-12. *The spectra of the luminous gases of various elements vaporized in the electric arc.* (Yerkes Observatory)

vibrations do not last long. Thus a multitude of atoms forming a gas will resemble a good radio station if they are emitting, and a good or "selective" wireless set if they are receiving.

If then a ray of light containing all frequencies passes through a crowd of such "selective" atoms there will be a sharp absorption or selection of particular frequencies, and the spectrum will show a number of narrow gaps, as illustrated in Figure 6-11. If the gas is emitting, it will send out light of the same frequencies as that which it absorbs. This effect is shown in Figure 6-12.

THE SPECTRUM OF THE SUN

The spectrum of the sun shows an immense number of these sharply defined gaps or "lines" as they are generally called. When the temperatures of various substances, iron, calcium, hydrogen, etc., are raised by heating until they become luminous vapours, it is found that they emit, each of them, light of a certain number of sharply defined frequencies. It is also found that these correspond line by line to the lines in the spectrum of the sun. The conclusion is obvious: the light from the sun must have gone through clouds of these atoms somewhere, and in respect to such substances as iron or calcium, or most other elements, this must have happened on the sun because in no other part of the path of the light has there been enough heat to bring the substances to the state of a luminous gas. If, therefore, the spectra of the sun and stars and other luminous bodies be analysed and compared with the spectra emitted in the laboratory by the various elements brought to the condition of luminous gases, it becomes possible at once to say whether or no these elements exist in the heavenly bodies as well as on the earth. These investigations were first carried out on the great scale by the enthusiasm of Huggins, Lockyer and their contemporaries. From that time to this they have increased in interest and importance because it has been found that there is far more in them than mere correspondences between the positions of definite frequencies in the two classes of spectra. The frequencies characteristic of each element depend on the condition of the element, which may be stripped of one or more of its electrons and at each such alteration may alter its notes: the condition depends on the temperature and the density of the gas, and thus the analysis of stellar spectra gives information of the state of the star as well as of its composition.

The elements that are found on the earth are found also in the stars. In one notable case the existence of an element was proved by Lockyer's observation of the sun's spectrum before it had been handled in the laboratory. Certain well marked lines were found which corresponded to nothing then known, and Lockyer deduced the presence of some new substance which he called "helium": it was not until long afterwards that it was extracted from cleveite and other minerals. It is now well known as the best gas for the inflation of dirigible balloons, since it is almost as much lighter than air as hydrogen is, and it is not inflammable. In the laboratory it is an object of extraordinary interest, since the atom of helium is the alpha particle which radioactive substances emit. . . .

But now there is one very important question to which some answer must be found. We have supposed that a radiation is emitted by the sun, which would have given a complete spectrum, were it not that on its way to us it has passed through an atmosphere surrounding the sun and containing the various elements in the form of gases. These gases themselves must be luminous and be emitting light of the very frequencies which we suppose them to have been absorbing and therefore causing black lines to appear in the spectrum. What is the origin of the original radiation? It must be so plentiful that its absorption in the sun's atmosphere makes a difference which the luminosity of the atmosphere itself does not make good. And its energy is distributed all over the spectrum: it is not limited to a number of definite frequencies.

We have already seen that independent atoms would issue their proper frequencies exactly. But they are not wholly independent: they spend a certain fraction of their time in each other's neighbourhood when their motions cause them to collide. During those moments they do not issue their proper frequencies so strictly. If they are so crowded together as to spend a large part of their time under each other's influence the ranges of proper frequency will spread until they meet and overlap and the whole spectrum is emitted. In a roughly analogous way, if a number of tuning forks, each on its stand, were thrown anyhow into a box which was then shaken violently, a medley of jangling sound would issue which would no longer be confined to the particular notes of the tuning forks. In the body of the sun the atoms are crowded together and violently agitated by the tremendous temperature: and it is in this way that we can account for the continuity of the spectrum. The somewhat cooler atmosphere surrounding the sun removes certain frequencies and hence the spectrum lines. . . .

When there is a total eclipse of the sun the incandescent vapours of its atmosphere can be seen to extend far beyond the obscured disc [Figure 6-13]: and their spectra consists of a series of bright lines as we should expect.

It is an interesting and important point that a luminous gas which gives a "bright line" spectrum will give a full spectrum if there is enough of the gas. There is always a certain amount of general radiation as well as that of particular frequencies of the atoms of which the gas is made. Now the latter is absorbed in going through the gas more than the former, and so it happens that it comes only from the external layers, while the former, the general radiation, can come from greater

Figure 6-13. *The sun's corona.* (Yerkes Observatory. Figure added by editor.)

depths. If the depth of the gas is great enough the two effects compensate so as to give equal weight to the two varieties of radiation, and the spectrum is complete. It becomes that which is characteristic of the temperature and is independent of the nature of the emitting body. Thus the completeness of the spectrum of a heavenly body does not mean that it is a solid, but merely that there is enough of it.

THE MOTIONS OF THE STARS TO AND FROM THE EARTH

By another form of careful observation it is possible to determine the rate at which a star is approaching or receding from the earth. For if, say, the star is rushing towards the earth it is crushing up the waves from the rear and all the wave-lengths are artificially shortened. . . . Just so an approaching motorcar crushes up the sound waves, and, as it recedes, draws them out [Figure 6-14]. All day the motorcars go by and the pitches of all the noises made by a car drop at the moment of passing. If the car is travelling at 25 miles an hour the drop is nearly a semi-tone. If the star gave a complete spectrum free from lines we should not be able to detect this shift, because the wave-lengths that gave the extreme violet would have been shortened and become invisible, while at the red end wave-lengths just too long to be visible would now be shortened and enter the red. All the wave-lengths would be shortened somewhat, both visible and invisible: but the visible colours would be supplied to our eyes just the same, and we should see no change. There is however an observable shift in the lines of the spectrum, an effect which was first described by Doppler. And as the positions of the lines can be very exactly measured, and as the shift is proportional to the speed of the star relative to the earth, it becomes possible to measure the relative motion of the two bodies along the line joining them. . . .

These measurements are independent of distance and so arises the curious situation that, while most of the stars are too far away to be seen moving across the sky, motions of approach or recession can be measured with considerable accuracy. In this way, for example, it is found that the distant nebulae, or at least the few that give enough light for the experiment, are receding: and apparently those that are furthest away are going the fastest, whence the modern conception of an expanding universe. Many stars approach and recede at regular intervals: and it is clear that each such star has a dark companion about which it is revolving. . . . [We] have seen how full of information are the characteristics of the light that comes to us from all parts of the universe. . . .

THE WAVE AND THE CORPUSCLE

A range of ether waves is at our disposal for experiment. Certain of them which lie within a narrow range are visible to our eyes: others may be detected by their action on a photographic plate. . . .

There is one particular phenomenon which is shown markedly by the

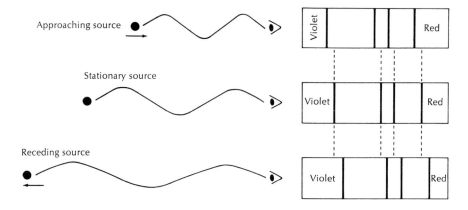

Figure 6-14. *The Doppler effect. The lines in the spectrum shift toward the violet end* (left) *when the light source is approaching. When the source recedes, the spectral lines shift toward the red end* (right). *(Figure added by editor.)*

radiations of the shortest wave-lengths, and to a less observable extent by the longer waves: it is known as the photo-electric effect. At this point the wave theory which has helped us so far fails to suggest an explanation. It is this fact and others associated with it that show our wave hypothesis to be incomplete, and lead to the curious position of physics at the present time. In 1905 Einstein suggested that the corpuscular theory had been set aside too hastily.

The Two Theories of Light (continued) *
by Edward Uhler Condon (1955)

Before considering detailed effects, let it be observed that on the wave view the intensity of light is connected with the amplitude of swing of the waves; to be specific, it is proportional to the square of the wave amplitude. As a wave spreads out this becomes continuously smaller with consequent reduction in the intensity of the light (Figure 6-15). On the corpuscular view the energy in the light beam is the total energy of a large number of independently moving corpuscles, called light quanta in modern terminology. The energy carried by each quantum is always the

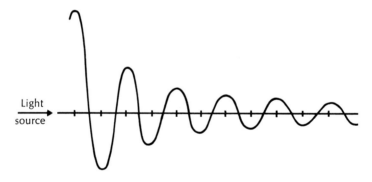

Figure 6-15. *The amplitude of a wave dies off inversely to the distance from the source as the wave spreads out in all directions. The intensity or brightness of the light is proportional to the square of the amplitude and so falls off with the inverse square of the distance from the source.*

* From Edward Uhler Condon, "Physics," in James R. Newman (ed.), *What Is Science?*, Simon and Schuster, Inc., New York. Copyright © 1955 by James R. Newman. Reprinted by permission of Simon and Schuster, Inc.

same—the decrease in intensity of the beam is a consequence only of the decrease in the number of particles crossing unit area in unit time. On one view the energy is continuously spread out over all the space occupied by the light beam; on the other it is localized in small bundles occupying a part of the total space, just as the molecules of a gas at ordinary pressure occupy only about one billionth of the total space which is filled by the gas.

The photoelectric effect gave the first clue to the corpuscular nature of light emission and absorption. If a piece of metal, insulated from the ground, is charged with negative electricity, the charge will leak off through the air when light, especially violet and ultraviolet light, shines on it. In this simple version of the experiment the effects are complicated by the action of the air layers which are absorbed on the metal.[1] Consequently, physicists who wish to study the photoelectric effect enclose the metal in a high-vacuum tube from which the gas can be removed to an extremely low pressure.

The amount of charge which is set free from a metal surface is proportional to the intensity of the light falling on it. This fact is explainable either on wave or corpuscular views. But it is found that the leakage of negative charge from the metal comes about through the ejection by the light of negatively charged electrons. Measurements of the speed of ejection of these electrons show that the energy with which they are ejected is approximately inversely proportional to the wave length of the light used. Actually the energy of motion of the liberated electrons is a little less than such a rule would imply. Einstein in 1905 pointed out that this experimental result could be most simply understood if one supposed that the light energy was absorbed in definite quanta of energy, the energy of one quantum going to the individual act of liberating one electron. The energy of one quantum was taken to be hc/λ, where h is a coefficient known as Planck's constant, c is the velocity of light and λ is the wave length of the light. The observed energy of motion of the freed electron is less than this by an approximately constant amount because a part of the energy supplied by the light quantum is used up in releasing the electron from the attractive forces which normally bind it within the metal.

Similar effects are observed when X-rays are used instead of light to

1. Although qualitative effects are easily demonstrated with the experiment done in air, exact measurements of the motion of the electrons ejected by the light require that the metal be enclosed in a vacuum tube, both because of chemical actions of the gas of the air, especially moisture, on the metal which strongly change its ability to emit electrons, and also because if gas is present the electrons very quickly become scattered and attached to gas molecules, making it difficult to observe the speed with which they are thrown out.

cause the ejection of electrons. The quantum effects are even more pronounced because the X-ray wave lengths are approximately ten thousand times smaller than those of visible light and therefore the energy in an X-ray quantum is some ten thousand times greater than in a visible light quantum. But because the wave lengths are so much smaller, it was for a long time impossible to observe, in the case of X-rays, the diffraction and interference effects which give us our most decisive criterion for the wave nature of visible light.

It happens, however, that the wave length of X-rays is just about the same as the distance between the layers of atoms in the regular arrangements that occur in crystals such as those of quartz or rocksalt [Figure 6-16]. This means that interference effects such as those described for light going through the silk umbrella are observed when a beam of X-rays is scattered by falling on a crystal. Instead of the scattered radiations going out from the crystal more or less equally in all directions, they are scattered only in certain specific directions. These directions are determined by relations between the X-ray wave length and the spacing of atoms in the crystal. Thus the effect can be used with a known crystal to determine the wave length of X-rays from a variety of sources. Or it can be used the other way round—to discover with X-rays of known wave lengths exact information about the arrangement of atoms in crystals whose geometry is not known. This diffraction of X-rays by the orderly arrangement of atoms in crystals was discovered in 1912 by Max Von Laue. In such work the scattering is effected by elec-

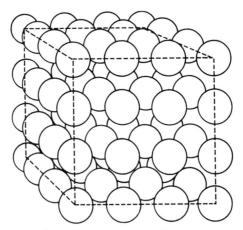

Figure 6-16. *Schematic drawing of atoms in a crystal.* (Figure added by editor.)

trons that are relatively securely bound to the atoms of which the crystal is made.

In 1922 Arthur H. Compton discovered another characteristic of the scattering of X-rays by matter. He noticed, particularly when high energy X-rays are scattered by lighter elements such as carbon, that the wave length of the deflected X-ray is somewhat longer than that of the X-ray before deflection, and that the amount of this Compton-shift toward longer wave length increases steadily from zero as the angle of deflection increases. Compton showed that this behavior was exactly what is to be expected from the quantum or corpuscular theory. If the X-rays are tied up in quanta of energy, it is natural to suppose that these quanta also carry definite amounts of momentum. When they strike an almost free or loosely bound electron and are scattered, they rebound from it, Compton supposed, according to the same mechanical laws as those governing the collisions of billiard balls. At a very glancing blow the X-ray is only slightly deflected in direction and gives up very little energy to the electron with which it collided. When the blow is more central it is deflected through a wider angle and gives up more energy to the struck electron, losing thereby more of its own energy and thus having its wave length increased.

This completed the duality. X-rays which—in respect of their being scattered by the tightly bound electrons in the regularly spaced atoms of a crystal—behave like a train of waves of a definite wave length, also behave—in respect of their scattering by loosely bound electrons—as if they were little bundles of energy and momentum rather than continuously spread out waves, these little bundles or quanta obeying the ordinary laws of collision mechanics when they collide with an electron. In order to describe all of the phenomena it is necessary to talk some of the time as if the X-rays are a wave motion and some of the time as if they are a stream of particles.

There used to be a joke current among physicists that X-rays were to be regarded as waves on Mondays, Wednesdays and Fridays and as quanta on the other days. In point of fact, however, the two aspects are so closely intertwined that no such neat separation of viewpoints can be effected; in Compton's experiments, for example, the scattering in carbon which is described by quantum or corpuscular language is followed by another scattering of the same X-rays by a crystal for the purpose of demonstrating that their wave length was altered in the first scattering.

Probability and Uncertainty *
by Sir James Jeans (1943)

THE PARTICLE- AND WAVE-PICTURES

We now have two distinct pictures of the nature of radiation, one depicting it as particles and the other as waves. The particle-picture is obviously the more suitable when the radiation is falling on matter, and the wave-picture when it is travelling through space. For a time there was a disposition to conclude that light must consist of two parts, a wave part and a particle part, but it is now clear that this is not so. The wave-picture and the particle-picture do not show two different things, but two aspects of the same thing. They are simply partial pictures which are appropriate to different sets of circumstances— . . . —and so are complementary but not additive. As soon as light shows the properties of particles, its wave properties disappear, and vice versa; the two sets of properties are never in evidence at the same time. Thus as we follow a beam of light, or even a single quantum, in its course, we must imagine the wave- and particle-pictures taking control of the situation alternately.

The wave-picture explains much in its own proper province, but it brings its own difficulties with it. In particular, it is not easy to pass back from the wave-picture to the particle-picture. For all waves scatter as they travel through space, and it is difficult to imagine how waves which have once scattered as the undulatory theory directs can recombine and concentrate their attack on single molecules or electrons in the way they are observed to do as soon as they encounter matter.

WAVES OF PROBABILITY

Let us [consider the] . . . imaginary experiment in which a single quantum of radiation is emitted from a source of light to fall on one point or another of a system of distant screens. We know that the whole energy of the quantum will concentrate on a single point of the screens, but which point will it be?

The obvious answer is that sometimes it will be one point, and sometimes another. It cannot always be the same point, or else when quanta

* From Sir James Jeans, *Physics and Philosophy*. Cambridge University Press, New York, 1943. Reprinted by permission.

were being emitted in millions, this one specially favoured point would be intensely bright and all others completely dark. Actually when quanta are being emitted in millions, there are some places on the screens at which the illumination is very bright, these indicating regions in which many photons have struck, and also places of less illumination, these indicating regions in which few photons have struck. Even the most faintly illuminated parts of the screen must have been struck by some photons.

If we now fix our attention on a single quantum of radiation of which we know nothing except that it belonged to the original beam, we may say that the extent to which either screen is illuminated at any point gives a measure of the *probability* that the quantum shall condense into a photon at this point. In this way we may interpret the waves of the undulatory theory as waves of probability; the extension of the wave system in space marks out the region within which a photon may be supposed to be travelling, while the intensity of the waves at each point within this region gives a measure of the probability that a photon will occur at that point if matter is placed there.

When half a million babies are born in England in a year, we may say that 20 per cent of them are born in London, 2 per cent in Manchester, 1 per cent in Bristol, and so on. But when we think of the one baby born in a single minute of time, we cannot say that 20 per cent of it is born in London, 2 per cent in Manchester, and so on. We can only say that there is a 20 per cent probability of its being born in London, a 2 per cent probability of its being born in Manchester, and so on. If we disregard variations of birth-rate with locality, a map exhibiting the density of population in different parts of England will also act as a chart showing the number of births per annum; but with reference to the birth occurring at any one instant, it merely shows the relative probabilities of the baby appearing in different areas. As soon as the waves of the undulatory theory fall on matter, they provide a precisely similar chart for the probability of photons appearing in the different areas of the matter. The waves, then, are again mental constructs—not enabling us to see what *will* happen, but what *may* happen.

WAVES OF KNOWLEDGE

The waves may equally well be interpreted as representations of our knowledge. In the experiment with the single photon, we do not know where the photon is, but the wave-picture gives a sort of diagrammatic representation of what we do know. We know that the photon must be within a certain region of space, this being the region mapped out by the waves at each instant. We may know that it is much more likely to be in a region A than in some other region B; if so, the waves represent this knowledge by being much more intense in the region A than in the region B, and so on. . . .

Before the quantum theory appeared, the principle of the uniformity of nature—that like causes produce like effects—had been accepted as a universal and indisputable fact of science. As soon as the atomicity of radiation became established, this principle had to be discarded.

In the experiment [just] described, the uniformity of nature would have required that every photon should hit the screen at the same point. Actually we have seen that they hit at different points, so that if a single quantum is discharged from the source several times in succession, different experiments will be found to give different results, and this although the conditions before the experiments were, so far as we could tell, precisely identical.

In this way we find that the atomicity of radiation destroys the principle of the uniformity of nature, and the phenomena of nature are no longer governed by a causal law—or at least if they are so governed, the causes lie beyond the series of phenomena as known to us. If, then, we wish to picture the happenings of nature as still governed by causal laws, we must suppose that there is a substratum, lying beyond the phenomena and so also beyond our access, in which the happenings in the phenomenal world are somehow determined. . . .

. . . Physics sets before itself the task of coordinating the various sense-data which reach us from the world beyond our sense-organs. If our senses could receive and measure infinitely delicate sense-data, we should be able in principle to form a perfectly precise picture of this outer world. Our senses have limitations of their own, but these can to a large extent be obviated by instrumental aid; telescopes, microscopes, etc. exist to make good the deficiencies of our eyes. But there is a further limitation which no instrumental aid can make good; it arises from the circumstances that we can receive no message from the outer world smaller than that conveyed by the arrival of a complete photon. As these photons are finite chunks of energy, infinite refinement is denied us; we have clumsy tools at best, and these can only make a blurred picture. It is like the picture a child might make by sticking indivisible wafers of colour onto a canvas. We might think we could avoid this complication by using radiation of infinite wave-length. For the quanta of this radiation have zero energy, and so might be expected to provide infinitely sensitive probes with which to explore the outer world. And so they do, so long as we only want to measure energy, but a true picture of the outer world will depend also on the exact measurement of lengths and positions. For this, long-wave quanta are useless. To measure a length accurately to within a millionth of an inch, we must have a measure graduated to millionths of an inch; a yardstick graduated only to inches is useless. Now quanta of one inch wave-length are, in a sense, graduated only to inches, while quanta of infinite wave-length are not graduated at all. Passing from quanta of short wave-length to quanta of long wave-length only shifts, but does not remove, the difficulty.

A rough analogy is to be found in the problem of photographing a rapidly moving object. A sensitized film can record no detail on a scale which is smaller than the grain of the film, so that if we use a large-grained film, all the fine detail of our picture will be blurred. If we try to escape this difficulty by using a film of very small grain, we merely cross over from Scylla to Charybdis; the speed of the film is now reduced so much that the picture is blurred through its subject having moved appreciably during the time necessary for exposure.

SUBJECT AND OBJECT

It used to be supposed that in making an observation on nature, as also in the more general activities of our everyday life, the universe could be supposed divided into two detached and distinct parts, a perceiving subject and a perceived object. Psychology provided an obvious exception, because the perceiver and perceived might be the same; subject and object might be identical, or might at least overlap. But in the exact sciences, and above all in physics, subject and object were supposed to be entirely distinct, so that a description of any selected part of the universe could be prepared which would be entirely independent of the observer as well as of the special circumstances surrounding him.

The theory of relativity (1905) first showed that this cannot be entirely so; the picture which each observer makes of the world is in some degree subjective. . . .[1]

The theory of quanta carries us further along the same road. For every observation involves the passage of a complete quantum from the observed object to the observing subject, and a complete quantum constitutes a not negligible coupling between the observer and the ob-

1. The theory of relativity will be taken up in Part 7. Ed.

served. We can no longer make a sharp division between the two; to try to do so would involve making an arbitrary decision as to the exact point at which the division should be made. Complete objectivity can only be regained by treating observer and observed as parts of a single system; these must now be supposed to constitute an indivisible whole, which we must now identify with nature, the object of our studies. It now appears that this does not consist of something we perceive, but of our perceptions; it is not the object of the subject-object relation, but the relation itself. But it is only in the small-scale world of atoms and electrons that this new development makes any appreciable difference; our study of the man-sized world can go on as before.

For instance, when an astronomer is observing the motion of a planet in the solar system, it is emitting millions of quanta every second, some of which pass through the telescope of the astronomer and into his eye. By noting the directions from which these arrive, he can follow and describe the motion of the planet across the sky. With the departure of each quantum, the planet suffers a recoil which changes its motion, but the changes are so minute that they may properly be disregarded. But it is different when a physicist tries to follow the motion of an electron inside an atom. He can only obtain knowledge of the internal state of the atom by causing it to discharge a full quantum of radiation, and . . . the emission of a quantum of radiation is so atom-shaking an event that the whole motion of the atom is changed, and the result is practically a new atom. A succession of quanta may give scraps of information about various stages of the atom, but can give no record of continuous change. In fact there can be no continuous change to record, since every departure of a quantum breaks the continuity.

For this reason it is futile to discuss whether the motion of the atom conforms to a causal law or not. The mere formulation of the law of causality presupposes the existence of an isolated objective system which an isolated observer can observe without disturbing it. The question is whether he, noticing that such a system is in a certain state at one instant, can or cannot foretell that it will be in some other specifiable state at some future instant. But if there is no sharp distinction between observer and observed, this becomes meaningless since any observation he makes must influence the future course of the system.

In more general terms, we may say that the law of causality acquires a meaning for us only if we have infinitesimals at our disposal with which to observe the system without disturbing it. When the smallest instruments at our disposal are photons and electrons, the law of causality becomes meaningless for us, except with reference to systems containing immense numbers of photons and electrons. For such systems the classical mechanics has already told us that causality prevails; for other systems causality becomes meaningless so far as our knowledge of the system is concerned if it controls the pattern of events, we can never know it.

We have now seen that six important consequences follow from the mere fact of the atomicity of radiation, coupled with those well-established facts of the undulatory theory of light that have been mentioned. These are:

(1) So far as the phenomena are concerned, the uniformity of nature disappears.

(2) Precise knowledge of the outer world becomes impossible for us.

(3) The processes of nature cannot be adequately represented within a framework of space and time.

(4) The division between subject and object is no longer definite or precise; complete precision can only be regained by uniting subject and object into a single whole.

(5) So far as our knowledge is concerned, causality becomes meaningless.

(6) If we still wish to think of the happenings in the phenomenal world as governed by a causal law, we must suppose that these happenings are determined in some substratum of the world which lies beyond the world of phenomena, and so also beyond our access.

Have We Reached the Limit of Vision?

Uncertainty °

by Lincoln Barnett (1948, 1957)

While quantum physics thus defines with great accuracy the mathematical relationships governing the basic units of radiation and matter, it seems to obscure our picture of the true nature of both. Most modern physicists, however, consider it rather naïve to speculate about the true nature of anything. They are "positivists"—or "logical empiricists"—who contend that a scientist can do no more than report his observations. And so if he performs two experiments with different instruments and one seems to reveal that light is made up of particles and the other that light is made up of waves, he must accept both results, regarding them not as contradictory but as complementary. By itself neither concept suffices to explain light, but together they do. Both are necessary to describe reality and it is meaningless to ask which is really true. For in the abstract lexicon of quantum physics there is no such word as "really."

It is futile, moreover, to hope that the invention of more delicate tools may enable man to penetrate much farther into the microcosm. There is an indeterminacy about all the events of the atomic universe which refinements of measurement and observation can never dispel. The element of caprice in atomic behavior cannot be blamed on man's coarse-grained implements. It stems from the very nature of things, as shown by Heisenberg in 1927 in a famous statement of physical law known as the "Principle of Uncertainty." To illustrate his thesis Heisenberg pictured an imaginary experiment in which a physicist attempts to observe the position and velocity [1] of a moving electron by using an immensely powerful supermicroscope. Now, as has already been suggested, an individual electron appears to have no definite position or velocity. A physicist can define electron behavior accurately enough so long as he is

° From Lincoln Barnett, *The Universe and Dr. Einstein*, 2d ed., William Sloane Associates, New York, 1957. Copyright 1948 by Harper & Row, Publishers, New York; copyright 1957 by Lincoln Barnett. Reprinted by permission of William Sloane Associates.

1. In physics the term "velocity" connotes direction as well as speed.

dealing with great numbers of them. But when he tries to locate a particular electron in space the best he can say is that a certain point in the complex superimposed wave motions of the electron group represents the *probable* position of the electron in question. The individual electron is a blur—as indeterminate as the wind or a sound wave in night —and the fewer the electrons with which the physicist deals, the more indeterminate his findings. To prove that this indeterminacy is a symptom not of man's immature science but of an ultimate barrier of nature, Heisenberg presupposed that the imaginary microscope used by his imaginary physicist is optically capable of magnifying by a hundred billion diameters—i.e., enough to bring an object the size of an electron within the range of human visibility. But now a further difficulty is encountered. For inasmuch as an electron is smaller than a light wave, the physicist can "illuminate" his subject only by using radiation of shorter wave length. Even X-rays are useless. The electron can be rendered visible only by the high-frequency gamma rays of radium. But the photoelectric effect, it will be recalled, showed that photons of ordinary light exert a violent force on electrons; and X-rays knock them about even more roughly. Hence the impact of a still more potent gamma ray would prove disastrous.

The Principle of Uncertainty asserts therefore that it is impossible with any of the principles now known to science to determine the position and the velocity of an electron at the same time—to state confidently that an electron is "right here at this spot" and is moving at "such and such a speed." For by the very act of observing its position, its velocity is changed; and, conversely, the more accurately its velocity is determined, the more indefinite its position becomes. And when the physicist computes the mathematical margin of uncertainty in his measurements of an electron's position and velocity he finds it is always a function of that mysterious quantity—Planck's Constant, *h*.

Quantum physics thus appears to shake two pillars of the old science, causality and determinism. For by dealing in terms of statistics and probabilities it abandons all idea that nature exhibits an inexorable sequence of cause and effect between individual happenings. And by its admission of margins of uncertainty it yields up the ancient hope that science, given the present state and velocity of every material body in the universe, can forecast the history of the universe for all time. One by-product of this surrender is a new argument for the existence of free will. For if physical events are indeterminate and the future is unpredictable, then perhaps the unknown quantity called "mind" may yet guide man's destiny among the infinite uncertainties of a capricious universe. Another conclusion of greater scientific importance is that in the evolution of quantum physics the barrier between man, peering dimly through the clouded windows of his senses, and whatever objective reality may exist has been rendered almost impassable. For whenever he attempts to penetrate and spy on the "real" objective world, he changes and distorts its workings by the very process of his observation. And when he tries to divorce this "real" world from his sense perceptions he is left with nothing but a mathematical scheme. He is indeed somewhat in the position of a blind man trying to discern the shape and texture of a snowflake. As soon as it touches his fingers or his tongue it dissolves. A wave electron, a photon, a wave of probability, cannot be visualized; they are simply symbols useful in expressing the mathematical relationships of the microcosm.

To the question, why does modern physics employ such esoteric methods of description, the physicist answers: because the equations of quantum physics define more accurately than any mechanical model the fundamental phenomena beyond the range of vision. In short, *they work*, as the calculations which hatched the atomic bomb spectacularly proved. The aim of the practical physicist, therefore, is to enunciate the laws of nature in ever more precise mathematical terms. Where the nineteenth century physicist envisaged electricity as a fluid and, with this metaphor in mind, evolved the laws that generated our present electrical age, the twentieth century physicist tries to avoid metaphors. He knows that electricity is not a physical fluid, and he knows that such pictorial concepts as "waves" and "particles," while serving as guideposts to new discovery, must not be accepted as accurate representations of reality. In the abstract language of mathematics he can describe how things behave though he does not know—or need to know—what they are.

Yet there are present-day physicists to whom the void between science and reality presents a challenge. Einstein more than once expressed the hope that the statistical method of quantum physics would prove a temporary expedient. "I cannot believe," he wrote, "that God plays dice with the world." He repudiated the positivist doctrine that science can only report and correlate the results of observation. He believed in a universe of order and harmony. And he believed that questing man may yet attain a knowledge of physical reality.

Editor's Note. The picture of the universe as a giant mechanism inexorably obeying the laws of motion and gravity was a direct consequence of Newton's work and dominated nineteenth-century thought. Pierre de Laplace crystallized this mechanistic world view in a passage which has had great influence on both science and philosophy.

Concerning Probability *
by Pierre Simon de Laplace (1814)

All events, even those which on account of their insignificance do not seem to follow the great laws of nature, are a result of it just as necessarily as the revolutions of the sun. In ignorance of the ties which unite such events to the entire system of the universe, they have been made to depend upon final causes or upon hazard, according as they occur and are repeated with regularity, or appear without regard to order; but these imaginary causes have gradually receded with the widening bounds of knowledge and disappear entirely before sound philosophy, which sees in them only the expression of our ignorance of the true causes.

Present events are connected with preceding ones by a tie based upon the evident principle that a thing cannot occur without a cause which produces it. This axiom, known by the name of *the principle of sufficient reason,* extends even to actions which are considered indifferent; the freest will is unable without a determinative motive to give them birth; if we assume two positions with exactly similar circumstances and find that the will is active in the one and inactive in the other, we say that its choice is an effect without a cause. It is then, says Leibnitz, the blind chance of the Epicureans. The contrary opinion is an illusion of the mind, which, losing sight of the evasive reasons of the choice of the will in indifferent things, believes that choice is determined of itself and without motives.

We ought then to regard the present state of the universe as the effect of its anterior state and as the cause of the one which is to follow. Given for one instant an intelligence which could comprehend all the forces by which nature is animated and the respective situation of the beings who compose it—an intelligence sufficiently vast to submit these data to analysis—it would embrace in the same formula the movements of the greatest bodies of the universe and those of the lightest atom; for it, nothing would be uncertain and the future, as the past, would be present to its eyes. The human mind offers, in the perfection which it has been able to give to astronomy, a feeble idea of this intelligence. Its dis-

* From Pierre Simon de Laplace, *Philosophical Essays on Probability,* John Wiley & Sons, Inc., New York, 1956. Reprinted by permission.

coveries in mechanics and geometry, added to that of universal gravity, have enabled it to comprehend in the same analytical expressions the past and future states of the system of the world. Applying the same method to some other objects of its knowledge, it has succeeded in referring to general laws observed phenomena and in foreseeing those which given circumstances ought to produce. All these efforts in the search for truth tend to lead it back continually to the vast intelligence which we have just mentioned, but from which it will always remain infinitely removed.

The Principle of Uncertainty *
by Jacob Bronowski (1953)

The principle of uncertainty . . . shook us all a good deal. After all, it said that nature could not be described as a rigid mechanism of causes and effects. And I recall again that all the successes of science, Newton's success and those of the nineteenth century, seemed to have been won hitherto by fitting nature with just this kind of machine. To say suddenly that at bottom these causal chains are not true, that the whole thing cannot be done—that seemed a strange discovery, and a disagreeable one.

It was a discovery, and it has had a profound effect. But it does not seem nearly so strange or unsettling now. On the contrary, to my generation the principle of uncertainty seems the most natural and sensible remark in the world. It does not seem to us to have taken the order out of science. It has taken out the metaphysics and left what had long been forgotten, the scientific purpose.

The purpose of science is to describe the world in an orderly scheme or language which will help us to look ahead. We want to forecast what we can of the future behaviour of the world; particularly we want to forecast how it would behave under several alternative actions of our own between which we are usually trying to choose. This is a very limited purpose. It has nothing whatever to do with bold generalisations about the universal workings of cause and effect. It has nothing to do with cause and effect at all, or with any other special mechanism. Nothing in this purpose, which is to order the world as an aid to decision and action, implies that the order must be of one kind rather than another. The order is what we find to work, conveniently and instructively. It is not

* From Jacob Bronowski, *The Common Sense of Science*, Harvard University Press, Cambridge, 1953. Reprinted by permission of Harvard University Press.

something we stipulate; it is not something we can dogmatise about. It is what we find; it is what we find useful.

In order to act, it is not necessary to have a metaphysical belief that the rules by which we are acting are universal and that all other rules are just like them. On the contrary, at bottom all general beliefs of this kind are at odds with the principles of science. Laplace believed that if we knew the present completely, we could completely determine the future. This belief had some political and religious force for Frenchmen of the Revolution. But it has no scientific meaning at all. It does not resemble a scientific statement, or for that matter a literary one, because it is not a statement about reality, either now or in the future. There simply is no sense in asserting what would happen if we knew the present completely. We do not, and plainly we never can.

This is precisely what the principle of uncertainty says to modern physics. It makes no assertions at all about whether we could or could not predict the future of an electron, supposing that we knew this or that about its present. It simply points out that we cannot completely know its present. For instance, we can know either its whereabouts or its speed with high precision; but we cannot know both. And, in consequence, we cannot predict its future.

At bottom then, the principle of uncertainty states in special terms what was always known, which is this. Science is a way of describing reality; it is therefore limited by the limits of observation; and it asserts nothing which is outside observation. Anything else is not science; it is scholastics. The nineteenth century was dominated by Laplace's belief that everything can be described by its causes. But this is no less scholastic than the medieval belief, that everything is contained in the First Cause.

At this stage those to whom causality is second nature are tempted to open a new line of retreat. Why, they say, should we not go on believing in a strictly determined nature anyway? Why must we say that some future event is not determined, simply because science says that it cannot be predicted? Even suppose, as science now insists, that this is not merely a momentary gap. Even suppose that scientists are right and that they will never discover new laws which will make them able to predict these small events. Granted all this, say the doubters: granted that there are material events which can be shown to be unpredictable by any scientific method at all now or in the future. Is that so profound a discovery? Is it indeed a discovery about anything but science itself? Is it more than a demonstration that the methods of science have shortcomings and are limited in scope? Why must we assume that because science cannot uncover the network of cause and effect in nature, therefore this network is not there? After all, even Laplace never supposed that any human being really could calculate the future conclusively from the present, in practice. He was perfectly aware of the practical limitations to scientific prediction. Why cannot we go on holding his view then, that the future is determined in theory, whether scientists in practice can predict it or not?

Alas, these winning and ingenuous remarks quite miss the point. Of course Laplace did not believe that the future could be punched out from the present on any calculating machine which men could build in practice. But he believed that in principle it could be done, if not by a human then by a superhuman computer. He believed that the future is fully and finally determined. The future as it were already exists in the mathematics; and the world itself is precisely a machine which calculates it by strict mechanical processes.

This is quite different from our own picture of the relation between present and future. Indeed, we would not be able to begin to set up the present on any universal machine like Laplace's, for two reasons. One is that Relativity has pointed out the difficulties in defining the present instant at two points which are far apart in space. And the other is that the principle of uncertainty has made it clear that even at one point the present cannot be defined with unlimited accuracy.

Does Physics Need Pictures? *

by Erwin Schrödinger (1935 and 1947)

Discoveries in physics cannot in themselves—so I believe—have the authority of forcing us to put an end to the habit of picturing the physical world as a reality.

I believe the situation is this. We have taken over from previous theory the idea of a particle and all the technical language concerning it. This idea is inadequate. It constantly drives our mind to ask for information which has obviously no significance. Its imaginative structure exhibits features which are alien to the real particle. An adequate picture must not trouble us with this disquieting urge; it must be incapable of picturing more than there is; it must refuse any further addition. Most people seem to think that no such picture can be found. One may, of course, point to the circumstantial evidence (which I am sorry to say is

* From Erwin Schrödinger, *Science, Theory and Man*. Dover Publications, Inc., New York, 1958, and *On the Peculiarity of the Scientific World View*, Cambridge University Press, New York, 1947. Reprinted by permission.

not changed by this essay) that in fact none has been found. I can, however, think of some reasons for this, apart from the genuine intricacy of
the case. The palliative, taken from positivist philosophy and purporting
to be a reasonable way out, was administered early and authoritatively.
It seemed to relieve us from the search for what I should call real understanding; it even rendered the endeavour suspect, as betraying an
unphilosophical mind—the mind of a child who regretted the loss of its
favourite toy (the picture or model) and would not realize that it was
gone for ever. . . .

To-day we are told . . . that we must not expect more from our science than those ad nauseam reiterated prophecies. Away with "image
worship." Let us only have differential equations or other mathematical
procedures and a recipe for deriving from them and from a set of actually performed observations all statements about all future observations of which foreknowledge is in principle at all possible. The desire
to visualize, so we are told, means wanting to know how Nature is
really constituted, and that is metaphysics—an expression that in present-day science is mainly used as an insult.

As is well known, Heisenberg's uncertainty assertion is strongly involved in the discussion. . . . It leads to the view that even straightforward observation of an experimentation with ordinary inanimate matter all at once confronts us with the whole profound perplexity of the
subject-object relationship. If those who believe that were right, it
would mean for us here that the intelligible objectified world-picture
. . . fails . . . not only in some specific form under consideration, but in
any one at all that we might try to give it.

Perhaps this possibility cannot be ruled out. I do not think it is likely.
So far it seems to me that the only reason for the iconoclastic uproar is
the following: the corpuscle concept has, it is true, become the unques-
tioned and inalienable possession of the physicist who continuously uses
it as a mental construct in his laboratory and at his desk, but on the
other hand, it leads to considerable embarrassment because we have not
yet succeeded in fusing it with the wave concept which, as one knows
to-day, ought to be applied not to different phenomena, but rather to exactly the same ones as the corpuscle concept, i.e., both have to be applied to simply everything. Some people believe that they have in
Mach's principle [1] found a wonderful way out of this dilemma which
frees from the obligation to search for clear conceptions of Nature by
condemning the belief in them as gross superstition. . . .

1. Mach believed that the laws of physics should be formulated so that they contain only concepts which can be defined by direct observations of nature or by a
short chain of reasoning from these observations. Ed.

Why should we here be allowed to take recourse to an epistemological principle as an excuse for our failures?

It seems to me that what we are striving for here (in physics) . . . is a comprehensive picture of the subject under investigation, a picture which becomes ever more distinct, lucid, and clearly understood in its interrelations. . . . the coherence would be utterly destroyed if we felt bound by pangs of conscience to omit all that is not directly ascertained or cannot, if so desired, be confirmed by sense perceptions; if we felt bound to formulate all propositions in such a way that their relations to sense perceptions were immediately manifest.

The Limit of Vision °
by Percy Williams Bridgman (1950)

I come to what it seems to me may well be from the long-range point of view the most revolutionary of the insights to be derived from our recent experiences in physics, more revolutionary than the insights afforded by the discoveries of Galileo and Newton, or of Darwin. This is the insight that it is impossible to transcend the human reference point. . . . The new insight comes from a realization that the structure of nature may eventually be such that our processes of thought do not correspond to it sufficiently to permit us to think about it all. We have already had an intimation of this in the behavior of very small things in the quantum domain . . . there can be no difference of opinion with regard to the dilemma that now confronts us in the direction of the very small. We are now approaching a bound beyond which we are forever stopped from pushing our inquiries, not by the construction of the world, but by the construction of ourselves. The world fades out and eludes us because it becomes meaningless. We cannot even express this in the way we would like. We cannot say that there exists a world beyond any knowledge possible to us because of the nature of knowledge. The very concept of existence becomes meaningless. It is literally true that the only way of reacting to this is to shut up. We are confronted with something truly ineffable. We have reached the limit of the vision of the great pioneers of science, the vision, namely, that we live in a sympathetic world, in that it is comprehensible by our minds.

° From Percy Williams Bridgman, "Philosophical Implications of Physics," *Bulletin of the American Academy of Arts and Sciences,* Vol. 3, No. 5, February 1950. Reprinted by permission.

Suggestions for Further Reading

° Born, Max: *The Restless Universe,* rev. ed., Dover Publications, Inc., New York, 1951, chap. III. An excellent description of modern concepts of light. Chapters I and II should be read first as background for Chapter III.

° Bragg, William: *The Universe of Light,* Dover Publications, Inc., New York, 1933. In addition to the selection that we have used in this part, we recommend this entire Bragg book for a clear and understandable account of the phenomena of reflection, refraction, polarization, and many other aspects of the nature of light that we did not have the space to include in these readings.

° Bronowski, Jacob: *The Common Sense of Science,* Vintage Books, Random House, Inc., New York, 1953. A very well-written discussion of the historical development and significance of the concepts of order, chance, and causality.

° Gamow, George, *Thirty Years That Shook Physics,* Anchor Books, Doubleday & Company, Inc., Garden City, New York, 1966. The story of the development of the quantum theory enlivened with personal recollections, sprightly drawings, and a modern version of Faust with some poetic license. A Science Study Series book.

° Jaffe, Bernard: *Michelson and the Speed of Light,* Anchor Books, Doubleday & Company, Inc., Garden City, N.Y., 1960. A Science Study Series biography of Michelson, including descriptions of his contributions to optics, the determination of the speed of light, and the Michelson-Morley experiment.

° Weisskopf, Victor F.: *Knowledge and Wonder,* Anchor Books, Doubleday & Company, Inc., Garden City, New York, 1963. A survey of important scientific ideas. Chapters Three, Four, and Five are especially recommended in connection with these readings. Another Science Study Series book.

° Paperback edition

7

Why Do Theories Change?

The changing and progressive aspect of science is illustrated in these readings which trace the evolution of the theory of gravity from Newton through Einstein and bring out the reasons why a theory that was true yesterday may be only partially true today.

Why Do Theories Change?

Introduction

THE READINGS IN THIS PART describe the evolution of a scientific theory. Bronowski said in Part 1 that a theory is the product of a creative imagination, resulting from the perception of a hidden likeness between apparently diverse phenomena. In this part we see how Newton created a new synthesis from Galileo's mechanics and Kepler's three laws of planetary motions. He united these apparently isolated principles into a broad working hypothesis and drew deductions from it that were confirmed in so many instances that the conceptual scheme became known as the law of gravity, one of the most comprehensive laws discovered by science. This law had great influence for many years and served as the foundation for the mechanistic and materialist thought of the eighteenth and nineteenth centuries.

However, toward the beginning of the twentieth century there began to appear several experimental facts that did not fit the Newtonian physics. This experimental evidence was of an extremely refined order of magnitude. We remember that Kepler threw out his first idea of planetary motion because of a discrepancy of eight minutes of arc in the orbit of Mars. But it was a difference of forty-three *seconds* of arc per century in the orbit of Mercury that was one of the observational facts not accounted for by Newton's law of gravity. In the Michelson-Morley experiment, it was necessary to be able to measure a time difference of 5×10^{-17} seconds in order that the results would be significant. This was the kind of experimental evidence that called for a revision in thinking and led to Einstein's broader synthesis, the theory of relativity.

Conant, in his book *On Understanding Science*, says, "We can put it down as one of the principles learned from the history of science that a theory is only overthrown by a better theory, never merely by contradictory facts. Attempts are first made to reconcile the contradictory facts to the existing conceptual scheme by some modification of the concept. Only the combination of a new concept with facts contradictory to the old ideas finally brings about a scientific revolution. When once this has taken place, then in a few short years discovery follows discovery and the branch of science in question progresses by leaps and bounds." [1]

The revolution brought about by Einstein has been so profound that all educated people today should have some knowledge of the ideas that

1. From James B. Conant, *On Understanding Science*, Mentor Books, New American Library of World Literature, Inc., New York, 1951. Reprinted by permission.

have emerged from his theory: ideas like the equivalence of mass and energy, which has been dramatically confirmed by the atom bomb; the definition of the speed of light as the top limiting velocity in the universe; and the fundamental reexamination of our measurements of space and time. These are some of the new concepts that are discussed in the readings in this part. In addition, the selections describe some of the reasons that led to the new theory and the experimental confirmations of it. The selection from Einstein is included to give the reader a feeling for the type of thinking he used in developing his theory. Within the limitations of a short introductory book with a nonmathematical approach, it is not possible to give a complete account of Einstein's theory. Readers who feel interested in gaining a better grasp of the theory itself will find suggestions for further reading at the end of this part. In the context of these readings, the most important points to understand are how theories are conceived, how they evolve, and how they are often incorporated into broader conceptual schemes. The story of Newton and Einstein is illustrative of this process. Studying the life history of a theory will help toward an appreciation of the nature of scientific theories in general.

In Part 2 and again in Part 4 we discussed the fact that scientific method cannot prove a theory to be definitely and absolutely true. At best it can show that there is a good probability that the theory represents a true description of reality. Yet, at any time, facts may appear that disprove it. Thus a scientific theory is forever "on trial," and it is this feature of the scientific process that prompted Gilbert Lewis to say that the concept of ultimate truth ". . . does not seem useful to science except in the sense of a horizon toward which we may proceed." [2] Since this is true, should theories serve only as aids to scientific progress? Are they too transitory to have any deeper significance, or does each new synthesis discover a greater unity in nature and lead mankind to a new level of understanding?

2. See p. 9.

Newton's Theory of Gravity

Rules of Correct Reasoning in Philosophy °

by Sir Isaac Newton (1686)

Rule I. We are to admit no more causes of natural things than such as are both true and sufficient to explain their appearances.

To this purpose the philosophers say that nature does nothing in vain, and more is in vain when less will serve; for Nature is pleased with simplicity, and affects not the pomp of superfluous causes.

Rule II. Therefore to the same natural effects we must, as far as possible, assign the same causes.

As to respiration in a man and in a beast; the descent of stones in Europe and in America; the light of our culinary fire and of the sun; the reflection of light in the earth, and in the planets.

Rule III. The qualities of bodies, which admit neither intensification nor remission of degrees, and which are found to belong to all bodies within the reach of our experiments, are to be esteemed the universal qualities of all bodies whatsoever.

For since the qualities of bodies are only known to us by experiments, we are to hold for universal all such as universally agree with experiments; and such as are not liable to diminution can never be quite taken away. We are certainly not to relinquish the evidence of experiments for the sake of dreams and vain fictions of our own devising; nor are we to recede from the analogy of Nature, which is wont to be simple, and always consonant to itself.

° From Sir Isaac Newton, *Principia Mathematica*, edited by Florian Cajori, from the translation by Andrew Motte. University of California Press, Berkeley, 1934. Reprinted by permission.

The Dawn of Universal Mechanics *
by W. P. D. Wightman (1951)

On Christmas Day of the year 1642 (in which Galileo died) there was born, near Grantham in Lincolnshire, a delicate babe who, during a vigorous life of nearly eighty-five years, extended the laws of motion of Galileo into realms undreamed of by their discoverer; and, triumphing over the immense mathematical difficulties, reduced all the phenomena of motion to manifestations of one simple law. Young Isaac Newton quickly threw off the weakness of infancy, and at an early age took advantage of a stormy night to measure the force of the gale by pacing out the lengths of leaps taken with the aid of the wind and in opposition to it. The story is interesting in that it indicates clearly the germ of the work—the application of calculation to actual experiments with the familiar but little understood forces of nature—by which he gained imperishable fame.

. . . But we must pass over these early years and picture him, as the French philosopher Voltaire, bids us, sitting in his garden at Woolsthorpe where, driven from Cambridge by threat of the Great Plague, he was passing the summer of 1665 in speculating concerning the motions of the heavenly bodies. Suddenly his attention was arrested by the fall of an apple to the ground. To no one before had such a familiar event seemed in any way connected with the motion of the heavenly bodies; to speak plainly, we cannot be sure that it was the fall of an *apple* which set Newton's mind to work, though his friend Stukeley also said that Newton told him so. In any case we have Newton's own written testimony that it was in 1665 that he began to think of "gravity extending to ye orb of the moon," [1] and further, that before he set about the calculations he marked the significance of the fact that this force of gravity extended to the deepest mine shafts and the highest mountains—why not then, his bold imagination hinted, to the moon itself?

Before we attempt to trace in outline the use he made of the materials which others had provided, and to learn something of the new intellectual weapons which he himself forged, it is imperative that we clearly discern what it was that Newton was seeking. One thing it certainly was not, though in our childhood we may often have been told so, namely

* From W. P. D. Wightman, *The Growth of Scientific Ideas,* Yale University Press, New Haven. Copyright © 1951 by Yale University Press. Reprinted by permission.

1. It is significant for the history of thought that his youthful attempts to master Euclid were prompted by his inability to solve a problem in *astrology!*

why an apple falls to the ground. Nor was it even why the moon moves as it does. His immediate purpose was in fact just that which in another connexion Galileo had proclaimed as the business of science, namely the demonstration that whatever may be the nature of the cause of an apple's fall to the ground, it is a cause of the *same nature* which determines the motion of the moon. Later, when he came to write his Preface to the First Edition of the *Principia,* he was to announce his belief that "the whole burden of philosophy [of nature] seems to consist in this—from the phenomena of motions to investigate the forces of nature, and then from these forces to demonstrate the other phenomena."

We must further understand that the search itself had been made possible only by Galileo's far-reaching studies; for so long as with the Greeks it was assumed that the motion of the moon in a circle round the earth was the most natural thing imaginable, there was clearly no problem to solve. But unless the moon is governed by influences entirely different from those apparently determining the behaviour of terrestrial bodies, its motion is anything but "natural"; for it should move in a straight line. Now Galileo taught us to distinguish as *forces* those natural influences which cause a body to diverge from rectilinear motion; hence some force must be acting on the moon. It was the genius of Newton to assume that since we know of no influences acting on the moon and the planets other than those acting on terrestrial bodies, it is worth while trying to prove that these will suffice to account for the behaviour of the former. Later in life he expressed this ideal—the tacitly accepted ideal of all *modern* science—in the following words: "We are to admit no more causes of natural things than such as are both true and sufficient to explain their appearances." . . . To accomplish this task alone would have enshrined him among the immortals; actually it was but a small part of his ultimate achievement.

To follow Newton step by step on his arduous task would take us far too long. . . . Indeed modern research has made it seem very probable that Newton himself was far from "moving in a straight line" towards a clearly envisaged goal. He worked on the problem at least three times in his life, and on two of these occasions seems to have lost interest in it for years at a time. It will nevertheless be most profitable to appreciate what the central problems were; for in so doing we can gain an insight into the true nature of mathematical physics such as entirely eludes anyone who is content to hear merely the bald statement of the results.

Let us then examine the *facts* as they were known at the time of Newton's first attack on the problem of the moon's orbit. The moon was known to describe an orbit—approximately circular—about the earth as centre. This was not of course a direct fact of observation, but an interpretation of such facts, and of whose truth no competent astronomer doubted. It was also known that the radius of the moon's orbit was about sixty times the radius of the earth itself. . . . The only other facts needed by Newton were the period of revolution of the moon—rather more than twenty-seven days—and the distance fallen in one second by a body near the earth's surface when unimpeded by the air—a distance of approximately sixteen feet.

These were the only *facts* used by Newton; to extract from them the relation which he believed to hold between the motion of the moon and that of an object near the earth's surface, he had to propose a method of *measuring* forces. To do this he assumed the truth of Galileo's law of inertia, and explicitly stated what was only *implicit* in Galileo's law of falling bodies. It is a strange thing that it never seems to have occurred to Galileo that if bodies left to themselves remain in a state of rest or uniform rectilinear motion, then it is natural to *measure* forces by the changes of motion which these produce in a given body: in other words, by the accelerations which the body undergoes. . . . In taking this next step Newton completed the work which Galileo had begun.

Newton's statement of this Second Law of Motion was to the effect that forces are proportional to the changes of motion which they produce, whether that change is in the speed with which the body "covers the ground" or of the direction in which it is moving; for it is a simple fact of experience that a weight can be twirled on the end of a string at a steady "speed" only if a constant force be applied. The modern statement of the law is therefore that "forces are proportional to the accelerations they produce," acceleration being defined as rate of change of *velocity,* which in turn means *speed in a given direction.*

Motion in a circle therefore may be regarded as equivalent to one in a straight line with which is compounded an acceleration towards the centre. Newton easily proved, as had his contemporary Huygens, that this acceleration is in fact proportional to v^2/r, where v is the speed with which the body is moving in the circle, and r the radius of the circle. Thus if we accept Galileo's and Newton's conception of force, it follows inevitably that any body moving in a circle must be acted upon by a force which accelerates it towards the centre of the circle. Now it is a fact that the moon is constantly moving round the earth in a path which is approximately a circle centred at the earth, therefore there must be a force acting on the moon accelerating it towards the earth; is it not the most natural, though not of course the only possible, assumption that the earth is itself the origin of the force? The only convincing manner of answering this question in the affirmative is to show that there is a precise numerical relationship between the supposed force of the earth on the moon and that on the apple. For numbers, like money, "talk"; that is, there is no disputing their testimony.

The next question therefore is the form of this supposed relationship. Here Newton made use of a hypothesis, probably first asserted by Bulli-aldus a few years before Newton's birth, and certainly familiar to New-ton's contemporaries, Halley, Wren and Hooke, that the power of one body, say the sun, to attract another, say a planet, would, if it exists at all, be inversely proportional to the square of the distance between them. That this suggestion was put forward by Bullialdus, of whom probably most people—even some men of science—have never even heard, only goes to show how mistaken is the oft-expressed opinion that the great advances in scientific thought are the exclusive products of in-dividual geniuses.

In order to show that his hypothesis was true, Newton had therefore to solve the comparatively easy problem (certain "simplifying assump-tions" being made) of showing that if f_m, f_a, are respectively the forces acting on the moon and on the apple, which are respectively at dis-tances d_m, d_a from the earth, then $f_m/f_a = d_a^2/d_m^2$. The first simplifying assumption is that the moon's orbit is circular, the second is that the dis-tance of each body "from the earth" is to be taken as its distance from the centre of the earth; for neither the moon nor the earth can be re-garded as a geometrical point, that is as negligibly small compared with the distance between them. And the difficulty naturally becomes far greater when bodies near the earth's surface are being considered. This is probably one of the subsidiary problems which Newton later referred to as being "more difficult than I was aware of." He referred to it in a slightly different connexion in III, 8 of the *Principia;* and his inability to solve it at that time, rather than the possible inaccuracy of his informa-tion regarding the diameter of the earth, probably accounted for his

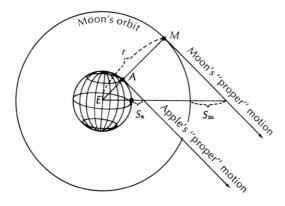

Figure 7-1

abandoning the formal proof of the whole project. Nevertheless it appears that he *did* make a provisional calculation, probably on such lines as these (Figure 7-1):

If a be the acceleration of the moon towards the earth, then $a = v^2/r$ and by Galileo's law (p. 188), s, the distance fallen $= \frac{1}{2}at^2 = \frac{1}{2}a = \frac{1}{2}v^2/r$ in one second. Now

$$v = \frac{\text{length of orbit}}{\text{time taken to describe it}}$$

$$= \frac{2\pi r}{t}$$

$$\therefore s = \frac{1}{2}\frac{v^2}{r}$$

$$= \frac{1}{2} \cdot \frac{4\pi^2 r^2}{t^2} \cdot \frac{1}{r}$$

$$= \frac{2\pi^2 r}{t^2}$$

$$= \frac{2 \times 3 \cdot 14^2 \times 240{,}000 \times 1760 \times 3}{(27.3 \times 24 \times 60 \times 60)^2} \text{ feet,}$$

27.3 being the number of *days* taken by the moon to describe its orbit of 240,000 *miles* radius.

The above fraction (in which the quantities used are those of to-day approximated to three significant figures) works out at .00449 ft.

Now if $f_m/f_a = d_a^2/d_m^2$, then, since force is proportional to acceleration, $s_m/s_a = d_a^2/d_m^2$, where s_m, s_a, are the respective distances fallen in one second by the moon and the apple. But the moon is known to be almost exactly sixty times as far from the centre of the earth as in the apple, so $s_a/s_m = 60^2/1$, that is, the distance fallen by the apple *should* be 3,600 times the distance fallen by the moon, i.e. $0.00449 \times 3600 = 16.2$ ft. The actual value found by experiment is 16.1 ft. Newton probably got 15.5 ft. for the former quantity, which would justify his remark that he "found the answer pretty nearly."

While Newton, perhaps dissatisfied with the number of simplifying assumptions he had been compelled to adopt, took no further steps to make his view public, a similar problem was taking shape in the minds of many of his contemporaries both in England and on the Continent. This was whether it could be proved that, on the assumption that the force between the sun and a planet is inversely proportional to the square of the distance between them, the planet would necessarily move according to Kepler's First Law. By 1684 several members of the newly formed Royal Society seem to have proved independently that if it could be assumed, as was unfortunately only roughly true, that the orbits of the planets are circles, then the inverse square hypothesis is implicit in Kepler's *Third* law. . . .

Of course no one pretended that this result *proved* the truth of the inverse square hypothesis, but the agreement was too striking to be merely a "fluke." One of these brilliant men, Robert Hooke, who in the intervals of assisting in the rebuilding of London after the fire, found time to invent carriage springs, to make one of the first efficient compound microscopes and to come very near to discovering oxygen, claimed that he could deduce a similar relationship for elliptic orbits, but on being asked to deliver it for inspection, temporised in a manner which suggested that the task was in reality beyond his mathematical powers. The events following on this unworthy display of duplicity were in their way as dramatic as anything in the history of human endeavour; for had it not been for the resolution of another of this group, Halley, the greatest single discovery ever made might never have been known to any but its discoverer. Halley however determined to consult Newton at Cambridge. So in August 1684 he called on Newton and, without mentioning the tentative views of Hooke and the others, bluntly asked him what would be the path of a planet on the assumption that the force acting on it decreased in the proportion of the square of its distance from the sun. To his astonishment and delight Newton replied that the planet would describe an ellipse. His delight was somewhat tempered when Newton failed to find the papers on which he had, so he said, worked out the proof. However in November the papers arrived, giving two proofs of the theorem that, in his own words, "by a centrifugal force reciprocally as the square of the distance, a planet must revolve in an ellipsis about the centre of the force placed in the lower umbilicus [focus] of the ellipsis and with a radius drawn to that centre describe areas proportional to the times." Newton later claimed to have obtained a proof in 1676, but had apparently lost interest in the subject! Now he set to work again and wrote a small treatise, *De Motu Corporum*, which Halley saw on a second visit to Cambridge. Fortunately for the world Halley realised that here was a "scoop" of the first order and, overcoming Newton's disinclination for publicity, persuaded him to send it to the Royal Society for publication.

The next year, under the stimulus of Halley's encouragement, Newton successfully proved that a spherical body such as the earth exercises its force of gravitation, whether on the moon or on an apple at its surface, in a manner which is mathematically equivalent to the force which would be exerted by the same body regarded as a mathematical point situated at its centre. To effect this wonderful result Newton had been faced by the task of calculating the combined effects of all the particles of matter of which the earth is composed. Such a task appeared impossible of fulfillment; in fact we may say that it was impossible until Newton himself had invented a new branch of mathematics—akin to what we now call the "integral calculus"—by which such problems involving the effect of an apparently infinite number of infinitesimal (that is "vanishingly" small) particles could be tackled. . . .

The Royal Society was delighted with what it saw of Newton's work, and at once ordered it to be completed and printed at the Society's expense. When the extent of the undertaking was realised however there seems to have been a cooling-off on the part of the well-to-do who largely composed the Society's membership. The exact reason for the delay will perhaps never be known, but the fact remains that Halley, a man of modest fortune and with a young family to bring up, undertook the expense and much of the labour of seeing the work through the press. The Royal Society took the credit for the publication, but posterity knows that it was to the vision and self-sacrifice of Halley that it owes the final publication in 1687 of the greatest single monument of human learning—*Philosophiae Naturalis Principia Mathematica.*

THE NEWTONIAN REVOLUTION

. . . We must now take note of an important implication of the law of gravitation. We have seen that the moon's path round the earth can be accounted for on the assumption that the earth draws the moon towards itself with exactly the same force as it does an apple on its own surface, only providing that magnitude of the force falls off in the proportion of the square of the distance from the centre of the earth. The moon being about sixty times as far from the earth's centre as is the apple, it will in unit time "fall" towards the centre about one three-thousand-six-hundredth part of the distance fallen by the apple. But what enabled Newton to enunciate his Second Law of Motion where Galileo had failed was the fact that the former had grasped the part played by the *mass*—the unchanging degree of resistance to motion, which characterises every material body; what, indeed, constitutes its "materiality." If of two garden rollers standing on the same level path, the one requires exactly twice the effort to give it the same acceleration as the other, we say that the former has twice the *mass* of the latter. Ordinarily we work the other way round: knowing that one *weighs* twice as much as the other, we at once *infer* that it will take twice as much effort to give it the same acceleration. It is so much easier to weigh garden rollers than to set up experiments by which may be measured the forces necessary to give them equal accelerations. Indeed it is one of the most difficult problems in mechanics to contrive the measurement of "naked" forces unassociated with lumps of matter. But it cannot be too strongly emphasised that though in practice we measure masses by weighing them, the two quantities refer to entirely distinct characteristics of matter.

Now in all that we have said so far about the force controlling the motions of the heavenly bodies, we have been able to get on perfectly well without any knowledge of their *masses*. How can we say that the force acting on the moon differs from that acting on the apple only in respect of the squares of their respective distances from the earth which attracts them? It is to be hoped that the watchful reader has been troubled with a feeling of uneasiness about this very point. Bearing in mind that the magnitude of a force causing motion is always proportional to the product of the mass of the body moved and its acceleration (by suitable choice of units it is *equal* to this product), he must have felt that to compare the forces acting on moon and apple by comparing their accelerations is a mere trifling with the problem, seeing that the *mass* of the moon is so enormously greater than that of the apple. . . . the argument holds equally well for a grain of sand and an aerial bomb. The whole of terrestrial mechanics rests on the observed fact that all bodies at a given point near the earth's surface are accelerated towards the centre at the same rate. The mass has nothing to do with it. Obviously if we are going to retain our original definition of a unit force (as that which causes unit acceleration in unit mass) we must ascribe to the particular force of gravity the curious property of being always proportional to the masses it is acting upon, providing of course that they are equidistant from the centre of the earth; for only thus can it come about that different masses are equally accelerated at the earth's surface. This is the course which Newton took, his actual words being: "There is a power of gravity tending to all bodies, proportional to the several quantities of matter they contain." (*Principia*, Bk. III, Prop. VII.)

It is a curious fact that the men of science of Newton's day and for nearly two centuries thereafter never seem to have been particularly struck by this odd characteristic of the "force" of gravity. . . . The law of gravitation as applied by Newton is enunciated in our text books in the following words: "Every particle of matter attracts every other particle with a force which varies directly as the product of their masses, and inversely as the square of the distance between them." The implication is clearly that "matter" itself somehow "generates" this "force" and exerts it in some mystical way across millions of miles of empty space. It is significant that Newton himself is much more guarded in his views than modern text books give him credit for. In speaking of this force he avoids the word "attraction," and speaks instead of "bodies gravitating towards one another." And in a letter to a friend, Richard Bentley, he writes: "You sometimes speak of gravity as essential and inherent to matter. Pray do not ascribe that notion to me; for the cause of gravity is what I do not pretend to know. . . . It is inconceivable that inanimate brute matter should, without the mediation of something else, which is not material, operate upon, and effect other matter without mutual contact." The widely held belief that Newton regarded gravitation as an *innate* characteristic of matter seems to have been due to one or two less cautious sentences in the *Principia* and in Query 31 of the *Opticks* (which however he is careful to correct at the end of the sentence). But more particularly was it due to the remarks made in the "popular" Preface to the Second Edition of the *Principia*, written by Roger Cotes, the editor. It is probable that Newton all along was struggling against the pseudo-Aristotelian tendency to invoke "occult powers" to "explain" everything—the antithesis to what Newton recognised the nature of scientific enquiry to be. His final views on the nature of matter appear in the last pages of the *Opticks:* "It seems probable to me that God in the beginning form'd matter in solid, massy, hard impenetrable, moveable Particles . . . and that these primitive Particles, being Solids, are incomparably harder than any porous Bodies compounded of them; even so very hard as never to wear or break in pieces; no ordinary power being able to divide what God himself made one in the first Creation." The innate qualities enumerated here do not include gravitation.

When following long established custom we speak of the earth's attraction for the apple and for the moon, we have in fact not a shred of evidence that any "force" such as is exerted by a tow-rope on a barge exists between them. If, however, we decide, as is implied by Galileo, that whenever a body suffers an acceleration, there must be a force acting on it, then we must inevitably deduce the existence of a force between earth and apple. And if further we decide, as did Newton, to measure forces by the acceleration they produce in unit masses, then we must infer that since all bodies fall with constant and equal accelerations at the same point on the earth's surface, the "force" which we have inferred to act on them must be proportional to their masses: this conclusion is inescapable *on the assumptions that we have made.*

This aspect of Newton's work is of great importance in assessing its ultimate validity, that is, as "science's last word." . . .

So far our discussion has been simplified by regarding the sun as "attracting" the planets, and the earth as "attracting" the moon and the apple. But Newton made it perfectly clear that the power of gravitation extends to every particle of matter; it is only for convenience that we have singled out one body—the greater—from the pair under consideration. The moon "attracts" the earth with the same force as that which the earth exerts upon it. To an observer on the sun, we may believe, the earth would not appear indifferent to this action: its path in space would not remain the same were the moon suddenly to be blotted out of existence.

The calculation of this mutual effect is made possible by Newton's Third Law of Motion: "To every action there is always opposed an equal reaction: or the mutual actions of two bodies upon each other are always equal, and directed to contrary parts." The operation of this law is a fact of daily experience. It is impossible to step from an unmoored rowing boat on to a landing stage without moving the boat in the opposite direction. We may believe that the same would be equally true of a man disembarking from the "Queen Mary"; but the effect is masked, even apart from the mooring ropes and rigid gangways, by the enormous mass of the ship compared with that of the man. By "action" Newton clearly meant what in the Second Law he had called "quantity of motion"—what we now call "momentum," that is the product of mass and velocity. Thus if any body is to alter its momentum it can do so only by altering that of some other body or bodies to an equal amount "directed to contrary parts." The propellent charge which sets the shell in motion can do so only by giving the gun an equal and opposite "motion." By using the mathematical device of a minus sign for a "contrary" motion—a convention which became familiar in later years—we are able to express Newton's Third Law in concise quantitative terms: "The algebraic sum of the momenta of an isolated system of bodies is a constant quantity." This is the first of the "conservation laws"—. . . .

Let us now pause for a moment to see what exactly was Newton's contribution to the growth of scientific ideas about matter and motion, and in what way it was in advance of all previous ideas in the same connexion. The simple idea which in Newton's skilled hands wove together the movements of all observable bodies was no more than this: All moving objects, whether planets or cannon balls, are composed of particles of something distinguished by two characteristics, namely an innate resistance to change of motion, the degree of which resistance is called the *mass* of the particle, and a tendency to gravitate together as if impelled by a force which varies directly as the product of the masses and inversely as the square of the distance between them. The "stuff" thus characterised is what we still call "matter." [2]

Thus stated, the Newtonian revolution sounds prosaic enough. Why

2. The critical reader may wonder how the unit of mass is to be measured without dragging in force and so making the argument circular. Newton left this point in an unsatisfactory state. It is too difficult to deal with in detail here.

then do we hail it as probably the greatest single achievement of the human intellect? It leaves us completely ignorant as to why an apple falls to earth; for to suppose this problem solved by saying "because the earth attracts the apple" is grievously to err. Such a "solution" is no better than that of Aristotle to the effect that the "earthy nature of the apple is seeking its natural place." The law of gravitation, like the laws of motion of Galileo, does not tell us why bodies move towards each other, but precisely how their movement is determined. And it tells us more; for it discovers . . . what is essential in the motion of all material bodies, namely the mere fact that they *are* material. Whether we keep strictly to Newton's own words or whether we simplify the *verbal expression* of the law by speaking of an actual "attraction," matters but little. Provided we bear in mind what we mean by a "force" in the science of physics, we are compelled to say that all material bodies move as if impelled by a force operating according to Newton's law. Whether the operation of that force is "caused" by matter itself, or whether with Newton we prefer to await the results of further research, the consequences predicted by the law remain completely unchanged. And that is after all what "science" since Galileo's day has stood for. . . . The future was to reveal that nothing but the refinement of the mathematical technique was necessary to enable man to calculate the amounts of the known disturbances and thus to seek for, as yet unobserved, causes of residual disturbances. The law of gravitation not only explains, it also foretells.

Although a century had to pass after the death of Newton before the law of gravitation was used as an instrument of discovery, it must be realised that Newton himself was able to show that many carefully observed but as yet unexplained phenomena were just as inevitable consequences of this universal characteristic of matter as was the nearly circular path of the moon. Thus he began the proof of the extremely complex problem of showing that the oceanic tides are mainly the result of the varying "pulls" of the sun and moon on the mass of water as they change their stations relative to one another. But perhaps the most beautiful interpretation of all was the demonstration that the precession of the equinoxes [3] (discovered by Hipparchos nearly two thousand years earlier— . . .), the so-called "inequalities" of the moon's motion, some already known to astronomers, but one of which was foretold by Newton; and the motions of the tides—all are manifestations of the same phenomenon, namely a disturbance of the simple gravitational reaction between two bodies by the action of a . . . *third* body. . . . Thus the moon instead of moving in an ellipse round the earth, according to Kepler's three

3. See footnote p. 124 and Fig. 3-36. Ed.

laws, really "wobbles" to a small but precisely determinable extent. The plane containing the earth's orbit about the sun is inclined to that of the moon about the earth, otherwise there would be an eclipse of the sun at every "new moon" and of the moon at every "full moon." Now the sun is about four hundred times as far from the moon as is the earth, but it is vastly more massive; so its "pull" on the moon imposes an additional periodic motion on that body. The actual nature of this motion was *proved* by Newton to involve a gradual revolution of the *axis* of the moon's *orbit*, occupying about eighteen years; this is the principal cause of the recurring cycles of eclipse known to the Chaldean astronomers as the "Saros" at least 2,500 years ago. By a superb use of the "principle of continuity" Newton extended this argument to the case of a ring of satellites, then to a ring of fluid surrounding the central body. . . . The last case of this sequence was that of a *solid ring* such as exists in the equatorial region of the earth. There will be similar rotation of the axis of this ring which, being rigidly connected to the sphere, will cause the axis of the latter also to revolve—and so the pole of the heavens will describe a circle and the equinoxes will precess! . . .

In Newton's hands the law of gravitation not only interpreted all the known motions of material bodies but actually foretold the existence of several others which were later observed by astronomers. In this Newton was helped, rather more than he gave him credit for, by the Astronomer Royal, Flamsteed. But the greatest triumph subsequent to Newton's death came towards the end of the eighteenth century, when by its application it became increasingly more probable that a hitherto unknown planet existed in the solar system. The whole story is one of the great romances of science; we have space here only to show how it illustrates one of the most valuable scientific methods.

On 13th March 1781, Sir William Herschel, by profession a musician but who became one of the most brilliant observational astronomers, noticed among a group of small stars one whose size increased progressively with the magnification employed in his telescope (the almost "infinitely," distant stars actually become smaller but at the same time *brighter* when viewed through an instrument of higher magnifying power). On succeeding nights its changed position relative to the "fixed" stars confirmed his view that the body might be a comet. Later however, other observations showed its movements to be those not of a comet but of a planet; and in due course it was established that the body had been observed many times during the previous hundred years. This fact enabled mathematicians to determine . . . the constants by which its orbit about the sun could be determined. But before very long Uranus, for such was the name given to the new planet, began to wander seriously from the strait and narrow path which men had assigned to it. Here was a chance to put Newton's law to the most crucial test by, as it were, "turning it inside out."

To Newton, . . . the problems of celestial mechanics were always of the form: "Given such and such bodies, to deduce their relative motions." But in the case of Uranus the problem was: "Given such and such irregularities in the motion of one body to deduce the position and magnitude of some unknown body which is the cause thereof." Such a piece of investigation is an illustration of what logicians call "the method of residues"; for before any attempt could be made to seek the position of the unknown body, all the possible effects of known bodies, such as Jupiter and Saturn, had to be calculated and allowed for: the residual effect must then be due to some undetermined cause.

To this task the young Cambridge mathematician, J. C. Adams, and the experienced Frenchman, Leverrier, applied themselves, each in complete ignorance of the fact that the other was seeking the same goal. On 23rd September 1846, the Berlin astronomer, Galle, found the planet which was the cause of the trouble very near the spot where Leverrier told him to look. Within a few days the Cambridge Astonomer, Challis, was able to prove that he had observed the planet in the same region of the sky but in different positions relative to the stars on 30th July, 4th August and 12th August. The new planet was called Neptune. It is perhaps too much to say that but for a purely abstract dynamical law it would never have been recognised by human eye; but it is certain that its discovery would have been indefinitely postponed. The light of Neptune is thus a reflection of the light of Newton's thought which shone with such power as actually to bring new material bodies within the range of human ken.

By assuming the truth of the law of gravitation it had been possible to deduce the existence of a massive body far beyond the orbit of Saturn but still within the confines of the solar system. In 1862 however an American astronomer, Alvan Clark, observed a new star whose existence, unknown to him, had been deduced from the observed irregularities in the motion of another star. In all that we have said about the heavens thus far we have assumed the immobility of the stars—Newton . . . took the "frame of the fixed stars" as his frame of reference to which all motions could be referred as absolute. . . .

The most sensational triumph however was still to come. The German mathematician, Bessel, . . . had in 1834 made so many observations on Sirius, the brightest star in the heavens, as to suspect that its proper motion (that is, its own motion among all the other stars which by this time had come to be recognised as moving at high speeds) was irregular. So just as a new *planet* had been "invented" to account for the irregular behaviour of Uranus which was troubling astonomers at the same time, Bessel, having in ten years failed to detect any cause for the "wandering" of Sirius, "invented" a new and invisible *star*. In a letter to Sir John Herschel (William's son) he wrote: "I adhere to the conviction that Procyon [the 'Little Dog' star, another wanderer, quite separate from Sirius] and Sirius form real binary systems, consisting of a visible and an invisible star. There is no reason to suppose luminosity an essential quality of cosmical bodies. The visibility of countless stars is no argument against the invisibility of countless others."

What a change from the Ionian conception of the stars as being jets of flame bursting through, or fixed to, the crystal sphere which rotates about the earth! Bessel's remarks exemplify in the clearest manner the Newtonian revolution in which "bodies celestial" differ not at all in their effects from "bodies terrestrial"; their presence being deducible from the assumption that, being fashioned from "matter," they will be subject to the law of gravitation now seen as *universally* valid . . . in 1862 Alvan Clark, knowing nothing of the predictions of Bessel and the more accurate ones of Peters, stumbled by accident upon a faint star in the proximity of Sirius *just where the mathematicians said it ought to be at that time to account for the irregularity in the motion of Sirius itself.* This dark companion, though as massive as our sun, has only one five-hundredth of the intrinsic brightness of the latter.

It is interesting to note that the . . . greatest advances in the secure establishment of the law of gravitation were the results of "accidents." William Herschel, examining the smaller stars, discovers a "comet" which turns out to be a planet, whose irregular behaviour leads to the discovery of another planet. Once again the same indefatigable "amateur," . . . discovers the physical connexion between certain double stars, and so paves the way for the precise verification of the *universality* of the law of gravitation, and of the discovery of dim bodies thousands of millions of miles beyond the reach of our solar system.

These "accidental" discoveries are gravely misunderstood by people not trained in scientific method. William Herschel was an *"amateur"* only in the literal French sense of the word: he *loved* astronomy so passionately as to conquer physical weakness to the extent of sitting at his telescope or polishing his mirrors all night after an ordinary day's work in his profession of music. He was a man of prodigious determination and unusual talent. "He studied science when he could" as George Forbes says in his *History of Astronomy;* but he studied to such good effect that his mind was prepared to make the best use of any "accidental" observation which he might happen to make. "Those that have eyes to see, let them see." . . .

So ends our study of the greatest sweep of human imagination, from the fall of an apple to the revolution of an invisible star! Every movement in the universe determined by the application of one simple law, so that "given the distribution of the masses and velocities of all the material particles of the universe at any one instant of time, it is *theoretically* possible to foretell their precise arrangement at any future time." "Theoretically," because the mathematical difficulties would far surpass the compass of any human mind however gifted. But even with that qualification this dictum of Laplace has raised philosophical doubts which seem to strike at the root of man's whole conception of the universe; for since every man's body consists of material particles, *their* motion, no less than that of the particles of inanimate matter, must inevitably follow a course already fully predictable to anyone possessed of sufficient knowledge and mathematical skill. And if this be so, then must all human efforts be in vain, all appearance of choice illusory; for the fabric of history which time weaves must be as completely determined as the melody which will emerge from a given gramophone record. Laplace and his contemporaries overlooked the fact that his deduction, like the gramophone record, bears within itself no mark to indicate why it should not be played backwards; but such a possibility makes nonsense of history.

The gradual realisation by the cultured world of this implication of the Newtonian revolution had, and is still having, the profoundest effect on the morals and ideals of Western civilisation. Thus the march of science leads us to positions whence we gain new views of spiritual realms as well as changing catastrophically our physical environment.

In the next chapter we shall briefly consider whether in the light of modern knowledge and criticism the Newtonian world-picture is quite as rigid in its outlines as it appeared to the thinkers of the eighteenth century.

IS THE LAW OF GRAVITATION TRUE?

In the summer of 1919 two topics produced a great deal of discussion in the more serious newspapers: one was the signing of the so-called "Peace Treaties," which left Europe on the brink of war for twenty years, and the other was the announcement that a handful of men of science had returned from a Pacific island with some photographs of a solar eclipse, from which conclusions had been drawn casting grave doubts on the absolute truth of the law of gravitation. Once again it seems the reader has been asked to spend an unconscionable amount of time and trouble being initiated into a scientific discovery which is not true. It is the purpose of this chapter to try to clear up this unsatisfactory situation.

At the present day most people fall into three groups in respect of their attitude towards science. The first group regard science as a messy business, whose study is undesirable for "nice" people; they are dying out fast, though unfortunately not quite fast enough. Of the other two groups, one accepts every statement of "Science," as our forefathers accepted "Holy Writ," as absolute truth, to doubt which is a special kind of blasphemy; while the other, bewildered by the constant rewriting of text books which are out of date almost before publication, regard "Science" as their forefathers did the "Evil One," as the spirit of confusion and deceit.

As we have already seen in connexion with the motions of the heavenly bodies only the study of its history can give us the clue to the unravelling of the apparent confusion in scientific knowledge. A method of interpretation which was perfectly true at the time of its enunciation has to be remodelled—often drastically—in the light of new knowledge. For knowledge is for ever *growing*—not merely accumulating like a heap of pebbles, but changing in outline and perspective like a living thing; the old writer knew what he was talking about when he wrote of the "tree of knowledge." The kind of knowledge we call science is a system of laws of nature. Laws are *concise* statements of how we believe nature *works*. To achieve this conciseness we have to make use of *general ideas.* The law of gravitation is of universal application simply because it purposely *excludes from consideration* all those properties of natural bodies except the universal property, that is, mass. The sun is so hot that the earth would be completely evaporated were it to approach it; the moon is so cold that its atmosphere is largely liquified and partly solidified. But by regarding these bodies as mere lumps of nondescript "stuff," each having a certain measurable resistance to change of motion (i.e. mass) and a certain orientation in space with regard to the other two, we find that their subsequent *motions* can be calculated by means of the law of gravitation. The temperature has nothing to do with it.

The law of gravitation therefore boils down to this: Given certain masses at certain known distances from each other, and whose relative velocities are known, we can calculate their subsequent motions.[4] Put in this way, the law of gravitation was absolutely true when it was enunciated, and is very nearly true to-day when the precision of measurement has greatly improved. Nor is it conceivable that it will ever be otherwise.

How can a law be "true" in 1687 and only "very nearly true" in 1948? It is not a question of a possible alteration in the way nature "works"; whether this be so or not is beside the point. What we are here con-

---

4. The mathematical difficulties become very grave for all possible cases of three bodies and, for four or more, almost beyond the power of man to handle.

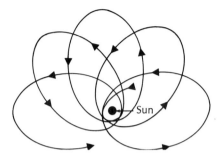

Figure 7-2. The rotation of Mercury's elliptical orbit, greatly exaggerated here, was an irregularity not explained by Newton's theory of the universe. The ellipse actually advances 43 seconds of arc per century. (After Lincoln Barnett. Figure added by editor.)

cerned with is the fact that we have "grown out of" the law of gravitation in two ways.

In the first place there was an irregularity in the motion of the planet Mercury which Newton's theory of the solar system failed to clear up [Figure 7-2]. This irregularity was observed only by virtue of an increase in the precision of instruments; latter-day astronomers "knew more" than Newton *could* have known. That was the first growing pain. But there was a more fundamental one.

If we refer to the "boiled down" gravitation law stated above, we shall observe that from this law we can calculate *motions* from *motions*. To do this we assume the truth of "laws of motion" connecting forces and masses; from which assumption it inevitably follows that the "force" of gravitation on any body *is always proportional to the mass of that body.* . . . Newton, in order to express his theory in words, had had to invent new general ideas, such as "mass," and adapt old ones such as "force" and "motion"; and while the new ones were clear and adequate, the old ones passed into his use with all the confused associations of the ages. We have already seen that "force" has a very unscientific vagueness about it, but "motion" is in an even worse position.

After all, what do we mean by "motion"? We shall at once be told "change of position." But when, seated in a stationary train, we see a neighbouring train slowly begin to move, how do we know which has "really" moved? Only by checking our position with respect to the platform. This Newton fully realised: a body "really" moves when it has changed its position with reference to the "fixed" stars. He realised moreover that this matter of a *fixed frame of reference* to which all motions are to be referred is so essential a part of his whole mechanical system that he dealt with it in the famous Introductory Scholium to the First Book of the *Principia.* To avoid any possibility of misunderstanding it is necessary to quote the whole Scholium here:

Hitherto I have laid down the definitions of such words as are less known, and explained the sense in which I would have them to be understood in the following discourse. I do not define time, space, place and motion, as being well known to all. Only I must observe that the vulgar conceive those quantities under no other notions but from the relation they bear to sensible objects. And thence arise certain prejudices, for the removing of which it will be convenient to distinguish them into absolute and relative, true and apparent, mathematical and common.

(1) Absolute, true, and mathematical time, of itself, and from its own nature flows equably without regard to anything external, and by another name is called duration: relative, apparent, and common time, is some sensible and external (whether accurate or unequable) measure of duration by the means of motion, which is commonly used instead of true time: such as an hour, a day, a month, a year.

(2) Absolute space, in its own nature, without regard to anything external, remains similar and immovable. Relative space is some movable dimension or measure of the absolute spaces; which our senses determine by its position to bodies; and which is vulgarly taken for immovable space; such is the dimension of a subterraneous, an aerial, or celestial space, determined by its position in respect of the earth. Absolute and relative space, are the same in figure and magnitude; but they do not remain always numerically the same. For if the earth, for instance, moves, a space of our air, which relatively and in respect of the earth remains always the same, will at one time be one part of the absolute space into which the air passes; at another time it will be another part of the same, and so, absolutely understood, it will be perpetually mutable.

Although a full discussion of this difficult matter would take us too long and too deep, yet since the answer to the question set at the head of this chapter depends on a clear understanding of the implications of Newton's assumptions, a brief exposition must be attempted.

The most important point to be noted is Newton's tacit departure from his own ideal of scientific enquiry: he *assumes* the existence of "absolute" time and space "without regard to anything external"—in other words, without any evidence from sensory experience, such as the "vulgar" base their ideas of time on. It is equally important to notice that Newton *had* to assume just this if he was to construct his mechanical System of the World. Finally we must not blame Newton for failing to haul himself up by his braces out of the climate of opinion into which he was born, and so to realise that "when once you tamper with your basic concepts," as Whitehead puts it, "philosophy is merely the marshalling of one main source of evidence and cannot be neglected." The liquidation of the Aristotelian physics involved the liquidation of the Aristotelian metaphysics of the immutable heavens; but Newton was not the man to do it. He was too great a man of science to attempt the impossible. The scientific advance has to proceed like that of an army—by jumps to tactical strong points, sometimes without the clearance of strategic minefields.

As soon as men of science realised that the "fixed" stars, so far from being "fixed," are moving relative to each other they began to look for another "platform." The most promising candidate was the "ether" whose undulations were supposed to constitute light. Unfortunately Michelson and Morley could find no evidence for the existence of the "ether" under conditions which should have rendered it easily observable. So the search had to start again. A few optimists are still searching, but most have given up the task as meaningless. Now if we can never be sure that a body has "really" moved, we can never be sure on which of two "moving" bodies a force is acting, for which reason it is hardly worth bothering about "force" at all. Following this train of thought, and taking as a fundamental property of nature the fact that weight and mass, though different properties are always equal in the same place, Professor Albert Einstein developed a system of dynamics (i.e. of the nature of motion) from which a new law of gravitation emerged. This law precisely accounted for the irregularity in the motion of Mercury which Newton's law had failed to do. But physicists rightly held that the "truth" of the law could not be finally established until, like Newton's and Galileo's laws, it had been shown that consequences *deduced* from it by pure reasoning were actually to be observed in nature. That is why the British eclipse expedition under the late Sir Arthur Eddington set out in 1919 to photograph some stars whose light was passing close to the eclipsed sun. The data they obtained were within the limits calculated on the basis of Einstein's theory.

Newton's law of gravitation is more "true" than Aristotle's for two reasons, because it enables us to control the movements of earthly bodies and to foretell the positions of heavenly bodies with a marvellous degree of exactitude, which the latter's was powerless to attempt, and also because Newton's concept of the relation between mass and acceleration was more precisely definable (and consequently "usable") than Aristotle's "natures."

Einstein's law of gravitation was derived from a consideration of the metrical properties of the "space" in which bodies move. As Professor Bridgman puts it, he was enquiring "into the meaning of simultaneity, and he was finding the meaning by analysing the physical operations employed in applying the concept in any concrete instance." [5] Newton had taken such ideas as simultaneity for granted. With the knowledge of

5. See P. W. Bridgman, *The Nature of Physical Theory,* Princeton, 1936.

the critical importance of the nature of light signals for the timing of physical events, Einstein saw that such an assumption was too simpleminded. His "analysis of the physical operations" yielded first the Special Theory of Relativity, concerned with measurement between bodies moving at *different* but *uniform* velocities. Later came the General Theory which included the case of bodies moving with velocities mutually *accelerated*. The result of this was to render the concept of *force* redundant.

Einstein's law of gravitation is "truer" than Newton's for reasons similar to those which made Newton's "truer" than Aristotle's. It is more than likely that history will repeat itself and Einstein's beautiful theory be found after all to be not quite "true." But no theory which can reveal to us new secrets of the universe can be "false," though it may be "out of date." We do not regard the "Cutty Sark" as falsely contrived merely because it could not equal the performance of the "Queen Mary." The latter ship is in the main a development of the older; but the "steam wind" which presses upon her turbine plates by being more controllable, is more economically applied. The "Cutty Sark" was the wonder of her age; the "Queen Mary" can do all that she did and more; but thousands of people still derive pleasure in using the graceful methods of the old "sea bird" rather than the more powerful ones of the modern machine. Similarly for all our everyday uses, whether for the design of engines or for the writing of the "time table" of the planets, we use and shall long continue to use Newton's law, which is sufficiently exact to meet all purposes except those in which we are probing more deeply into the heart of the universe.

One question remains on which we may close this rather difficult discussion. If Einstein's dynamics is "truer" than Newton's, why do we waste so much time in teaching the latter in our schools? As well ask, why waste a child's time with instruction in the use of the foot rule, when the vernier calipers give us a much more accurate estimate of length? Because over-refinement tends merely to obscure the simple ideas of length and number which every human being must acquire. Einstein's dynamics are Newtonian dynamics subjected to intense refinement. . . . It will be the chief service of the modern critical study of the history of science to present scientific knowledge as a growing tree which depends for its vigour as much on its roots in the past as upon its fresh young branches; but from which for true health all dead and rotten wood needs to be drastically pruned. In other words it will teach us that new theories do not replace old ones because they are more *true;* but grow out of and embrace the older because they are more adequate.

Not that we may not, to explain any phenomena of nature, make use of any probable hypotheses whatsoever: hypotheses, if they are well made, are at least great helps to the memory, and often direct us to new discoveries. But my meaning is, that we should not take up any one too hastily (which the mind, that would always penetrate into the causes of things, and have principles to rest on, is very apt to do) till we have very well examined particulars, and made several experiments, in that thing which we would explain by our hypothesis, and see whether it will agree to them all; whether our principles will carry us quite through, and not be as inconsistent with one phenomenon of nature, as they seem to accommodate and explain another. And at least that we take care that the name of *principles* deceive us not, nor impose on us, by making us receive that for an unquestionable truth, which is really at best but a very doubtful conjecture; such as are most (I had almost said all) of the hypotheses in natural philosophy.

John Locke
Concerning Human Understanding (1690)

From the point of view of the physicist, a theory . . . is a policy rather than a creed; its object is to connect or coordinate apparently diverse phenomena and above all to suggest, stimulate, and direct experiment.

J. J. Thomson
The Corpuscular Theory of Matter (1907)

Einstein's Theory of Relativity

The Universe and Dr. Einstein °

by Lincoln Barnett (1957)

In his great treatise *On Human Understanding* philosopher John
Locke wrote three hundred years ago: "A company of chessmen standing on the same squares of the chessboard where we left them, we say,
are all in the same place or unmoved: though perhaps the chessboard has
been in the meantime carried out of one room into another. . . . The
chessboard, we also say, is in the same place if it remain in the same
part of the cabin, though perhaps the ship which it is in sails all the
while; and the ship is said to be in the same place supposing it kept the
same distance with the neighboring land, though perhaps the earth has
turned around; and so chessmen and board and ship have every one
changed place in respect to remoter bodies."

Embodied in this little picture of the moving but unmoved chessmen
is one principle of relativity—relativity of position. But this suggests another idea—relativity of motion. Anyone who has ever ridden on a railroad train knows how rapidly another train flashes by when it is traveling in the opposite direction, and conversely how it may look almost
motionless when it is moving in the same direction. A variation of this
effect can be very deceptive in an enclosed station like Grand Central
Terminal in New York. Once in a while a train gets under way so gently
that passengers feel no recoil whatever. Then if they happen to look out
the window and see another train slide past on the next track, they have
no way of knowing which train is in motion and which is at rest; nor
can they tell how fast either one is moving or in what direction. The
only way they can judge their situation is by looking out the other side
of the car for some fixed body of reference like the station platform or a
signal light. Sir Isaac Newton was aware of these tricks of motion, only
he thought in terms of ships. He knew that on a calm day at sea a sailor

can shave himself or drink soup as comfortably as when his ship is lying motionless in harbor. The water is his basin, the soup in his bowl, will remain unruffled whether the ship is making five knots, 15 knots, or 25 knots. So unless he peers out at the sea it will be impossible for him to know how fast his ship is moving or indeed if it is moving at all. Of course if the sea should get rough or the ship change course abruptly, then he will sense his state of motion. But granted the idealized conditions of a glass-calm sea and a silent ship, nothing that happens below decks—no amount of observation or mechanical experiment performed *inside* the ship—will disclose its velocity through the sea. The physical principle suggested by these considerations was formulated by Newton in 1687. "The motions of bodies included in a given space," he wrote, "are the same among themselves, whether that space is at rest or moves uniformly forward in a straight line." This is known as the Newtonian or Galilean Relativity Principle. It can also be phrased in more general terms: mechanical laws which are valid in one place are equally valid in any other place which moves uniformly relative to the first.[1]

The philosophical importance of this principle lies in what it says about the universe. Since the aim of science is to explain the world we live in, as a whole and in all its parts, it is essential to the scientist that he have confidence in the harmony of nature. He must believe that physical laws revealed to him on earth are in truth universal laws. . . .

In the next two centuries it appeared probable that Newton's view would prevail. For with the development of the wave theory of light scientists found it necessary to endow empty space with certain mechanical properties—to assume, indeed, that space was some kind of substance. Even before Newton's time the French philosopher, Descartes, had argued that the mere separation of bodies by distance proved the existence of a medium between them. And to eighteenth and nineteenth century physicists it was obvious that if light consisted of waves, there must be some medium to support them, just as water propagates the waves of the sea and air transmits the vibrations we call sound. Hence when experiments showed that light can travel in a vacuum, scientists evolved a hypothetical substance called "ether" which they decided must pervade all space and matter. Later on Faraday propounded another kind of ether as the carrier of electric and magnetic forces. When Maxwell finally identified light as an electromagnetic disturbance the case for the ether seemed assured.

A universe permeated with an invisible medium in which the stars wandered and through which light traveled like vibrations in a bowl of jelly was the end product of Newtonian physics. It provided a mechani-

1. See p. 182. Ed.

cal model for all known phenomena of nature, and it provided the fixed frame of reference, the absolute and immovable space, which Newton's cosmology required. Yet the ether presented certain problems, not the least of which was that its actual existence had never been proved. To discover once and for all whether there really was any such thing as ether, two American physicists, A. A. Michelson and E. W. Morley, performed a classic experiment in Cleveland in the year 1881.

The principle underlying their experiment was quite simple. They reasoned that if all space is simply a motionless sea of ether, then the earth's motion through the ether should be detectable and measurable in the same way that sailors measure the velocity of a ship through the sea. As Newton pointed out, it is impossible to detect the movement of a ship through calm waters by any mechanical experiment performed *inside* the ship. Sailors ascertain a ship's speed by throwing a log overboard and watching the unreeling of the knots on the log line. Hence to detect the earth's motion through the ether sea, Michelson and Morley threw a "log" overboard, and the log was a beam of light. For if light really is propagated through the ether, then its velocity should be affected by the ether stream arising from the earth's movement. Specifically a light ray projected in the direction of the earth's movement should be slightly retarded by the ether flow, just as a swimmer is retarded by a current when going upstream. The difference would be slight, for the velocity of light (which was accurately determined in 1849) is 186,284 miles a second, while the velocity of the earth in its orbit around the sun is only 20 miles a second. Hence a light ray sent *against* the ether stream should travel at the rate of 186,264 miles a second, while one sent *with* the ether stream should be clocked at 186,304 miles a second. With these ideas in mind Michelson and Morley constructed an instrument of such great delicacy that it could detect a variation of even a fraction of a mile per second in the enormous velocity of light (Figure 7-3). This instrument, which they called an "interferometer" consisted of a group of mirrors so arranged that a light beam could be split in two and flashed in different directions at the same time. The whole experiment was planned and executed with such painstaking precision that the result could not be doubted. And the result was simply this: there was no difference whatsoever in the velocity of the light beams regardless of their direction.

The Michelson-Morley experiment confronted scientists with an embarrassing alternative. On the one hand they could scrap the ether theory which had explained so many things about electricity, magnetism, and light. Or if they insisted on retaining the ether they had to abandon the still more venerable Copernican theory that the earth is in

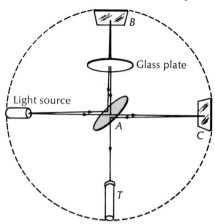

Figure 7-3. The Michelson-Morley interferometer was an arrangement of mirrors designed so that a beam transmitted from the light source was divided and sent in two directions at the same time. This was done by a mirror, A, whose face was only thinly silvered, so that part of the beam could pass through to the mirror C, the remainder being reflected at right angles toward mirror B. Mirrors B and C then reflected the rays back to mirror A where, reunited, they proceeded to the observing telescope T. Since the beam ACT had to pass through the thickness of glass behind the reflecting face of mirror A three times, a clear glass plate of equal thickness was placed between A and B to intercept beam ABT and compensate for the resulting retardation. The whole apparatus was rotated in different directions so that the beams ABT and ACT could be sent with, against, and at right angles to the postulated ether stream. At first it might appear that a trip "downstream," for example, from B to A, should compensate in time for an "upstream" trip from A to B. But this is not so. To row a boat one mile upstream and another mile downstream takes longer than rowing two miles in still water or across current, even with allowance for drift. Had there been any acceleration or retardation of either beam by the ether stream, the optical apparatus at T would have detected it.

motion. To many physicists it seemed almost easier to believe that the earth stood still than that waves—light waves, electromagnetic waves—could exist without a medium to sustain them. It was a serious dilemma and one that split scientific thought for a quarter century. Many new hypotheses were advanced and rejected. The experiment was tried again by Morley and by others, with the same conclusion; the apparent velocity of the earth through the ether was zero.

THE SOLUTION

Among those who pondered the enigma of the Michelson-Morley experiment was a young patent office examiner in Berne, named Albert

Einstein. In 1905, when he was just twenty-six years old, he published a short paper suggesting an answer to the riddle in terms that opened up a new world of physical thought. He began by rejecting the ether theory and with it the whole idea of space as a fixed system or framework, absolutely at rest, within which it is possible to distinguish absolute from relative motion. The one indisputable fact established by the Michelson-Morley experiment was that the velocity of light is unaffected by the motion of the earth. Einstein seized on this as a revelation of universal law. If the velocity of light is constant regardless of the earth's motion, he reasoned, it must be constant regardless of the motion of any sun, moon, star, meteor, or other system moving anywhere in the universe. From this he drew a broader generalization, and asserted that the laws of nature are the same for all uniformly moving systems. This simple statement is the essence of Einstein's Special Theory of Relativity. It incorporates the Galilean Relativity Principle which stated that mechanical laws are the same for all uniformly moving systems. But its phrasing is more comprehensive; for Einstein was thinking not only of mechanical laws but of the laws governing light and other electromagnetic phenomena. So he lumped them together in one fundamental postulate: all the phenomena of nature, all the laws of nature, are the same for all systems that move uniformly relative to one another.

On the surface there is nothing very startling in this declaration. It simply reiterates the scientist's faith in the universal harmony of natural law. It also advises the scientist to stop looking for any absolute, stationary frame of reference in the universe. The universe is a restless place: stars, nebulae, galaxies, and all the vast gravitational systems of outer space are incessantly in motion. But their movements can be described only with respect to each other, for in space there are no directions and no boundaries. It is futile moreover for the scientist to try to discover the "true" velocity of any system by using light as a measuring rod, for the velocity of light is constant throughout the universe and is unaffected either by the motion of its source or the motion of the receiver. Nature offers no absolute standards of comparison; and space is—as another great German mathematician, Leibnitz, clearly saw two centuries before Einstein—simply "the order or relation of things among themselves." Without things occupying it, it is nothing.

Along with absolute space, Einstein discarded the concept of absolute time—of a steady, unvarying, inexorable universal time flow, streaming from the infinite past to the infinite future. Much of the obscurity that has surrounded the Theory of Relativity stems from man's reluctance to recognize that sense of time, like sense of color, is a form of perception. Just as there is no such thing as color without an eye to discern it, so an instant or an hour or a day is nothing without an event to mark it. And just as space is simply a possible order of material objects, so time is simply a possible order of events. The subjectivity of time is best explained in Einstein's own words. "The experiences of an individual," he says, "appear to us arranged in a series of events; in this series the single events which we remember appear to be ordered according to the criterion of 'earlier' and 'later.' There exists, therefore, for the individual, an I-time, or subjective time. This in itself is not measurable. I can, indeed, associate numbers with the events, in such a way that a greater number is associated with the later event than with an earlier one. This association I can define by means of a clock by comparing the order of events furnished by the clock with the order of the given series of events. We understand by a clock something which provides a series of events which can be counted."

By referring our own experiences to a clock (or a calendar) we make time an objective concept. Yet the time intervals provided by a clock or a calendar are by no means absolute quantities imposed on the entire universe by divine edict. All the clocks ever used by man have been geared to our solar system. What we call an hour is actually a measurement in space—an arc of 15 degrees in the apparent daily rotation of the celestial sphere. And what we call a year is simply a measure of the earth's progress in its orbit around the sun. An inhabitant of Mercury, however, would have very different notions of time. For Mercury makes its trip around the sun in 88 of our days, and in that same period rotates just once on its axis. So on Mercury a year and a day amount to the same thing. But it is when science ranges beyond the neighborhood of the sun that all our terrestrial ideas of time become meaningless. For Relativity tells us there is no such thing as a fixed interval of time independent of the system to which it is referred. There is indeed no such thing as simultaneity, there is no such thing as "now," independent of a system of reference. For example a man in New York may telephone a friend in London, and although it is 7:00 P.M. in New York and midnight in London, we may say that they are talking "at the same time." But that is because they are both residents of the same planet, and their clocks are geared to the same astronomical system. A more complicated situation arises if we try to ascertain, for example, what is happening on the star Arcturus "right now." Arcturus is 38 light years away. A light year is the distance light travels in one year, or roughly six trillion miles. If we should try to communicate with Arcturus by radio "right now" it would take 38 years for our message to reach its destination and another 38 years for us to receive a reply.[2] And when we look at Arcturus and say that we see it "now," in 1957, we are actually seeing a ghost

2. Radio waves travel at the same speed as light waves.

—an image projected on our optic nerves by light rays that left their source in 1919. Whether Arcturus even exists "now" nature forbids us to know until 1995.

Despite such reflections it is difficult for earthbound man to accept the idea that *this very instant* which he calls "now" cannot apply to the universe as a whole. Yet in the Special Theory of Relativity Einstein proves by an unanswerable sequence of example and deduction that it is nonsense to think of events taking place simultaneously in unrelated systems.

Relativity: The Special Theory °
by Albert Einstein (1916)

In your schooldays most of you who read this book made acquaintance with the noble building of Euclid's geometry, and you remember —perhaps with more respect than love—the magnificent structure, on the lofty staircase of which you were chased about for uncounted hours by conscientious teachers. By reason of your past experience, you would certainly regard everyone with disdain who should pronounce even the most out-of-the-way proposition of this science to be untrue. But perhaps this feeling of proud certainty would leave you immediately if someone were to ask you: "What, then, do you mean by the assertion that these propositions are true?" Let us proceed to give this question a little consideration.

Geometry sets out from certain conceptions such as "plane," "point," and "straight line," with which we are able to associate more or less definite ideas, and from certain simple propositions (axioms) which, in virtue of these ideas, we are inclined to accept as "true." Then, on the basis of a logical process, the justification of which we feel ourselves compelled to admit, all remaining propositions are shown to follow from these axioms, i.e., they are proven.

A proposition is then correct ("true") when it has been derived in the recognized manner from the axioms. The question of the "truth" of the individual geometrical propositions is thus reduced to one of the "truth" of the axioms. Now it has long been known that the last question is not only unanswerable by the methods of geometry, but that it is in itself entirely without meaning. We cannot ask whether it is true that only

° From Albert Einstein, *Relativity: The Special and General Theory*, translated by Robert W. Lawson, 1931, Crown Publishers, Inc., New York. Copyright © 1961 by the Estate of Albert Einstein. Reprinted by permission of Crown Publishers, Inc.

one straight line goes through two points. We can only say that Euclidean geometry deals with things called "straight lines," to each of which is ascribed the property of being uniquely determined by two points situated on it. The concept "true" does not tally with the assertions of pure geometry, because by the word "true" we are eventually in the habit of designating always the correspondence with a "real" object; geometry, however, is not concerned with the relation of the ideas involved in it to objects of experience, but only with the logical connection of these ideas among themselves.

Every description of the scene of an event or of the position of an object in space is based on the specification of the point on a rigid body (body of reference) with which that event or object coincides. This applies not only to scientific description, but also to everyday life. If I analyze the place specification "Times Square, New York," [1] I arrive at the following result. The earth is the rigid body to which the specification of place refers; "Times Square, New York," is a well-defined point, to which a name has been assigned, and with which the event coincides in space.

This primitive method of place specification deals only with places on the surface of rigid bodies, and is dependent on the existence of points on this surface which are distinguishable from each other. But we can free ourselves from both of these limitations without altering the nature of our specification of position. If, for instance, a cloud is hovering over Times Square, then we can determine its position relative to the surface of the earth by erecting a pole perpendicularly on the Square, so that it reaches the cloud. The length of the pole measured with the standard measuring rod, combined with the specification of the position of the foot of the pole, supplies us with a complete place specification. On the basis of this illustration, we are able to see the manner in which a refinement of the conception of position has been developed.

We thus obtain the following result: Every description of events in space involves the use of a rigid body to which such events have to be referred. The resulting relationship takes for granted that the laws of Euclidean geometry hold for "distances," the "distance" being represented physically by means of the convention of two marks on a rigid body.

"The purpose of mechanics is to describe how bodies change their position in space with time." I should load my conscience with grave sins against the sacred spirit of lucidity were I to formulate the aims of mechanics in this way, without serious reflection and detailed explanations. Let us proceed to disclose these sins.

1. We have chosen this as being more familiar to the American reader than the "Potsdamer Platz, Berlin," which is referred to in the original.

It is not clear what is to be understood here by "position" and "space." I stand at the window of a railway carriage which is traveling uniformly, and drop a stone on the embankment, without throwing it. Then, disregarding the influence of the air resistance, I see the stone descend in a straight line. A pedestrian who observes the misdeed from the footpath notices that the stone falls to earth in a parabolic curve. I now ask: Do the "positions" traversed by the stone lie "in reality" on a straight line or on a parabola? Moreover, what is meant here by motion "in space"? From the considerations of the previous section the answer is self-evident. In the first place, we entirely shun the vague word "space," of which, we must honestly acknowledge, we cannot form the slightest conception, and we replace it by "motion relative to a practically rigid body of reference." The positions relative to the body of reference (railway carriage or embankment) have already been defined in detail in the preceding section. If instead of "body of reference" we insert "system of co-ordinates," which is a useful idea for mathematical description, we are in a position to say: The stone traverses a straight line relative to a system of co-ordinates rigidly attached to the carriage, but relative to a system of co-ordinates rigidly attached to the ground (embankment) it describes a parabola. With the aid of this example it is clearly seen that there is no such thing as an independently existing trajectory (literally, "path-curve" [2]), but only a trajectory relative to a particular body of reference.

In order to have a *complete* description of the motion we must specify how the body alters its position *with time;* i.e., for every point on the trajectory it must be stated at what time the body is situated there. These data must be supplemented by such a definition of time that, in virtue of this definition, these time values can be regarded essentially as magnitudes (results of measurements) capable of observation. If we take our stand on the ground of classical mechanics, we can satisfy this requirement for our illustration in the following manner. We imagine two clocks of identical construction; the man at the railway carriage window is holding one of them, and the man on the footpath the other. Each of the observers determines the position on his own reference body occupied by the stone at each tick of the clock he is holding in his hand. In this connection we have not taken account of the inaccuracy involved by the finiteness of the velocity of propagation of light. With this and with a second difficulty prevailing here we shall have to deal in detail later.

As long as one was convinced that all natural phenomena were capable of representation with the help of classical mechanics, there was no need to doubt the validity of this principle of relativity. But in view of

2. That is, a curve along which the body moves.

the more recent development of electrodynamics and optics, it became more and more evident that classical mechanics affords an insufficient foundation for the physical description of all natural phenomena. At this juncture the question of the validity of the principle of relativity became ripe for discussion, and it did not appear impossible that the answer to this question might be in the negative.

Now in virtue of its motion in an orbit round the sun, our earth is comparable with a railway carriage traveling with a velocity of about 30 kilometers per second. If the principle of relativity were not valid we should therefore expect that the direction of motion of the earth at any moment would enter into the laws of nature, and also that physical systems in their behavior would be dependent on the orientation in space with respect to the earth.

However, the most careful observations have never revealed such anisotropic properties in terrestrial physical space, i.e., a physical nonequivalence of different directions.[3] This is very powerful argument in favor of the principle of relativity.

There is hardly a simpler law in physics than that according to which light is propagated in empty space. Every child at school knows, or believes he knows, that this propagation takes place in straight lines with a velocity $c = 300,000$ km./sec. At all events we know with great exactness that this velocity is the same for all colors, because if this were not the case, the minimum of emission would not be observed simultaneously for different colors during the eclipse of a fixed star by its dark neighbor. By means of similar considerations based on observations of double stars, the Dutch astronomer De Sitter was also able to show that the velocity of propagation of light cannot depend on the velocity of motion of the body emitting the light. The assumption that this velocity of propagation is dependent on the direction "in space" is in itself improbable.

At this juncture the theory of relativity entered the arena. As a result of an analysis of the physical conceptions of time and space, it became evident that *in reality there is not the least incompatibility between the principle of relativity and the law of propagation of light,* and that by systematically holding fast to both these laws a logically rigid theory could be arrived at. This theory has been called the *special theory of relativity* to distinguish it from the extended theory, with which we shall deal later.

Lightning has struck the rails on our railway embankment at two places A and B far distant from each other. I make the additional asser-

3. The reference here is to the negative results of the Michelson-Morley experiment described on p. 341. Ed.

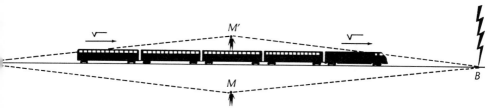

Figure 7-4

tion that these two lightning flashes occurred simultaneously. If I ask you whether there is sense in this statement, you will answer my question with a decided "Yes." But if I now approach you with the request to explain to me the sense of the statement more precisely, you find after some consideration that the answer to this question is not so easy as it appears at first sight.

Are two events (e.g., the two strokes of lightning A and B) which are simultaneous with reference to the *railway embankment* also simultaneous *relatively to the train?* We shall show directly that the answer must be in the negative (Figure 7-4).

When we say that the lightning strokes A and B are simultaneous with respect to the embankment, we mean: the rays of light emitted at the places A and B where the lightning occurs, meet each other at the mid-point M of the length A→B of the embankment. But the events A and B also correspond to positions A and B on the train. Let M′ be the mid-point of the distance A→B on the traveling train. Just when the flashes [4] of lightning occur, this point M′ naturally coincides with the point M, but it moves toward the right, with the velocity v of the train. If an observer sitting in the position M′ in the train did not possess this velocity, then he would remain permanently at M, and the light rays emitted by the flashes of lightning A and B would reach him simultaneously, i.e., they would meet just where he is situated. Now in reality (considered with reference to the railway embankment) he is hastening toward the beam of light coming from B, whilst he is riding on ahead of the beam of light coming from A. Hence the observer will see the beam of light emitted from B earlier than he will see that emitted from A. Observers who take the railway train as their reference body must therefore come to the conclusion that the lightning flash B took place earlier than the lightning flash A. We thus arrive at the important result:

Events which are simultaneous with reference to the embankment are

4. As judged from the embankment.

not simultaneous with respect to the train, and vice versa (relativity of simultaneity).

Our train of thought in the foregoing pages can be epitomized in the following manner. Experience has led to the conviction that, on the one hand, the principle of relativity holds true, and that on the other hand the velocity of transmission of light *in vacuo* has to be considered equal to a constant *c*. By uniting these two postulates we obtained the law of transformation for the rectangular co-ordinates *x, y, z,* and the time *t* of the events which constitute the processes of nature. In this connection we did not obtain the Galilei transformation, but, differing from classical mechanics, the *Lorentz transformation* [see following selection]. General laws of nature are covariant with respect to Lorentz transformations.

It is clear from our previous considerations that the (special) theory of relativity has grown out of electrodynamics and optics. In these fields it has not appreciably altered the predictions of theory, but it has considerably simplified the theoretical structure, i.e., the derivation of laws, and—what is incomparably more important—it has considerably reduced the number of independent hypotheses forming the basis of theory. The special theory of relativity has rendered the Maxwell-Lorentz theory so plausible that the latter would have been generally accepted by physicists even if experiment had decided less unequivocally in its favor.

Classical mechanics required to be modified before it could come into line with the demands of the special theory of relativity. For the main part, however, this modification affects only the laws for rapid motions, in which the velocities of matter *v* are not very small as compared with the velocity of light. We have experience of such rapid motions only in the case of electrons and ions; for other motions the variations from the laws of classical mechanics are too small to make themselves evident in practice. We shall not consider the motion of stars until we come to speak of the general theory of relativity.

Let me add a final remark of a fundamental nature. The success of the Faraday-Maxwell interpretation of electromagnetic action at a distance resulted in physicists becoming convinced that there are no such things as instantaneous actions at a distance (not involving an intermediary medium) of the type of Newton's law of gravitation. According to the theory of relativity, action at a distance with the velocity of light always takes the place of instantaneous action at a distance or of action at a distance with an infinite velocity of transmission.

The Universe and Dr. Einstein (continued) *
by Lincoln Barnett (1957)

The paradox of the lightning flashes dramatizes one of the subtlest and most difficult concepts in Einstein's philosophy: the relativity of simultaneity. It shows that man cannot assume that his subjective sense of "now" applies to all parts of the universe. For, Einstein points out, "every reference body (or coordinate system) has its own particular time; unless we are told the reference body to which the statement of time refers, there is no meaning in a statement of the time of an event." The fallacy in the old principle of the addition of velocities lies therefore in its tacit assumption that the duration of an event is independent of the state of motion of the system of reference. In the case of the man pacing the deck of a ship, for example, it was assumed that if he walked three miles in one hour as timed by a clock on the moving ship, his rate would be just the same timed by a stationary clock anchored somehow in the sea. It was further assumed that the distance he traversed in one hour would have the same value whether it was measured relative to the deck of the ship (the moving system) or relative to the sea (the stationary system). This constitutes a second fallacy in the addition of velocities—for distance, like time, is a relative concept, and there is no such thing as a space interval independent of the state of motion of the system of reference.

Einstein asserted, therefore, that the scientist who wishes to describe the phenomena of nature in terms that are consistent for all systems throughout the universe must regard measurements of time and distance as variable quantities. The equations comprising the Lorentz transformation do just that. They preserve the velocity of light as a universal constant, but modify all measurements of time and distance according to the velocity of each system of reference.[1]

---

* From Lincoln Barnett, *The Universe and Dr. Einstein*, 2nd ed., William Sloane Associates, New York, 1957. Copyright 1948 by Harper & Row, Publishers, New York; copyright 1957 by Lincoln Barnett. Reprinted by permission of William Sloane Associates.

1. The Lorentz transformation relates distances and times observed on moving systems with those observed on systems relatively at rest. Suppose, for example, that a system, or reference body, is moving in a certain direction, then *according to the old principle of the addition of velocities,* a distance or length x', measured with respect to the moving system along the direction of motion, is related to length x, measured with respect to a relatively stationary system, by the equation $x' = x \pm vt$, where v is the velocity of the moving system and t is the time. Dimensions y' and z', mea-

So although the Dutch physicist H. A. Lorentz had originally developed his equations to meet a specific problem, Einstein made them the basis of a tremendous generalization and to the edifice of Relativity another axiom: the laws of nature preserve their uniformity in all systems when related by the Lorentz transformation. Stated thus, in the abstract language of mathematics the significance of this axiom can scarcely be apparent to the layman. But in physics an equation is never a pure abstraction; it is simply a kind of shorthand expression which the scientist finds convenient to describe the phenomena of nature. Sometimes it is also a Rosetta Stone in which the theoretical physicist can decipher secret realms of knowledge. And so by deduction from the message written in the equations of the Lorentz transformation, Einstein discovered a number of new and extraordinary truths about the physical universe.

These truths can be described in very concrete terms. For once he had evolved the philosophical and mathematical bases of Relativity, Einstein had to bring them into the laboratory, where abstractions like time and space are harnessed by means of clocks and measuring rods. And so translating his basic ideas about time and space into the language of the laboratory, he pointed out some hitherto unsuspected properties of clocks and rods. For example: a clock attached to any moving system runs at a different rhythm from a stationary clock; and a measuring rod attached to any moving system changes its length according to

sured with respect to the moving system at right angles to x' and at right angles to each other (i.e., height and breadth), are related to dimensions y and z on the relatively stationary system by $y' = y$, and $z' = z$. And finally a time interval t', clocked with respect to the moving system, is related to time interval t, clocked with respect to the relatively stationary system, by $t' = t$. In other words, distances and times are not affected, *in classical physics*, by the velocity of the system in question. *But it is this presupposition which leads to the paradox of the lightning flashes.* The Lorentz transformation reduces the distances and times observed on moving systems to the conditions of the stationary observer, keeping the velocity of light c a constant for all observers. Here are the equations of the Lorentz transformation which have supplanted the older and evidently inadequate relationships cited above:

$$x' = \frac{x - vt}{\sqrt{1 - (v^2/c^2)}}$$
$$y' = y$$
$$z' = z$$
$$t' = \frac{t - (v/c^2)x}{\sqrt{1 - (v^2/c^2)}}$$

It will be noted that, as in the old transformation law, dimensions y' and z' are unaffected by motion. It will also be seen that if the velocity of the moving system v is small relative to the velocity of light c, then the equations of the Lorentz transformation reduce themselves to the relations of the old principle of the addition of velocities. But as the magnitude of v approaches that of c, then the values of x' and t' are radically changed.

the velocity of the system. Specifically the clock slows down as its velocity increases, and the measuring rod shrinks in the direction of its motion. These peculiar changes have nothing to do with the construction of the clock or the composition of the rod. The clock can be a pendulum clock, a spring clock, or an hour glass. The measuring rod can be a wooden ruler, a metal yardstick, or a ten-mile cable. The slowing of the clock and the contraction of the rod are not mechanical phenomena; an observer riding along with the clock and the measuring rod would not notice these changes. But a stationary observer, i.e., stationary relative to the moving system, would find that the moving clock has slowed down with respect to his stationary clock, and that the moving rod has contracted with respect to his stationary units of measurement.

This singular behavior of moving clocks and yardsticks accounts for the constant velocity of light. It explains why all observers in all systems everywhere, regardless of their state of motion, will always find that light strikes their instruments and departs from their instruments at precisely the same velocity. For as their own velocity approaches that of light, their clocks slow down, their yardsticks contract, and all their measurements are reduced to the values obtained by a relatively stationary observer. The laws governing these contractions are defined by the Lorentz transformation and they are very simple: the greater the speed, the greater the contraction. A yardstick moving with 90 per cent the velocity of light would shrink to about half its length; thereafter the rate of contraction becomes more rapid; and if the stick should attain the velocity of light, it would shrink away to nothing at all. Similarly a clock traveling with the velocity of light would stop completely. From this it follows that nothing can ever move faster than light, no matter what forces are applied. Thus Relativity reveals another fundamental law of nature: the velocity of light is the top limiting velocity in the universe.

At first meeting these facts are difficult to digest but that is simply because classical physics assumed, unjustifiably, that an object preserves the same dimensions whether it is in motion or at rest and that a clock keeps the same rhythm in motion and at rest. Common sense dictates that this must be so. But as Einstein has pointed out, common sense is actually nothing more than a deposit of prejudices laid down in the mind prior to the age of eighteen. Every new idea one encounters in later years must combat this accretion of "self-evident" concepts. And it is because of Einstein's unwillingness ever to accept any unproven principle as self-evident that he was able to penetrate closer to the underlying realities of nature than any scientist before him. Why, he asked, is it any more strange to assume that moving clocks slow down and moving rods contract, than to assume that they don't? The reason classical physics took the latter view for granted is that man, in his everyday experience, never encounters velocities great enough to make these changes manifest. In an automobile, an airplane, even in a V-2 rocket, the slowing down of a watch is immeasurable. It is only when velocities approximate that of light that relativistic effects can be detected. The equations of the Lorentz transformation show very plainly that at ordinary speeds the modification of time and space intervals amounts practically to zero. Relativity does not therefore contradict classical physics. It simply regards the old concepts as limiting cases that apply solely to the familiar experiences of man.

Einstein thus surmounts the barrier reared by man's impulse to define reality solely as he perceives it through the screen of his sense. . . . Relativity shows that we cannot foretell the phenomena accompanying great velocities from the sluggish behavior of objects visible to man's indolent eye. Nor may we assume that the laws of Relativity deal with exceptional occurrences; on the contrary they provide a comprehensive picture of an incredibly complex universe in which the simple mechanical events of our earthly experience are the exceptions. The present-day scientist, coping with the tremendous velocities that prevail in the fast universe of the atom or with the immensities of sidereal space and time, finds the old Newtonian laws inadequate. But Relativity provides him in every instance with a complete and accurate description of nature.

Whenever Einstein's postulates have been put to test, their validity has been amply confirmed. Remarkable proof of the relativistic retardation of time intervals came out of an experiment performed by H. E. Ives of the Bell Telephone Laboratories in 1936. A radiating atom may be regarded as a kind of clock in that it emits light of a definite frequency and wave-length which can be measured with great precision by means of a spectroscope. Ives compared the light emitted by hydrogen atoms moving at high velocities with that emitted by hydrogen atoms at rest, and found that the frequency of vibration of the moving atoms was reduced in exact accordance with the predictions of Einstein's equations. Someday science may devise a far more interesting test of the same principle. Since any periodic motion serves to measure time, the human heart, Einstein has pointed out, is also a kind of clock. Hence, according to Relativity, the heartbeat of a person traveling with a velocity close to that of light would be relatively slowed, along with his respiration and all other physiological processes. He would not notice this retardation because his watch would slow down in the same degree. But judged by a stationary timekeeper he would "grow old" less rapidly.

In order to describe the mechanics of the physical universe, three quantities are required: time, distance, and mass. Since time and distance are relative quantities one might guess that the mass of a body also varies with its state of motion. And indeed the most important practical results of Relativity have arisen from this principle—the relativity of mass.

In its popular sense, "mass" is just another word for "weight." But as used by the physicist, it denotes a rather different and more fundamental property of matter: namely, resistance to a change of motion. A greater force is necessary to move or stop a freight car than a velocipede; the freight car resists a change in its motion more stubbornly than the velocipede because it has greater mass. In classical physics the mass of any body is a fixed and unchanging property. Thus the mass of a freight car should remain the same whether it is at rest on a siding, rolling across country at 60 miles an hour, or hurtling through outer space at 60,000 miles a second. But Relativity asserts that the mass of a moving body is by no means constant, but increases with its velocity relative to an observer. The old physics failed to discover this fact simply because man's senses and instruments are too crude to note the infinitesimal increases of mass produced by the feeble accelerations of ordinary experience. They become perceptible only when bodies attain velocities relatively close to that of light. (This phenomenon, incidentally, does not conflict with the relativistic contraction of length.[2] One is tempted to ask: how can an object become smaller and at the same time get heavier? The contraction, it should be noted, is only in the direction of motion; width and breadth are unaffected. Moreover mass is not merely "heaviness" but resistance to a change in motion.)

Einstein's equation giving the increase of mass with velocity is similar in form to the other equations of Relativity but vastly more important in its consequences:

$$m = \frac{m_0}{\sqrt{1 - v^2/c^2}}$$

Here m stands for the mass of a body moving with velocity v, m_0 for its mass when at rest, and c for the velocity of light. Anyone who has ever studied elementary algebra can readily see that if v is small, as are all the velocities of ordinary experience, then the difference between m_0 and m is practically zero. But when v approaches the value of c then the increase of mass becomes very great, reaching infinity when the velocity of the moving body reaches the velocity of light. Since a body of infinite mass would offer infinite resistance to motion the conclusion is once again reached that no material body can travel with the speed of light.

Of all aspects of Relativity the principle of increase of mass has been most often verified and most fruitfully applied by experimental physi-

2. See "Fitzgerald-Lorentz contraction," in Glossary. Ed.

cists. Electrons moving in powerful electrical fields and beta particles ejected from the nuclei of radioactive substances attain velocities ranging up to 99 per cent that of light. For atomic physicists concerned with these great speeds, the increase of mass predicted by Relativity is no arguable theory but an empirical fact their calculations cannot ignore. In fact the mechanics of the proton-synchroton and other new super-energy machines must be designed to allow for the increasing mass of particles as their speed approaches the velocity of light, in order to make them operate at all.

By further deduction from his principle of Relativity of mass, Einstein arrived at a conclusion of incalculable importance to the world. His train of reasoning ran somewhat as follows: since the mass of a moving body increases its motion increases, and since motion is a form of energy (kinetic energy), then the increased mass of a moving body comes from its increased energy. In short, energy has mass! By a few comparatively simple mathematical steps, Einstein found the value of the equivalent mass m in any unit of energy E and expressed it by the equation $m = E/c^2$. Given this relation a high school freshman can take the remaining algebraic step necessary to write the most important and certainly the most famous equation in history: $E = mc^2$.

The part played by this equation in the development of the atomic bomb is familiar to most newspaper readers. It states in the shorthand of physics that the energy contained in any particle of matter is equal to the mass of that body multiplied by the square of the velocity of light, the proper units being chosen in each case. This extraordinary relationship becomes more vivid when its terms are translated into concrete values: i.e., one kilogram of coal (about two pounds), if converted *entirely* into energy, would yield 25 billion kilowatt hours of electricity or as much as all the power plants in the U.S. could generate by running steadily for two months.

$E = mc^2$ provides the answer to many of the long-standing mysteries of physics. It explains how radioactive substances like radium and uranium are able to eject particles at enormous velocities and to go on doing so for millions of years. It explains how the sun and all the stars can go on radiating light and heat for billions of years; for if our sun were being consumed by ordinary processes of combustion, the earth would have died in frozen darkness eons ago. It reveals the magnitude of the energy that slumbers in the nuclei of atoms, and forecasts how many grams of uranium must go into a bomb in order to destroy a city. Finally it discloses some fundamental truths about physical reality. Prior to Relativity scientists had pictured the universe as a vessel containing two distinct elements, matter and energy—the former inert, tangible, and characterized by a property called mass, and the latter active, invisible, and without mass. But Einstein showed that mass and energy are equivalent: the property called mass is simply concentrated energy. In other words matter is energy and energy is matter, and the distinction is simply one of temporary state.

In the light of this broad principle many puzzles of nature are resolved. The baffling interplay of matter and radiation, which appears sometimes to be a concourse of particles and sometimes a meeting of waves, becomes more understandable. The dual role of the electron as a unit of matter and a unit of electricity, the wave electron, the photon, waves of matter, waves of probability, a universe of waves—all these seem less paradoxical. For all these concepts simply describe different manifestations of the same underlying reality, and it no longer makes sense to ask what any one of them "really" is. Matter and energy are interchangeable. If matter sheds its mass and travels with the speed of light we call it radiation or energy. And conversely if energy congeals and takes on a different form we call it matter. Heretofore science could only note their ephemeral properties and relations as they touched the perceptions of earth-bound man. But since July 16, 1945 man has been able to transform one into the other. For on that night at Alamogordo, New Mexico, man for the first time transmuted a substantial quantity of matter into the light, heat, sound, and motion which we call energy.

Yet the fundamental mystery remains. The whole march of science toward the unification of concepts—the reduction of all matter to elements and then to a few types of particles, the reduction of "forces" to the single concept "energy," and then the reduction of matter *and* energy to a single basic quantity—leads still to the unknown.

Relativity: The General Theory *
by Albert Einstein (1916)

The non-mathematician is seized by a mysterious shuddering when he hears of "four-dimensional" things, by a feeling not unlike that awakened by thoughts of the occult. And yet there is no more commonplace statement than that the world in which we live is a four-dimensional space-time continuum.

That we have not been accustomed to regard the world in this sense

* From Albert Einstein, *Relativity: The Special and General Theory*, translated by Robert W. Lawson, 1931, Crown Publishers, Inc., New York. Copyright © 1961 by the Estate of Albert Einstein. Reprinted by permission of Crown Publishers, Inc.

as a four-dimensional continuum is due to the fact that in physics, before the advent of the theory of relativity, time played a different and more independent role, as compared with the space co-ordinates. It is for this reason that we have been in the habit of treating time as an independent continuum. As a matter of fact, according to classical mechanics, time is absolute, i.e., it is independent of the position and the condition of motion of the system of co-ordinates.

Since the introduction of the special principle of relativity has been justified, every intellect which strives after generalization must feel the temptation to venture the step toward the general principle of relativity. But a simple and apparently quite reliable consideration seems to suggest that, for the present at any rate, there is little hope of success in such an attempt. Let us imagine ourselves transferred to our old friend the railway carriage, which is traveling at a uniform rate. As long as it is moving uniformly, the occupant of the carriage is not sensible of its motion, and it is for this reason that he can without reluctance interpret the facts of the case as indicating that the carriage is at rest but the embankment in motion. Moreover, according to the special principle of relativity, this interpretation is quite justified also from a physical point of view.

If the motion of the carriage is now changed into a non-uniform motion, as for instance by a powerful application of the brakes, then the occupant of the carriage experiences a correspondingly powerful jerk forward. The retarded motion is manifested in the mechanical behavior of bodies relative to the person in the railway carriage. The mechanical behavior is different from that of the case previously considered, and for this reason it would appear to be impossible that the same mechanical laws hold relatively to the non-uniformly moving carriage, as hold with reference to the carriage when at rest or in uniform motion.

"If we pick up a stone and then let it go, why does it fall to the ground?" The usual answer to this question is: "Because it is attracted by the earth." Modern physics formulates the answer rather differently.

The action of the earth on the stone takes place indirectly. The earth produces in its surroundings a gravitational field, which acts on the stone and produces its motion of fall.

In contrast to electric and magnetic fields, the gravitational field exhibits a most remarkable property, which is of fundamental importance for what follows. Bodies which are moving under the sole influence of a gravitational field receive an acceleration, *which does not in the least depend either on the material or on the physical state of the body.* For instance, a piece of lead and a piece of wood fall in exactly the same manner in a gravitational field (*in vacuo*) when they start off from rest or with the same initial velocity.

We obtain a new result of fundamental importance when we carry out the analogous consideration for a ray of light. We conclude *that, in general, rays of light are propagated curvilinearly in gravitational fields.* In two respects this result is of great importance.

In the first place, it can be compared with the reality. Although a detailed examination of the question shows that the curvature of light rays required by the general theory of relativity is only exceedingly small for the gravitational fields at our disposal in practice, its estimated magnitude for light rays passing the sun at grazing incidence is nevertheless 1.7 seconds of arc. This ought to manifest itself in the following way. As seen from the earth, certain fixed stars appear to be in the neighborhood of the sun, and are thus capable of observation during total eclipse of the sun. At such times these stars ought to appear to be displaced outward from the sun by an amount indicated above, as compared with their apparent position in the sky when the sun is situated at another part of the heavens. The examination of the correctness or otherwise of this deduction is a problem of the greatest importance, the early solution of which is to be expected of astronomers.[1]

To establish a relationship between time and space in the physical world it is necessary to consider the behavior of the clocks and measuring rods with which we might measure them. The problem of accurate and meaningful measurement turns out to be especially complicated when the timepieces and measuring rods are moving rapidly—say rotating at the speed of light—from the point of view of the observer. Hence it becomes necessary to specify the position of objects and events in the physical world by means of a mathematical system which does not depend on the rigid-body geometry of Euclid for definition of length or the straight-line co-ordinate system established by Descartes (Cartesian co-ordinates) for marking their position. Fortunately, the great mathematician, Gauss, had worked out his own system of co-ordinates (Gaussian co-ordinates) by which positions (that is, the distances between points) can be established by saying in effect that they are to be found at the intersections of two or more named curves. Since the four-dimensional space-time continuum of relativity is not rigidly Euclidean in character, it is extremely convenient to use Gaussian co-ordinates instead of a rigid

1. By means of the star photographs of two expeditions equipped by a Joint Committee of the Royal and Royal Astronomical Societies, the existence of the deflection of light demanded by theory was first confirmed during the solar eclipse of May 29, 1919. [See Figures 7-5 and 7-6. Ed.]

body of reference to determine the time and place of an event or object in nature.

The following statement corresponds to the fundamental idea of the general principle of relativity: "*All Gaussian co-ordinate systems are essentially equivalent for the formulation of the general laws of nature.*"

In gravitational fields there are no such things as rigid bodies with Euclidean properties; thus the fictitious rigid body of reference is of no avail in the general theory of relativity. The motion of clocks is also influenced by gravitational fields, and in such a way that a physical definition of time which is made directly with the aid of clocks has by no means the same degree of plausibility as in the special theory of relativity.

For this reason non-rigid reference bodies are used, which are as a whole not only moving in any way whatsoever, but which also suffer alterations in form *ad lib,* during their motion. Clocks, for which the law of motion is of any kind, however irregular, serve for the definition of time. We have to imagine each of these clocks fixed at a point on the non-rigid reference body. These clocks satisfy only the one condition, that the "readings" which are observed simultaneously on adjacent clocks (in space) differ from each other by an indefinitely small amount.

If we confine the application of the theory to the case where the gravitational fields can be regarded as being weak, and in which all masses move with respect to the co-ordinate system with velocities which are small compared with the velocity of light, we then obtain as a first approximation the Newtonian theory. Thus the latter theory is obtained here without any particular assumption, whereas Newton had to introduce the hypothesis that the force of attraction between mutually attracting material points is inversely proportional to the square of the distance between them. If we increase the accuracy of the calculation, deviations from the theory of Newton make their appearance, practically all of which must nevertheless escape the test of observation owing to their smallness.

We must draw attention here to one of these deviations. According to Newton's theory, a planet moves round the sun in an ellipse, which would permanently maintain its position with respect to the fixed stars, if we could disregard the motion of the fixed stars themselves and the action of the other planets under consideration. Thus, if we correct the observed motion of the planets for these two influences, and if Newton's theory be strictly correct, we ought to obtain for the orbit of the planet an ellipse, which is fixed with reference to the fixed stars. This deduction, which can be tested with great accuracy, has been confirmed for all the planets save one, with the precision that is capable of being obtained by the delicacy of observation attainable at the present time. The sole exception is Mercury, the planet which lies nearest the sun. Since the time of Leverrier, it has been known that the ellipse corresponding to the orbit of Mercury, after it has been corrected for the influences mentioned above, is not stationary with respect to the fixed stars, but that it rotates exceedingly slowly in the plane of the orbit and in the sense of the orbital motion. The value obtained for this rotary movement of the orbital ellipse was forty-three seconds of arc per century, an amount ensured to be correct to within a few seconds of arc. This effect can be explained by means of classical mechanics only on the assumption of hypotheses which have little probability, and which were devised solely for this purpose.

On the basis of the general theory of relativity, it is found that the ellipse of every planet round the sun must necessarily rotate in the manner indicated above; that for all the planets, with the exception of Mercury, this rotation is too small to be detected with the delicacy of observation possible at the present time; but that in the case of Mercury it must amount to forty-three seconds of arc per century, a result which is strictly in agreement with observation [Figure 7-2].

If we ponder over the question as to how the universe, considered as a whole, is to be regarded, the first answer that suggests itself to us is surely this: As regards space (and time) the universe is infinite. There are stars everywhere, so that the density of matter, although very variable in detail, is nevertheless on the average everywhere the same. In other words: However far we might travel through space, we should find everywhere an attenuated swarm of fixed stars of approximately the same kind of density.

This view is not in harmony with the theory of Newton. The latter theory rather requires that the universe should have a kind of center in which the density of the stars is a maximum, and that as we proceed outward from this center the group density of the stars should diminish, until finally, at great distances, it is succeeded by an infinite region of emptiness. This stellar universe ought to be a finite island in the infinite ocean of space.

This conception is in itself not very satisfactory. It is less satisfactory because it leads to the result that the light emitted by the stars and also individual stars of the stellar system are perpetually passing out into infinite space, never to return, and without ever again coming into interaction with other objects of nature. Such a finite material universe would be destined to become gradually but systematically impoverished.

But speculations on the structure of the universe also move in quite another direction. The development of non-Euclidean geometry led to the recognition of the fact that we can cast doubt on the *infiniteness* of our space without coming into conflict with the laws of thought or with experience (Riemann, Helmholtz).

In the first place, we imagine an existence in two-dimensional space. Flat beings with flat implements, and in particular flat rigid measuring rods, are free to move in a *plane* [Figure 7-8, top]. For them nothing exists outside of this plane: that which they observe to happen to themselves and to their flat "things" is the all-inclusive reality of their plane. In particular, the constructions of plane Euclidean geometry can be carried out by means of the rods. In contrast to ours, the universe of these beings is two-dimensional; but, like ours, it extends to infinity. In their universe there is room for an infinite number of identical squares made up of rods, i.e., its volume (surface) is infinite. If these beings say their universe is "plane," there is sense in the statement, because they mean that they can perform the constructions of plane Euclidean geometry with their rods. In this connection the individual rods always represent the same distance, independently of their position.

Let us consider now a second two-dimensional existence, but this time on a spherical surface instead of on a plane [Figure 7-8, bottom]. The flat beings with their measuring rods and other objects fit exactly on this surface and they are unable to leave it. Their whole universe of observation extends exclusively over the surface of the sphere. Are these beings able to regard the geometry of their universe as being plane geometry and their rods withal as the realization of "distance"? They cannot do this. For if they attempt to realize a straight line, they will obtain a curve, which we "three-dimensional beings" designate as a great circle, i.e., a self-contained line of definite finite length, which can be measured up by means of a measuring rod. Similarly, this universe has a finite area that can be compared with the area of a square constructed with rods. The great charm resulting from this consideration lies in the recognition of the fact that *the universe of these beings is finite and yet has no limits.*

But the spherical-surface beings do not need to go on a world tour in order to perceive that they are not living in a Euclidean universe. They can convince themselves of this on every part of their "world," provided they do not use too small a piece of it.

To this two-dimensional sphere universe there is a three-dimensional analogy, namely, the three-dimensional spherical space which was discovered by Riemann. Its points are likewise all equivalent. It possesses a finite volume, which is determined by its "radius."

According to the general theory of relativity, the geometrical properties of space are not independent, but they are determined by matter. Thus we can draw conclusions about the geometrical structure of the universe only if we base our considerations on the state of the matter as being something that is known. We know from experience that, for a suitably chosen co-ordinate system, the velocities of the stars are small as compared with the velocity of transmission of light. We can thus as a rough approximation arrive at a conclusion as to the nature of the universe as a whole if we treat the matter as being at rest.

We already know from our previous discussion that the behavior of measuring rods and clocks is influenced by gravitational fields, i.e., by the distribution of matter. This in itself is sufficient to exclude the possibility of the exact validity of Euclidean geometry in our universe. But it is conceivable that our universe differs only slightly from a Euclidean one, and this notion seems all the more probable, since calculations show that the metrics of surrounding space is influenced only to an exceedingly small extent by masses even of the magnitude of our sun [Figures 7-5 and 7-6]. We might imagine that, as regards geometry, our universe behaves analogously to a surface which is irregularly curved in its individual parts, but which nowhere departs appreciably from a plane: something like the rippled surface of a lake [Figure 7-7, top]. Such a universe might fittingly be called a quasi-Euclidean universe. As regards its space it would be infinite. But calculation shows that in a quasi-Euclidean universe the average density of matter would necessarily be nil. Thus such a universe could not be inhabited by matter everywhere; it would present to us that unsatisfactory picture which we portrayed.

If we are to have in the universe an average density of matter which differs from zero, however small may be that difference, then the universe cannot be quasi-Euclidean. On the contrary, the results of calculation indicate that if matter be distributed uniformly, the universe would necessarily be spherical (or elliptical) [Figure 7-7, middle]. Since in reality the detailed distribution of matter is not uniform, the real universe will deviate in individual parts from the spherical, i.e., the universe will be quasi-spherical [Figure 7-7, bottom]. But it will be necessarily finite. In fact, the theory (of relativity) supplies us with a simple connection between the space expanse of the universe and the average density of matter in it.[2]

2. After publishing this essay in 1916, Einstein worked on several other models of the universe, illustrated in Figure 7-7 and discussed further in Part 8. Ed.

Illustrations of Relativity *

by George Gamow (1947)

Einstein came to the remarkable conclusion that the phenomenon of gravity is merely the effect of the curvature of the four-dimensional space-time world (Figures 7-5 and 7-6). In fact we may now discard as

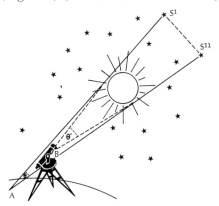

Figure 7-5. The light rays from two stars S $_I$ and S $_{II}$ located (at the moment of observation) at opposite sides of the sun disk converge into a theodolite, which measures the angle between them. The experiment is then repeated later when the sun is out of the way, and the two angles are compared. If they are different we have proof that the mass of the sun changes the curvature of the space around it, deflecting the rays of light from their original paths. Such an experiment was originally suggested by Einstein to test his theory. The reader may understand the situation somewhat better by comparing it with its two-dimensional analogy shown in Figure 7-6. Obviously there was a practical barrier to carrying out Einstein's suggestion under ordinary conditions: because of the brilliance of the solar disk, you cannot see the stars around it; but during a total solar eclipse the stars are clearly visible in the daytime. In 1919, taking advantage of this fact, a British astronomical expedition to the Principe Islands (West Africa), from which the total solar eclipse of that year could best be observed, actually made the test. The difference of angular distances between the two stars with and without the sun between them was found to be 1.61 in. ±0.30 in. as compared with 1.75 in. predicted by Einstein's theory. Similar results were obtained by various expeditions at later dates. (After George Gamow)

* From George Gamow, *One, Two, Three . . . Infinity*. The Viking Press, Inc., New York. Copyright 1947 by George Gamow. Reprinted by permission of The Viking Press, Inc.

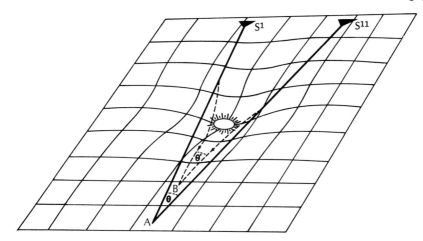

Figure 7-6 (After George Gamow)

inadequate the old statement that the sun exercises a certain force that acts directly on the planets making them describe circular orbits around it. It would be more accurate to say that the mass of the sun curves the space-time world around it, and the world-lines of planets look the way

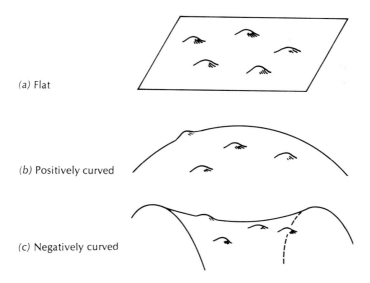

(a) Flat

(b) Positively curved

(c) Negatively curved

Figure 7-7. *Three views of space.* (After George Gamow)

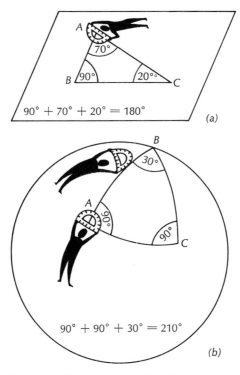

$$90° + 70° + 20° = 180°$$

(a)

$$90° + 90° + 30° = 210°$$

(b)

*Figure 7-8. Two-dimensional scientists of the flat and curved "surface worlds"
check the Euclidian theorem about the sum of the angles in a triangle. (After
George Gamow)*

they do only because they are geodesical lines running through the
curved space.

In Figure 7-7 we give a two-dimensional illustration of the flat space
with "pimples," and two possible types of curved spaces. The so-called
"positively curved" space corresponds to the surface of a sphere or any
other closed geometrical figure, and bends "the same way" regardless of
the direction in which one goes. The opposite type of "negatively
curved" space bends up in one direction, but down in another, and resembles closely the surface of a western saddle. The difference between
the two types of curvatures can be clearly realized if you cut out two
pieces of leather, one from a football, another from the saddle, and
try to straighten them out on a table. You will notice that neither can be
straightened out without stretching or shrinking it, but, whereas the periphery of the football piece must be stretched, that of the saddle piece
must be shrunk. The football piece has not enough material around the

center to flatten it out; the saddle piece has too much, and it gets folded whenever we try to make it flat and smooth.

We can state the same point in still another way. Suppose we count the number of pimples located within one, two, three, etc. inches (counted along the surface) from a certain point. On the flat uncurved surface the number of pimples will increase as the square of the distances, that is, as 1, 4, 9, etc. On a spherical surface the number of pimples will increase more slowly than that, and on a "saddle" surface more rapidly. Thus the two-dimensional shadow scientists living in the surface, and thus having no way of looking at it from outside to notice its shape, will be still able to detect the curvature by counting the number of pimples that fall within the circles of different radii (Figure 7-8). It may be also noticed here that the difference between the positive and negative curvatures shows itself in the measurements of the angles in corresponding triangles. As we have seen in the previous section the sum of angles of a triangle drawn on the surface of a sphere is always larger than 180°. If you try to draw a triangle on the saddle surface you will find that the sum of its angles is always less than 180°.

The above results obtained specifically in regard to curved surfaces can be generalized in regard to curved three-dimensional spaces according to the following table:

Type of space	Behavior at large distances	Angle-sum of a triangle	Volume of a sphere increases
Positively curved (sphere analogue)	Closes on itself	$> 180°$	Slower than cube of the radius
Flat (plane analogue)	Extends into infinity	$= 180°$	As the cube of the radius
Negatively curved (saddle analogue)	Extends into infinity	$< 180°$	Faster than cube of the radius

Theory and Common Sense

"I can't believe *that*," said Alice.

"Can't you?" the Queen said, in a pitying tone. "Try again: draw a long breath, and shut your eyes."

Alice laughed. "There's no use trying," she said: "One can't believe impossible things."

"I daresay you haven't had much practice," said the Queen. "When I was your age, I always did it for half-an-hour a day. Why, sometimes I've believed as many as six impossible things before breakfast."

Lewis Carroll
Through the Looking Glass (1871)

Common Sense and the Universe *

by Stephen Leacock (1942)

I

Speaking last December [1941] at the annual convention of the American Association for the Advancement of Science, and speaking, as it were, in the name of the great 100-inch telescope under his control, Professor Edwin Hubble, of the Mount Wilson Observatory, California, made the glad announcement that the universe is not expanding.[1] This was good news indeed, if not to the general public who had no reason to suspect that it was expanding, at least to those of us who humbly attempt to "follow science." . . .

Now it appears that that distant universe has *not* been receding at all; in fact, it isn't away out there. Heaven knows where it is. Bring it back. Yet not only did the astronomers assert the expansion but they proved it, from the behavior of the red band in the spectrum, which blushed a deeper red at the revelation of it, like the conscious water that "saw its God and blushed" at Cana in Galilee long ago. One of the most distinguished and intelligible of our astronomers, Sir Arthur Eddington, had written a book about it, *The Expanding Universe*, to bring it down to

---

* From Stephen Leacock, *Last Leaves*. Dodd, Mead & Company, Inc., New York. Copyright 1945 by Dodd, Mead & Company, Inc. Reprinted by permission.
1. This conclusion has subsequently been reversed. See Part 8. Ed.

our level. Astronomers at large accepted this universal explosion in all directions as calmly as they once accepted the universal fall of gravitation, or the universal death in the cold under Carnot's Second Law of Thermodynamics.

But the relief brought by Professor Hubble is tempered on reflection by certain doubts and afterthoughts. It is not that I venture any disbelief or disrespect toward science, for that is as atrocious in our day as disbelief in the Trinity in the days of Isaac Newton. But we begin to doubt whether science can quite keep on believing in and respecting itself. If we expand today and contract tomorrow; if we undergo all the doubled-up agonies of the curvature of space only to have the kink called off . . . if we get reconciled to dying a martyr's death at one general, distributed temperature of 459 degrees below zero, the same for all, only to find that the world is perhaps unexpectedly warming up again —then we ask, where are we? To which, of course, Einstein answers "Nowhere," since there is no place to be. So we must pick up our little book again, follow science, and wait for the next astronomical convention.

Let us take this case of the famous Second Law of Thermodynamics, that inexorable scroll of fate which condemned the universe—or at least all life in it—to die of cold. I look back now with regret to the needless tears I have wasted over that, the generous sympathy for the last little band of survivors, dying at 459 degrees below our zero ($-273°$ centigrade), the absolute zero of cold when the molecules cease to move and heat ends. No stove will light at that, for the wood is as cold as the stove, and the match is as cold as both, and the dead fingers motionless.

I remember meeting this inexorable law for the first time in reading, as a little boy, a piece of "popular science" entitled *Our Great Timepiece Running Down.* It was by Richard Proctor, whose science-bogeys were as terrifying as Mrs. Crow's *Night Thoughts,* only slower in action. The sun, it appeared, was cooling; soon it would be all over. Lord Kelvin presently ratified this. Being Scotch, he didn't mind damnation and he gave the sun and the whole solar system only ninety million years more to live.

This famous law was first clearly enunciated in 1824 by the great French physicist, Nicolas Carnot. It showed that all bodies in the universe kept changing their temperature—hot things heated cold, and cold things chilled hot. Thus they pooled their temperature. Like the division of a rich estate among a flock of poor relations, it meant poverty for all. We must all share ultimately the cold of absolute space.

It is true that a gleam came when Ernest Rutherford and others, working on radioactivity, discovered that there might be a contrary process of "stoking up." Atoms exploding into radioactivity would keep the home fires burning in the sun for a long time. This glad news meant that the sun was both much older and much younger than Lord Kelvin had ever thought it was. But even at that it was only a respite. The best they could offer was 1,500,000,000 years. After that we freeze. . . .

At this point I may perhaps pause to explain that the purpose of this article is not to make fun of science, nor to express disbelief in it, but only to suggest its limits. What I want to say is that when the scientist steps out from recording phenomena and offers a general statement of the nature of what is called "reality," the ultimate nature of space, of time, of the beginning of things, of life, of a universe, then he stands exactly where you and I do, and the three of us stand where Plato did—and long before him, Rodin's primitive thinker.

Consider this. Professor Hubble, like Joshua, has called upon the universe to be still. All is quiet. The universe rests, motionless, in the night sky. The mad rush is over. Every star in every galaxy, every island universe, is at least right where it is. But the old difficulty remains: Does it go forever, this world in the sky, or does it stop? Such an alternative has posed itself as a problem for every one of us, somewhere about the age of twelve. We cannot imagine that the stars go on forever. It's unthinkable. But we equally cannot imagine that they come to a stop and that beyond them is nothing, and then more nothing. Unending nothing is as incomprehensible as unending something. This alternative I cannot fathom, nor can Professor Hubble, nor can anyone ever hope to.

Let me turn back in order to make my point of view a little clearer. I propose to traverse again the path along which modern science has dragged those who have tried to follow it for about a century past. It was at first a path singularly easy to tread, provided that one could throw aside the inherited burden of superstition, false belief, and prejudice. For the direction seemed verified and assured all along by the corroboration of science by actual physical results. Who could doubt electricity after the telegraph? Or doubt the theory of light after photography? Or the theory of electricity when read under electric light? At every turn, each new advance of science unveiled new power, new mechanism of life—and of death. To "doubt science" was to be like the farmer at the circus who doubted the giraffe. Science, of course, had somehow to tuck into the same bed as Theology, but it was the theologian who protested. Science just said, "Lie over."

Let us follow then this path.

II

When mediaeval superstition was replaced by the new learning, mathematics, astronomy, and physics were the first sciences to get organized and definite. By the opening of the nineteenth century they were well set; the solar system was humming away so drowsily that Laplace was able to assure Napoleon that he didn't need God to watch over it. Gravitation worked like clockwork and clockwork worked like gravitation. Chemistry, which, like electricity, was nothing but a set of experiments in Benjamin Franklin's time, turned into a science after Lavoisier had discovered that fire was not a thing but a process, something happening to things—an idea so far above the common thought that they guillotined him for it in 1794. Dalton followed and showed that all things could be broken up into a set of very, very small atoms, grouped into molecules all acting according to plan. With Faraday and Maxwell, electricity, which turned out to be the same as magnetism, or interchangeable with it, fell into its place in the new order of science.

By about 1880 it seemed as if the world of science was fairly well explained. Metaphysics still talked in its sleep. Theology still preached sermons. It took issue with much of the new science, especially with geology and the new evolutionary science of life that went with the new physical world. But science paid little attention.

For the whole thing was so amazingly simple. There you had your space and time, two things too obvious to explain. Here you had your matter, made up of solid little atoms, infinitely small but really just like birdseed. All this was set going by and with the Law of Gravitation. Once started, the nebulous world condensed into suns, the suns threw off planets, the planets cooled, life resulted and presently became conscious, conscious life got higher up and higher up till you had apes, then Bishop Wilberforce, and then Professor Huxley.

A few little mysteries remained, such as the question of what space and matter and time and life and consciousness really were. But all this was conveniently called by Herbert Spencer the *Unknowable,* and then locked in a cupboard and left there.

Everything was thus reduced to a sort of Dead Certainty. Just one awkward skeleton remained in the cupboard. And that was the peculiar, mysterious aspect of electricity, which was not exactly a thing and yet more than an idea. There was also, and electricity only helped to make it worse, the old puzzle about "action at a distance." How does gravitation pull all the way from here to the sun? And if there is *nothing* in space, how does light get across from the sun in eight minutes, and even all the way from Sirius in eight years?

Even the invention of "ether" as a sort of universal jelly that could have ripples shaken across it proved a little unconvincing.

Then, just at the turn of the century the whole structure began to crumble.

The first note of warning that something was going wrong came with the discovery of X-rays. Sir William Crookes, accidentally leaving round tubes of rarefied gas, stumbled on "radiant matter," or "matter in the fourth state," as accidentally as Columbus discovered America. The British Government knighted him at once (1897) but it was too late. The thing had started. Then came Guglielmo Marconi with the revelation of more waves, and universal at that. Light, the world had learned to accept, because we can see it, but this was fun in the dark.

There followed the researches of the radioactivity school and, above all, those of Ernest Rutherford which revolutionized the theory of matter. I knew Rutherford well as we were colleagues at McGill for seven years. I am quite sure that he had no original intention of upsetting the foundations of the universe. Yet that is what he did and he was in due course very properly raised to the peerage for it.

When Rutherford was done with the atom all the solidity was pretty well knocked out of it.

Till these researches began, people commonly thought of atoms as something like birdseed, little round solid particles, ever so little, billions to an inch. They were small. But they were there. You could weigh them. You could apply to them all the laws of Isaac Newton about weight and velocity and mass and gravitation—in other words, the whole of first-year physics. . . .

One must not confuse Rutherford's work on atoms with Einstein's theories of space and time. Rutherford worked all his life without reference to Einstein. Even in his later days at the Cavendish Laboratory at Cambridge when he began, ungratefully, to smash up the atom that had made him, he needed nothing from Einstein. I once asked Rutherford—it was at the height of the popular interest in Einstein, in 1923—what he thought of Einstein's relativity. "Oh, that stuff!" he said. "We never bother with that in our work!" His admirable biographer, Professor A. S. Eve, tells us that when the German physicist Wien told Rutherford that no Anglo-Saxon could understand relativity Rutherford answered, "No, they have too much sense."

But it was Einstein who made the real trouble. He announced in 1905 that there was no such thing as absolute rest. After that there never was. But it was not till just after the Great War that the reading public caught on to Einstein and little books on "relativity" covered the bookstalls.

Einstein knocked out space and time as Rutherford knocked out matter. The general viewpoint of relativity toward space is very simple. Einstein explains that there is no such place as *here*. "But," you answer, "I'm here; here is where I am right now." But you're moving, you're spinning round as the earth spins; and you and the earth are both spinning round the sun, and the sun is rushing through space towards a distant galaxy, and the galaxy itself is beating it away at 26,000 miles a second. Now where is that spot that is here! How did you mark it? You remember the story of the two idiots who were out fishing, and one said, "We should have marked that place where we got all the fish," and the other said, "I did, I marked it on the boat." Well, that's it. That's *here*.

You can see it better still if you imagine the universe swept absolutely empty: nothing in it, not even *you*. Now put a *point* in it, just one point. Where is it? Why, obviously it's nowhere. If you say it's right there, where do you mean by there? In which direction is there? In *that* direction? Oh! hold on, you're sticking yourself in to make a direction. It's in *no* direction; there aren't any directions. Now put in another point. Which is which? You can't tell. They *both* are. One is on the right, you say, and one on the left. You keep out of that space! There's no right and no left.

The discovery by Einstein of the curvature of space was greeted by the physicists with the burst of applause that greets a winning home-run at baseball. That brilliant writer just mentioned, Sir Arthur Eddington, who can handle space and time with the imagery of a poet, and even infiltrate humor into gravitation, as when he says that a man in an elevator falling twenty stories has an ideal opportunity to study gravitation —Sir Arthur Eddington is loud in his acclaim. Without this curve, it appears, things won't fit into their place. The fly on the globe, as long as he thinks it flat (like Mercator's map), finds things shifted as by some unaccountable demon to all sorts of wrong distances. Once he gets the idea of a sphere everything comes straight. So with our space. The mystery of gravitation puzzles us, except those who have the luck to fall in an elevator, and even for them knowledge comes too late. They weren't falling at all: just curving. "Admit a curvature of the world," wrote Eddington in his Gifford Lectures of 1927, "and the mysterious agency disappears. Einstein has exorcised this demon."

But it appears now, fourteen years later, that Einstein doesn't care if space is curved or not. He can take it either way. A prominent physicist of today, head of the department in one of the greatest universities of the world, wrote me on this point: "Einstein had stronger hopes that a general theory which involved the assumption of a property of space, akin to what is ordinarily called curvature, would be more useful than he now believes to be the case." Plain talk for a professor. Most people just say Einstein has given up curved space. It's as if Sir Isaac Newton years after had said, with a yawn, "Oh, about that apple—perhaps it wasn't falling."

Science as Foresight [*]
by Jacob Bronowski (1955)

Nature is more intricately organized and cross-linked than our theories, so that each model or likeness that we try scores some striking successes and then falls short. All that we can do, at any state of our factual knowledge, is to prefer that code which makes what we know most orderly. That is, we choose those concepts which organize the messages of nature most coherently. We achieve this, in my view, by maximizing their content of information or meaning. This is the test for the inductions by which we try to generalize from our particular experience and to gain a basis for foresight.

Charles Darwin said sadly at the end of his life that

My mind seems to have become a kind of machine for grinding general laws out of large collections of facts.

Darwin was too modest. If such a machine can indeed be made, we do not know how to make it. Indeed, we do not know anything like it, except a man. For the discovery of laws is a complex induction like the solving of a cryptogram which, so far as we know, a machine procedure can help but not complete. It depends on an imaginative act, seeing the structure of the solution in a likeness, and seizing a likeness where none was expected.

[*] From Jacob Bronowski, "Science as Foresight," in James R. Newman (ed.), *What Is Science?*, Simon & Schuster, Inc., New York. Copyright © 1955 by James R. Newman. Reprinted by permission of Simon & Schuster, Inc.

Suggestions for Further Reading

On Newton

[*] Bronowski, Jacob: *The Common Sense of Science*, Vintage Books, Random House, Inc., New York, 1953, chap. III. An interesting discussion of Newton's model and its effect on the concepts of cause and order in the universe.

[*] Burtt, E. A.: *The Metaphysical Foundations of Modern Science*, Anchor Books, Doubleday & Company, Inc., Garden City, N.Y., 1954, chap. VII. A detailed and philosophical analysis of Newton's methods and ideas.

[*] Butterfield, Herbert: *The Origins of Modern Science*, The Macmillan Company, New York, 1957, chap. 8. The history of the modern theory of gravitation.

° Lodge, Oliver: *Pioneers of Science,* Dover Publications, Inc., New York, 1960, lectures VII, VIII, and IX. These three chapters deal in detail with Newtonian physics. Some mathematics is used but the treatment is not difficult.

On Relativity

° Barnett, Lincoln: *The Universe and Dr. Einstein,* Mentor Books, New American Library of World Literature, Inc., New York, 1958. In addition to the selection we have used, the reader may find the rest of this book helpful in explaining relativity.

Bergmann, Peter G.: *The Riddle of Gravitation,* Charles Scribner's Sons, New York, 1968. An explanation of Einstein's general theory and its relationship to new ideas in cosmology.

° Durell, Clement V.: *Readable Relativity,* Harper Torchbooks, Harper & Row, Publishers, New York, 1960. One of the best simple mathematical accounts of relativity. It explains both the special and general theory using only elementary algebra and geometry.

° Einstein, A., and L. Infeld: *The Evolution of Physics,* Simon and Schuster, Inc., New York, 1961, chap. III. An excellent popular presentation of relativity by Einstein and one of his coworkers in research. It is told in simple language avoiding all highly technical terms and mathematical formulae.

° Gamow, George: *One, Two, Three . . . Infinity,* Mentor Books, New American Library of World Literature, Inc., New York, 1947, part II, chaps. III–V. A vivid and interesting account for the layman of the principles of relativity and the fourth dimension.

———: *Matter, Earth and Sky,* Prentice-Hall, Inc., Englewood Cliffs, N.J., 1958, part 1. Another very good description for the general reader of the concepts of relativity (Special Theory, chap. 7, and General Theory, chap. 12).

° Russell, Bertrand: *The A B C of Relativity,* Mentor Books, New American Library of World Literature, Inc., New York, 1959. One of the best simple but authoritative books on relativity for the layman.

° Paperback edition

8

How Was the Universe Created?

Did creation occur as a single event billions of years ago, or is it going on around us all the time? This question, of great philosophical significance, is debated by two opposing schools of modern cosmology. The selections in this part show a scientific idea in the making and offer a new appreciation of the impact that scientific theories can have on other areas of human thought.

How Was the Universe Created?

Introduction

THE READINGS IN THIS PART bring us up to the frontiers of modern cosmology in the description of the two principal theories of the origin of the universe. They begin with a selection from Gamow, which fills in the background concerning present scientific knowledge of the galaxies: measurements of their size, rotation, and speed of recession, the geometry of space-time to which the reader was introduced in Part 7. One of the most revolutionary and reorienting concepts since the time of Copernicus is that our planet has been further displaced from the center of the universe and is now believed to occupy a very peripheral position in the spiral arm of an ordinary galaxy. This discovery, George Gamow says, comes as ". . . a slap in the face of human pride." [1] This peripheral position hampers our investigation of the universe. Dark clouds prevent us from seeing the center of our galaxy and the part of the universe that lies beyond it. In these readings we see how science has overcome this difficulty and continues to find methods of gaining new knowledge of the universe.

One of the newest tools in astronomical research—the radio telescope—was discovered just at the moment when we seemed to have approached the theoretical limits of the older methods. The maximum magnification of the optical telescope is limited by the unsteadiness of our atmosphere, but this does not affect the radio telescope to the same degree because the longer wavelengths that it receives are less affected by slight fluctuations in the atmosphere. Also the long radio wavelengths are not deflected as much by small dust particles as are the waves of visible light, and therefore the radio telescope is able to penetrate the clouds that lie between us and the center of the galaxy.

Another limitation of the optical telescope is due to the faintness of the signals from the most distant sources. This difficulty, too, has been overcome by the radio telescope because of the unusually strong radio signals coming from sources that are believed to be beyond the range of our optical telescopes. For these reasons the radio telescopes may widen our horizons several-fold.

It is important to remember that we are seeing further back in time as we look farther out into space, so that our knowledge of the universe has been increased in all four dimensions as the new radio telescopes

1. See p. 384.

began scanning the skies and as the information was evaluated by astronomers. Telescopes in satellites orbiting above the earth's atmosphere can achieve much greater clarity and resolution than the most powerful earth-based telescopes. The full exploitation of this latest possibility is expected greatly to increase our knowledge of the universe. However, we may encounter another barrier out where the galaxies are receding so fast that their radiation can never reach us. "Then," Lovell tells us, "we must be content. No further strivings or inventions of man will enable us to probe the conditions which existed in the epochs of history beyond these few thousand million years. They are gone for ever beyond the fundamental limits of observability." [2]

"At this point we reach the second stage of our inquiry, where we appeal to cosmological theory. The question is this. Can we formulate a theory in terms of known physical laws whose predictions agree so well with the present observable universe that we can predict the past and future?" Lovell goes on to describe the two principal theories of cosmology that have been brought forward in recent years: the evolutionary theory, sometimes referred to as the "big bang" theory and the steady-state or continuous creation theory. A decade ago both these theories seemed to fit the framework of known facts. The situation was reminiscent of cosmological theory at the time of Ptolemy and Aristarchus, when two very different theories could be made to explain the observed facts.

We have also seen one case in which science was forced to accept simultaneously two opposing theories. The facts demonstrated that no choice could be made between the two theories of light; they were both necessary to explain the observations and experiments. In Part 7 we saw how one theory, Newton's law of gravitation, was shown to be a special case of the larger synthesis of Einstein. Now in these readings we see a scientific decision actually in the making, an argument hotly debated by both sides, and we realize that it is not always possible for scientists to remain aloof and impartial. Scientists are men who have convictions and are willing to do battle for them.

In the absence of decisive facts, the arguments for the two rival theories of cosmology revolved around a discussion about the nature of the assumptions on which they are based. However, with the advent of radio telescopes the measurements on distribution of distant galaxies and the discovery of quasars produced evidence favoring the big bang theory. In fact, Robert Dicke has identified radiation which he believes to be waves of energy from the big bang itself!

The scientists who supported the steady-state theory have conceded

2. See p. 409.

that these most recent discoveries cannot be fitted into the framework of their original theory. However, there still remain many unanswered questions. Within the next few years observations from scientific satellites may provide conclusive evidence in favor of the evolutionary theory or evidence that may demand an entirely new conceptual scheme. The unraveling of this mystery will be exciting and deeply significant for modern man, having far-reaching implications for our religious and philosophical beliefs. As Fred Hoyle once said, "The Universe is everything; both living and inanimate things; both atoms and galaxies; and if the spiritual exists as well as the material, of spiritual things also . . . for by its very nature the Universe is the totality of all things."

The Galaxies

The Infinite [*]

by Giacomo Leopardi (1798–1837)

Dear to me always was this lonely hill
And this hedge that excludes so large a part
Of the ultimate horizon from my view.
But as I sit and gaze, my thought conceives
Interminable vastnesses of space
Beyond it, and unearthly silences,
And profoundest calm; whereat my heart almost

Becomes dismayed. And as I hear the wind
Blustering through these branches, I find myself
Comparing with this sound that infinite silence;
And then I call to mind eternity,
And the ages that are dead, and this that now
Is living, and the noise of it. And so
In this immensity my thought sinks drowned:
And sweet it seems to shipwreck in this sea.

Toward the Limits of the Unknown †

by George Gamow (1962)

More than a century ago the famous British astronomer, William Herschel, observing the stellar sky through his large self-made telescope,
was struck by the fact that most of the stars that are ordinarily invisible
to the naked eye appear within the faintly luminous belt cutting across

* From R. C. Trevelyan, *Translations from Leopardi*, Cambridge University Press,
New York, 1941. Reprinted by permission.
† From George Gamow, *One, Two, Three . . . Infinity*. The Viking Press, Inc.,
New York. Copyright 1947 by George Gamow. Reprinted by permission of The Vi-
king Press, Inc.

Figure 8-1. Great nebula in Andromeda. (Hale Observatories. Figure added by editor.)

the night sky and known as the Milky Way. And it is to him that the science of astronomy owes the recognition of the fact that the Milky Way is not ordinary nebulosity or merely a belt of gas clouds spreading across space, but is actually formed from a multitude of stars that are so far away and consequently so faint that our eye cannot recognize them separately.

Using stronger and stronger telescopes we have been able to see the Milky Way as a larger and larger number of separate stars, but the main bulk of them still remains in the hazy background. It would be, however, erroneous to think that in the region of the Milky Way the stars are distributed any more densely than in any other part of the sky. It is, in fact, not the denser distribution of stars but the greater depth of stellar distribution in this direction that makes it possible to see what seems to be a larger number of stars in a given space than anywhere else in the sky. In the direction of the Milky Way the stars extend as far as the eye (strengthened by telescopes) can see, whereas in any other direction the distribution of stars does not extend to the end of visibility, and beyond them we encounter mostly the almost empty space.

Looking in the direction of the Milky Way it is as though we are looking through a deep forest where the branches of numerous trees overlap each other forming a continuous background, whereas in other directions we see patches of the empty space between the stars, as we would see the patches of the blue sky through the foliage overhead.

Thus the stellar universe, to which our sun belongs as one insignificant member, occupies a flattened area in space, extending for large distances in the plane of the Milky Way, and being comparatively thin in the direction perpendicular to it.

A more detailed study by generations of astronomers led to the conclusion that our stellar system includes about 40,000,000,000 individual stars, distributed within a lens-shaped area about 100,000 light-years in diameter and some 5,000 to 10,000 light-years thick. And one result of this study comes as a slap in the face of human pride—the knowledge that our sun is not at all at the center of this giant stellar society but rather close to its outer edge.

In Figure 8-2 we try to convey to our readers the way this giant beehive of stars actually looks. By the way, we have not mentioned yet that in more scientific language the system of the Milky Way is known as the *Galaxy* (Latin of course!). The size of the Galaxy is here reduced by a factor of a hundred billion billions, though the number of points that represent separate stars are considerably fewer than forty billions, for, as one puts it, typographical reasons.

One of the most characteristic properties of the giant swarm of stars

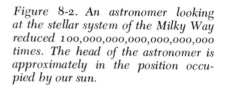

Figure 8-2. An astronomer looking at the stellar system of the Milky Way reduced 100,000,000,000,000,000,000 times. The head of the astronomer is approximately in the position occupied by our sun.

forming the galactic system is that it is in a state of rapid rotation similar to that which moves our planetary system. Just as Venus, Earth, Jupiter, and other planets move along almost circular orbits around the sun, the billions of stars forming the system of the Milky Way move around what is known as the galactic center. This center of galactic rotation is located in the direction of the constellation of Sagittarius (the Archer), and in fact if you follow the foggy shape of the Milky Way across the sky you will notice that approaching this constellation it becomes much broader, indicating that you are looking toward the central thicker part of the lens-shaped mass. (Our astronomer in Figure 8-2 is looking in this direction.)

What does the galactic center look like? We do not know that, since unfortunately it is screened from our sight by heavy clouds of dark interstellar material hanging in space. In fact, looking at the broadened part of the Milky Way in the region of Sagittarius [1] you would think first that the mythical celestial road branches here into two "one-way traffic lanes." But it is not an actual branching, and this impression is given simply by a dark cloud of interstellar dust and gases hanging in space right in the middle of the broadening between us and the galactic center. Thus whereas the darkness on both sides of the Milky Way is due to the background of the dark empty space, the blackness in the middle is produced by the dark opaque cloud. A few stars in the dark central

1. Which can be best observed on a clear night in early summer.

patch are actually in the foreground, between us and the cloud (Figure 8-3).

It is, of course, a pity that we cannot see the mysterious galactic center around which our sun is spinning, along with billions of other stars. But in a way we know how it must look, from the observation of other stellar systems or galaxies scattered through space far beyond the outermost limit of our Milky Way. It is not some supergiant star keeping in subordination all the other members of the stellar system, as the sun reigns over the family of planets. The study of the central parts of other galaxies (which we will discuss a little later) indicates that they also consist of large multitudes of stars with the only difference that here the stars are crowded much more densely than in the outlying parts to which our sun belongs. If we think of the planetary system as an autocratic state where the Sun rules the planets, the Galaxy of stars may be likened to a kind of democracy in which some members occupy influential central places while the others have to be satisfied with more humble positions on the outskirts of their society [Figure 8-1].

As said above, all the stars including our sun rotate in giant circles around the center of the galactic system. How can this be proved, how large are the radii of these stellar orbits, and how long does it take to make a complete circuit?

All these questions were answered a few decades ago by the Dutch astronomer Oort, who applied to the system of stars known as the Milky Way observations very similar to those made by Copernicus in considering the planetary system.

Let us remember first Copernicus' argument. It had been observed by the ancients, the Babylonians, the Egyptians, and others, that the big planets like Saturn or Jupiter seemed to move across the sky in a rather peculiar way. They seemed to proceed along an ellipse in the way the

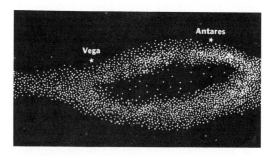

Figure 8-3. *Looking toward the galactic center it would seem that the mythical celestial road branches into two one-way traffic lanes.*

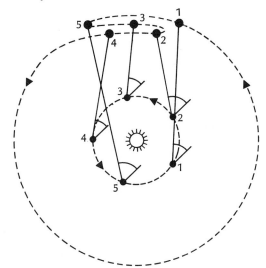

Figure 8-4

sun does, then suddenly to stop, to back, and after a second reversal of motion, to continue their way in the original direction. In the illustration, Figure 8-4, we show schematically such a loop as described by Saturn over a period of about two years. (The period of Saturn's complete circuit is 29½ years.) Since, on account of religious prejudices that dictated the statement that our Earth is the center of the universe, all planets and the sun itself were believed to move around the Earth, the above described peculiarities of motion had to be explained by the supposition that planetary orbits have very peculiar shapes with a number of loops in them.

But Copernicus knew better, and by a stroke of genius, he explained the mysterious looping phenomenon as due to the fact that the Earth as well as all other planets move along simple circles around the Sun. This explanation of the looping effect can be easily understood after studying the schematic picture in Figure 8-4.

The sun is in the center, the Earth (small sphere) moves along the smaller circle, and Saturn (with a ring) moves along the larger circle in the same direction as the Earth. Numbers 1, 2, 3, 4, 5 represent different positions of the Earth in the course of a year, and the corresponding positions of Saturn which, as we remember, moves much more slowly. The parts of vertical lines from the different positions of the Earth represent the direction to some fixed star. By drawing lines from the various Earth positions to the corresponding Saturn positions we see that the angle formed by the two directions (to Saturn and to the fixed star) first in-

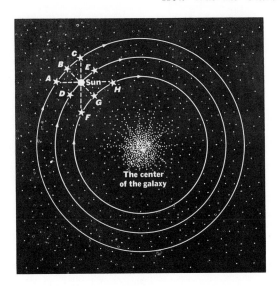

Figure 8-5

creases, then decreases, and then increases again. Thus the seeming phenomenon of looping does not represent any peculiarity of Saturn's motion but arises from the fact that we observe this motion from different angles on the moving Earth.

The Oort argument about the rotation of the Galaxy of stars may be understood after inspection of Figure 8-5. Here in the lower part of the picture we see the galactic center (with dark clouds and all!) and there are plenty of stars all around it through the entire field of the figure. The three circles represent the orbits of stars at different distances from the center, the middle circle being the orbit of our sun.

Let us consider eight stars (shown with rays to distinguish them from other points), two of which are moving along the same orbit as the sun, but one slightly ahead and one slightly behind it, the others located on somewhat larger and somewhat smaller orbits as shown in the figure. We must remember that owing to the laws of gravity the outer stars have lower and the inner stars higher velocity than the stars on solar orbits (this is indicated in the figure by the arrows of different lengths).

How will the motion of these eight stars look if observed from the sun, or, what is of course the same, from the Earth? We are speaking here about the motion along the line of sight, which can be most conveniently observed by means of the so-called Doppler effect.[2] It is clear, first of all, that the two stars (marked D and E) that move along the same orbit and with the same speed as the sun will seem stationary to a

2. See the discussion of the Doppler effect on p. 395. [Also pp. 290–91. Ed.]

solar (or terrestrial) observer. The same is true of the other two stars (B and G) located along the radius, since they move parallel to the sun, so that there is no component of velocity along the line of sight.

Now what about the stars A and C on the outer circle? Since they both move more slowly than the sun we must conclude, as clearly seen in this picture, that the star A is lagging behind, whereas the star C is being overtaken by the sun. The distance to the star A will increase, while the distance to C will decrease, and the light coming from the two stars must show respectively the red and violet Doppler effect. For the stars F and H on the inner circle the situation will be reversed, and we must have a violet Doppler effect for F and a red one for H.

It is assumed that the phenomenon just described could be caused only by a circular motion of the stars, and the existence of that circular motion makes it possible for us not only to prove this assumption but also to estimate the radius of stellar orbits and the velocity of stellar motion. By collecting the observational material on the observed apparent motion of stars all over the sky, Oort was able to prove that the expected phenomenon of red and violet Doppler effect really exists, thus proving beyond any doubt the rotation of the Galaxy. . . .

The exact measurements of the Oort effect of stellar motion now makes it possible to measure the orbits of stars and determine the period of rotation. Using this method of calculation it has been learned that the radius of the solar orbit having its center in Sagittarius is 30,000 light-years, that is, about two thirds the radius of the outermost orbit of the entire galactic system. The time necessary for the sun to move a complete circle around the galactic center is some 200 million years. It is a long time, of course, but remembering that our stellar system is at least 5 billion years old, we find that during its entire life our sun with its family of planets has made 25 or more complete rotations. If, following the terminology of the terrestrial year, we call the period of solar rotation "a solar year" we can say that our universe is only 25 years old. Indeed things happen slowly in the world of stars, and a solar year is quite a convenient unit for time measurements in the history of the universe!

As we have already mentioned above, our Galaxy is not the only isolated society of stars floating in the vast spaces of the universe. Telescopic studies reveal the existence, far away in space, of many other giant groups very similar to that to which our sun belongs. The nearest of them, the famous Andromeda Nebula, can be seen even by the naked eye. It appears to us as a small, faint, rather elongated nebulosity. In Figures 8-6 and 8-7 are shown photographs, taken through the large telescope of the Mt. Wilson Observatory, of two such celestial objects. The

Figure 8-6. *Spiral nebula in Coma Berenices.* (Hale Observatories)

two objects shown in these photographs are: the Nebula in Coma Berenices seen straight on the edge, and the Nebula in Ursus Major seen from
the top. We notice that, as a part of the characteristic lens-shape ascribed to our Galaxy, these nebulae possess a typical spiral structure;
hence the name "spiral nebulae." There are many indications that our
own stellar structure is similarly a spiral, but it is very difficult to determine the shape of a structure when you are inside it. As a matter of fact,
our sun is most probably located at the very end of one of the spiral
arms of the "Great Nebula of the Milky Way."

For a long time astronomers did not realize that the spiral nebulae
are giant stellar systems similar to our Milky Way, and confused them
with ordinary diffuse nebulae like that in the constellation of Orionis,
which represent the large clouds of interstellar dust floating between the
stars inside our Galaxy. Later, however, it was found that these foggy
spiral-shaped objects are not fog at all, but are made of separate stars,
which can be seen as tiny individual points when the largest magnifica-
tions are used. But they are so far away that no parallactic measurements can indicate their actual distance.

Figure 8-7. A spiral nebula in Ursa Major, a distant island universe, seen from above. (Hale Observatories)

Thus it would seem at first that we had reached the limit of our means for measuring celestial distances. But no! In science, when we come to an insuperable difficulty the delay is usually only temporary; something always happens that permits us to go still farther. In this case a quite new "measuring rod" was found by the Harvard astronomer Harlow Shapley in the so-called pulsating stars or Cepheids.[3]

There are stars and stars. While most of them glow quietly in the sky, there are a few that constantly change their luminosity from bright to dim, and from dim to bright in regularly spaced cycles. The giant bodies of these stars pulsate as regularly as the heart beats, and along with this pulsation goes a periodic change of their brightness.[4] The larger the star, the longer is the period of its pulsation, just as it takes a long pendulum more time to complete its swing than a short one. The really

3. So called after the star δ-Cepheus, in which the phenomenon of pulsation was first discovered.

4. One must not confuse these pulsating stars with the so-called eclipsing variables, which actually represent systems of two stars rotating around each other and periodically eclipsing one another.

small ones (that is, small as stars go) complete their period in the course of a few hours, whereas the real giants take years and years to go through one pulsation. Now, since the bigger stars are also the brighter, there is an apparent correlation between the period of stellar pulsation, and the average brightness of the star. This relation can be established by observing the Cepheids, which are sufficiently close to us so that their distance and consequently actual brightness may be directly measured.

If now you find a pulsating star that lies beyond the limit of parallactic measurements, all you have to do is to watch the star through the telescope and observe the time consumed by its pulsation period. Knowing the period, you will know its actual brightness, and comparing it with its apparent brightness you can tell at once how far away it is. This ingenious method was successfully used by Shapley for measuring particularly large distances within the Milky Way and has been most useful in estimating the general dimensions of our stellar system.

When Shapley applied the same method to measuring the distance to several pulsating stars found imbedded in the giant body of the Andromeda Nebula, he was in for a big surprise! The distance from the Earth to these stars, which, of course, must be the same as the distance to the Andromeda Nebula itself, turned out to be . . . much larger than the estimated diameter of the stellar system of Milky Way. And the size of Andromeda Nebula came out only a little smaller than the size of our entire Galaxy. The two spiral nebulae shown in our Plates (Figures 8-6 and 8-7) are still farther away and their diameters are comparable to that of the Andromeda.[5]

This discovery dealt the death blow to the earlier assumptions that the spiral nebulae are comparatively "small things" located within our Galaxy, and established them as independent galaxies of stars very similar to our own system, the Milky Way. No astronomer would now doubt that to an observer located on some small planet circling one of the billions of stars that form the Great Andromeda Nebula, our own Milky Way would look much as the Andromeda Nebula looks to us [Figures 8-1 and 8-8].

The further studies of these distant stellar societies, which we owe mostly to Dr. E. Hubble, the celebrated galaxy-gazer of Mt. Wilson Observatory, reveal a great many facts of great interest and importance. It was found first of all that the galaxies, which appear more numerous through a good telescope than the ordinary stars do to the naked eye, do not all have necessarily spiral form, but present a great variety of different types. There are *spherical galaxies,* which look like regular discs

5. The distance to Andromeda is estimated at about 2.2 million light years. Ed.

Figure 8-8. *Edge of Andromeda.* (Hale Observatories. Figure added by editor.)

with diffused boundaries; there are *elliptical galaxies* [Figure 8-9] with different degrees of elongation. The spirals themselves differ from each other by the "tightness with which they are wound up." There are also very peculiar shapes known as "barred spirals." . . . although up to the present we do not have a completely satisfactory explanation of why and how these spiral forms are formed and what causes the difference between the simple and the barred spirals.

Much is still to be learned from further study of the structure, motion, and stellar content in the different parts of galactic societies of stars. A very interesting result was, for example, obtained . . . by a Mt. Wilson astronomer, W. Baade, who was able to show that, whereas the central bodies (nuclei) of spiral nebulae are formed by the same type of stars as the spherical and elliptic galaxies, the arms themselves show a rather different type of stellar population. This "spiral-arm" type of stellar population differs from the population of the central region by the presence of very hot and bright stars, the so-called "Blue Giants," which are absent in the central regions as well as in the spherical and elliptic galax-

Figure 8-9. *An elliptical galaxy.* (Yerkes Observatory. Figure added by editor.)

ies. Since . . . the Blue Giants most probably represent the most recently formed stars, it is reasonable to assume that the spiral arms are so to speak the breeding grounds for new stellar populations. One could imagine that a large part of the material ejected from the equatorial bulge of a contracting elliptic galaxy is formed by primordial gases that come out into the cold intergalactic space and condense into the separate large lumps of matter, which through subsequent contraction become very hot and very bright. . . .

We must state here, first of all, that the method of distance measurements based on pulsating stars, though giving excellent results when applied to quite a number of galaxies that lie in the neighborhood of our Milky Way, fails when we proceed into the depth of space, since we soon reach distances at which no separate stars may be distinguished and the galaxies look like tiny elongated nebulosities even through the strongest telescopes. Beyond this point we can rely on the visible size, since it is fairly well established that, unlike stars, all galaxies of a given type are of about the same size. If you know that all people are of the same height, that there are no giants or dwarfs, you can always say how far a man is from you by observing his apparent size.

Using this method for estimating distances in the far-outflung realm of galaxies, Dr. Hubble was able to prove that the galaxies are scattered more or less uniformly through space as far as the eye (fortified by the most highly powered telescope) can see. We say "more or less" because there are many cases in which the galaxies cluster in large groups containing sometimes many thousands of members, in the same way as the separate stars cluster in galaxies.

Our own galaxy, Milky Way, is apparently one member of a comparatively small group of galaxies numbering in its membership three spirals (including ours, and the Andromeda Nebula) and six elliptical and four irregular nebulae (two of which are Magellanic clouds).

However, save for such occasional clustering, the galaxies, as seen through the 200-inch telescope of the Palomar Mountain Observatory, are scattered rather uniformly through space . . .

Studying the spectra of the light coming from these distant galaxies, Mt. Wilson's astronomer E. Hubble noticed that the spectral lines are shifted slightly towards the red end of the spectrum, and that this so-called "red shift" is stronger in the more distant galaxies. In fact, it was found that the "red shift" observed in different galaxies is directly proportional to their distance from us.

The most natural way to explain this phenomenon is to assume that *all galaxies recede from us with a speed that increases with their distance from us*. This explanation is based on the so-called "Doppler effect," which makes the light coming from a source that is approaching us change its color toward the violet end of the spectrum, and light from a receding source change toward the red. Of course, to obtain a noticeable shift the relative velocity of the source in relation to the position of the observer must be rather large. When Prof. R. W. Wood was arrested for going through a red traffic signal in Baltimore and told the judge that, because of this phenomenon, the light he saw looked green to him, since he was approaching it in the car, the professor was simply pulling the judge's leg. Had the judge known more about physics, he would have asked Professor Wood to calculate the speed with which he must have been driving in order to see green in a red light, and then would have fined him for speeding.

Returning to the problem of the "red shift" observed in galaxies, we come to what is at first sight a rather awkward conclusion. It looks as though all of the galaxies in the universe were running away from our Milky Way as if it were a galactic Monster of Frankenstein! What then are the horrible properties of our own stellar system, and why does it

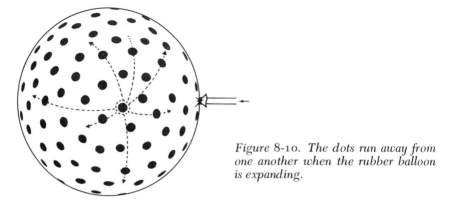

Figure 8-10. The dots run away from one another when the rubber balloon is expanding.

seem to be so unpopular among all other galaxies? If you think a little about this question, you will easily come to the conclusion that there is nothing particularly wrong with our Milky Way, and that, in fact, other galaxies do not run away from it exclusively but rather that all run away from one another. Imagine a rubber balloon with a polka-dot pattern painted on its surface (Figure 8-10). If you will begin to inflate it, gradually stretching its surface to a large and larger size, the distances between individual dots will continually increase so that an insect sitting on any one of the dots would receive the impression that all other dots are "running away" from it. Moreover the recession velocities of different dots on the expanding balloon will be directly proportional to their distance from the insect's observation point.

This example must make it quite clear that the recession of galaxies observed by Hubble has nothing to do with the particular properties or position of our own galaxy, but must be interpreted simply as due to *the general uniform expansion of the system of galaxies scattered through the space of the universe.*

From the observed velocity of the expansion and the present distances between the neighboring galaxies one easily calculates that *this expansion must have started at least five billion years ago.* . . .

We are now prepared to answer the fundamental question concerning the size of our universe. *Shall we consider the universe as extending into infinity and conclude that bigger and better telescopes will always reveal to the inquiring eye of an astronomer new and hitherto unexplored regions of space, or must we believe, on the contrary, that the universe occupies some very big but nevertheless finite volume, and is, at least in principle, explorable down to the last star?*

When we speak of the possibility that our universe is of "finite size," we do not mean, of course, that somewhere at a distance of several billion light-years the explorer of space will encounter a blank wall on which is posted the notice "No trespassing."

In fact, . . . *space can be finite without being necessarily limited by a boundary.* It can simply curve around and "close on itself," so that a hypothetical space explorer, trying to steer his rocket ship as straight as possible will describe a geodesic line in space and come back to the point from which he started.

The situation would be, of course, quite similar to an ancient Greek explorer who travels *west* from his native city of Athens, and, after a long journey, finds himself entering the eastern gates of the city.

And just as the curvature of the Earth's surface can be established without a trip around the world, simply by studying the geometry of only a comparatively small part of it, the question about the curvature of the three-dimensional space of the universe can be answered by similar measurements . . . one must distinguish between two kinds of curvature: the positive one corresponding to the closed space of finite volume, and the negative one corresponding to the saddle-like opened infinite space [Figure 7-7]. The difference between these two types of space lies in the fact that, whereas in the *closed space* the number of uniformly scattered objects falling within a given distance from the observer increases more slowly than the cube of that distance, the opposite is true in *opened space.* . . .

Now we can ask ourselves what kind of forces are responsible for the expansion of the universe, and whether this expansion will ever stop or even become contraction. Is there any possibility that the expanding masses of the universe will turn back on us and squeeze our stellar system, the Milky Way, the sun, the Earth, and the humanity on earth into a pulp with nuclear density? . . .

It seems that at present the kinetic energy of receding galaxies is several times greater than their mutual potential gravitational energy, from which it would follow that *our universe is expanding into infinity without any chance of ever being pulled more closely together again by the forces of gravity.* It must be remembered, however, that most of the numerical data pertaining to the universe as a whole are not very exact,[6] and it is possible that future studies will reverse this conclusion.

6. It must be mentioned here that the above-described method of estimating the age of the universe is based on the knowledge of the distances of faraway galaxies. Since at large distances no stars, and in particular no Cepheids, can be seen, even through the largest telescopes, these distances are estimated on the basis of the inverse square law for the observed brightness of light sources. Thus, if one galaxy seems 25 times fainter than another, it is considered to be 5 times farther away. This method presupposes, however, that the intrinsic brightness of a galaxy does not

But even if the expanding universe does suddenly stop in its tracks, and turn back in a movement of compression, it will be billions of years before that terrible day envisioned by the Negro Spiritual, "when the stars begin to fall," and we are crushed under the weight of collapsing galaxies!

Early Years of Radio Astronomy *
by A. C. B. Lovell (1959)

The succession of telescopes of ever-increasing size which have been constructed has continued to reveal more and more distant nebulae of stars. However, it seems likely that we have nearly reached the culmination of this line of development with the 200-inch telescope on Palomar Mountain in America which began its programme of work about 15 years ago. Culmination, because it seems possible that optical telescopes of this order of size may represent the largest which it is worth while building on earth. Even under the excellent atmospheric conditions which exist on Palomar, the unsteadiness of the atmosphere limits the realization of the maximum penetration of this telescope to a few dozen nights during the year.

This telescope can photograph star systems which are so far away that the light has taken over two thousand million [1] years to reach us, but when I discuss the problem of the origin of the universe you will realize that this distance is a most tantalizing limit. The desire to find out what lies beyond is very great—some knowledge of the star systems at twice this distance might well give us the key to unlock the secrets of the evolution of the entire universe. You will therefore understand that those of us who seek this knowledge seize with a particular passion any prospects of surmounting the hindrances of our earthbound environment. And indeed we have been born in a fortunate and exciting epoch. Out of the cataclysm of the world war have emerged two technical develop-

change with its age, which may not be true. In fact, younger galaxies probably contain many very bright fast-evolving stars which mostly disappear in older galaxies. Looking far out in space, we look far back in time and see distant galaxies as they were in the distant past. This may lead to an underestimate of their intrinsic brightness, resulting in an underestimate of their distance.

* Abridged from A. C. B. Lovell, *The Individual and the Universe,* Harper & Row, Publishers, Inc., New York. Copyright © 1958 by A. C. B. Lovell. Reprinted by permission of Harper & Row, Publishers, Inc.

1. Throughout these selections, Professor Lovell uses the British "billion." The British "billion" is a million million, whereas the American "billion" is a thousand million. Ed.

ments which are creating a revolution in astronomical observations—
radio astronomy and the earth satellite.

When I think of the enormous scientific and technological problems
which had to be solved I still stand in awe when I reflect that at this
moment[2] at least four objects launched by man are relentlessly circling this earth. They have carried scientific instruments above the
dense regions of the earth's atmosphere and are sending back by radio
the information about the conditions in the environment of the earth
which hitherto we have only been able to probe remotely and by inference. The astronomer whose interest lies in the problems of the solar
system has already gathered a rich harvest of results about the complex
of radiations and particles which exist in space and which are normally
absorbed or transformed by processes in the upper regions of the atmosphere. Soon, telescopes will be carried into these regions and man's vision will be freed from the disturbing effects of the atmosphere. Limitless possibilities are now emerging; indeed, we seem to be on the verge
of another epoch of discovery which may well parallel that of the radio
telescope.

In the ordinary course of events our knowledge of the universe comes

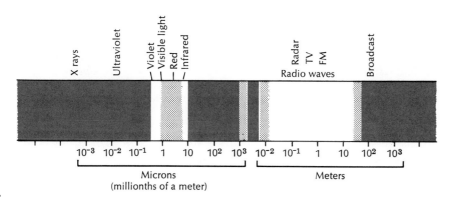

*Figure 8-11. The electromagnetic spectrum showing the regions in which
the earth's atmosphere is normally transparent. The most important of these
"windows" are the visible wavelengths and the radio window that has made
possible the new science of radio astronomy.* (After Arthur Clarke, The
Promise of Space. Figure added by editor.)

2. The moment in question was at the time of the original broadcast on 23 November, 1958, when the Vanguard satellite, Explorer IV, Sputnik 3, and its carrier
rocket were all in orbit. Ed.

Figure 8-12. The 250-ft paraboloid at the Jodrell Bank Radio Astronomy Observatory in Cheshire. This great instrument remains the largest fully steerable radio telescope in operation and it has been responsible for fundamental advances in research into radio astronomy. It has even been used occasionally for tracking satellites and space-probes, though this is certainly not the purpose for which it was designed, and it is not now used in this way.

to us because the sun and the stars emit light. Less than 30 years ago Jansky, an American engineer, discovered almost by accident that radio waves were reaching the earth from outer space. But although his proof that they were coming from regions of space outside the solar system was quite decisive, he could not find out much more about these radio waves. Radio waves and light waves travel through space with the same speed, 186,000 miles a second, but are distinguished by the difference in wavelength. [See Figure 8-11.] The light which reaches us from the stars has a wavelength measurable in millionths of a centimetre whereas the wavelength of the radio waves is measured in metres. To detect these radio waves we need a special kind of telescope, now commonly known as a radio telescope, which is, in effect, a very large version of the common television aerial. The urge to build big radio telescopes is the same as the driving force behind the construction of large optical telescopes, namely, the desire to penetrate far into space. Because of the long wavelength of the radio waves the radio telescopes have to be very much larger than optical telescopes. Even with the enormous instruments now in operation the radio-astronomer's view of the heavens is ill-defined compared with that given by a small optical telescope [Figure 8-12]. This factor greatly increases the difficulty of correlating the universe which we study with radio telescopes with the stars and galaxies visible to our eyes.

It is hardly surprising that Jansky and those who followed him were unable to relate their radio signals with the stellar objects perceived by our normal senses, and for a long time the idea existed that these radio waves were coming from the hydrogen gas which fills the space between the stars. For years Jansky's discovery remained almost unknown, and this unexpected gift of nature to man lay disregarded by the world's astronomers. My own introduction to this work was accidental—a casual remark by a colleague during the endless wartime discussions on how to get a few more miles of range out of our radar. Yet within four years I was wanting most desperately the telescope which now towers over the Cheshire plain. You see, the wartime developments in radar had placed in our hands equipment of a sensitivity and excellence far greater than anything previously available. As soon as these new techniques were used to study the radio waves from the sky one realized that Jansky's discovery had opened an entirely new avenue for the exploration of space.

The improved definition and sensitivity of the equipment soon showed that some of the radio emission was concentrated in particular parts of the sky. The idea that the interstellar hydrogen gas emitted the radio waves turned out to be only a small part of the story. The real situation was much more exciting. The concentration of the radio emissions into localized sources does not seem to occur in any regions of the heavens in which there are prominent visible objects. On the other hand, if we turn our radio telescopes towards the bright stars like Sirius or Capella we cannot record any radio waves from them. Indeed, we face a paradox which has sometimes made us reflect on those heavenly bodies of Lucretius 'which glide devoid of light forevermore.' However, we do not believe that these newly discovered radio sources are dark stars. Our difficulty in relating them to the universe of our ordinary vision arises, we think, partly because the objects which emit radio waves are very faint and generally peculiar, and partly because the radio waves come from distant regions which are beyond the range of the optical telescopes.

Before I talk further about this situation it would, I think, be helpful if I described the results of our attempts to detect radio emissions from the more common stars and galaxies. To begin with, the sun is a very strong radio source—so strong indeed that it sometimes hinders the observations of the more distant signals, in the same way that its light blocks out stars from the ordinary telescopes. One of the earliest of the post-war surprises was the discovery by Appleton and Hey that the sun spots and flares which occasionally appear on the solar surface are associated with great and irregular increases in the solar radio emissions. These solar eruptions are often accompanied by disturbing terrestrial events such as the appearance of the aurora borealis and the fadeouts in transatlantic radio-communications. When the sun is quiet, or undisturbed by spots, then the radio waves from the sun are much less intense. They are generated in the solar corona, that highly tenuous region of the sun extending far outside the photosphere which is the disk of the sun which we see normally [Figure 6-13]. For the radio astronomer the size of the sun depends on the wavelength which he uses to study it. At a radio wavelength of a few centimetres we find that the sun is about the same size as the sun which we see with our eyes, but as the wavelength we use increases so does the size of the sun as seen by the radio astronomer. At a few metres' wavelength its radius is several times greater than the optical radius. If our eyes were sensitive to these radio wavelengths the sun would appear to be enormous and probably flattened, not spherical as it now appears. The subjective basis for our knowledge of the sun introduces a nice point for philosophical argument.

Of the planets in our solar system, Jupiter at least behaves abnormally as far as the radio astronomer is concerned. Several years ago scientists in Washington got some unexpected deflections on the chart connected to their radio telescope. For a long time they thought it was some local interference but after a few months the same kind of interference began to appear in the middle of the night and eventually they realized that the signals were coming from the planet Jupiter. Now these results are very surprising, since, unlike the sun, Jupiter does not have the kind of hot atmosphere conducive to the generation of radio waves. It now seems possible that these signals come from the planetary surface, and recent work has indicated that if this is the case only a few places on the surface are responsible. It is interesting to speculate on the events which might generate these signals, since the energies involved are enormous. In fact, it is difficult to establish a parallel with anything which might happen on earth. One has to think in terms of the energy of ten or more hydrogen bombs, or giant volcanic eruptions like the explosion of Krakatau.[3] We may well have to wait for the close approach of a space probe to Jupiter before this problem can be settled.

After these remarks about the radio emissions from the sun and Jupiter you may be somewhat puzzled by my earlier reference to the difficulty of detecting radio waves from the common visible objects in the universe. The sun is a common star, and one might therefore expect other stars in the Milky Way to emit similar radio waves. Indeed, they might well do so, but even the nearest star is over 200,000 times farther from us than the sun. If the sun were at that distance its radio waves would be 50,000 million times more difficult to detect and that presents us with an almost insuperable problem. . . .

Between the stars it has for long been realized that there must exist great clouds of hydrogen gas, but this gas is in a neutral state and does not emit light and so cannot be seen with ordinary telescopes. However, the neutral hydrogen atom emits radio waves on a particular wavelength of 21 centimetres, and these interstellar hydrogen clouds can be detected and measured with the radio telescopes. The story of the detection of these 21-centimetre emissions is an epic of modern science. In the terrible circumstances of the German occupation Van de Hulst, a young research student in Holland, made the calculations which led him to predict the existence of these radio waves. He showed that although for any one hydrogen atom the process of emission is only likely to take place once in eleven million years, the numbers of atoms in the interstellar clouds are so great that the emission should be detectable. Years of peace were required to develop the right equipment, and then in 1951 these weak radio waves were detected—an event which represented a triumph of technical skill and a brilliant vindication of Van de Hulst's prediction.

The astronomers in Leiden under their director, Jan Oort, seized on this unparalleled opportunity of piercing the secrets of the structure of the Milky Way. The large optical telescopes are powerless to penetrate the interstellar dust clouds which obscure the structural details of the system—and you must remember that we are placed in an unprivileged position on the edge of the galactic disk. The dust presents little handicap to the passage of the 21-centimetre waves, and during the last few years Oort and his staff have produced results of almost unbelievable detail and elegance describing parts of the Milky Way which men will never see.

3. The theory has been proposed more recently that Jupiter has a very strong magnetic field (some ten times stronger than the earth's) and that the microwave emission is caused by particles trapped in the magnetic field and spiraling back and forth along the lines of magnetic force. This is called synchrotron radiation. Ed.

The second feature of the radio waves which come from within the confines of the Milky Way is that the smooth background of radio emission is punctuated by very strong sources which stand out for the radio telescope in the same way that the bright visual stars stand out from the diffuse light of the Milky Way. I have already emphasized that the common bright stars do not emit radio waves—at least they have not yet been detected—and the solution to the origin of these strong sources of radio waves cannot be found in the common stars. There are a few cases where these radio sources have been identified with unusual objects which can be seen in the optical telescopes. The most spectacular is the Crab Nebula [Figure 8-13], 4,000 light years away. This nebula is the gaseous remains of an exploding star or supernova outburst seen by Chinese astronomers 900 years ago in A.D. 1054. The millions of tons of hydrogenous material of the star disintegrated in a catastrophic explosion, and at present we see the gaseous shell of the explosion still moving out through space at the rate of 70 million miles a day. This gas is at a high temperature and in a great state of turmoil, and we believe that

Figure 8-13. *Crab nebula.* (Hale Observatories. Figure added by editor.)

Figure 8-14. *Galaxy NGC 5128, which is associated with an intense radio source in Centaurus.* (Hale Observatories. Figure added by editor.)

these conditions are responsible for the radio emission. There are two other well-established cases of supernova outbursts in the galactic system—those observed by Tycho Brahe in 1572 and by Kepler in 1604. The visible remnants of these are difficult to see in the telescope, but both have been identified as radio sources. . . .

From these remarks you will see that the radio emission generated in the Milky Way system has complex origins, such as the diffuse regions of turbulent gas which give us localized radio sources, of which some at least are supernova; the clouds of neutral hydrogen gas; the corona around the system where no matter can be seen; and probably emission from ionized hydrogen which contributes to the background radiation. But when we inspect our records we find many thousands of radio sources which seem to be distributed uniformly over the sky. Most of these are much weaker than those which I have just described in relation to the Milky Way, and their uniform distribution leads us to believe that they are external and lie at great distances from the local galaxy. . . .

Ryle and his colleagues in Cambridge made a very accurate measurement of the position of one of the stronger of these unidentified sources which lies in the constellation of Cygnus. These measurements were suf-

ficiently accurate for the American astronomers to use the 200-inch telescope on Palomar to make an exhaustive search of this region of the sky.
They found there a remarkable event in which two great extragalactic
nebulae seem to have collided with one another. This collision, which is
at a distance of 700 million light years, is nearly at the limit of clear definition of the world's biggest optical telescope, and yet the radio waves
are relatively strong—so strong that even if the nebulae were ten times
farther away we could still detect them as a radio source. We do not yet
understand why two nebulae in collision like this produce radio waves
which are far more intense than those they would produce separately.[4]
For our present argument the important fact is that the strength of the
radio emission is out of all proportion to the visibility of the nebulae in
the optical sense [Figure 8-14]. . . .

. . . The precise interpretation of the current results from Cambridge
and Sydney, where so much of this work has been done, is hotly disputed. The problem is of the very greatest significance to cosmology,
and I must reserve further discussion until we consider the question of
the origin of the universe.

4. As more observations were made it became increasingly difficult to interpret
these radio sources as galaxies in collision. Many astronomers believe now that these
sources represent explosions of single galaxies. However, the difficulties of explaining
the radio galaxies have not been entirely resolved. Ed.

Theories of the Origin of the Universe

The Evolutionary Theory *
by A. C. B. Lovell (1959)

I want to talk to you about the problem of the origin of the universe. I suppose it would hardly be an exaggeration to say that this is the greatest challenge to the intellect which faces man, and I cannot pretend that I have any new solution to offer you. However, . . . the air is alive with a new hope and expectancy, because our new instruments may be reaching out so far into space that we may soon be able to speak with more confidence. I am going to set out the problem as I see it, and I hope you will get an idea of these vast cosmological issues and of the implications of the alternative solutions which lie ahead. At the end I shall tell you what I think about it all as an ordinary human being.

We have seen that observational astronomy tells us about the universe as it exists out to distances of about two thousand million light years. At that distance we are seeing the universe as it existed two thousand million years ago. Within this vast area of space and time we can study the innumerable stars and galaxies, and from these observations we can attempt to infer the probable nature and extent of the cosmos beyond the range of observations.

I think there are three stages in which we might consider this problem. The first stage is to inquire whether the observations are likely to be extended in the future to even greater distances and thereby penetrate even farther into past history than the present two thousand million years. The second stage is an appeal to cosmological theory, an inquiry as to the extent to which the present observations agree with any particular cosmology and the nature of the past and future as predicted by these theories. Finally, we shall reach a stage where theories based on our present conceptions of physical laws have nothing further to say. At this point we pass from physics to metaphysics, from astronomy to

* Abridged from A. C. B. Lovell, *The Individual and the Universe.*

theology, where the corporate views of science merge into the beliefs of the individual.

The vast region of space and time enclosed by the present observations includes several hundred million galaxies of stars. As far as we can see, the overall large-scale structure of the universe within these limits has a high degree of uniformity. When we look at these distant regions we find that the light is reddened, indicating that the galaxies are receding from us. As far as we can see, the red shift of the most distant nebulae is still increasing linearly with distance. There is no indication that we are seeing anything but a small part of the total universe. However, in the second stage of our inquiry we shall see that an observational test between rival cosmological theories demands a still further penetration, and an extension of the present observational limit is a matter of some urgency in cosmology. Unfortunately there are fundamental difficulties introduced by the recession of the galaxies which no device of man will ever surmount. At the present observable limit of the large optical telescopes the galaxies are receding with a speed of about one-fifth of the velocity of light. From this aspect alone we face a limit to future progress. Even if no other effects intervened we could never obtain information about those farther regions of space where the velocities of recession of the galaxies reach the speed of light. The light from the more distant galaxies will never reach us. In Eddington's phrase 'Light is like a runner on an expanding track with the winning post receding faster than he can run.'

There are, moreover, further difficulties which will hinder the approach to this fundamental limit. If the remote galaxies were stationary, then all the light emitted, say, in one second would reach our telescopes. But the galaxies are moving away with speeds which are an appreciable fraction of the velocity of light, and as the speed increases less and less of the light actually emitted by the galaxies in one second reaches our instruments. This degradation of the intensity of the light coupled with the accompanying shift in wavelength to the red end of the spectrum worsens still further the technical difficulties of these observations.

The radio telescopes may well be in a stronger position with respect to these hindrances. To begin with, the [radio] galaxies, which I described in an earlier lecture, generate very powerful radio emissions, and the shifts in wavelength which accompany the recession do not present the same observational difficulties as in the optical case. In fact, the present belief is that many of the objects already studied by their radio emissions lie at distances which exceed considerably the present two thousand million light years' limit of the optical telescopes. Therefore, we can, I think, answer the first stage of our inquiry with some degree of certainty in the following way. The present observable horizon of the universe will be pushed back by a limited amount in the near future, perhaps to a few thousand million light years. Then we must be content. No further strivings or inventions of man will enable us to probe the conditions which existed in epochs of history beyond these few thousand million years. They are gone for ever beyond the fundamental limits of observability.

At this point we reach the second stage of our inquiry, where we appeal to cosmological theory. The question is this. Can we formulate a theory in terms of known physical laws whose predictions agree so well with the present observable universe that we can predict the past and future?

Indeed, when we turn to the cosmological theories which are today seriously considered by astronomers we find a most absorbing state of affairs. Not one, but several theories can explain from acceptable postulates the present observable state of the universe. These predictions bring us face to face with the ultimate problem of the origin of the universe in ways which are startlingly different. But the new techniques in astronomy may be on the verge of producing observational data which may be decisively in favour of one or other of these cosmologies. At least one of these alternatives would, I think, present theology with a very serious dilemma. In fact, if the full implications of the theory eventually receive the support of astronomical observations it is difficult to see how certain fundamental doctrines could be maintained in their present harmonious relation with our physical knowledge of the universe.

First of all, though, I want to discuss the cosmological theories which are generally classed as the evolutionary models of the universe. I think it would be correct to say that these theories, which are a consequence of Einstein's general theory of relativity, are regarded with the most favour by the majority of contemporary astronomers. In passing, perhaps I should add that in the light of our present knowledge it does not seem worth while discussing for our present purpose any of the cosmological theories which preceded the introduction of the theory of general relativity in 1915. The application of Newton's theory of gravitation, in which the attraction between bodies varies inversely as the square of their distance apart, to the large-scale structure of the universe would require that the universe had a centre in which the spatial density of stars and galaxies was a maximum. As we proceed outwards from this centre the spatial density should diminish, until finally at great distances it should be succeeded by an infinite region of emptiness. The observed uniformity in the large-scale structure of the universe is clearly at variance with these ideas. Neither does any theory based on Newton's laws of universal attraction and conservation of mass offer hope of explaining the observed expansion and recession of the nebulae. On the other hand, in Einstein's theory of general relativity gravitation is not explained in terms of a force but of the deformation of space near massive bodies. In our ordinary life we treat space as though it were flat, or Euclidean, in the sense that the geometrical properties obey the axioms of Euclid as we were taught in school. For example, the three angles of a triangle add up to two right angles. According to Einstein's theory, however, these simple conceptions must be abandoned, and though in ordinary circumstances the differences are insignificant, nevertheless when we consider the properties of space near a massive star, for example, the conceptions of a flat space no longer apply . . . [Figure 7-6].

Einstein attempted to apply his new ideas of the gravitational curvature of space-time to the universe as a whole. In this case the curvature of space would be influenced not only by one star, but by countless stars and galaxies. However, in the large-scale view, as we have seen, the distribution has a high degree of uniformity, and the problem of the overall curvature of space can be related to the average density of the matter in the universe. In working out the equations Einstein was unable to find any solution which described a static universe. We must remember that this was a decade before the discovery of the recession of the nebulae, and any cosmological theory which did not provide for a static cosmos could have been little more than a curiosity.

Faced with this dilemma, Einstein realized in 1917 that the difficulties could be surmounted by the introduction of a new term in his equations. This is the famous λ term, or the cosmical constant, over which there was to be so much future dispute. This new term appears in the equations as an arbitrary universal constant. Its interpretation in terms of a physical model of the universe is that it introduces an effect analogous to repulsion. This cosmic repulsion increases with the distance between bodies, and is to be regarded as superimposed on the usual forces of Newtonian attraction. Thus at great distances the repulsion outweighs the attraction, and in the equilibrium condition the Newtonian attraction and cosmical repulsion are in exact balance.

We cannot follow in detail the subsequent developments, which are of the utmost complexity. Some years after the introduction of these ideas the whole situation was altered in dramatic fashion by the discovery that the universe was non-static but was expanding. At about the same time the Russian mathematician, Friedman, found other solutions of Einstein's equations which predicted either an expanding or contracting universe. In fact, now it has for long been realized that the equations of general relativity cannot define an unique universe because there are three unknowns in the equations, whereas observationally we have only two sets of data. The possible types of non-static universes fall into three main families determined by the various possible combinations of the sign of the cosmical constant and the space curvature. These are a universe which starts from a point origin at a finite time in the past and expands continuously to become infinitely large after an infinite time, a universe whose radius has a certain finite value at the initial moment of time, and thence expands to become infinite after an infinite time, and lastly a universe which expands from zero radius to a certain maximum and then collapses to zero again, this process of oscillation being capable of indefinite repetition. Within each of these three main categories a large number of possible models can be constructed differing in various points of detail. For the past thirty years cosmologists have sought for arguments based on the observed characteristics of the universe which would identify the actual universe with one of the theoretical models.

All that I propose to do here is to give some examples of these evolutionary models, one of which is today believed by many cosmologists to describe the past history with some degree of certainty. The first example is a solution discovered by the Abbé Lemaître in 1927 and developed by Eddington. I have already said that by introducing the cosmical constant Einstein was able to specify a static condition of the universe in which the Newtonian attraction and cosmical repulsion are in exact balance. However, this equilibrium is unstable. If something upsets the balance so that the attraction is weakened, then cosmical repulsion has the upper hand and an expansion begins. As the material of the universe separates, the distance between the bodies becomes greater, the attraction still further weakens, the cosmical repulsion ever increases, and the expansion becomes faster. On the other hand, if the equilibrium was upset in the other way so that the forces of attraction became superior, then the reverse would occur and the system would contract continuously. Eddington's view was that in the initial stage the universe consisted of a uniform distribution of protons and electrons, by our standards very diffuse. This proton-electron gas comprised the entire primeval universe, which would have had a radius of about a thousand million light years. At some stage an event or series of events must have occurred in this diffuse gas which determined that the universe was launched on a career of expansion and not contraction. There were many views as to how this might have happened. Eddington held that the accumulation of irregularities in the gas started the evolutionary tendency. Soon, condensations formed in the gas and those ultimately became the galaxies of stars. On these views the present radius of the universe must be about five times that of the initial static primeval universe.

In the light of modern knowledge this theory receives little support. The time scale of its evolution is too short, and one cannot find a compelling reason why the primeval gas should have been disturbed in such a way as to determine that the universe was launched on a career of expansion rather than contraction. The initial condition is a special case, ephemeral and fortuitous. As far as the laws of physics are concerned, one can only say that by chance the initial disturbances were such as to determine the history of the universe. One cannot feel very happy that such a chance occurrence some thousands of millions of years ago should have determined the fundamental features of the universe. Moreover, although originally the theory as expounded by Jeans and Eddington undoubtedly had attractive features for some theologians, I feel now that this might well have been enhanced by feelings of relief that the vastness, uniformity, and organization of the universe which had just been revealed still remained outside the conceivable laws of physics in its initial state. Indeed, when considering these initial conditions Jeans spoke in terms of 'the finger of God agitating the ether,' implying a divine intervention at a predictable time in past history after which the laws of physics became applicable. This degree of familiarity with divine processes is, I think, undesirable theologically, and for science it evades the problem by obscuring the ultimate cosmological issue.

Moreover, there is another problem which must be faced. The event which we have considered in the unstable static assemblage of primeval gas predetermined the subsequent history of the universe. One must still inquire how long the gas existed in this condition of unstable equilibrium and how the primeval gas originated. Science has nothing to say on this issue. Indeed, it seems that the theory requires the exercise of yet another divine act at some indeterminate time before the occurrence which set off the gas on its career of condensation and expansion.

Of course, this particular model is now of little more than historical interest as being one of the first of the evolutionary theories based on general relativity to receive serious attention. It provides, however, a remarkable example of the influence in cosmology of the predilection of the individual. When faced with the various possible cosmological models which we have outlined, Eddington said this: 'Since I cannot avoid introducing this question of a beginning, it has seemed to me that the most satisfactory theory would be one which made the beginning not too unaesthetically abrupt. This condition can only be satisfied by an Einstein universe with all the major forces balanced.' He continues, 'Perhaps it will be objected that, if one looks far enough back, this theory does not really dispense with an abrupt beginning, the whole universe must come into being at one instant in order that it may start in balance. I do not regard it in that way. To my mind undifferentiated sameness and nothingness cannot be distinguished philosophically.' In this way Eddington attempted to rationalize the basis on which to build the universe.

I have already mentioned the Abbé Lemaître. His original work in 1927, published in a little-known journal, was discovered by Eddington. Although Eddington remained faithful to this idea that the universe evolved from the static but unstable Einstein universe, the conception was soon abandoned by Lemaître himself. For the past twenty-five years Lemaître's name has been associated with another model whose origin recedes even farther back in time than the static Einstein state. Of all cosmologies, it is, perhaps, by far the most thoroughly studied. We shall see later that during the last few years a tremendous clash has occurred with other opinions, but at the present time there are no known features of the observable universe which are incompatible with Lemaître's evolutionary cosmology. Lemaître's model is typical of one of the groups of theories inherent in general relativity, according to which the universe originated at a finite time in the past and expands to an infinite size at an infinite future time.

Perhaps we can most easily visualize this conception by taking the universe as we see it now and inquiring quite simply what might have been the situation long ago. The observations of the distant galaxies show that their light and radio emission is shifted in wavelength so that as received on the earth the light is redder and the radio waves longer in wavelength than those which are actually emitted. The interpretation of this shift is that we are separating from the galaxies at a very high speed, and that the speed of recession increases as we move out into space. At the limits of present-day observation the speed of recession is about thirty-seven thousand miles per second, which is a fifth of the velocity of light.[1] The observation which gives us this figure is of a cluster of galaxies in Hydra photographed in the two-hundred-inch telescope. The so-called cosmological principle which is inherent in Lemaître's theory implies that if human beings equipped with similar instruments existed on a planet in this Hydra cluster of galaxies, then they would see the cluster of galaxies to which we belong at the limit of their powers of observation, and the velocity of recession would also be thirty-seven thousand miles per second. It is important to rid ourselves of any idea that because all around us we find galaxies in recession, then we are the

1. With the discovery of quasars these figures have very dramatically increased. See selection beginning on p. 437. Ed.

centre of the recessional movement. This is not the case. It is an impression which we obtain because we can see only a small part of the total universe.

To return to this cluster of galaxies in Hydra. We are now seeing it as it was two thousand million years ago moving away at a rate of thirty-seven thousand miles a second. What is the likely past history of this and all other similar galaxies? Up to a point this question is not too difficult to answer. For example, a minute ago we were two million miles closer to this cluster than we are now. A year ago we were over a billion miles closer. If we recede back into history in this manner we realize that the galaxies such as Hydra which are now almost beyond our view must have been very much closer to us in the remote past. In fact, if we proceed in this way, then we reach a time of about eight or nine thousand million years ago when all the galaxies must have been very close together indeed. Of course, the galaxies themselves have evolved during this time, but the primeval material from which they were formed must have existed in a space which is very small compared with the universe today.

With important reservations which I shall deal with now, this in essence is the fundamental concept of Lemaître's theory, namely, that the universe originated from a dense and small conglomerate which Lemaître calls the primeval atom. I shall return in a moment to the conditions which might have existed at the beginning, and to the possible events which might have initiated the disruption and expansion of the primeval atom. It is in fact necessary to emphasize that the theory does not demand the formation of the galaxies in the first phase of the expansion. The primeval atom contained the entire material of the universe, and its density must have been inconceivably high—at least a hundred million tons per cubic centimetre. The initial momentum of the expansion dispersed this material, and after thousands of millions of years the conditions applicable to the so-called Einstein universe would have been reached. Then the size of the universe was about a thousand million light years and the density would have been comparable to that with which we are familiar on earth. According to Lemaître, at this stage the initial impetus of the expansion was nearly exhausted and the universe began to settle down into the nearly static condition which we have previously considered, where the forces of gravitational attraction and cosmical repulsion were in balance. The mathematical treatment indicates that the universe must have stayed for a long time in this condition. It is during this phase that the great clusters of galaxies began to form from the primeval material. Then the conditions of near equilibrium were again upset, the forces of cosmical repulsion began to win over those of gravitational attraction, and the universe was launched on the career of expansion which after nine thousand million years brought it to the state which we witness today.

The time scale determined by tracing back the past history of the galaxies brings us not to the beginning of time and space, but merely to a condition which existed a few thousand million years ago when the universe was probably about one-tenth of its present size and consisted of the original gaseous clouds from which the clusters of galaxies began to form. The processes of the formation and evolution of the galaxies from this early stage are the subject of very detailed mathematical treatment. There is, at present, every reason to believe that a satisfactory explanation of the evolution of the universe from that condition can be given in terms of the known laws of physics. But when we pass on to consider the even earlier stages, difficulties and uncertainties appear. How much farther do we have to go back in time to the condition of the primeval atom? The theory does not determine this with any precision, because the delay which the universe suffered during the equilibrium phase when the gaseous clouds were forming into galaxies cannot be specified. One can, however, say this—that the explosion or disintegration of the primeval atom must have occurred between twenty thousand million and sixty thousand million years ago. In other words the period of about nine thousand million years ago, when the galaxies began to form and the present period of expansion began, represents a comparatively recent phase in the history of the universe. . . .

I shall talk about the alternative view which science can offer on the origin of the universe, but before doing this I want to dwell a moment on the implications of this evolutionary theory. The time scale, although vast, is conceivable in human terms. From the initial moment of time when the primeval atom disintegrated, astronomy and mathematics can attempt to describe the subsequent history of the universe to the state which we observe today. Moreover, there is every chance that in the foreseeable future man will produce experimental tests which will either substantiate or destroy this picture. But when we inquire what the primeval atom was like, how it disintegrated and by what means and at what time it was created we begin to cross the boundaries of physics into the realms of philosophy and theology. The important thing at that stage is what you and I think about this situation, this beginning of all time and space.

As a scientist I cannot discuss this problem of the creation of the primeval atom because it precedes the moment when I can ever hope to infer from observations the conditions which existed. If, indeed, the universe began in this way, then the concepts of space and time with which we deal originated at some moment between twenty thousand million and sixty thousand million years ago. Time, in the sense of being measured by any clock, did not exist before that moment, and space, in the sense of being measured by any yardstick, was contained entirely within the primeval atom. The vast regions of space which we survey today are just a small part of those which were originally the space of that small conglomerate.

We can, of course, speculate on the issues of the creation of the primeval atom and its initial condition, but it is the philosopher who must first build a scheme which is self-consistent and which leads us smoothly into beginning of space-time where the mathematician can take over. Or one can simply refuse to discuss the question. If we wish to be materialistic, then we adopt the same attitude of mind as the materialist adopts in more common situations. The materialist will begin in the present case at the initiation of space-time when the primeval atom disintegrated. . . .

With an effort of imagination the human mind can trace its way back through the thousands of millions of years of time and space, and we can attempt to describe in common concepts the condition of the primeval atom. The primeval atom was unstable and must have disintegrated as soon as it came into existence. There we reach the great barrier of thought because we begin to struggle with the concepts of time and space before they existed in terms of our everyday experience. I feel as though I've suddenly driven into a great fog barrier where the familiar world has disappeared.

I think one can say that philosophically the essential problem in the conception of the beginning of the universe is the transfer from that state of indeterminacy to the condition of determinacy, after the beginning of space and time when the macroscopic laws of physics apply. When viewed in this way we see that the problem bears a remarkable similarity to one with which we are familiar. This is the indeterminacy which the quantum theory of physics introduces into the behaviour of individual atoms, compared with the determinacy which exists in events where large numbers of atoms are involved. The process of thought by which we reduce the multiplicity of the entire universe to its singular condition of the primeval atom is equivalent in principle to the reduction of the chair in which you are sitting to one of its individual atoms. Not in the evolutionary sense of course, but in the sense that quantum theory and the principle of uncertainty explains why the behaviour of the individual atom is indeterminate and why it is impossible for you to find out the condition of the atom with any precision, because you will disturb it in the very process of investigation. In fact, the application of

the fundamental concepts of quantum theory to the cosmological problem enables us to begin the struggle with the barrier which arises whenever we think about the beginning of space and time.

The primeval atom was a singular state of the universe, as incapable
of precise specification by physical methods as the familiar individual
particle in the uncertainty principle of modern physics. When the primeval atom disintegrated the state of multiplicity set in and the universe
became determinate in a macroscopic sense. Philosophically, space and
time had a natural beginning when the condition of multiplicity occurred, but the beginning itself is quite inaccessible. In fact, in the beginning the entire universe of the primeval atom was effectively a single
quantum unit in the sense that only one of the future innumerable potential states existed. I am aware that this discussion is merely a line of
metaphysical thought. Its importance lies in the parallel with the fundamental difficulties and basic indeterminacies in modern quantum theory.
If future advances should occur in these directions, then it may become
possible to speak with more certainty about the condition of the original
cosmological quantum. In the light of our present knowledge of atomic
physics it is possible only to surmise the kind of condition which might
have existed. I suggested earlier that the density of matter in this primeval atom was inconceivably high. This is arrived at by a simple arithmetical deduction from the probable total mass of the universe as we
see it now, and by assuming that the radius of the primeval atom was
not greater than, say, a few million miles. However, it is possible that
the primeval atom was not like this, but that it consisted of intense radiation and corpuscular rays which formed the primeval gas during the
first phases of the expansion. In fact, it is a fundamental concept of Lemaître's theory that the cosmic radiation which we observe today is a
relic of this early state. A characteristic of this picture of evolution is the
long time-scale involved in the transformation of the intense energy of
the original primeval atom, first into the gaseous clouds of hydrogen and
then by processes, which awaited the high temperatures and pressures
which arose when stars began to form, into the other elements with
which we are familiar today. If pressed to describe this primeval atom
in conventional terms one would, I think, refer to a gigantic neutron. By
radioactive decay this neutron suffered a tremendous explosion. Protons,
electrons, alpha particles, and other fundamental particles emerged from
it at great velocity and continued to fill all space nearly uniformly as
this basic material expanded for many thousands of millions of years
until the clusters of galaxies began to form.

An alternative picture of the condition of the primeval atom has been
given by Gamow, who believes that it consisted entirely of high-temper-

ature thermal radiation. Five minutes after the expansion began the temperature of the universe was a thousand million degrees, after a day it had fallen to forty million degrees—say, nearly to the temperature of the centre of the sun; after ten million years it had fallen to an average temperature, which we call room temperature. On this theory of Gamow all the chemical elements which we deal with today must have been formed within the first thirty minutes of the life of the universe.

Gamow differs from Lemaître in other important respects. In Lemaître's theory the force of the initial disintegration was exhausted after a few thousand million years, and the expansion which we witness today came into play only as a result of the forces of cosmical repulsion which developed when the galaxies began to form. In Gamow's theory the force of the initial explosion was so great that the expansion of the universe is attained without invoking the force of cosmical repulsion. In other words, the beginning in the Gamow theory is close to the nine thousand million years [2] which we deduce by tracing back the history of the galaxies, and there is no protracted period in the state of diffuse gas with all the major forces balanced as in Lemaître's theory.

The most distinguished living exponent of the evolutionary theory of the origin of the universe is himself in Holy Orders. For him and for all who associate their universe with God, the creation of the primeval atom was a divine act outside the limits of scientific knowledge and indeed of scientific investigation. The probable condition of intense radiation in the primeval atom is entirely consistent with the divine command "Let there be light." It would, of course, be wrong of me to suggest that this view of the origin of the universe demands necessarily the possibility of creation of matter by a divine act. On the contrary, those who reject God adopt a strictly materialistic attitude to the problem of the creation of the primeval atom. They would argue that the creation of the primeval material had no explanation within the framework of contemporary scientific knowledge, but would escape from the dilemma by reserving the possibility that science would, if given the opportunity of studying these initial conditions, find a satisfactory solution. Or they would evade the problem of a beginning altogether by following a further line of thought due to Gamow that the primeval atom was not the beginning, but merely a state of maximum contraction of a universe which had previously existed for an eternity of time.[3] I think, how-

2. More recent estimates are higher, 10,000 million years or more. Ed.
3. Gamow in his book, *One, Two, Three . . . Infinity* states it this way: "The universe is now expanding because in some previous period of its history (of which, of course, no record has been left), it contracted from infinity into a very dense state and then rebounded, as it were, propelled by the strong elastic forces inherent in compressed matter. . . . We can now send our imagination flying beyond any limits,

ever, that for theology, there is one important observation to make. If the universe was created and evolved in the manner just described, then the conception that the creation of the primeval material was a divine act can never be attacked by scientific investigation. A set of conditions which existed over twenty thousand million years ago, and which can never return again, is for ever beyond investigation.

The theory which we have discussed envisages a once for all creation in the remote past followed by a steady evolution to the present conditions. The alternative to this theory is that the creation of matter is taking place continuously and that although stars and galaxies evolve from this basic material, the universe, when considered as a large-scale structure, is in a steady state. We can illustrate this view by considering the future history of the galaxies which are now near the limit of observation. We are receding at great speed from these galaxies. In a billion years' time the galaxies will have passed for ever from our field of view and other galaxies which are now closer to us will have moved out to our observable horizon. So much is common ground on both the evolutionary and steady-state theories. The sharp distinction arises when we compare the picture of the universe within the observable horizon now and in a billion years' time. On the evolutionary theory more and more galaxies move out of our field of view, and the number of galaxies which we can see with our instruments will for ever decrease. In other words, the average spatial density of the universe is decreasing. On the steady-state theory this is not the case. Although individual galaxies recede beyond the observable horizon, others are always being created to take their place. In a billion years' time the universe will look to us very much as it does now. The individual galaxies will have changed, but their average spatial density remains the same, because matter is always in creation throughout all of space. The cosmological principle of the evolutionary theory in which the universe would appear to be the same to any observer, wherever he was situated in space, has become the perfect cosmological principle according to which the universe is the same throughout all space and time.

The implications of this point of view are, of course, profound. For example, there cannot have been a beginning in any scale of time at all. If we trace back in time the history of the galaxies, they dissolve into gas

and ask ourselves whether during the precompressive stages of the universe everything that is now happening was happening in reverse order. Were you reading this book from the last page to the first some six or eight billion years ago? . . . Interesting as they are, such questions cannot be answered from the purely scientific point of view, since the maximum compression of the universe, which squeezed all matter into a uniform nuclear fluid, must have completely obliterated all the records of the earlier compressive stages." Ed.

and then into uncreated matter as they move in towards us, whereas others come into view from beyond the observable horizon. At a time of twenty thousand million years ago the evolutionary models picture the universe as a concentrated conglomerate of gas, whereas the steady-state universe would have appeared as it does today. Indeed, however far we go back in time, there is no stage at which we can say that the universe, as a whole, had a beginning. In the only language at our command we can say that the history of the universe on the steady-state theory extends to an infinite time in the past.

Whereas there is hope that we can put our inferences about the past to an experimental test, we can discuss the future only in terms of the predictions of cosmological theory. Here again there are great differences between the evolutionary and steady-state models. The predictions of the steady-state theory are quite clear. The universe has an infinite extent in space and an infinite future in time. There is, of course, a limit to the observable universe from any one place in it determined by the speed of expansion. But if an intelligent being exists at our observable limit he would find himself surrounded by a similar universe of galaxies and so on without end. Neither does the theory of continuous creation place any limitation on the future extent in time. In the same way that a billion years ago the universe would look the same as it does now, so in a billion years of future existence the overall large-scale picture will be unchanged.

The future on the evolutionary models is quite different. The total content of matter was fixed once and for all at the time of creation. The expansion is thinning out the galaxies, and in a billion years our view of space would indeed be vastly different from what it is today. In some variations of the evolutionary theory the process of expansion is expected to reverse when the spatial density has fallen to a certain value, and then the contraction of space would bring the ageing galaxies into view again. But even in such variations of the evolutionary models the ultimate death of the universe seems inescapable, because the energy with which the universe was imbued at its creation is relentlessly becoming less available.

The finite limitations of space, time, and content in some of the evolutionary models lead one to ask whether our universe is, in fact, the entire cosmos. It is a question which at present cannot be discussed with profit. There is no feature of the theory which would preclude the existence of other universes created at different times, but unless we are hopelessly wrong in our interpretation of our observations of the universe we see, there is no conceivable way in which we can ever penetrate the regions of time and space where they might exist.

The conflict between the steady-state and evolutionary theories is of the very greatest significance to cosmology and to human thought. The evolutionary theory places the creation of matter at a definite moment in the remote past, beyond human investigation. Although the steady-state theory has no solution to the problem of the creation of matter,[4] it is important to appreciate that if this theory is correct, then the primeval gas is being created now, at this moment, and hence is open to human investigation. On the whole, I think it must be incontestable that the steady-state theory is more materialistic than the evolutionary theory. It could be said that the creation process is a divine act which is proceeding continuously, and which is beyond the conception of the human mind. On the other hand, it cannot be denied that this may be a somewhat perilous attitude for the simple reason that the tools of science can probe the regions of space where this creation is occurring. In fact, in the equations of the cosmologists a creation term already exists. Philosophically, it is, I think, important to emphasize the approachability of the creation of hydrogen which is inherent in these modern theories of continuous creation. Otherwise the metaphysical impact would not be severe. In this sense the concept was stated long ago by Kant in these words ". . . the remaining part of the succession of eternity is always infinite," he said, "and that which has flowed is finite, the sphere of developed nature is always but an infinitely small part of that totality which has the seed of future worlds in itself, and which strives to evolve itself out of the crude state of chaos through longer or shorter periods. The creation," he went on, "is never finished or complete. It has, indeed, once begun, but it will never cease," But, of course, Kant's doctrine was egocentric, in the sense that God had completed the creation in the part of the cosmos which we can see. In the contemporary theories of continuous creation the processes of formation should still be occurring all around us, and are therefore open to human investigation.

I think it is true to say that during the last few years the cosmological issue has crystallized into a conflict between these evolutionary and steady-state theories of the origin of the universe. The variations in detail within these two broad principles are numerous. Many of these dif-

---

4. Fred Hoyle in his book *Frontiers of Astronomy* says on this point: "There is an impulse to ask where originated material comes from. But such a question is entirely meaningless within the terms of reference of science. Why is there gravitation? Why do electric fields exist? Why is the Universe? These queries are on a par with asking where newly originated matter comes from and they are just as meaningless and unprofitable. . . . We can ask questions quite freely about the consequences of the laws of physics. But if we ask why the laws of physics are as they are, we shall receive only the answer that the laws of physics have consequences that agree with observation." Ed.

ferences are highly abstract, but in so far as the stream of human thought is concerned these internal variations are of small consequence compared with the major issue as to whether creation is occurring now and throughout all time in the past and in the future, or whether the fundamental material of the universe was created in its entirety some billions of years ago.

It seems possible that we may be on the verge of settling by experimental observation which of these two principles is correct. In fact, the group of young cosmologists[5] who have promulgated the theories of continuous creation have always emphasized that, as distinct from the theoretical arguments which have surrounded the variations of evolutionary cosmology in the last thirty years, the new theories should be capable of direct experimental test. . . .

If time and space had a beginning, then when the universe was only a few thousand million years old it would be much more compact than it is today. The galaxies would be in existence, but they would be packed closer together compared with their spatial density today. The spatial density today—by which I mean the number of galaxies within, say, fifty or a hundred million light years of the Milky Way—can be determined by the large telescopes. If we could count the number in a similar volume of space at a distance of several thousand million light years we should in effect be making a count of the galaxies as they existed several thousand million years ago. If creation is still taking place, then on the steady-state theories this number should be the same as today. If the evolutionary model is correct, then the spatial density at this distance in time and space will be much greater [Figure 8-15].

The possibility of carrying out this decisive observational test excites the imagination. Unfortunately it seems likely that the hindrances introduced by the atmosphere of the earth will prevent the great optical telescopes from penetrating to the required regions of space. It may well be that only when optical telescopes can be carried in earth satellites or erected on the moon will it be possible to look back into the past to this extent. Before the advent of such futuristic enterprises it seems likely that the great radio telescopes will give us the answer we require. . . .

As individuals we must therefore face the possibility that within the next few years astronomers may be able to speak with unanimity about the ultimate cosmological problem. Only the materialist can turn aside unmoved by this prospect. For others, a settlement of this cosmological issue might mean an affirmation or rejection of deeply embedded philosophical and theological beliefs.

5. Fred Hoyle, Hermann Bondi, and Thomas Gold have been the most prominent advocates of this theory. Ed.

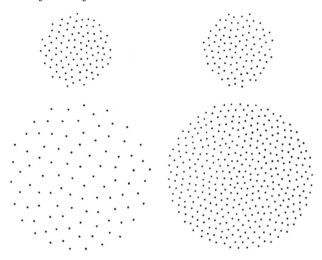

Figure 8-15. *Evolutionary and steady-state views are compared in these diagrams. At left is a schematic view of an evolutionary universe. At the top is a sample of the universe with the galaxies represented by dots. At the bottom is a picture of the same galaxies after the passage of time. The galaxies have merely receded from one another. At right is the same kind of view of a steady-state universe. At the top is a sample of the universe. At the bottom is a picture of the volume occupied by the same galaxies after the passage of time. In that time, however, new galaxies have been created, maintaining the density of the galaxies as before.* (From "The Steady-State Universe" by Fred Hoyle. Copyright © 1956 by Scientific American, Inc. All rights reserved. Figure added by editor.)

So far I have tried to present the contemporary background without prejudice, but no doubt before I finish you will expect me to say a word about my own personal views. At the moment our outlook in astronomy is optimistic. A new epoch has been opened by the development of radio telescopes, and we are perhaps within a generation of an even more astonishing one because of the inherent possibilities of astronomical observations from earth satellites or the moon. We can only guess as to the nature of the remote regions which might be photographed by telescopes removed from their earth-bound environment. In the case of radio telescopes this development is still very young. Three hundred years elapsed between Galileo's small telescope and the inauguration of the 200-inch telescope on Palomar Mountain. In the development of radio telescopes we have not covered a tenth of that time-span. I think therefore that our present optimism may well be of the kind which comes from the initial deployment of great new instruments and techniques. I have no doubt that within a few years these instruments will enable us to resolve the conflict which I have described between the evolutionary and steady-state models. In this process new difficulties will certainly appear, and these might make my present description of the universe as out of date as the static egocentric description which was in vogue in the first twenty years of this century. When we are dealing with time-spans of thousands of millions of years it would be sheer impudence to suggest that the views of the cosmos which have evolved from the techniques developed in our age possess any degree of finality. My present attitude to the scientific aspects of the problem is therefore neutral in the sense that I do not believe that there yet exist any observational data which are decisively in favour of any particular contemporary cosmology. The optimism with which I believe that we are on the verge of producing the necessary observational data is tempered with a deep apprehension, born of bitter experience, that the decisive experiment nearly always extends one's horizon into regions of new doubts and difficulties.

On the question of the creation of the primeval material of the universe it seems to me unlikely that there can ever be a scientific description, whether in terms of the evolutionary or steady-state theories. If the idea of continuous creation is substantiated, then science will have penetrated very far indeed into the ultimate processes of the universe. It might then appear that a completely materialistic framework would have been established, but it does not seem to me that this is the case. If one imagines a scientific device which is so perfect that it could record the appearance of a single hydrogen atom as demanded by the continuous creation theory, then the scientific description of the process would still be imperfect. The same basic and quite fundamental difficulty would appear, as I have described in the case of the primeval atom, in the further effort to obtain information about the nature of the energy input which gave rise to the created atom.

If I were pressed on this problem of creation I would say, therefore, that any cosmology must eventually move over into metaphysics for reasons which are inherent in modern scientific theory. The epoch of this transfer may be now and at all future time, or it may have been twenty thousand million years ago. In respect of creation the most that we can hope from our future scientific observations is a precise determination of this epoch. I must emphasize that this is a personal view. The attitudes of my professional colleagues to this problem would be varied. Some would no doubt approve of this or a similar line of metaphysical thought. Others would not be willing to face even this fundamental limit to scientific knowledge, although, as I have said, an analogous limitation occurs in modern scientific theory which describes the well-known processes of atomic behaviour. Some, I am afraid, will be aghast at my temerity in discussing the issues at all. As far as this group is concerned, all that I say is that I sometimes envy their ability to evade by neglect such a problem which can tear the individual's mind asunder.

On the question of the validity of combining a metaphysical and physical process as a description of creation, this, as I said earlier, is the individual's problem. In my own case, I have lived my days as a scientist, but science has never claimed the whole of my existence. Some, at least, of the influence of my upbringing and environment has survived the conflict, so that I find no difficulty in accepting this conclusion. I am certainly not competent to discuss this problem of knowledge outside that acquired by my scientific tools, and my outlook is essentially a simple one. Simple in the sense that I am no more surprised or distressed at the limitation of science when faced with this great problem of creation than I am at the limitation of the spectroscope in describing the radiance of a sunset or at the theory of counterpoint in describing the beauty of a fugue.

When I began my talks I mentioned the mixture of fear and humility with which I approached the task. Now you see the irony of the modern astronomer's life in its entirety. The devices of a world war have been forged, with the help of the fear of another, into a system of scientific experiments which take us back through time and space to deal with the problems of the origin of the universe.

Why Is It Dark at Night? *
by Hermann Bondi (1959)

One of the bases of modern cosmology is known as Olbers' paradox, which makes the darkness of the night sky appear as a curious phenomenon. The argument leading up to this is so simple and attractive and beautiful that it may not be out of place to consider it here in full.

When one looks at the sky at night one notices that there are some very bright stars, more medium bright ones and very large numbers of faint ones. It is easy to see that this phenomenon might be accounted for by the fact that the bright-looking stars happen to be near; the medium bright ones rather farther away, and the faint ones a good deal farther away still. In this way one would not only account for the variations in brightness but also for the fact that there are more of the faint ones than of the medium bright ones, and more of the medium bright ones than of the very bright ones, for there is more space farther away than nearby. One can now speculate about stars yet farther away, so far away, in fact, that they cannot be seen individually, not by the naked eye, nor even by the telescope. The question then arises of whether these very distant stars, though they would individually be too faint to be seen, might not be so exceedingly numerous as to provide an even background illumination of the night sky? This is the question that the German astronomer Olbers asked 130-odd years ago. The argument will now be presented in the light of the astronomical knowledge of 1826, without considering any of the phenomena discovered by modern astronomy.

OLBERS' PARADOX

Olbers then attempted to calculate what the brightness of the background of the sky should be on this basis. He immediately realized that in trying to consider effects from regions too far away to be seen in detail, he was forced to make assumptions about what the depths of the universe were like. He then made a set of assumptions which looks so plausible even nowadays that they may well serve as a model of what the beginning of a scientific investigation should be like. He first assumed, in the light of the knowledge of his day (1826), that the distant regions of the universe would be very much like our own. He expected there would be stars there, with the same average distance between them as between near stars. He expected that while each star would have an intrinsic brightness of its own, there would be an average brightness of stars very much like that in our astronomical neighborhood. In other words, he assumed that we get a typical view of the universe. This is in full accord with the ideas that have been current since the days of Copernicus, that there is nothing special, nothing pre-selected about our position in the scheme of things. This is a convenient assumption from a scientific point of view and a very fruitful one because we can assume that what goes on around us holds elsewhere as well, if not in detail at least on the average.

Unfortunately, this assumption is not sufficient for the calculation Olbers wished to make. For light travels at a finite speed: at a high speed, it is true, but a finite one nevertheless. The light we now receive from many distant regions was sent out by the objects there a long time ago, having spent the intervening period on its journey from there to here.

What is important for us, therefore, in trying to calculate the amount of light we get from the depths of the universe, is not how much the stars there radiate *now*, but how much they radiated at the time when the light which we receive now was sent out by them. We have to make a guess about the variation of astronomical conditions, not only with space, but also with time. And here, again, Olbers made the simplest of all possible assumptions, for he assumed time to matter as little as space. In other words, he supposed that not only in other parts of the universe, but also at other times, there would be stars, that their brightness would be the same as it is in our astronomical neighborhood, and similarly, their average distance apart would be the same as it is near us. Next, Olbers assumed, very naturally, that the laws of physics, as we know them from here, apply elsewhere and at other times. In particular, he assumed that the laws of the propagation of light—the way light spreads out after leaving its source—applied over these vast regions just as much as they apply in our rooms here. This, again, is the most obvious, most convenient and most fruitful assumption one can make. It would seem a stupid thing to set out on a voyage of discovery into the depths of the universe by first throwing away all the knowledge we have gained in our vicinity. Finally, Olbers made an assumption which is of the utmost importance, but he made it implicitly. He was not aware of the fact that he was making an assumption at all. Scientists know very well that this is the most dangerous kind of assumption. This assumption was that there were no large, systematic motions in the universe; that the universe was static.

THE MATHEMATICS OF STARLIGHT

On the basis of these four assumptions, it is easy to work out the background light of the sky. Imagine a vast, spherical shell surrounding us (Figure 8-16). The thickness of the shell is supposed to be small compared with its radius; but the whole shell is supposed to be so enormous that there are vast numbers of stars within the shell. How many stars are there in this shell? In order to work this out we have to know the volume of the shell. If we call the radius of the shell R and its thickness H, then we see readily that the surface of the sphere on which the shell is built is $4\pi R^2$ and thus the volume of the shell is, to a sufficient approximation, $4\pi R^2 H$. If, now, N is the number of stars per unit volume, then the number of stars in the shell of volume $4\pi R^2 H$ will be $4\pi R^2 HN$. How much light will all the stars in the shell send out? If the average rate at which an individual star sends out light is L, then all the stars in the shell put together will send out $4\pi R^2 HNL$. However, what interests us is

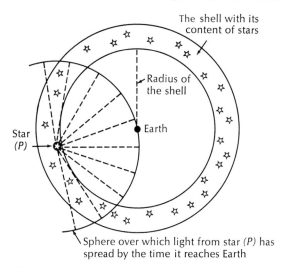

The shell with its
content of stars

Radius of
the shell

Earth

Star
(P)

Sphere over which light from star (P) has
spread by the time it reaches Earth

Figure 8-16. *Each star gives off a large amount of light, but only a small fraction reaches our earth. But there are so many millions of stars, a large amount of light should reach us. If we consider shells of equal thickness, equal amounts of light should reach us from each shell. If this were so, the sky should be bright and there would be neither light nor day nor any life on earth.*

not how much light all these stars send out, but how much light we receive from them. Consider the light of an individual star in the shell. By the time the light from it reaches us, it will have traveled through a distance R: and so it will have spread out over a sphere of surface $4\pi R^2$. That is to say, the light of each individual star has to be divided by $4\pi R^2$ to tell us the intensity of light from it which is received here. This is true of all the stars in the shell, and therefore, the total light we receive from all the stars in the shell is the total light they send out divided by $4\pi R^2$. This division leads to the cancellation of the factor $4\pi R^2$ and we are left with HNL.

It will be seen that this does not involve the radius of the shell at all. The amount of light we receive from any shell of equal thickness is the same irrespective of the radius of the shell. If, therefore, we add shell after shell, then, since we get the same amount of light from each shell, the amount received will go up and up without limit. On this basis, we should be receiving an infinite amount of light from all the shells stretching out to infinity. However, this argument not only leads to an absurd conclusion, but is not quite right. For each star, in addition to sending out light, obstructs the light from the stars beyond it. In other words, we will not be receiving light from stars in the very distant shells because there will generally be a star in between us and there—a nearer star—which will intercept the light. Of course, it will be realized that stars send out very much light, considering how small a surface they have. Therefore, this obscuring or shadowing effect is not very strong; it will prevent the sum from going up to infinity, but it still leads to our getting from all these shells of stars a flood of light equal to 50,000 times sunlight when the sun is in the zenith. On this basis, then, it should be incredibly bright both day and night. Everything would be burned up; it would correspond to a temperature of over 10,000 degrees Fahrenheit. Naturally, this remarkable result astonished Olbers, and he tried to find a way out. He thought that this flood of light might be stopped by obscuring clouds of matter in space between us and these distant stars. However, this way of escape does not work. For, if there were such a cloud, it would be getting hot owing to the very fact that it was absorbing light from stars; and it would go on getting hotter and hotter until it radiated by its glow as much light as it received from the stars. And then it would not be a shield worth having any longer. Other ways out have been tried, but none of them works. We are, therefore, inevitably led to the result that, on the basis of Olbers' assumptions, we should be receiving a flood of light which is not, in fact, observed.

EXPANSION AND DIMNESS

This little argument may well serve as a prototype of scientific arguments. We start with a theory, the set of assumptions that Olbers made. We have deduced from them by a logical argument consequences that are susceptible to observation, namely, the brightness of the sky. We have found that the forecasts of the theory do not agree with observation, and thus the assumptions on which the theory is based must be wrong. We know, as a result of Olbers' work, that whatever may be going on in the depths of the universe, they cannot be constructed in accordance with his assumptions. By this method of empirical disproof, we have discovered something about the universe and so have made cosmology a science.

In order to escape from this paradox, we have to drop at least one of his assumptions. In the light of modern knowledge described in the previous chapter, the reader will have no difficulty in spotting the assumption that has to be dropped. It is the one that the universe is static. If the universe is expanding, then the distant stars will be moving away from us at highest speeds, and, it is well known from ordinary physics that light emitted by a receding source is reduced in intensity compared with light emitted by a source at rest. With an expanding universe, such as the one we live in, it may indeed be dark at night, for the light from the distant shells is tremendously weakened by the fact that the luminous objects in them are rushing away from us at high speed. Thus, the darkness of the night sky, the most obvious of all astronomical observations, leads us almost directly to the expansion of the universe, this remarkable and outstanding phenomenon discovered by modern astronomy.

Other changes made by modern astronomy in Olbers' assumptions are relatively minor. It is true that we know that our stars do not go on and on, but form a large stellar system, our galaxy; but we also know that beyond our galaxy there are millions and millions of other galaxies, all more or less like ours. We could, therefore, put Olbers' argument into modern language by changing the reference to stars going on and on in space to galaxies going on and on. The substance of the argument would not be affected.

The Steady-State Theory of the Universe [*]
by Hermann Bondi (1960)

In cosmology, one is considering an extrapolation of physics as we know it here to a very much larger scale of phenomenon. What we have learnt in the laboratory is to be applied to the universe at large. Clearly, there are dangers and difficulties in making such an enormous extrapolation. . . .

Can we really suppose that physical processes go on elsewhere as they do in our neighbourhood? Clearly, the answer to this question will depend on whether we are in a very special place in the universe or whether ours is a typical one. If our position in the universe in space and time is typical, then we can feel confident that our locally acquired knowledge is applicable elsewhere. If, on the other hand, the universe were very different elsewhere or at other times from what it is here and now, then we would need to know which aspects of our physical knowledge were truly permanent and which of them had just caught a mood of the moment of the universe. If we assume that all the physics we know is unchangeable, although the universe is changing, then we make a possible but quite arbitrary assumption.

[*] From Hermann Bondi, "The Steady-State Theory of the Universe" in Hermann Bondi *et al.*, *Rival Theories of Cosmology*. London, Oxford University Press, 1960. Reprinted by permission.

Of course, it may be necessary to consider the very difficult problems of the variation of physics in a varying universe; but before we enter the enormous complication of this question, we first try to see whether our universe might not happen to be one that is the same everywhere and at all times when viewed on a sufficiently large scale. In examining this possibility, we by no means claim that this must be the case; but we do say that this is so straightforward a possibility that it should be disproved before we begin to consider more complicated situations.

We are thus led to consider a model of the universe which is uniform on a large scale, both in space and in time. This model is known as the steady-state model. It is a useful model because in it we can be sure that physics, as we know it here, applies everywhere else. Moreover, as I shall explain later in this talk, it is a model that makes many forecasts that can be checked by experiment and observation. Therefore, it is a testable and, accordingly, a useful scientific theory. It follows immediately from the assumptions of the steady-state theory that the universe must be expanding, for otherwise, as a simple argument shows, we would be drowned by a flood of light from the most distant regions. In order to be consistent with the assumption of uniformity the motion of expansion must be such that there is a velocity of recession proportional to distance. The effect of the recession will then prevent the flood of light. This indeed is the type of motion that is being observed.

Next, if we have such a motion, then it would seem at first sight that the mean density in the universe must be diminishing, because if the distances between the galaxies are increasing all the time, it follows that the same matter now fills a larger volume. However, this would be in flagrant contradiction with the postulate that the universe is the same at all times. The only way out of this difficulty is to suppose that there is a process of continual creation going on—a process by which, in the enormous spaces between the galaxies, new matter constantly appears. This new matter condenses and forms new galaxies to fill the increasing spaces between the older ones.

Furthermore, every star ages since it converts hydrogen into helium in order to supply the energy the star radiates into space. As each star in the galaxy goes through these changes, the galaxy itself ages. However, the *average* age of galaxies is kept down since new galaxies constantly form in the increasing spaces between the old ones. It is for this reason, in order to keep the average age constant, that we require the new matter to be laid down in the vast intergalactic spaces. Only in this way can new galaxies be formed so that the average distance between galaxies stays constant, although, because of the expansion of the universe, the distance between existing galaxies is all the time increasing. Old galaxies, as they move farther and farther away, become less and less observable.

The whole picture of the steady-state universe is, therefore, very much like a picture of a stationary human population. Each individual is born, grows up, grows old and dies, but the average age stays the same owing to the fact that, all the time, new individuals are being born. We have, in the steady-state theory, a very similar picture of the universe of galaxies. Old galaxies die by drifting into regions where they are harder and harder to observe, and new galaxies are formed all the time in the spaces between the old ones. In this way, we arrive at a universe that is on the large scale uniform and unchanging. Moreover, it is the only model of this type. Of course, it deviates from ordinary physics in assuming this phenomenon of continual creation of matter which is, indeed, a major infringement of present formulations of physics. Dr. Bonnor has argued that this process of continual creation violates the principle of conservation of energy which has withstood all the revolutions in physics in the last sixty years and which most physicists would be prepared to give up only if the most compelling reasons were presented; but this seems to me to be unsound. The principle of conservation of mass and energy, like all physical principles, is based on observation. These observations, like all experiments and observations, have a certain measure of inaccuracy in them. We do not know from the laboratory experiments that matter is absolutely conserved; we only know that it is conserved to within a very small margin. The simplest formulation of this experimental result seems to be to claim that matter must be absolutely conserved. But this is purely a mathematical abstraction from certain observational results that may contain, indeed are bound to contain, errors.

Now, in fact, the mean density in the universe is so low, and the time scale of the universe is so large, by comparison with terrestrial circumstances, that the process of continual creation required by the steady-state theory predicts the creation of only one hydrogen atom in a space the size of an ordinary living-room once every few million years. It is quite clear that this process, therefore, is in no way in conflict with the experiments on which the principle of the conservation of matter and energy is based. It is only in conflict with what was thought to be the simplest formulation of these experimental results, namely that matter and energy were precisely conserved. The steady-state theory has shown, however, that much simplicity can be gained in cosmology by the alternative formulation of a small amount of continual creation, with conservation beyond that. This may, therefore, be the formulation with the greatest overall simplicity. There is thus no reason whatever, on the basis of any available evidence, to put the steady-state theory out of court because it requires this process of continual creation. This would be indeed a prejudice, and not a scientific argument.

Finally, as I said at the beginning, we must see how testable this theory is. How many forecasts does it make that can be checked by observation and experiment? There is a whole class of observations based on a very simple consideration. When we see the most distant galaxies that we can observe, then we look at them, not as they are now, but as they were a long time ago, for the light that travelled from them to us took a long time to cover the distance between them and us.

In the case of the most distant galaxies visible in optical telescopes this time is probably 5,000,000,000 years. If the universe as a whole is evolving, . . . then, presumably, all the galaxies originated at more or less the same time. In particular, we can definitely say that in such a universe no galaxies originated very recently. According to relativistic theories, then, we see the distant galaxies at an earlier stage in their evolution than the near ones which we see as they are now, more or less. Therefore, one would expect some variations with distance in the appearance of the galaxy, or the colour of light that it sends out, or in the degree of clustering, or possibly in the likelihood that it is a strong emitter of radio waves observable by radio astronomy. Accordingly, if one looks out into space and compares the shapes of distant galaxies with those of near galaxies, or compares in the same manner any other of the characteristics I mentioned, then either one will or one will not find a variation with distance. On the basis of the steady-state theory, time does not matter. A long time ago the universe looked just the same as it does now. Accordingly, no such variation can occur in the picture of the steady-state theory. In the evolutionary pictures one would expect precisely such a variation. Therefore, if these observations are made and any variation is found, then the steady-state theory is stone dead. If no such variation is found, it does not necessarily mean that the evolutionary theories are wrong, because one can always say that the period of time into which we can look back is too short for any such changes to show themselves. Some such observations are within the range of existing equipment, or equipment now in process of being built. Indeed, from the point of view of the steady-state theory we have the very satisfactory situation that although two different observations of this type have been claimed to disprove the steady-state theory, in both cases it has since been shown that they involved far greater observational errors than had originally been believed. In one case the absence of any such variation has now been established, in the other no definite conclusions can be drawn at present. However, many of these tests may be practicable in the near future. . . .

Enough has happened in the twelve years that this theory has existed to show that it gives us a useful way of looking at the universe, a way that inspires new observations and is vulnerable to them.

Evidence from the Radio Telescope *

by George Gamow (1962)

Radio emission from the colliding galaxies [1] is so intensive that the radio telescope can observe them at distances four times as large as those at which the galaxies themselves can be seen through the largest optical telescope. Within that volume one expects to find about one thousand colliding galactic pairs, a sample large enough for statistical studies. The British radio astronomer, Martin Ryle, undertook the construction of such a giant radio telescope, which occupies nearly five acres of ground near Cambridge University, and published his results early in 1961 in the *Monthly Notices* of the Royal Astronomical Society. The main point of investigation was the study of the distribution of colliding galaxies in the space of the Universe up to a distance of about 8 billion light years. The farther they are away from us the fainter is the radio signal, and a basic law of physics states that the observed intensity decreases in inverse proportion to the square of the distance. (Thus, being moved twice as far away, a candle will look four times fainter, and 10 times away it will look 100 times fainter.) On the other hand, if the sources are distributed uniformly through space, their number within a sphere of a given radius will increase as the volume of the sphere, i.e., as the cube of its radius (twice the radius, eightfold the volume; 10 times the radius, 1,000 times the volume). Thus if we plot the number of sources with the observed intensity below a certain upper limit against that limit, the numbers should increase in inverse proportion to the square root of the cube of that limit. Indeed, the sources which appear to us 100 times fainter must be located 10 times farther away in space. But if the radius of a sphere is 10 times larger its volume, and also the number of sources within it, must be 1,000 times larger. Similarly, 10,000 times fainter sources must be located 100 times farther

* From George Gamow, *The Great Ideas Today*. Encyclopedia Britannica, Inc., Chicago, 1962. Reprinted by permission.
1. See footnote 4 on p. 406. Ed.

away, so that the total number of sources located closer to us than this limit must be 1,000,000 times larger. This theoretically expected relation is shown by a broken line (marked: slope—1.5) in the diagram (Figure 8-17) which is taken from the article of Ryle and his collaborator P. F. Scott. The readers familiar with logarithms will notice that the plot is constructed on the logarithmic scale, i.e. by making on both axes the marks for 10, 100, 1,000, at equal distances. In this case the expected relation should be a straight line sloping to the right as shown in the figure. The circles and crosses represent the direct results of the observation and a few vertical lines indicate the limits of the possible observational error.

We notice at once that the observed points do not fall on the expected line (broken) but rather on another line (continuous) with a larger slope. It seems that the sources are located closer to each other at

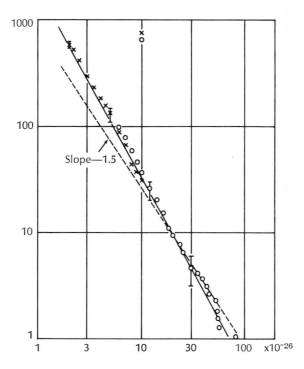

Figure 8-17. The number of observed galactic collisions plotted against their intensity (inverse square of the distance). If the universe were in a steady state, the observed points (circles and crosses) would fall on the broken line marked slope −1.5.

greater distances (lower observed intensities) and become less crowded closer to us. . . . What is true for colliding galaxies must also be true for all other galaxies which are not in collision at present. . . .

If the [steady state] theory is correct, the regions of the Universe located far away from us must show the same picture as the nearby regions. Ryle's work seems to indicate that this is not the case, and speaks in favor of the [evolutionary] theory according to which the Universe originated from a highly compressed state of matter many billions of years ago.

Cosmic Dispute *
by Barbara P. Gamow (1967)

"Your years of toil,"
Said Ryle to Hoyle,
    "Are wasted years, believe me.
The Steady State
Is out of date.
    Unless my eyes deceive me,

My telescope
Has dashed your hope;
    Your tenets are refuted.
Let me be terse:
Our Universe
    Grows daily more diluted!"

Said Hoyle, "You quote
Lemaître, I note,
    And Gamow. Well, forget them!
That errant gang
And their Big Bang—
    Why aid them and abet them?

You see, my friend,
It has no end
    And there was no beginning,
As Bondi, Gold,
And I will hold
    Until our hair is thinning!"

"Not so!" cried Ryle
With rising bile
    And straining at the tether;
"Far galaxies
Are, as one sees,
    More tightly packed together!"

"You make me boil!"
Exploded Hoyle,
    His statement re-arranging;
"New matter's born
Each night and morn.
    The picture is unchanging!"

"Come off it, Hoyle!
I aim to foil
    You yet." (The fun commences.)
"And in a while,"
Continued Ryle,
    "I'll bring you to your senses!"

* From George Gamow, *Mr. Tompkins in Paperback*. Cambridge University Press, New York, 1947. Reprinted by permission of Barbara Gamow and the publisher.

The Referee: New Facts [*]
by Isaac Asimov (1967)

Ryle's work was not completely convincing however. It rested upon the detection and measurement of very faint radio sources, and even slight errors, which could easily have occurred, would have sufficed to wipe out completely the trend upon which Ryle had based his conclusion. The backers of "continuous creation" grimly clung, therefore, to their own view of the universe.

As radio-wave sources continued to be pinned down into narrower areas, several in particular attracted attention. The sources seemed to be so small that it was possible they might be individual stars rather than galaxies. If so, they would have to be quite close (individual stars cannot be made out at very great distances) and Ryle's assumption that all radio sources were distant galaxies would be upset; and with it his conclusion. "Continuous creation" would then gain a new lease on life.

Among the compact radio sources were several known as 3C48, 3C147, 3C196, 3C273, and 3C286. The "3C" is short for "Third Cambridge Catalog of Radio Stars," a listing compiled by Ryle and his group, while the remaining numbers represent the placing of the source on that list.

Every effort was made to detect the stars that might be giving rise to these 3C sources. In America, Allan Sandage was meticulously searching the suspected areas with the 200-inch telescope at Mount Palomar ready to pounce on any suspicious-looking star. In Australia, Cyril Hazard kept his radio telescope focused on 3C273 while the moon bore down in its direction. As the moon moved in front of 3C273, the radio-wave beam was cut off. At the instant of cutoff, the edge of the moon had obviously cut across the exact location of the source.

By 1960, the stars had been found. They were not new discoveries at all; they had been recorded on previous photographic sweeps of the sky but had always been taken to be nothing more than faint members of our own galaxy. A new painstaking investigation, spurred by their unusual radio-wave emission, now showed, however, that they were not ordinary stars after all. Faint clouds of matter seemed to hover about a couple of them, and 3C273 showed signs of a tiny jet of something or other emerging from it.

What's more, their spectra, when obtained by two American astronomers, Jesse L. Greenstein and the Dutch-born Maarten Schmidt, proved to be most peculiar. The few lines that were present were in locations that couldn't be identified with any known elements. It was a most puzzling mystery and was abandoned in frustration for a time.

In 1963, Schmidt returned to the spectrum of 3C273. Six lines were present and it suddenly occurred to him that four of these were spaced in such a way as to resemble a well-known series of lines that should be in a far different part of the spectrum. In order for these four lines to be in the place they were actually observed, they would have had to have undergone a red-shift of unprecedented size. Could that be? He turned to the other spectra. If he allowed very large red-shifts, he could identify every single one of the lines involved.

Within the next two or three years, the result of a concentrated search of the skies was to uncover about forty of these objects altogether. The spectra of more than half were obtained and all showed enormous red-shifts. One, in fact, is receding at a record velocity of 150,000 miles per second and is estimated to be about nine billion light-years away (fifty billion trillion miles).

And yet, if such red-shifts are allowed, then these apparent "stars" had to be very distant because, on the basis of the expanding universe, a large red-shift is always associated with huge distances. In fact, these queer objects had to be farther away than any other known bodies in the universe.

At such distances, what looked like stars certainly could not be stars. No ordinary star could possibly be seen at such huge distances. The objects were therefore called "quasi-stellar" ("star-like") radio sources, and quasi-stellar soon came to be shortened to "quasar."

The quasars are a rich source of puzzlement for astronomers. If the red-shift is interpreted in the light of the expanding universe and if the quasars are indeed billions of light-years away, then they have very unusual properties. To appear as bright as they do at such enormous distances, they must be glowing with the luminosity of ten to a hundred galaxies. And yet, there are many reasons for supposing that they are not very large in size. They may be only one to ten light-years in diameter rather than the hundred-thousand-light-year span of an ordinary galaxy.

What kind of a body can it be that has its substance crowded into so tiny a fraction of a galactic volume and yet blazes with the light of dozens of galaxies? There are almost as many theories as there are astronomers—but as far as the fate of the "continuous creation" view of the universe is concerned, the theories don't matter. The mere fact that quasars exist might be enough.

The key point is that there are many quasars far away and no quasars within a billion light-years of ourselves. This means that there were many quasars in the long-gone youthful universe and none now.[1] The number of quasars (which may be the source of all or almost all the radio-wave beams studied by Ryle) may increase with distance and, therefore, with the youthfulness of the universe. This means that we have detected one important change in the universe with advancing time—the number of quasars diminishes. That is enough to eliminate "continuous creation."

It is enough, that is, *if*, indeed, the quasars are far, far distant objects. The belief that they are rests on the assumption that the gigantic redshifts they display are part of the expansion of the universe—but what if they aren't?

Suppose that quasars are small portions of nearby galaxies, hurled outward from the core of those galaxies by means of coresized explosions. Examples of "exploding galaxies" have indeed been detected in recent years and astronomers are now carefully tracking down galaxies which for one reason or other—odd shapes, wisps of fogginess, signs of internal convlusion—look unusual. A few quasars have been detected not far from such "peculiar galaxies."

Is this coincidence? Do the quasars happen to be in the same line of sight as the peculiar galaxies? Or were they cast outward with monstrous velocities from those galaxies as a result of explosions involving millions of stars? If so, the quasars might not all be unusually far away from us after all. Some might be close, some far, and their distribution might not force us to give up the "continuous creation" theory after all.

This is possible, but there are arguments against it. Suppose that quasars are objects hurled out of galaxies with such force as to be trav-

---

1. "Sandage, in 1965, announced the discovery of objects that may indeed be aged quasars. They seemed like ordinary bluish stars, but they possessed huge red shifts as quasars did. They were as distant, as luminous, as small as quasars, but they lacked the microwave emission. He called them "blue stellar objects" which can be abbreviated as BSO's.

"The BSO's seem to be more numerous than quasars, perhaps fifty times as numerous, and one estimate places the total number of BSO's within reach of our telescopes at 400,000, though others suggest a considerably larger figure. If they develop out of quasars, then they are fifty times as numerous because they endure in the BSO form fifty times as long—say 50,000,000 years. Still older quasars must dim to the point where they can be detected neither by microwave emission nor by light emission. What shape they may take or how they may then be recognized is as yet unknown." (Isaac Asimov, *The Universe*, Walker and Company, New York, 1966.)

eling at large fractions of the speed of light. Some of them would indeed be hurled away from us and would show a gigantic red-shift that would be misleading if it were interpreted as representing a recession caused by the general expansion of the universe rather than by a special explosion of a galaxy.

A roughly equal number would, however, be hurled toward us and would be approaching us at large fractions of the speed of light. They would then show a gigantic violet-shift.

Then, too, some would be hurled neither toward us nor away from us, but more or less across our line of sight in a sideways direction. Such quasars would show only a small (if any) red-shift or violet-shift, but, considering how close they might be and how rapidly they might be moving, they would alter their positions in the sky by a slight but measurable amount over the couple of years they have been observed.

The fact is, however, that no quasars have been found that show a violet-shift, and none that alter position. Only red-shifts have been observed, gigantic red-shifts. To suppose that comparatively nearby explosions have cast out quasars in such a way as to produce red-shifts only is to ask too much of coincidence.

So the weight of the evidence is in favor of the great distance of the quasars and of the elimination of the "continuous creation" theory—and Fred Hoyle gave up.

The elimination of "continuous creation" does not necessarily mean the establishment of the "big bang." Suppose there is some third possibility that is as yet unsuggested. To strengthen the "big bang" theory against the general field of unsuggested possibilities, it would be nice to consider some phenomenon that the "big bang" theory would predict, some phenomenon that could then actually be observed.

Suppose, for instance, that the universe *did* begin as an incredibly dense cosmic egg that exploded. At the moment of explosion, it must have been tremendously hot—possibly as hot as 10 billion degrees Centigrade (equivalent to 18 billion degrees Fahrenheit).

If so, then if our instruments could penetrate far enough, to nearly the very edge of the observable universe, they might reach far enough back in time to catch a whiff of the radiation that accompanied the "big bang."

At temperatures of billions of degrees, the radiation would be in the form of very energetic X rays. However, the expanding universe would be carrying that source of X rays away from us at nearly the speed of light. This incredible speed of recession would have the effect of vastly weakening the energy of the radiation; weakening it to the point where it would reach us in the form of radio waves with a certain group of properties. Through the 1960s, estimates of what those properties might be were advanced.

Then, early in 1966, a weak background of radio-wave radiation was detected in the skies; radiation that would just fit the type to be expected of the "big bang." This has been verified and it looks very much as though we have not only eliminated "continuous creation" but have actually detected the "big bang."

If so, then we have lost something. In facing our own individual deaths, it was possible after all, even for those who lacked faith in an afterlife, to find consolation. Life itself would still go on. In a "continuous creation" universe, it would even be possible to conceive of mankind as moving, when necessary, from an old galaxy to a young one and as existing eventually through all infinity and for all eternity. It is a colossal, godlike vision, that might almost make individual death a matter of no consequence.

In the "big bang" scheme of things, however, our particular universe has a beginning—and an ending, too. Either it spreads out ever more thinly while *all* the galaxies grow old and the individual stars die, one by one, or it reaches some maximum extent and then begins to collapse once more, returning after many eons to a momentary existence as a cosmic egg.

In either case, mankind, as we know it, must cease to exist, and the dream of godhood must end. Death has now been rediscovered and Homo sapiens, as a species, like men as individuals, must learn to face the inevitable end.

—Or, if the universe oscillates, and if the cosmic egg is reformed every hundred billion years or so, to explode once more; then, perhaps, in each of an infinite number of successive universes, a manlike intelligence (or a vast number of them) arises to wonder about the beginning and end of it all.

The pertinent facts of the natural sciences . . . are awaiting still further research and confirmation, and the theories founded on them are in need of further development and proof before they can provide a sure foundation for arguments which, of themselves, are outside the proper sphere of the natural sciences. This notwithstanding, it is worthy of note that modern scholars in these fields regard the idea of the creation of the universe as entirely compatible with their scientific conceptions and that they are even led spontaneously to this conclusion by their scientific research. Just a few decades ago, any such "hypothesis" was rejected as entirely irreconcilable with the present state of science.

Pope Pius XII
Address to the Pontifical Academy of Science
(November 22, 1951)

This picture of the universe as exploding fireworks which went off ten billion years ago invites us to consider the remarkable question of Miguel de Unamuno, whether the whole world—and we with it—be not possibly only a dream of God; whether prayer and ritual perhaps be nothing but attempts to make Him more drowsy, so that He does not awaken and stop our dreaming.

Pascual Jordan
Physics of the Twentieth Century (1944)

There was a monk indulging against the teaching of the Master in cosmological enquiries. In order to know where the world ends he began . . . interrogating the gods of the successive heavens. . . . Finally, the Great Brahma himself became manifest, and the monk asked him where the world ends. . . . The great Brahma took that monk by the arm, led him aside and said: "These gods, my servants, hold me to be such that there is nothing I cannot see, understand, realize. Therefore, I gave no answer in their presence. But I do not know where the world ends. . . ."

Dialogues of the Buddha (563–483 B.C.)

Suggestions for Further Reading

° Asimov, Isaac: *The Universe: From Flat Earth to Quasar*, Walker and Company, New York, 1966. A lucid description of the development of our understanding of the universe.

° Bondi, Hermann: *The Universe at Large*, Anchor Books, Doubleday & Company, Inc., Garden City, N.Y., 1960. A Science Study Series book giving a clear exposition of the theory of continuous creation, as well as a number of other subjects: radiation belts, tides, gravitation, motion, and magnetism.

° Bondi, Hermann, et al.: *Rival Theories of Cosmology*, Oxford University Press, London, 1960. An interesting collection of essays presenting different sides of the "cosmic dispute," within the context of the time of publication.

° Gamow, George: *The Creation of the Universe*, Viking Press, Inc., New York, 1961. A very interesting presentation of Gamow's version of the evolutionary theory of the origin of the universe.

° Hoyle, Fred: *The Nature of the Universe*, Harper & Row, Publishers, New York, 1960. Hoyle describes for the layman the steady-state theory as it was believed at that time.

Munitz, Milton K. (ed.): *Theories of the Universe*, The Free Press, Glencoe, Illinois, 1957. A very fine anthology of original writings from the myths of Babylonia to modern times.

° Schatzman, E. L., *The Structure of the Universe,* translated from the French by Patrick Moore, World University Library, McGraw-Hill Book Company, New York, 1968. An authoritative and up-to-date presentation, moderately technical.

Shklovskii, I. S., and Carl Sagan: *Intelligent Life in the Universe,* Holden-Day, In., San Francisco, London, Amsterdam, 1966. This outstanding book is the product of a collaboration between a Russian and an American astronomer. It covers a broad panorama of scientific knowledge and philosophical speculation about the universe. Several of the later chapters deal more specifically with the subject of other life in the universe.

° Paperback edition

9

Is There Other Life in the Universe?

Science is a leap of the imagination. The selections in this part take the reader along on a speculative search for other life in the universe. They show how a change in theoretical concepts can create a new possibility that scientific imagination jumps in to explore, with results that may be very exciting from our human point of view.

Is There Other Life in the Universe?

Introduction

A FEW DECADES AGO the question of whether there might be life on other planets would have been relegated to writers of science fiction. Now, as the readings in this part testify, astronomers are agreed that the probability of life existing elsewhere in the universe is very high. The difference in attitude has been brought about by a change in the scientific theory of the origin of our solar system. If the planets were caused by a chance collision of two stars, as astronomers believed until recently, the event would have been rare enough to make it unlikely that there would be other planets in the proper zone to support life. But when the collision hypothesis was overthrown and the nebular hypothesis, which had first been proposed by Kant in 1755 was re-established, the calculations concerning the uniqueness of planetary systems were drastically altered. According to the nebular hypothesis, all stars were formed in the same way as our sun and therefore could possess planetary systems. This theory makes the number of possible sites for life truly "astronomical," especially in view of the latest estimates of numbers of stars in the whole universe.

The change in concepts came about as the result of new evidence about the chemical constitution of matter in the universe—the discovery of larger amounts of hydrogen and helium in the bodies of the sun and stars. This evidence which was made available by increased skill and precision in analyzing the spectra of the stars, led to a sweeping change of theoretical concepts and, as a corollary, to the conclusion that there might be billions of sites for life throughout the universe. Thus we see how an idea of enormous significance from our human point of view has occurred because of the reversal of a decision between two rival scientific theories. This story may well serve as illustrative of the sudden way science can revolutionize our human views. It is advisable to remember also that the information and the theories in this field are still in a state of rapid development. Every year brings the discovery of new facts and these may at any time bring about new theoretical interpretations. We recall J. J. Thomson's statement that ". . . from the point of view of the physicist a theory is a policy rather than a creed." [1] In considering the human corollaries to scientific theories, it is well to keep in mind the dynamic and growing nature of science. Shapley says, "And as groping

1. See p. 338.

philosophers and scientists we are thankful for the mysteries that still lie beyond our grasp." [2]

The central question of this part is an especially tantalizing mystery that is almost within our grasp today. With the exploration of the other planets of our solar system, it may be settled before many years have gone by. The evidence from space exploration has been disappointingly negative so far. The landing on the moon and the analysis of the lunar rocks have shown no sign that life has ever existed there. Furthermore, the probes sent to Mars have returned pictures of a forbidding planet which seems less hospitable to life than some astronomers had expected. However, it is still possible that primitive forms of life may exist in some regions of that planet and the question will probably not be resolved until a landing on Mars has been achieved.

Scientifically, the problem of the existence of other life in the universe spans two fields of inquiry. One side of the problem involves the chemical and biological nature of life and the process of evolution. The other side is an astronomical problem, the existence of possible sites in which life could develop. Since our main focus in this book is on the broad aspects of the universe, the readings in this part concentrate on the second question. We have included only enough information on biology and chemistry to delineate the conditions necessary for the support of living things. Assuming the right site and enough time, the majority of scientists believe today that life would be inevitable. Some scientists believe that, given sufficient time, the process of evolution must lead eventually to the appearance of intelligent beings. Furthermore, there is a good probability that these beings may have reached a higher state of development than we have attained. There is enough support for this point of view that impulses received by radio telescope are now being analyzed for messages from outer space. If such a message were found, it would surely be one of the most dramatic discoveries of all time.

2. See p. 519.

A Reconsideration of Man's Place in the Universe

Sky, universe, all-embracing ether, and immeasurable space alive with movement—all these are of one nature. In space there are countless constellations, suns, and planets; we see only the suns because they give light; the planets remain invisible, for they are small and dark. There are also numberless earths circling around their suns, no worse and no less inhabited than this globe of ours. For no reasonable mind can assume that heavenly bodies which may be far more magnificent than ours would not bear upon them creatures similar or even superior to those upon our human Earth.

Giordano Bruno (1548–1600)

An Inquiry Concerning Other Worlds °

by Harlow Shapley (1964)

We shall start our account with the following question: Is there not at this time a justification for a revised look at mankind as a world factor? "Yes!" is our prompt answer. To the scientist, rich in new knowledge, and to the puzzled layman, and perhaps to some philosophers, the answer is decidedly affirmative. An elementary reason for a reconsideration lies in the recognition in recent years of the "displacement" of the sun, earth, and other planets from a central place, or even a significant place, in the sidereal universe—in the placing of the observer in a very undistinguished location in a faint spiral arm of an ordinary galaxy.

This reason is elementary but momentous, for it concerns the replacement of the earlier *geocentric* and *heliocentric* theories of the universe by the *eccentric* arrangement that we now all accept. By this move we have made a long forward step in cosmic adjustment—a step that is unquestionably irreversible. We must get used to the fact that we are peripheral, that we move along with our star, the sun, in the outer part of a galaxy that is one among billions of star-rich galaxies.

If there is some special grandeur in our position in space and time, I

° From Harlow Shapley, *Of Stars and Men*. The Beacon Press, Boston. Copyright © 1958, 1964 by Harlow Shapley. Reprinted by permission of The Beacon Press.

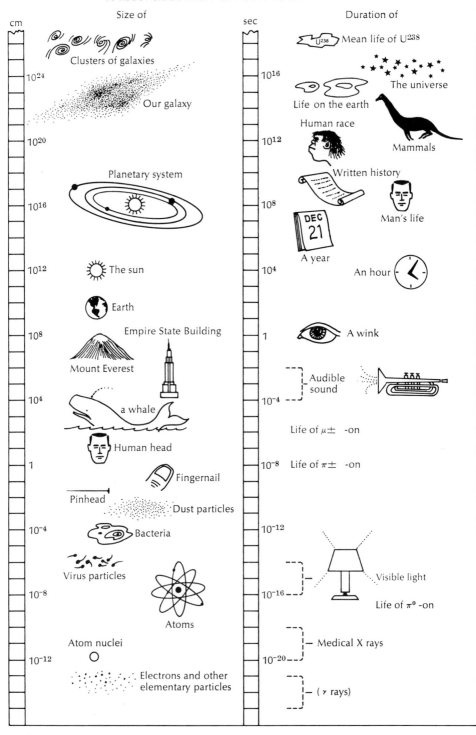

A RECONSIDERATION OF MAN'S PLACE IN THE UNIVERSE

fail to find it. Our glory if any must lie elsewhere. Also, should we not openly question the rather vain and tedious dogma that man somehow is something very remarkable, something superior? He may be. I hope that he is. But certainly it is not in his location in space, or in his times; not in his energy content or chemical composition. He is not at all outstanding in the four basic entities—space, time, matter, energy. There is nothing unique, and worthy of boast, in his size, activity, or even his epoch in cosmic chronology. He is not at the beginning of things, nor at the end. . . .

Before we can propose ourselves and our destiny as significant concerns of the universe, we should turn our attention to the probable existence and general spread of protoplasm throughout stellar spaces and cosmic times. We can no longer be content with the hypothesis that living organisms are of this earth only. But before we ponder on the life spread, we should inquire into the prevalence of suitable sites for biological operations. The initial question is not whether such sites are presently inhabited. First we ask: Are there other habitable celestial bodies—bodies that would be hospitable if life were there? No field of inquiry is more fascinating than a search for humanity, or something like humanity, in the mystery-filled happy lands beyond the barriers of interstellar space. But are there such happy lands? . . .

Human bodies are constructs of commonly known chemical elements, and nothing else. . . . The element oxygen accounts for about 65 percent of our bodies; 18 percent is carbon, 10 percent is hydrogen, 3 percent nitrogen, 2 percent calcium, and the remaining 2 percent includes silicon, phosphorus, sodium, sulphur, iron, and a dozen other elements —all common to the crust of the earth and to the flames of the sun. The percentages vary somewhat from rat to leech, from watery octopus to crusty coral. There is more than average calcium in the bony vertebrates, more silicon in the brachiopods, more H_2O in the jelly fish; but all animals are composed of the common atoms. The elements uncom-

Figure 9-1. Space and time scales of the universe. The relative sizes of various objects in everyday life, in the macrocosm, and in the microcosm are shown in a decimal logarithmic scale, i.e., in the scale in which each factor of 10 is represented by one division of the yardstick. The sizes range from the diameter of an electron, and other elementary particles that are about 0.00001 angstrom, to the diameter of giant stellar galaxies, which often measure 100,000 light-years across. It is interesting to notice that the size of the human head is just about halfway between the size of an atom and the size of the sun, or halfway between the size of an atomic nucleus and the diameter of the planetary system (on the logarithmic scale in both cases, of course). (After George Gamow)

mon to the rocks, like gold, platinum, and radium, are also uncommon to the bodies of men.

The stars are composed of the same stuff as that which constitutes the sun and the earth's crust. They are built of the same materials as those that compose terrestrial organisms. As far as we can tell, the same physical laws prevail everywhere. The same natural rules apply at the center of the Milky Way, in the remote galaxies, and among the stars and planets of the solar neighborhood.

In view of the common physics and chemistry, should we not also expect to find animals and plants everywhere? That seems completely reasonable; and soon we shall say that it seems inevitable. But to demonstrate the actual presence of organic life in other planetary systems is now impossible for us because the stars are so remote and we, as earth-bound searchers of the sky, are yet too feeble in the face of stellar realities. To establish through statistical analysis the high probability of planets suitable for living organisms is, however, not difficult. A statistical argument, as a matter of fact, can be more convincing than would be a marginal observation.

It will clarify the discussion if we start with two routine reminders: (1) By life we mean what we terrestrials recognize as life—a biochemical operation involving oxygen, carbon, and nitrogen, and making use of water in a liquid state. (Other kinds are imaginable; e.g., one where silicon replaces carbon, or where sulphur's participation is like that of oxygen, or where liquid ammonia replaces water. Such is imaginable, but unlikely.) (2) Mars and Venus are therefore the only other planets of our solar system that are at all suitable for living organisms.

The evidences are good that Martian life is low and lichen-like, if it exists at all. The surface of Venus was until recently an unsolved problem. Now we know that a blanket effect prevails on Venus and that the surface temperature is much too high for the existence of protoplasm.

Among the many definitions of life is the cold rigid version: "material organizations perpetuating their organization." We might put it better: "the perpetuation by a material organization of its organization." The definition can properly refer both to individuals and to species, and also to societies. They are all at times alive. They all die, if we suitably define death. The lively deathless atoms of our breaths, however, are not, in this defining, alive.

Life is tough, tenacious, and consequently persistent when we give it time to adjust to varying environments. We find it thriving in geysers and hot springs. Some flowers bloom under the snow. Both plants and animals on occasion endure for long periods on hot deserts. Some seeds and spores can withstand desiccation and extreme cold indefinitely. Life as we know it on the earth has wide adaptability; but there are limits, and one of these limits is the heat and radiation near a star's surface, where the molecules constituting protoplasm would be dissociated into atoms.

In our consideration of the spread of life throughout the universe, we must therefore immediately drop all thoughts of living organisms on the trillions of radiant stars. The flames of the sun are rich in the lively atoms of oxygen, carbon, nitrogen, hydrogen, and calcium—the principal constituents of living matter—but physical liveliness and organic livingness are quite different behaviors. At the surfaces of some of the cool stars, like Antares and Betelgeuse, and in the cooling sunspots, we find a few simple molecules in addition to the scores of kinds of atoms; but there is nothing in the stars that is as complicated and tender as the proteins—those molecular aggregates that underlie the simplest life. And of course the stars harbor no water in the liquid state.

The stars are out of it, therefore, and they probably contain more than half of all the material in the universe. Most of the non-stellar material is believed to be in the form of interstellar gas, with a bit of dust. The dust is of the sort that shows up as meteors, when in collision with the atoms and molecules of the earth's atmosphere; it appears also among the stars as the dark nebulosities that interrupt and make patchy the glow of the Milky Way.

No life exists on these minute meteor specks or on the relatively larger meteorites. It is absent for several reasons. The masses are too small to hold gravitationally an atmosphere. (Even our moon cannot retain the oxygen and carbon dioxide necessary for breathing animals and plants.) Moreover, the meteors out among the stars are too cold for liquid water, and they are unprotected against the lethal ultraviolet radiation from hot stars.

How about life on the comets? The same general argument holds as for meteors and meteorites, since the comets are simply assemblages of dusty and fragmented meteoric materials, infused with gases. In addition, most of the large comets of the solar system are when brightest too near the sun for living organisms, and the rest of the time in their orbital travels too frigidly remote in the outer parts of their paths.

In our search for life we are therefore left with the planets and those on which it can occur and survive must be neither too near the stars nor so remote from them that the cold is unrelieved. They should not be too small to hold an oxygen atmosphere, unless we are content to settle for primitive anaerobic life. (A few types of low organisms thrive in the absence of elemental oxygen.)

The life-bearing planets must also have non-poisonous atmospheres and salutary waters; but given time enough, organisms could no doubt become adjusted to environments that would be poisonous and impossible for life such as that now developed on the earth.

Finally, the propitious planets that are suitable in size, temperature, and chemistry must also move in orbits of low eccentricity. Highly eccentric orbits would bring their planets too near the star at periastron a part of the year, and then too far out at apastron. The resulting temperature oscillations would be too much for comfort, perhaps even too much for the origin and persistence of early life. Also, in the interest of avoiding too great differences in temperature from night to day, it would be best if the planets rotate rapidly and their rotational axes be highly inclined (as is the earth's) to their orbital planes.

With the foregoing requirements in mind we ask if there are many really suitable planetary systems, and ask also: How are planets born?

The Origin of the Solar System and Theories of Stellar Evolution

The Birth of Planets [*]
by George Gamow (1962)

For us, the people living in the seven parts of the World (counting Admiral Byrd for Antarctica) the expression "solid ground" is practically synonymous with the idea of stability and permanence. As far as [most of us] are concerned all the familiar features of the Earth's surface, its continents and oceans, its mountains and rivers could have existed since the beginning of time. True, the data of historical geology indicate that the face of the Earth is gradually changing, that large areas of the continents may become submerged by the waters of the oceans, whereas submerged areas may come to the surface.[1]

We also know that the old mountains are gradually being washed away by the rain, and that new mountain ridges rise from time to time as the result of tectonic activity, but all these changes are still only the changes of the solid crust of our globe.

It isn't, however, difficult to see that there must have been a time when no such solid crust existed at all, and when our Earth was a glowing globe of melted rocks. In fact, the study of the Earth's interior indicates that most of its body is still in a molten state, and that the "solid ground" of which we speak so casually is actually only a comparatively thin sheet floating on the surface of the molten magma. The simplest way to arrive at this conclusion is to remember that the temperature measured at different depths under the surface of the Earth increases at

1. Studies in oceanography in very recent years have provided strong evidence that the ocean floors have been spreading at a small but consistent rate over many millions of years. It is now believed that about 200 million years ago the Atlantic ocean did not exist and that the present continents were part of one (or two) large land masses. See the September 1969 issue of Scientific American for descriptions of the evidence that has led to this view and the various theories relating to it. Ed.

the rate of about 30°C per kilometer of depth (or 16°F per thousand feet) so that, for example, in the world's deepest mine (a gold mine in Robinson Deep, South Africa) the walls are so hot that an air-conditioning plant had to be installed to prevent the miners from being roasted alive.

At such a rate of increase, the temperature of the Earth must reach the melting point of rocks (between 1200°C and 1800°C) at a depth of only 50 km. (approx. 31 miles) beneath the surface, that is, at less than 1 percent of the total distance from the center. All the material farther below, forming more than 97 per cent of the Earth's body, must be in a completely molten state.

It is clear that such a situation could not have existed forever, and that we are still observing a certain stage in a process of gradual cooling that started once upon a time when the Earth was a completely molten body, and will terminate some time in the distant future with the complete solidification of the Earth's body all the way to the center. A rough estimate of the rate of cooling and growth of the solid crust indicates that the cooling process must have begun several billion years ago.

The same figure can be obtained by estimating the age of rocks forming the crust of the Earth. Although at first sight rocks exhibit no variable features, thus giving rise to the expression "unchangeable as a rock," many of them actually contain a sort of natural clock, which indicates to the experienced eye of a geologist the length of time that has passed since they solidified from their former molten state.

This age-betraying geological clock is represented by a minute amount of uranium and thorium, which are often found in various rocks taken from the surface and from different depths within the Earth. . . . the atoms of these elements are subject to a slow spontaneous radioactive decay ending with the formation of the stable element lead.

To determine the age of a rock containing these radioactive elements we need only to measure the amount of lead that has been accumulated over the centuries as the result of radioactive decay.

In fact, as long as the material of the rock was in the molten state, the products of radioactive disintegration could have been continuously removed from the place of their origin by the process of diffusion and convection in the molten material. But as soon as the material solidified into a rock, the accumulation of lead alongside the radioactive element must have begun, and its amount can give us an exact idea of how long it was going on, in exactly the same way as the comparative numbers of empty beer cans scattered between the palms on two Pacific islands could have given to an enemy spy an idea of how long a garrison of marines had stayed on each island.

The general result of the survey of the accumulation of lead reveals an extremely important fact that: *no rocks found on the surface of the Earth exhibit an age of more than 3.5 billion years,* from which we must conclude that *the solid crust of the Earth was formed from previously molten material 3.5 or more billion years ago.* Recent studies of Clare C. Peterson in Sweden and Harrison Brown of California, United States, who studied the age of meteorites, indicate that the planetary system may be as much as 4.5 billion years old.

Thus we can picture the Earth several billion years ago as a completely molten spheroid surrounded by a thick atmosphere of air, water-vapors, and probably other extremely volatile substances.

How did this hot lump of cosmic matter come into being, what kind of forces were responsible for its formation, and who supplied the material for its construction? These questions, pertaining to the origin of our Globe as well as to the origin of every other planet of our solar system, have been the basic inquiries of scientific *Cosmogony* (the theory of the origin of the universe), the riddles that have occupied the brains of astronomers for many centuries.

The first attempt to answer these questions by scientific means was made in 1749 by the celebrated French naturalist George-Louis Leclerc, Comte de Buffon, in one of the forty-four volumes of his *Natural History.* Buffon saw the origin of the planetary system as the result of a collision between the sun and a comet that came from the depth of interstellar space. His imagination painted a vivid picture of a "comèt fatale" with a long brilliant tail brushing the surface of our, at that time lonely, sun and tearing from its giant body a number of small "drops," which were sent spinning into space by the force of the impact (Figure 9-2, left).

A few decades later entirely different views concerning the origin of our planetary system were formulated by the famous German philosopher Immanuel Kant, who was more inclined to think that the sun made up its planetary system all by itself without the intervention of any other celestial body. Kant visualized the early state of the sun as a giant, comparatively cool, mass of gas occupying the entire volume of the present planetary system, and rotating slowly around its axis. The steady cooling of the sphere through radiation into the surrounding empty space must have led to its gradual contraction and to the corresponding increase of its rotational speed. The increasing centrifugal force resulting from such rotation must have led to the progressive flattening of the gaseous body of the primitive sun, and resulted in the ejection of a series of gaseous rings along its extended equator (Figure 9–2, right). Such a ring formation from the rotating masses can be demonstrated by the classical experiment performed by Plateau in which a large sphere of oil (not

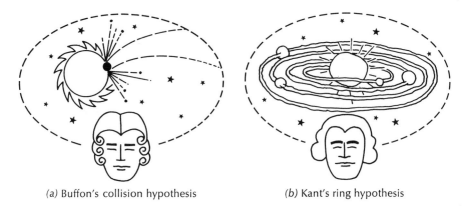

(a) Buffon's collision hypothesis   (b) Kant's ring hypothesis

Figure 9-2. Two schools of thought in cosmogony.

gaseous, as in the case of the sun) suspended within some other liquid with equal density and brought into rapid rotation by some auxiliary mechanical device begins to form rings of oil around itself when the speed of rotation exceeds a certain limit. The rings formed in this way were supposed to have broken up later and to have condensed into various planets circling at different distances around the sun.

These views were later adopted and developed by the famous French mathematician Pierre-Simon, Marquis de Laplace, who presented them to the public in his book *Exposition du système du monde,* published in 1796. Although a great mathematician, Laplace did not attempt to give mathematical treatment to these ideas, but limited himself to a semipopular qualitative discussion of the theory.

When such a mathematical treatment was first attempted sixty years later by the English physicist Clerk Maxwell, the cosmogonical views of Kant and Laplace ran into a wall of apparently insurmountable contradiction. It was, in fact, shown that if the material concentrated at present in various planets of the solar system was distributed uniformly through the entire space now occupied by it, the distribution of matter would have been so thin that the forces of gravity would have been absolutely unable to collect it into separate planets. Thus the rings thrown out from the contracting sun would forever remain rings like the ring of Saturn, which is known to be formed by innumerable small particles running on circular orbits around this planet and showing no tendency toward "coagulation" into one solid satellite.

The only escape from this difficulty would consist in the assumption that the primordial envelope of the sun contained much more matter (at least 100 times as much) than we now find in the planets, and that most of this matter fell on the sun, leaving only about 1 per cent to form planetary bodies.

Such an assumption would lead, however, to another no less serious contradiction. Indeed if so much material, which must originally have rotated with the same speed as the planets do, had fallen on the sun, it would inevitably have communicated to it an angular velocity 5000 times larger than that which it actually has. If this were the case, the sun would spin at a rate of 7 revolutions per hour instead of at 1 revolution in approximately 4 weeks.

These considerations seemed to spell death to the Kant-Laplace views, and with the eyes of astronomers turning hopefully elsewhere, Buffon's collision theory was brought back to life by the works of the American scientists T. C. Chamberlin and F. R. Moulton, and the famous English scientist Sir James Jeans. Of course, the original views of Buffon were considerably modernized by certain essential knowledge that had been gained since they were formulated. The belief that the celestial body that had collided with the sun was a comet was now discarded, for the mass of a comet was by then known to be negligibly small even when compared with the mass of the moon. And so the assaulting body was now believed rather to be another star comparable to the sun in its size and mass.

However, the regenerated collision theory, which seemed at that time to represent the only escape from the fundamental difficulties of the Kant-Laplace hypothesis, likewise found itself treading on muddy ground. It was very difficult to understand why the fragments of the sun thrown out as a result of the vigorous punch delivered by another star would move along the almost circular orbits followed by all planets, instead of describing elongated elliptical trajectories.

To save the situation it was necessary to assume that, at the time the planets were formed by the impact of the passing star, the sun was surrounded by a uniformly rotating gaseous envelope, which helped to turn the originally elongated planetary orbits into regular circles. Since no such medium is now known to exist in the region occupied by the planets, it was assumed that it was later gradually dissipated into interstellar space, and that the faint luminosity known as *Zodiacal Light,* which at present extends from the sun in the elliptical plane, is all that is left from that past glory. But this picture, representing a kind of hybrid between the Kant-Laplace assumption of the original gaseous envelope of the sun and Buffon's collision hypothesis was very unsatisfactory. However, as the proverb says, one must choose the lesser of two evils, and the collision hypothesis of the origin of the planetary system was accepted as the correct one, being used until very recently in all scientific treatises, textbooks, and popular literature (including the author's two books *The Birth and Death of the Sun*, 1940, and *Biography of the Earth*, 1941).

It was only in the fall of 1943 that the young German physicist C. von Weizsäcker cut through the Gordian Knot of the planetary theory. Using the new information collected by recent astrophysical research, he was able to show that all the old objections against the Kant-Laplace hypothesis can be easily removed, and that, proceeding along these lines, one can build a detailed theory of the origin of planets, explaining many important features of the planetary system that had not even been touched by any of the old theories.

The main point of Weizsäcker's work lies in the fact that during the last couple of decades astrophysicists have completely changed their minds about the chemical constitution of matter in the universe. It was generally believed before that the sun and all other stars were formed by the same percentage of chemical elements as those that we have learned from our Earth. Geochemical analysis teaches us that the body of the Earth is made up chiefly of oxygen (in the form of various oxides), silicon, iron, and smaller quantities of other heavier elements. Light gases such as hydrogen and helium (along with other so-called rare gases such as neon, argon, etc.) are present on the Earth in very small quantities.[2]

In the absence of any better evidence, astronomers had assumed that these gases were also very rare in the bodies of the sun and the other stars. However, the more detailed theoretical study of stellar structure led the Danish astrophysicist B. Stromgren to the conclusion that such an assumption is quite incorrect and that, in fact, at least 35 per cent of the material of our sun must be pure hydrogen. Later this estimate was increased to above 50 per cent, and it was also found that a considerable percentage of the other solar constituents is pure helium. Both the theoretical studies of the solar interior (which recently culminated in the important work of M. Schwarzschild), and the more elaborate spectroscopic analysis of its surface, led astrophysicists to a striking conclusion that: *the common chemical elements that form the body of the Earth constitute only about 1 per cent of the solar mass, the rest being almost*

---

[2] Hydrogen is found on our planet mostly in its union with oxygen in water. But everybody knows that although water covers three quarters of the Earth's surface the total water mass is very small compared with the mass of the entire body of the Earth.

evenly divided between hydrogen and helium with a slight preponder-ance of the former. Apparently this analysis also fits the constitution of the other stars.

Further, it is now known that *interstellar space is not quite empty,* but is filled by a mixture of gas and fine dust with a *mean density of about 1 mg matter in 1,000,000 cu miles space,* and this diffuse, highly rarefied material apparently has the same chemical constitution as have the sun and the other stars.

In spite of its incredibly low density the presence of this interstellar material can be easily proved, since it produces noticeable selective absorption of the light from stars so distant that it has to travel for hundreds of thousands of light-years through space before entering into our telescopes. The intensity and location of these "interstellar absorption lines" permits us to obtain good estimates of the density of that diffuse material and also to show that it consists almost exclusively of hydrogen and probably helium. In fact, the dust, formed by small particles (about 0.001 mm in diameter) of various "terrestrial" materials, constitutes not more than 1 per cent of its total mass.

To return to the basic idea of Weizsäcker's theory, we may say that this new knowledge concerning the chemical constitution of matter in the universe, plays directly into the hand of the Kant-Laplace hypothesis. In fact, if the primordial gaseous envelope of the sun was originally formed from such material, *only a small portion of it, representing heavier terrestrial elements, could have been used to build our Earth and other planets.* The rest of it, represented by noncondensible hydrogen and helium gases, must have been somehow removed, either by falling into the sun or by being dispersed into surrounding interstellar space. Since the first possibility would result, as it was explained above, in much too rapid axial rotation of the sun, we have to accept another alternative, namely, that the gaseous "excess-material" was dispersed into space soon after the planets were formed from the "terrestrial" compound.

This brings us to the following picture of the formation of the planetary system. When our sun was first formed by the condensation of interstellar matter a large part of it, probably about a hundred times the present combined mass of planets, remained on the outside forming a giant rotating envelope. (The reason for such behavior can easily be found in the differences between the rotational states of various parts of interstellar gas condensing into the primitive sun.) This rapidly rotating envelope should be visualized as consisting of *noncondensible gases* (hydrogen, helium, and a smaller amount of other gases) and *dust-particles* of various terrestrial materials (such as iron oxides, silicon compounds, water droplets and ice crystals) which were floating inside the gas and carried along by its rotational motion. The formation of big lumps of "terrestrial" material, which we now call planets must have taken place as the result of collisions between dust particles and their gradual aggregation into larger and larger bodies. In Figure 9-3 we illustrate the results of such mutual collisions which must have taken place at velocities comparable to that of meteorites.

One must conclude, on the basis of logical reasoning, that at such velocities the collision of two particles of about equal mass would result in their mutual pulverization (Figure 9-3, left), a process leading not to the growth but rather to the destruction of larger lumps of matter. On the other hand, when a small particle collides with a much larger one (Figure 9-3, middle) it seems evident that it would bury itself in the body of the latter, thus forming a new, somewhat larger mass.

Obviously these two processes would result in the gradual disappearance of smaller particles and the aggregation of their material into larger bodies. In the later stages the process will be accelerated due to the fact that the larger lumps of matter will attract gravitationally the smaller particles passing by and add them to their own growing bodies. This is illustrated in Figure 9-3, right, which shows that in this case the capture-effectiveness of massive lumps of matter becomes considerably larger.

Weizsäcker was able to show that *the fine dust originally scattered through the entire region now occupied by the planetary system must have been aggregated into a few big lumps to form the planets, within a period of about a hundred million years.*

As long as the planets were growing by the accretion of variously

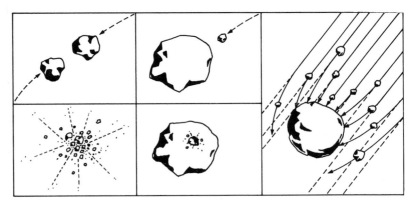

Figure 9-3

sized pieces of cosmic matter on their way around the sun, the constant bombardment of their surfaces by fresh building material must have kept their bodies very hot. As soon, however, as the supply of stellar dust, pebbles, and larger rocks was exhausted, thus stopping the process of further growth, the radiation into interstellar space must have rapidly cooled the outer layers of the newly formed celestial bodies, and led to the formation of the solid crust, which is even now growing thicker and thicker, as the slow internal cooling continues.

The next important point to be attacked by any theory of planetary origin is the explanation of the peculiar rule (known as the *Titus-Bode* rule) that governs the distance of different planets from the sun. In the table on this page these distances are listed for nine planets of the solar system, as well as for the belt of *planetoids,* which apparently corresponds to an exceptional case where separate pieces did not succeed in collecting themselves into a single big lump.

Name of the planet	Distance from the sun in terms of earth's distance from the sun	The ratio of the distance of each planet from the sun, to the distance from the sun of the planet listed above it
Mercury	0.387	
Venus	0.723	1.86
Earth	1.000	1.38
Mars	1.524	1.52
Planetoids	about 2.7	1.77
Jupiter	5.203	1.92
Saturn	9.539	1.83
Uranus	19.191	2.001
Neptune	30.07	1.56
Pluto	39.52	1.31

The figures in the last column are of especial interest. In spite of some variations, it is evident that none are very far from the numeral 2, which permits us to formulate the approximate rule: *the radius of each planetary orbit is roughly twice as large as that of the orbit nearest to it in the direction of the sun.*

It is interesting to notice that a similar rule holds also for the satellites of individual planets, a fact that can be demonstrated, for example, by the table below giving the relative distances of nine satellites of Saturn.

As in the case of the planets themselves, we encounter here quite large deviations (especially for Phoebe!) but again there is hardly any doubt that there is a definite trend for regularity of the same type.

Name of satellite	Distance in terms of Saturn's radius	The ratio of increase in two successive distances
Mimas	3.11	
Enceladus	3.99	1.28
Tethys	4.94	1.24
Dione	6.33	1.28
Rhea	8.84	1.39
Titan	20.48	2.31
Hyperion	24.82	1.21
Japetus	59.68	2.40
Phoebe	216.8	3.63

How can we explain the fact that the aggregation process that took place in the original dust cloud surrounding the sun did not result in the first place in just one big planet, and why the several big lumps were formed at these particular distances from the sun?

To answer this question we have to undertake a somewhat more detailed survey of motions that took place in the original dust cloud. We must remember first of all that every material body—whether it is a tiny dust particle, a small meteorite, or a big planet—that moves around the sun under the Newtonian law of attraction is bound to describe an elliptical orbit with the sun in the focus. If the material forming the planets was formerly in the form of separate particles, say, 0.0001 cm. in diameter, there must have been some 10^{45} particles moving along the elliptical orbits of all various sizes and elongations. It is clear that, in such heavy traffic, numerous collisions must have taken place between the individual particles, and that, as the result of such collisions, the motion of the entire swarm must have become to a certain extent organized. In fact, it is not difficult to understand that such collisions served either to pulverize the "traffic violators" or to force them to "detour" into less crowded "traffic lanes." . . . By careful analysis of the situation, Weizsäcker was able to show that for the stability of such a system . . . the entire picture of motion must have looked very much like Figure 9-4. Such an arrangement would assure "safe traffic" within each individual ring, but, since these rings rotated with different periods, there must have been "traffic accidents" where one ring touched another. The large number of mutual collisions taking place in these boundary regions between the particles belonging to one ring and those belonging to neighboring rings must have been responsible for the aggregation process and for the growth of larger and larger lumps of matter at these particular distances from the sun. Thus, through a gradual thinning process within

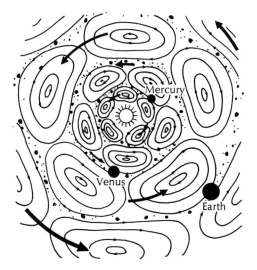

Figure 9-4. Dust-traffic lanes in the original solar envelope.

each ring, and through the accumulation of matter at the boundary regions between them, the planets were finally formed.

The above described picture of the formation of the planetary system gives us a simple explanation of the old rule governing the radii of planetary orbits. In fact, simple geometrical considerations show that in the pattern of the type shown in Figure 9-4, *the radii of successive boundary lines between the neighboring rings form a simple geometrical progression, each of them being twice as large as the previous one.* We also see why this rule cannot be expected to be quite exact. In fact, it is not the result of some *strict law* governing the motion of particles in the original dust cloud, but must be rather considered as expressing a certain *tendency* in the otherwise irregular process of dust traffic.

The fact that the same rule also holds for the satellites of different planets of our system indicates that the process of satellite formation took place roughly along the same lines. When the original dust cloud surrounding the sun was broken up into separate groups of particles that were to form the individual planets, the process repeated itself in each case with most of the material concentrating in the center to form the body of the planet, and the rest of it circling around condensing gradually into a number of satellites.

With all our discussion of mutual collisions and the growth of dust particles, we have forgotten to tell what happened to the gaseous part of the primordial solar envelope that, as may be remembered, constituted originally about 99 per cent of its entire mass. The answer to this question is a comparatively simple one.

While the dust particles were colliding, forming larger and larger lumps of matter, the gases that were unable to participate in that process were gradually dissipating into interstellar space. It can be shown by comparatively simple calculations that the time necessary for such dissipation was about 100,000,000 years, that is, about the same as the period of planetary growth. Thus by the time the planets were finally formed, most of the hydrogen and helium that had formed the original solar envelope must have escaped from the solar system, leaving only the negligibly small traces referred to above as Zodiacal Light.

One important consequence of the Weizsäcker theory lies in the conclusion that *the formation of the planetary system was not an exceptional event, but one that must have taken place in the formation of practically all of the stars.* This statement stands in sharp contrast with the conclusions of the collision theory, which considered the process by which the planets were formed as very exceptional in cosmic history. . . .

If, as it appears now, . . . there must be millions of planets within our galaxy alone, . . . it would be at least strange if life—even in its highest forms—had failed to develop in these "inhabitable" worlds.

In fact, . . . the simplest forms of life, such as different kinds of viruses, actually are merely rather complicated molecules composed mainly of carbon, hydrogen, oxygen, and nitrogen atoms. Since these elements must be presented in sufficient abundance on the surface of any newly formed planet, we must believe that sooner or later after the formation of the solid crust of earth and the precipitation of atmospheric vapors forming the extensive water reservoirs, a few molecules of such type must have appeared, owing to an accidental combination of the necessary atoms in the necessary order. To be sure, the complexity of living molecules makes the probability of their accidental formation extremely small, and we can compare it with the probability of putting together a jigsaw puzzle by simply shaking the separate pieces in their box with the hope that they will accidentally arrange themselves in the proper way. But on the other hand we must not forget that there were an immense number of atoms continuously colliding with one another, and also a lot of time in which to achieve the necessary result. The fact that life appeared on our Earth rather soon after the formation of the crust indicates that, improbable as it seems, the accidental formation of a complex organic molecule required probably only a few hundred million years. Once the simplest forms of life appeared on the surface of the newly formed planet, the process of organic reproduction, and the evolution would lead to the formation of more and more complicated forms of living organisms. There is no telling whether the evolution of life on different "inhabitable" planets takes the same track as it did on our Earth. The study of life in different worlds would contribute essentially to our understanding of the evolutionary process.

But whereas we may be able to study the forms of life that may have developed on Mars and Venus (the best "inhabitable" planets of the solar system) in the not too distant future by means of an adventuresome trip to these planets on a "nuclear-power propelled space ship," the question about the possible existence and the forms of life in other stellar worlds hundreds and thousands of light-years away, will probably remain forever an unsolvable problem of science.

The Private Life of the Stars *

by George Gamow (1958)

Having a more or less complete picture of how the individual stars give birth to their families of planets, we may now ask ourselves about the stars themselves.

What is the life history of a star? What are the details of its birth, through what changes does it go during its long life, and what is its ultimate end?

We can start studying this question by looking first at our own sun, which is a rather typical member among the billions of stars forming the system of the Milky Way. We know, first of all, that our sun is a rather old star, since according to the data of paleontology it has been shining with unchanged intensity for a few billion years supporting the development of life on the Earth. No ordinary source could supply so much energy for such a long period of time, and the problem of solar radiation remained one of the most puzzling riddles of science until the discovery of radioactive transformations and the artificial transformation of elements revealed to us tremendous sources of energy hidden in the depths of atomic nuclei. . . . practically every chemical element represents an alchemical fuel with a potentially tremendous output of energy, and that this energy can be liberated by heating up these materials to millions of degrees. . . .

* From George Gamow, *Matter, Earth and Sky*. Prentice-Hall, Inc., Englewood Cliffs, New Jersey. © 1958. Reprinted by permission.

THERMONUCLEAR REACTIONS

The discovery of radioactivity and the recognition of the fact that the energy stored within atomic nuclei exceeds, by a factor of a million, the energy liberated in ordinary chemical transformations, threw an entirely new light on the problem of solar energy sources. If the sun could have existed for several thousands of years fed by an ordinary chemical reaction (burning), nuclear energy sources are surely rich enough to supply an equal amount of energy for billions of years. The trouble is, however, that in natural radioactive decay the liberation of nuclear energy is extremely slow. In order to explain the observed mean rate of energy production in the sun, we would have to assume that the sun is composed almost entirely of uranium, thorium, and their decay products. Thus, we are forced to the conclusion that the liberation of nuclear energy inside the sun is not an ordinary radioactive decay, but rather some kind of induced nuclear transformation caused by the specific physical conditions in the solar interior. It is natural to expect that the factor responsible for the induced nuclear transformations is the tremendously high temperature existing in the solar interior. . . . And, indeed, the calculations carried out in this direction in 1929 by the Austrian physicist, F. G. Houtermans, and the British astronomer, R. Atkinson, led to the conclusion that at the temperatures existing in the solar interior, thermonuclear reactions between hydrogen nuclei (protons) and the nuclei of other light elements can be expected to liberate sufficient amounts of nuclear energy to explain the observed radiation of the sun.

CARBON CYCLE AND H-H REACTION

Although the work of Houtermans and Atkinson proved, beyond any doubt, that the energy production inside the sun is due to thermonuclear reactions between hydrogen and some light elements, the exact nature of these reactions remained obscure for another decade because of the lack of experimental knowledge concerning the result of nuclear bombardment by fast protons. However, with the pioneering work of Cockcroft and Walton on artificially accelerated proton beams and subsequent work in this direction, enough material has been collected in this field to permit the solution of the solar energy problem. One possible solution was proposed in 1937 by H. Bethe in the United States and C. von Weizsäcker in Germany (independently of one another) and is known as the *carbon cycle*, while the other possibility was conceived by an American physicist, Charles Critchfield, and is known as the *H-H re-*

action. The net total result of both reactions is the transformation of hydrogen into helium, but it is achieved in a different manner in each reaction.

In carbon cycles, the atom of carbon can be considered as a "nuclear catalyst" that helps unite four independent protons into a single α-particle by capturing them one by one and holding them together until the union is achieved. After 4 protons are caught and the newly formed α-particle is released, we get back the original carbon atom which can go again through the next cycle. The total period of the cycle is 6×10^6 years while the total energy liberated by the cycle is 4×10^{-5} erg, which yields for the rate of energy liberation per carbon atom the value of 2×10^{-19} erg/sec. Since, according to present data concerning the chemical composition of the sun, each gram of solar material contains 10^{-4} gm of carbon (5×10^{18} carbon atoms), this leads to the total rate of energy liberation of 1 erg/gm $\times$ sec, which is only one per cent of the observed production in the sun.

In the H-H reaction, the series of reactions takes 3×10^9 years and liberates 4×10^{-5} erg, leading to an energy production rate of 5×10^{-22} erg/sec per proton. Since hydrogen constitutes about 50 per cent of the solar material (2×10^{23} protons per gram), the total rate of energy liberation comes out to be 100 erg/gm $\times$ sec, in good agreement with the observed value.

The predominance of the H-H reaction over the carbon cycle in the sun, however, is not the general rule, and is reversed in many other stars. The point is that these two sets of thermonuclear reactions possess different sensitivities to temperature. While the rate of the H-H reaction increases comparatively slowly with increasing temperature, the rate of the carbon cycle goes up very rapidly. We shall see . . . that the brightness of various stars depends essentially on their mass, which determines the temperatures in their deep interior: the more massive the star, the hotter it is inside and the more violent are the energy-producing thermonuclear reactions. Thus, if we take Sirius, for example, we find that its central temperature is 2.3×10^7 °K (compared with only 2.0×10^7 °K in the center of the sun). At this higher temperature the rate of energy production by the carbon cycle becomes considerably larger than that of the H-H reaction, so that it plays here the principal role. On the other hand, in all stars less massive than the sun (and 95 per cent of all stars belong to this group), the carbon cycle becomes quite unimportant and these stars draw their energy entirely from the H-H reaction.

THE FUTURE OF OUR SUN

Since the energy radiated by the sun is due to the continuous transformation of hydrogen into helium in its interior, the sun evidently cannot shine for an eternity and is bound to run out of fuel sometime in the future. It is estimated that during the 5 billion years of its existence our sun has used about one-half of its original supply of hydrogen, so that it still has enough nuclear fuel for another 5 billion years. What will happen 5 billion years from now when our sun comes close to the end of its resources? To answer this question we have to remember that thermonuclear reactions proceed almost exclusively near the center of the sun, where the temperature is the highest. Thus, the shortage of nuclear fuel will be felt first in the central regions of the sun, where all the originally available hydrogen will have been transformed into (unburnable) helium. We can easily visualize that this will result in a rearrangement of things in the solar interior in such a way that the high-temperature region will move to the interface between the "burned-out core" and the outer layers that still contain enough hydrogen to maintain a nuclear fire. The internal structure of the sun, therefore, will be transformed from a so-called *point source model* (energy source in the center) to a *shell source model* in which the energy is liberated in a thin spherical shell that separates the burned-out core from the rest of the solar body

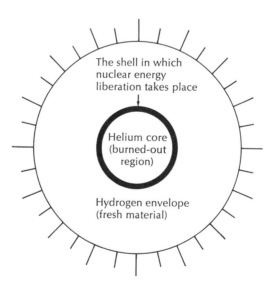

The shell in which nuclear energy liberation takes place

Helium core (burned-out region)

Hydrogen envelope (fresh material)

Figure 9-5. A shell source model of red giant stars.

(Figure 9-5). As more and more hydrogen is consumed, the "shell" will move outwards from the center, as a ring of fire does that has been started by a carelessly dropped match in a dry grass field.

It was suggested by Charles Critchfield and the author, and later confirmed by the more detailed calculations of M. Schwarzschild and his associates, that the formation of such a shell source inside the sun (or any other star) must result in a steady growth of the star's body and in a gradual increase of its luminosity. In fact, within a few hundred million years after the shell source is formed the diameter of the sun is expected to become as large as the orbit of Venus, and its luminosity will increase by a factor of between 10 and 20, making the oceans on the earth boil violently. After this last effort, the sun will begin to shrink and fade out again, until it becomes quite faint and insignificant. But there is no reason for immediate panic—we still have five billion years to go.

NORMAL STARS, STELLAR GIANTS, AND DWARFS

As we have mentioned before, different stars have different colors, ranging from a brilliant bluish hue to a dull red. This is, of course, due to the difference in their surface temperatures, and it makes it possible for us to learn about the physical conditions on their surfaces. More exactly, the surface temperatures of stars can be estimated by the study of the absorption lines in their spectra, in the same manner as it is done in the case of our sun. We find that, while the surface temperature of Capella is 6,500° K, i.e., about the same as that of our sun, the temperature of Rigel tops the 11,000° K mark. The reddish star, Antares, on the other hand, has a temperature of only about 3,000° K.

Knowing the surface temperature and the absolute brightness of a star, we can find its linear diameter, even though through the largest telescope the star appears no larger than a point. According to the Stefan-Boltzmann law, the temperature determines the amount of light emitted by a unit area of stellar surface. Thus, dividing the total light emission of a star by the emissivity per unit area, we can calculate the total surface, and, hence the diameter of the star. In this way we find that, whereas Sirius A is only 1.8 times larger than the sun, Capella and Antares are larger than the sun by factors of 12 and 450, respectively. On the other hand, Sirius B is much smaller than the sun, being, in fact, only slightly larger than the earth.

Plotting the magnitudes of a large number of stars against their surface temperatures (both in logarithmic scale) we obtain the very interesting diagram (Figure 9-6) associated with the name of a famous American astronomer, H. N. Russell. We find that most of the stars in this diagram fall within a narrow band known as the *main sequence*. Our sun, Sirius, Rigel, and many other stars belong to this sequence. As we go along this sequence towards brighter and brighter stars, the radius increases, but only moderately. If we check on the masses of the stars belonging to the main sequence (these masses can be estimated if the stars in question belong to a double system), we find that they also in-

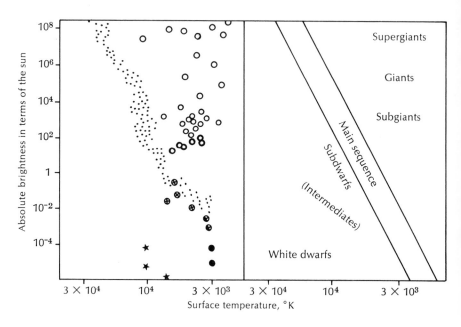

Figure 9-6. *The Russell diagram for the stars near the sun that form the spiral arms of our galaxy (see page 479).* (After Payne Gaposchkin)

crease along the sequence. Thus, we conclude that the increase of radius is simply due to the increase of stellar mass and that the mean density of stars belonging to the main sequence varies but little.

An entirely different picture emerges if we take stars such as *Antares,* or ϵ *Aurigae.* The points representing these stars in the Russell diagram fall far to the right of the regular main sequence, indicating that these stars must have entirely different internal structures. This class of stars is known under the general name, *Red Giants*—red because they all have a low surface temperature and thus a red color, and giants because of their abnormally large geometrical dimensions and their very high luminosity [Figure 9-7]. Taking into account the measured masses of these stars (30 sun masses for Antares, and for 15 ϵ Aurigae), we find

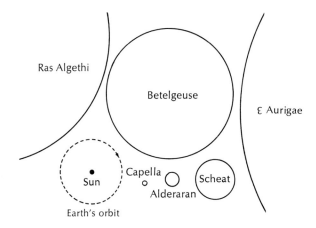

Figure 9-7. *Giant and supergiant stars compared with the size of our planetary system.* (Figure added by editor.)

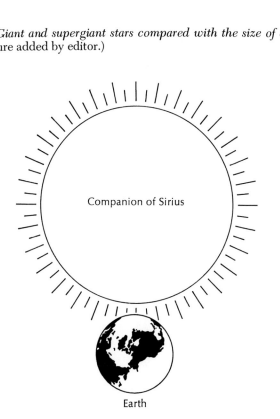

Figure 9-8. *White dwarf stars as compared with the earth.* (Figure added by editor.)

that their material is extremely rarefied. While the mean density of our sun is 1.6 with respect to water, the mean densities of Antares, and ε Aurigae are only 0.02, and 0.000003, respectively, as compared to atmospheric air.

In contrast to the Red Giants are the so-called *White Dwarfs*, which are located in the lower left corner of the Russell diagram. A typical representative of this class is the companion of Sirius (Sirius B) which having a mass almost equal to that of the sun, is only slightly larger than the earth [Figure 9-8]. Thus, its mean density must be 500,000 times greater than that of water!

PULSATING AND EXPLODING STARS

While most of the stars are seen as luminous points, shining quietly with a constant light, there are many wide deviations from this general rule. To this class of unusual stars belong the stars known as *Cepheid variables,* or simply *Cepheids,* which are subject to periodic pulsations because of some as yet unknown reason. The giant bodies of these stars expand and contract with rhythmic regularity, and this periodic change of dimensions is accompanied by corresponding periodic changes of color and luminosity. The luminosity curve of δ Cephei, after which all this class of stars is named, is shown in Figure 9-9 and is distinctly different from the luminosity curve of an eclipsing variable. The interesting point about the Cepheids is that their pulsation period, which varies for different stars from a fraction of a day to several months, is directly correlated with their intrinsic brightness: *the brighter the star, the longer is its pulsation period.* This gives astronomers an invaluable method for the measurement of distances that are too large to be tackled by the parallax method. Indeed, by observing the changes of luminosity of a distant Cepheid variable, we obtain directly its pulsation period from which we can determine its absolute brightness. Comparing this bright-

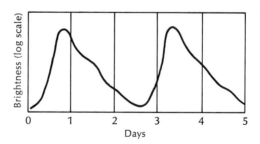

Figure 9-9. Luminosity changes of δ Cephei.

ness with the observed one, and using the inverse square law, we obtain a very exact measurement of the distance of that star from us.

Somewhat similar behavior is observed in the case of the so-called *U Geminorum stars*, named after a star in the constellation of Twins that was the first star of this type to be studied. Instead of going through smooth continuous changes of luminosity as Cepheids do, these stars remain completely tranquil for a longer or shorter period of time, and then burst in the process of a minor explosion. Some of these stars blow up every week and some others only every few months. It is interesting to note that the intensity of these periodic explosions increases in direct proportion to the interval between them.

The extreme case of the periodic explosions of the U Geminorum-type stars is probably represented by the so-called *novae* stellar explosions in which a star's brightness may increase overnight by a factor of many thousands, as compared with a factor of only 30 or 40 for typical U Geminorum stars. These violent stars were given the name, *novae*, i.e., "new ones," because it was originally believed that they were not in the sky before they were first seen. However, later observation proved that these "new stars" were actually not new at all, but rather old stars which, because of some catastrophic process, suddenly increased their luminosity. It is very likely, indeed, that the novae are rather similar to the U Geminorum-type stars except that they explode only once in a few thousand years and with correspondingly higher intensity. About a hundred novae appear in our stellar system of the Milky Way each year, but very few of them are close enough to us to be seen by the naked eye.

But the real standouts among exploding stars are the so-called *supernovae*, which stand in about the same relation to ordinary novae as the H-bomb does to a small A-bomb. When a star becomes a supernova its luminosity increases by a factor of a few hundred million and at its maximum the exploding star may become as bright as the rest of the galaxy to which it belongs. The first historically recorded explosion of this type took place in the constellation of Taurus on July 4, 1054, and is described in detail in the chronicles of the Peiping Observatory. We read:

In the first year of the period Chihha, the fifth moon, the day Chi-chou [i.e. July 4, 1054], a great star appeared approximately several inches southeast of T'ien-Kuan [i.e. ζ Tauri]. After more than a year, it gradually became invisible.

If we look now at the point in the sky where the Chinese astronomers observed a new star nine centuries ago, we see a very interesting object known as the *Crab Nebula* [Figure 8-13]. It does look like an expanding smoke cloud caused by some vigorous explosion, and, indeed, comparing the older photographs of the Crab Nebula with more recent ones, we find that this nebulosity *is* expanding at the rate of 0.18 angular seconds per year. Since the present angular radius of the Crab Nebula is about 160″, we conclude that the expansion must have started about 900 years ago, which is in perfect agreement with the date of the new star in Taurus as given in Chinese chronicles.

Careful observations reveal that in the center of the Crab Nebula is a faint, but very hot, star, apparently all that is left of the star that must have blown up nine centuries ago. The mass of that star is estimated to be about 25 per cent larger than that of the sun, while the mass of the expanding nebulosity is about 10 per cent of solar mass. It may be added that the Crab Nebula is 5,000 light-years away from us, which means that, whereas the explosion was observed on the earth only nine centuries ago, it actually took place during the beginning of human civilization.

Since the time of the Chinese supernova, only two explosions of comparable violence have taken place among the stars of our Milky Way system. One of them occurred in the constellation of Cassiopeia in 1572 and was described by Tycho Brahe in his book, *De Stella Nuova;* one of

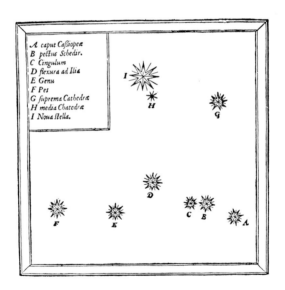

Figure 9-10. An illustration from Tycho Brahe's book "De Stella Nuova," indicating the position of a new star (marked by I) in the constellation of Cassiopeia.

the illustrations from this book is shown in Figure 9-10. Another similar explosion took place soon thereafter in the year 1604 and was observed by Tycho's assistant, Johannes Kepler. Since that time no new supernovae have appeared in our galaxy, and, considering that statistical data indicate that the average interval between the supernovae in a given galaxy is about 300 years, we may expect to see one in the sky pretty soon (i.e., within a month or within a century).

However, our stellar system of the Milky Way is only one of the billions of other similar stellar systems, known as *galaxies*, that float in space within the range of modern telescopes, and we have a much better chance to observe supernovae in these far-away stellar systems. As a matter of fact, a bright supernovae appeared in 1885 in our nearest cosmic neighbor, the Great Nebula in Andromeda. By constantly observing a few hundred of our neighboring galaxies, the Mount Wilson astronomers, W. Baade and F. Zwicky, were able to observe and to study the luminosities of a dozen supernovae going off in them.

THE EVOLUTION OF STARS

Since the stars forming our system of the Milky Way are similar to our sun, differing from it essentially only in their mass, their evolutionary history must be similar to that of the sun, described in the previous section. Like the sun, the stars must have condensed once upon a time from a dilute mixture of gas and dust, and have finally reached the steady (main-sequence) state in which they consume their original hydrogen supply by means of carbon cycles or H-H reactions.

However, stars of different masses go through their evolutionary cycles at different rates. According to observationally established, and theoretically confirmed, *mass-luminosity relations*, the brightness of the stars, and hence their rate of hydrogen consumption, increases as the cube of their mass. Since the original supply of hydrogen in a star is apparently proportional to its mass, we must conclude that *the total life spans of different stars must be inversely proportional to the squares of their masses.* Thus, we speak about the "calendar age" of a star, which simply refers to the number of years it has existed, and of its "genetic age," which represents its age as the fraction of its total life span. Stars of the same calendar age may have entirely different genetic ages. Imagine a human being, a dog, and a mouse—all three having been born five years ago. The human being is still a baby; the dog is already a mature dog; and the mouse is a *dead* mouse since mice do not live that long. A similar situation exists in the world of stars, and, by observing the stars that are seen in the sky now, we may expect to find among them repre-

Figure 9-11. *The horsehead in Orion.* (Hale Observatories)

sentative examples of all different stages of stellar evolution. Thus, our theoretical considerations concerning our sun's past and future can be directly compared with the observed properties of stars that are at the present time passing through these evolutionary stages.

A very important fact about the stellar population of our system that was emphasized by the Mount Wilson astronomer, W. Baade, a number of years ago is that there exist two different types of stellar population. In the neighborhood of the sun, which Baade calls the "local swimming hole," the space between the stars if filled by a very diluted mixture of gas and dust that forms giant interstellar nebulae obscuring the stars located behind them (Figure 9-11). The density of this interstellar material is about 10^{-24} gm/cm^3, which means that a cube of space 1,000 km on a side contains only one gram of matter. This material, the total mass of which is comparable to the total mass of the stars themselves, serves as the source for the formation of new stars by a slow condensation process. The continuous formation process of stars from the dilute interstellar material accounts for the fact that this type of stellar population contains very bright stars like Rigel, which, because of their very short life span, must be considered to have been born considerably later than our sun and its planetary system. It is interesting to notice that stars are apparently born in large groups that originate from large contracting nebulosities and then disperse in different directions as the children of one large family. This fact was . . . emphasized by a Russian astronomer, V. A. Ambarzumian, in his studies of the so-called *stellar associations* formed by dispersing groups of stars, all of the same comparatively small age.

In contrast to the above described stellar communities, known as *Population I*, there exists also stellar communities known as *Population II* that are characterized by the complete absence of any interstellar material. A typical example of such a stellar community is represented by the *globular clusters* (Figure 9-12) that float in space far away from us near the center of our stellar system. Apparently in this case the formation of the stars took place all at once in some distant past, with all the available material being completely utilized. All stars of Population II are of the same calendar age, and no new stars can be formed any more. Thus, while stellar Population I can be compared with the population of a city containing the members of all ages, from newborn babies to gray-haired veterans, stellar Population II resembles more the alumni of the class of 1912 (or any other year) of some large university. No new members are admitted, and the membership is gradually extinguished by the dying-out process.

Stellar Population II presents particular interest for the study of stel-

Figure 9-12. *Globular star cluster in Hercules.* (Hale Observatories)

lar evolution, since, being of the same calendar age, the stars of different masses belonging to it show a clear picture of their genetic age differences. In Figure 9–13 we give the Russell diagram of the Population II stars forming a typical globular cluster like the one shown in Figure 9-12. Comparing it with the corresponding diagram of the Population I stars (Figure 9-6), we notice that the main sequence is sharply limited from above and that the stars that are more than 3.5 times more luminous (and more than 50 per cent more massive) than our sun are completely absent. These are the "dead mice" stars which, because of their fast fuel consumption, were completely extinguished a long time ago.

✿   ✿   ✿

LATER STAGES OF STELLAR EVOLUTION

What happens to a star when all the hydrogen in its central regions is turned into helium? We have already mentioned in the previous chapter in relation to the future of our sun that such exhaustion of hydrogen is expected to result in the formation of an *energy-producing shell* at the base of the outer hydrogen-rich envelope, and that the growth of that shell should lead to a gradual increase of stellar radius and luminosity. The stars that come to that stage of their evolution are bound to take off from the main squence of the Russell diagram and to travel toward its upper right corner. And, in fact, that is exactly what one observes in the case of globular star clusters. In Figure 9-13 we notice a narrow band

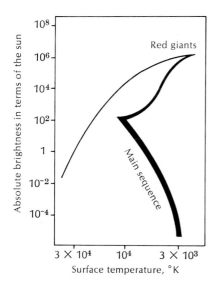

Figure 9-13. The Russell diagram of stars that form the central body of the Milky Way system and of globular clusters.

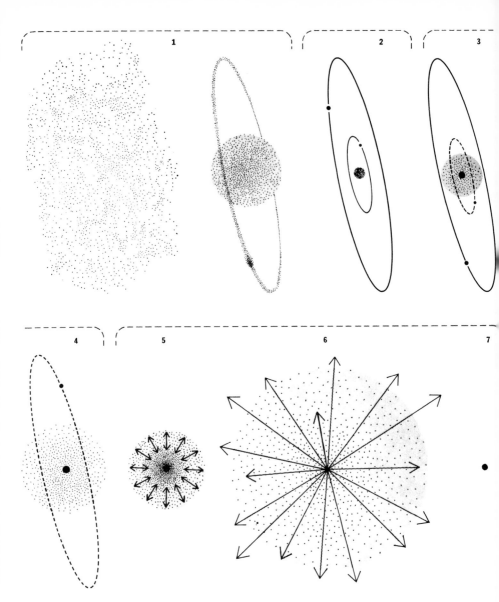

Figure 9-14. *Possible evolution of a star 1.2 times the mass of the sun begins with a dust cloud condensing into a star and protoplanets (1), a process taking perhaps 10 million years. The star enters the "main sequence" (2) and remains there for approximately 8 billion years. Then it expands (3) into the red giant stage (4), first destroying life on its inner planet, then burning up the planets in turn. This period may last about 100 million years. Then the star may pulsate in luminosity every few hours (5) for thousands of years, finally exploding into a nova (6) and eventually collapsing into white dwarf (7). The time period for final stages (5, 6, and 7) is not known.* (From "Life Outside the Solar System" by Su-Shu Huang. Copyright © 1960 by Scientific American, Inc. All rights reserved. Figure added by editor.)

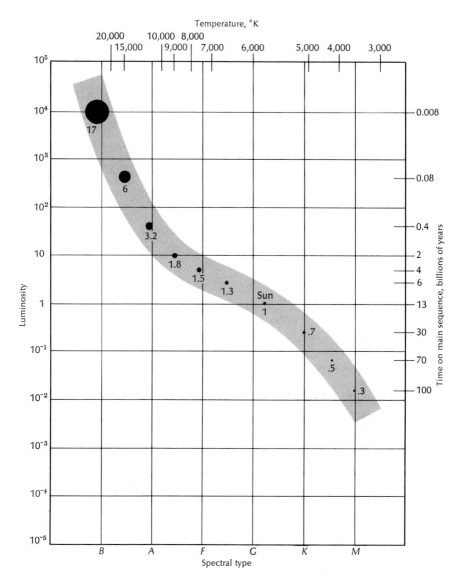

Figure 9-15. *Typical stars on the main sequence are represented on this diagram. Vertical scale at left indicates luminosity, the sun being unity. Scale at right shows time on the main sequence. The numbers at top denote temperature in degrees Kelvin; the letters at bottom, spectral type. Small figures beneath each star give its mass with respect to the mass of the sun.* (From "Life Outside the Solar System" by Su-Shu Huang. Copyright © 1960 by Scientific American, Inc. All rights reserved. Figure added by editor.)

that starts from the cut-off point of the main sequence and extends in the predicted direction; this band represents those stars that apparently have exhausted their central hydrogen supply and are in various stages of their shell-source development. Theoretical studies of this phase of stellar evolution indicate that a new kind of thermonuclear reaction begins to be of importance in addition to regular H $\rightarrow$ He transformations. In fact, the rapidly increasing central densities in shell-model stars begin to favor triple-collision processes in which the helium forming the isothermal core is transformed directly into carbon:

$$_2\mathrm{He}^4 + {}_2\mathrm{He}^4 + {}_2\mathrm{He}^4 = {}_6\mathrm{C}^{12} + \text{radiation}$$

This important reaction, first proposed by the American physicist E. E. Salpeter, is followed by other reactions, such as:

$$_6\mathrm{C}^{12} + {}_2\mathrm{He}^4 = {}_8\mathrm{O}^{16} + \text{radiation}$$
$$_8\mathrm{O}^{16} + {}_2\mathrm{He}^4 = {}_{10}\mathrm{Ne}^{20} + \text{radiation}$$
$$\text{etc.}$$

which gradually build up the original hydrogen of the stellar interior into more and more complex nuclei. This continuous build-up process is presumably terminated when nuclei of iron atoms are formed, since iron nuclei are the most stable nuclei among all chemical elements.

In order to maintain the high central temperature and density needed for the maintenance of the above described reactions, the star must begin to contract, and its luminosity gradually fades; this is responsible for the band in the Russell diagram for Population II that runs downward and to the left from the highest luminosity point reached by the evolving stars. During this descent into oblivion, stars pass through various unstable states and are subject to the pulsations and periodic minor explosions described earlier in this chapter.

Arriving at the end of their evolutionary track at the lower left corner of the Russell diagram, stars become internally unstable and blow up in the brilliant display of supernovae explosions.

Life in the Universe

Billions of Planetary Systems [*]
by Harlow Shapley (1964)

On the basis of our sampling census of stars in our own galaxy, and our sampling of the galaxy population out to the limit attainable by present telescopes, we can readily compute that there are more than 10^{20} stars in the universe, each one competent of course through radiation to maintain the photochemical reactions that are the basis of plant and animal life. Perhaps only a few percent of these are single stars with planetary potentialities. Perhaps only a few percent of these few developed in such a way (nebular contraction) or had such a suitable experience in the past (collisional) that they would now possess persisting planets. Perhaps only a few percent of these that succeed in having stable-orbited persisting planets would have one or more at the right distance from the central star; and of these rightly placed planets but 1 percent have an orbit of suitable circularity to maintain sufficiently equable temperatures. We could go on to a few of the few of the few, because non-poisonous airs and waters are also required, and that particular activity of oxygen, carbon, hydrogen, and nitrogen that we call "living" must get started. We could by such restrictions reduce the number of stars with livable and actually "inhabited" planets to nearly a nothing.

All of these restrictions, however, get us practically nowhere in isolating ourselves as something unique and special, for there are too many stars! Three undeniable factors have entered our consideration—the ordinariness of our sun which has accomplished the creation of life on this planet; the evidence for the universality of the kind of chemistry and physics we know here, and the existence of more than 10^{20} opportunities for life, that is, the existence of more than 100,000 million billion stars.

Let us look once more at large numbers and work this argument over again. Let us suppose that because of doubling, clustering, secondary

[*] From Harlow Shapley, *Of Stars and Men*. The Beacon Press, Boston. Copyright © 1958, 1964 by Harlow Shapley. Reprinted by permission of The Beacon Press.

collisions, and the like, only one star in 1000 has a planetary system. Personally I would think that 1 in 50 would be a better estimate, and many of those who believe in the nebular contraction theory of stellar formation would say that at least 1 out of 10 stars has planets. But to be conservative, we say that only 1 out of 1000 has a planetary system, and then assume that but 1 out of 1000 of those stars with systems of planets has one or more planets at the right distance from the star to provide the water and warmth that protoplasm requires. In our solar system we have two or three planets in such an interval of distance. Further, let us suppose that only 1 out of 1000 of those stars with planets suitably distant has one large enough to hold an atmosphere; in our system we have at least seven planets out of nine with atmospheres. That will reduce our suitable planet to a one in a billion chance.

Let us make one other requirement of our suitable planet: the chemical composition of air and water must be of the sort that would develop the inorganic molecules into the organic. Perhaps that biological evolution occurs but once in a thousand times.

Assembling all four of the one to a thousand chances (all grossly underestimated, I believe, but in the effort to establish our uniqueness in the world, and hence our "importance," we are making it as hard as possible to find other habitable planets), we come to the estimate that only one star out of 10^{12} meets all tests; that is, one star out of a million million. Where does that high improbability of livable planets leave us?

Dividing the million million into the total number of stars, $10^{20} \div 10^{12}$, we get 10^8; that is, 100 million planetary systems suitable for organic life. This number is a minimum; personally I would recommend . . . its multiplication by at least 1000, possibly by a million.

To state a conclusion: The scientific researches of recent times have enriched and clarified our concepts of habitable planets. Through discovering the true stellar nature of the spiral "nebulae," and through the deep sounding of star-and-galaxy populated space to such great depths that the number of knowable stars rises to billions of times the number formerly surmised, and also through the discovery of the expansion of the universe with its concomitant deduction that a few billion years ago the stars and planetary materials were much more densely and turbulently crowded together than in the present days of relative calm, we have strengthened our beliefs with respect to the existence of "other worlds." The present concept includes the identifying of our own world as the surface of Planet No. 3, in the family of a run-of-the-mill yellowish star, situated in the outer part of a typical spiral galaxy that contains billions of typical stars—this "home galaxy" being one item in an overall system, the Metagalaxy, that numbers its galaxies in the multibillions.

Worlds Around the Sun *
by Lee Edson (1969)

VENUS: EARTH'S MYSTERIOUS TWIN

From the sun to Pluto, the farthest planet, the solar system stretches almost four billion miles. The intervening space is filled with objects that can be conveniently divided into three categories: the terrestrial planets (Mercury, Venus, the earth, and Mars) and their moons, the Jovian planets (Jupiter, Saturn, Uranus, and Neptune) and their moons, and the objects that occupy interplanetary space (asteroids, comets, and miscellaneous debris). The terrestrial (or inner) planets are not only roughly comparable in size—ranging from about 3,000 to 7,900 miles in diameter—but they seem to be basically alike in chemical composition, if not in physical setting. The Jovians, on the other hand, are giant worlds far less dense than the terrestrial planets. (See Figure 9–16.)

Of the inner planets, Venus, closest to earth and its near twin in size and mass, has been—quite literally—shrouded in mystery. A heavy cloud cover masks its surface and until the recent development of radio

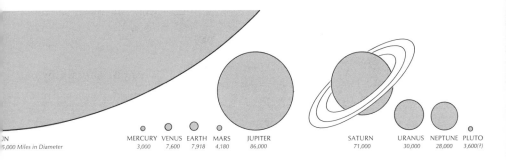

	MERCURY	VENUS	EARTH	MARS	JUPITER	SATURN	URANUS	NEPTUNE	PLUTO
5,000 Miles in Diameter	3,000	7,600	7,918	4,180	86,000	71,000	30,000	28,000	3,600(?)

Figure 9-16. The basics of planetary astronomy begin with visualizing the immense diameters of the planets and their great distances from the sun. At right, the sizes of the planets (but not their distances) are diagrammed in relation to a section of the sun. (Based on drawing from Worlds Around the Sun.)

* From Lee Edson (with Carl Sagan as consultant), *Worlds Around the Sun: The Emerging Portrait of the Solar System,* American Heritage Publishing Co., Inc., New York. © Copyright 1969 by American Heritage Publishing Co., Inc. Updated for this book by Carl Sagan and reprinted by permission.

and radar techniques to "see" through the clouds, little could be known with assurance about the planet itself. Even its size was only imprecisely known, and the length of the Venerean day (the planet's period of rotation about its axis) was anyone's guess. Such lack of data left men free to speculate on the "inhabitants" of Venus, and over the years the speculations ranged from a wistful picture of a world, much like our own, populated by intelligent creatures to the more dispassionate view of oceans swarming with primeval organic forms.

Scientific awareness of a potential life-supporting environment on Venus goes back to 1761 when the noted Russian scientist M. V. Lomonosov observed a transit of Venus—a rare occurrence in which the planet passes in front of the sun—and was amazed to see a ring around the disk as it passed one limb or edge of the sun and another ring as it passed the other limb. The ring must be due, the scientist thought, to light scattered or refracted by the Venerean atmosphere. Lomonosov concluded that Venus had an atmosphere "equal to if not greater than that which envelops our earthly sphere." Even the most careful telescopic observations of Venus showed no sign of any apparent surface detail whatever. It was early deduced, therefore, that the planet must be almost completely covered by clouds. Scientists speculated that, since earthly clouds are composed of water, Venerean clouds must also be. Since the supposed water clouds of Venus were much more extensive than those of the earth, it was thought that the planet's surface must also be very wet—perhaps a great Carboniferous swamp inhabited by reptiles. Subsequently, other ideas were put forth: that Venus is covered with carbonated water—an immense seltzer sea—or with pools of oil under clouds of smog.

Spectroscopic studies failed to settle the question of the habitability of Venus. In 1932, scientists detected carbon dioxide in the planet's atmosphere, but no one could find evidence of water vapor or oxygen. The failure to find oxygen was not crucial, since biologists know that some life forms on earth do not depend on oxygen, and many believe that entire life systems on other planets could be based on biochemical processes that do not involve oxygen at all.

Evidence against a life-supporting planet began to accumulate when scientists applied the new techniques of radio astronomy to study temperatures on Venus. In the 1920's Edison Pettit and Seth B. Nicholson of the Mount Wilson Observatory monitored the infrared radiation from Venus and deduced a temperature at some level in the atmosphere of about $-40°$ F. This is quite close to the temperature expected for Venus: its nearness to the sun's heat is more than compensated for by the high reflectivity of its clouds. But in 1956 Cornell Mayer and his associates at the U.S. Naval Research Laboratory in Washington, D.C., studied radiation of longer microwavelengths from Venus. At these wavelengths the planet emitted radiation as if it were at a temperature of about 600° F. —a searing heat beyond the melting points of lead and tin.

How could Venus be at a temperature so much higher than that expected theoretically and confirmed by infrared measurements? The question led to several hypotheses and to considerable controversy in scientific journals. One suggestion, put forth by Carl Sagan in 1960, was that the atmosphere of Venus acts as a kind of one-way filter to let solar radiation in but prevent heat in the form of infrared radiation from getting out—functioning like the glass of a greenhouse is supposed to. According to this "greenhouse" model, the higher temperature would apply at the surface, accounted for by the heat-trapping effects of carbon dioxide and, it was assumed, of water vapor in the atmosphere. The infrared temperature would apply to the cloud tops. A competing hypothesis, called the "aeolosphere" model (after Aeolus, the Greek god of the winds), was proposed by Ernst J. Öpik, an Estonian-born astronomer. This also located the source of microwave emission at the surface but attributed the high temperature to friction from wind-borne dust particles. Both models recognized a surface too hot to support life. . . .

Further evidence came in October, 1967, when the Soviet Union's Venera 4 approached the dark side of Venus and ejected a capsule that sent back temperature, pressure, and composition readings as it gently floated, on a parachute tether, toward the surface. Its last temperature communication was more than 600° F. It is not yet clear whether this represents Venus' surface temperature, but it is known that the temperatures should increase with depth, so that the planet's nighttime equatorial temperatures must be in excess of 600° F.

Venera 5 and 6 survived to greater depths in the Venus atmosphere and recorded even higher temperatures. Combining the Venera data, the Mariner 5 data and groundbased radar measurements of the radius of the solid body of Venus, it has become clear that the surface of the planet is an approximately uniform 750° K [or Fahrenheit equivalent if you like]. This means Venus is far too hot to support familiar forms of life, even at the poles . . . Venera 4 found traces of water vapor in the atmosphere below the clouds [and] evidence, surprising to many astronomers, that is [the atmosphere] is composed almost entirely of carbon dioxide—perhaps as much as 95 per cent. Many astronomers had thought nitrogen would be the major ingredient, but the gas analyzers aboard the Venera 4, 5, and 6 capsules found no nitrogen;

it may exist, but in an amount too small to be detected by these Venera instruments. . . .

Venus' period of rotation about its axis has become known for the first time because of radar's ability to penetrate the cloud cover. In the past, calculations varied wildly, ranging from 20 hours to 225 earth days. . . .

One method of analysis is to measure the Doppler broadening: when a single frequency of radio waves is beamed to Venus from the earth, it is intercepted by the entire planet. Part of the signal is reflected back to earth from the approaching limb (and is increased in frequency by the Doppler effect) and part of it is reflected from the receding limb of the planet (and is decreased in frequency by the Doppler effect). The center of the planet is moving at right angles to our line of sight and therefore produces no Doppler effect on the reflected radio waves. Measuring these frequency shifts makes possible an estimate of the planet's rate of spin, and in 1962, Roland L. Carpenter and Richard M. Goldstein at the Jet Propulsion Laboratory concluded that Venus makes one complete rotation every 250 days. Other astronomers, using a modification of the Doppler-shift method, later refined the figure to 243. Further radar data is beginning to reveal a radar map of the Venus surface.

If the length of the day and the length of the year were identical, then Venus would always present the same face to the sun—provided, however, that the planet spins on its axis in the same sense as does the earth (counterclockwise, as seen from our North Pole) and the other planets of the solar system. (Uranus, as we shall see, is a special case.) But the remarkable discovery from the radar data is that Venus rotates every 243 days, not every 225 days (the time the planet takes to go once around the sun), and, most surprisingly, that Venus spins around its axis in the opposite (clockwise, from our conventional standpoint) direction. A 243-day retrograde rotation period has another equally surprising meaning—Venus always keeps the same face turned to earth when the planets' orbits are closest. How the earth could influence its sister planet in this way is one of the most baffling riddles of planetary astronomy, and the ultimate answer probably lies hidden in the very origin of the solar system.

All told, the portrait of Venus emerging from new astronomical discoveries is a bleak one. Because the atmospheric pressure at the surface of Venus is far greater than on earth—perhaps one hundred times as great, the equivalent of the ocean two thirds of a mile down—liquid water could exist there at much higher temperatures. But the accumulated radio and radar observations of Venus now suggest that the average surface temperature is higher than 800° F. and no amount of pressure can hold water in liquid form at a temperature greater than about

700° F. There is little difference in temperature between the light and dark sides of Venus; the dark side manages to retain much of its daytime heat, though it probably also has its supply replenished by hot winds blowing across the planet.

If the temperature of the planet is as high as present measurements indicate, the prospects for familiar forms of life on Venus are zero. However, it is not out of the question that life forms might exist in the cooler clouds or even—based on biochemical structures quite unlike those known on the earth—near the planet's surface.

Of the many problems that remain, one of the most intriguing is why there is so little water on Venus in any form—by present measurements, barely one ten-thousandth the amount on earth. This difference between the sister planets remains one of the great puzzles of solar system astronomy. One theory is that ultraviolet radiation from the sun has been dissociating water molecules since the planet's beginning, and the weak gravitational attraction of Venus has been allowing the liberated hydrogen atoms to escape into space. Some argue that this process would account for only part of the loss, assuming Venus and the earth began with equal amounts of water.

Whatever the explanation, the dryness adds to the portrait of a planet that is arid and uninviting. Though Venus and the earth may be sisters, born almost as twins from the same primordial material, their histories have been enigmatically different.

NEW VIEWS OF MARS AND MERCURY

The possibility that Venus is a lifeless and uninhabited planet has made all the more enticing the prospect of searching for life on the planet Mars. It is a good place to look—as the fourth planet outward from the sun, and only thirty-five million miles from the earth at its closest approach, Mars has surface temperatures that would not be insupportable for some forms of life known on the earth. Summer equatorial noontime temperatures have been measured at about 80° F. At night, however, the temperature falls to below −100° F. at the same locale and season, while at latitudes equivalent to the so-called temperate latitudes on the earth Martian temperatures may never rise above −30° F., even in the daytime.

The idea of life on Mars, especially of intelligent creatures somewhat similar to ourselves, is almost as old as man's discovery that the planets are "other worlds." In Babylonian times the planets were thought to be ethereal abodes of the gods, but not real places. Lucian of Samosata in the second century A.D. was probably the first man to fantasize that a planet might be inhabited with beings that were not divine, and with the rediscovery of Greek learning in the Renaissance came fresh impetus to consider the likelihood of life on worlds other than our own. In 1600 the Italian scholar Giordano Bruno was burned at the stake for the heretical suggestion that God might have shown his infinite wisdom by peopling the universe with countless other beings like ourselves. The most influential proponent of this idea of the plurality of worlds was probably an imaginative Frenchman, Bernard de Fontenelle; in 1686 he revived the theme in a best-selling book that stocked the then known planets with interesting creatures and characters. Mars' sole inhabitants, he suggested, were luminous birds. . . .

The planet lends itself both to the fantasies of science fiction and to the serious speculation by scientists that there may be life on Mars.

The Martian day is close to earth's, about twenty-four and a half hours long; moreover, because its axis is tilted to the plane of its orbit by about the same sixty-seven degrees as that of the earth, Martian and terrestrial seasons are similar in pattern. On the other hand, the Martian year is almost twice as long as that on earth: 687 earth days, so that the length of each season is also twice as long.

During the seasons, however, there are marked changes in contrast on the planet's surface, that is, in the relative brightness of dark areas as compared with adjacent bright areas. These have suggested to some people a yearly cycle of blossoming, growth, and decay similar to that of earthly vegetation. Many astronomers—especially earlier visual observers—reported vivid color changes in the dark areas, a progression from a neutral gray to pastel greens and blues, accompanying the contrast changes, but the reality of this color alteration is now disputed. Brilliant white caps cover much of the area around the Martian poles during the winter, then recede as spring approaches, in a way that seems very much like the behavior of icecaps on earth. [See Figure 9-19.]

Then there are the remarkable "canals"—the long dark lines that run from one dark area to another. They were first noted by Giovanni Schiaparelli, an Italian astronomer, in 1877 and at once created a sensation. Schiaparelli called them *canali,* meaning channels, but in English the word came to be written "canals."

In the United States, Percival Lowell, a member of a prominent and wealthy Boston family, became interested in the Martian canals. He was a self-taught astronomer but an able one, and a man of considerable imagination. In 1906 and 1908 he published two startling books, *Mars and Its Canals* and *Mars as the Abode of Life,* in which he declared that the canals were inland waterways built by intelligent beings to transport

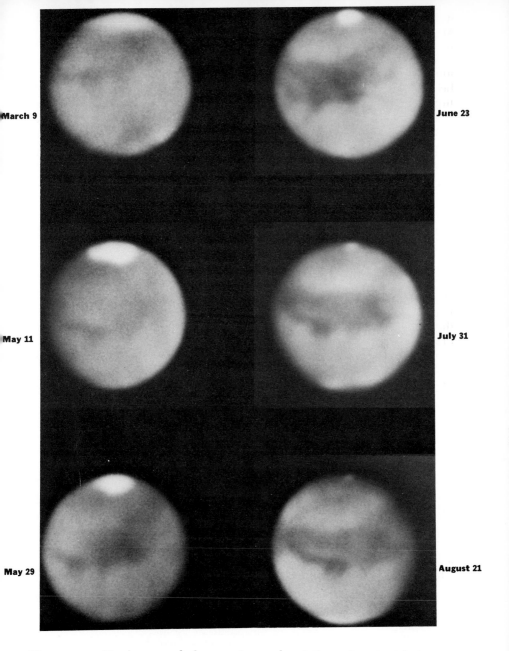

March 9

June 23

May 11

July 31

May 29

August 21

Figure 9-17. *Mars's seasonal changes. Seasonal variations of contrast between bright and dark areas on Mars between March 9 and August 21 Martian dates. The polar caps recede, and the contrast between bright and dark areas increases, reaching a maximum during Martian June and July. Such seasonal changes may be due to Martian vegetation inhabiting the dark areas, but may also be due to fine bright particles of windblown dust scoured off the dark regions of Mars by the fierce winds of winter.* (Lowell Observatory)

493

water from the melting polar caps to Martian cities. He believed that Mars was running out of water and the inhabitants had built the canals in a massive engineering project essential for their survival. . . .

The nature of the Martian "canals" is still disputed, although scientists today dismiss the notion of their being irrigation ditches. Even in the best telescopes Mars presents a shimmering, evanescent image, due to turbulence in the earth's atmosphere—what an astronomer calls "bad seeing." Some expert contemporary observers still see thin dark lines crossing the Martian bright areas. Other observers have noted that when the atmosphere becomes very steady and the seeing excellent, the canals seem to dissolve into a mélange of disconnected fine detail. In these observers' opinion, the canals may be an optical illusion—the result of the eye's penchant for order—rather than a true Martian surface feature.

On the other hand, there is some recent evidence that points to canals of a sort. Photographs radioed back to earth by Mariner 4, during the spacecraft's fly-by of Mars in July, 1965, revealed a number of straight lines, although shorter and thinner than those some observers had reported. In addition, Carl Sagan and James B. Pollack have found radar evidence leading to the conclusion that at least some of Lowell's broad classical canals are real topographic features, namely, ridges. In fact, the shadows cast by the lines in the Mariner 4 photographs indicate ridges in some cases, but grooves in others. Sagan and Pollack have suggested that the Martian canals may represent a system of fractures in the planet's crust, somewhat like the ridges at the bottom of the earth's oceans. . . .[1]

Scientists at the Jet Propulsion Laboratory (J.P.L.), watching the computer build up the photographs from code signals received from the spacecraft, were intrigued to see a crater-pocked surface. Although they superficially resemble those of our moon, Mars' craters are not entirely moon-like: they have a more filled-in appearance and their circular ramparts are often breached or worn away. Clearly, erosion is more prevalent on Mars than on the moon. This is to be expected, since, unlike the moon, Mars does have an atmosphere and water, at least in frozen form, is present on its surface. . . .

For all we know, shallow seas and rivers may once have existed on primitive Mars, but all traces of them may have been eroded subse-

---

1. Photographs relayed back by Mariners 6 and 7 in August 1970 (and still in a relatively rough form) show several previously identified canals as well-defined features. Examples include Agathodaemon and Cerberus. Other canals appear to be dissolved into a sequence of dark patches of varying size and contrast, suggesting that many canals involve the chance alignment of randomly distributed dark patches. The true physical nature of these features is still unknown. Ed.

quently by wind-blown dust, by the cratering process itself, and by the diurnal freezing and thawing of water, if any.

Certainly there is little or no liquid water on Mars now. In the cold, thin atmosphere of the planet, heated ice crystals would become water vapor, bypassing the liquid stage in the same way that "dry ice" on earth becomes gaseous carbon dioxide. Before the Mariner 4 flight, scientists had concluded that the atmosphere of Mars was very thin, but the spacecraft established that the atmosphere is thinner beyond most previous expectations—about equivalent to the earth's atmosphere at an altitude of twenty miles.

This finding also forced scientists to revise their earlier conclusions about the composition of the Martian atmosphere. Through spectroscopic studies at McDonald Observatory in 1948, Gerard M. Kuiper, now at the University of Arizona, detected the presence of carbon dioxide in the atmosphere of Mars. Like most astronomers, however, he had concluded that the major ingredient was nitrogen. But while Mariner 4's

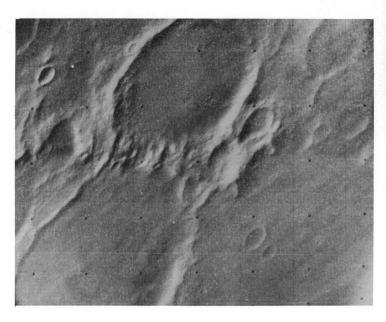

Figure 9-18. Two oblique craters near the south polar cap of Mars, jokingly described as a giant's footprint, first evidence of life on Mars. This picture is No. 20 in Mariner 7's sequence. Other pictures showed details of the cap with vast drifts that looked like snow but are believed to be frozen carbon dioxide. Very small amounts of water vapor were found to be present in the thin Martian atmosphere. (NASA. Figure added by editor.)

Figure 9-19. *This view shows the planet's South Pole cap of dry ice and Nix Olympica, a crater (circle) 300 miles wide amid unexplained white streaks.* (NASA. Figure added by editor.)

instruments showed a reduced total atmospheric pressure on Mars, the actual amount of carbon dioxide present did not differ from that recorded earlier. This implies that carbon dioxide is, in fact, a major constituent of the Martian atmosphere.

Recently, the spectroscopic identification of very small quantities of water vapor and oxygen has been claimed—and disputed—and ultraviolet spectroscopy, conducted during a brief rocket trip above the earth's atmosphere, found negligible amounts of ozone in the atmosphere of

Mars. The oxygen and ozone in the earth's atmosphere prevent solar ultraviolet light from reaching the surface of our planet. This light, artificially produced, is used on earth for sterilization purposes. The lack of oxygen and ozone on Mars means that this sterilizing ultraviolet light from the sun will reach the planet's surface unimpeded by the atmosphere, posing a serious hazard to most familiar forms of life as we know it.

If carbon dioxide is the major constituent of the Martian atmosphere, then the polar caps may not be ice crystals but frozen carbon dioxide or dry ice. This hypothesis is not new, but had fallen into disfavor after Gerard Kuiper and Audouin Dollfus each found independent evidence for icecaps of frozen water. Leighton and Murray have calculated that polar temperatures on Mars should be low enough (about $-210°$ F.) for carbon dioxide to condense out of the atmosphere, and infrared evidence supports this. Advocates of the dry ice cap do not, however, rule out the existence of ice particles mixed in with the frozen carbon dioxide, and many astronomers believe that ice is to be found beneath the Martian surface in the form of permafrost.[2]

Aside from the polar caps, the Martian surface is divided into two kinds of regions: bright areas and dark areas. A current astronomical debate concerns their relative heights: Are the dark areas higher than the light, or vice versa? In any case, there is quite convincing radar evidence that some areas of Mars are elevated as much as ten miles above their surroundings. These are probably not isolated mountain peaks, however, but more likely large land masses hundreds to thousands of miles across, perhaps similar to the continents of earth.

As is apparent even to the naked eye, Mars is red, presumably because of some material present in abundance on its surface that is reddish. On the earth all the abundant minerals that are colored red are iron oxides, a kind of rust. A natural suggestion is, therefore, that iron oxides are the cause of Mars' reddish coloration. But astronomers are still unsure how much iron oxide would be required to account for such noticeable surface properties. As on the earth, large amounts of silicates should also be present, because silicon is a very abundant cosmic element.

Radar evidence and the way in which Mars reflects sunlight suggest that its surface is powdery. Indeed, what look very much like dust storms are seen fairly frequently on the planet, arising in bright areas and temporarily blotting out dark ones. To raise dust in such a thin at-

2. Observations by Mariners 6 and 7 seemed to confirm the hypothesis of frozen carbon dioxide polar caps (in surprisingly deep drifts). They also reported very small amounts of atomic oxygen and water vapor. Ed.

mosphere, Martian winds must be very fierce; meteorological calculations indicate occasional speeds of up to two hundred miles per hour.

Since the winds vary with the seasons—and are strongest in the winter—there is some reason to think that the highlands will be more readily denuded of fine bright dust in winter than in summer. This phenomenon possibly explains the seasonal contrast changes. Some scientists still believe, however, in the appealing idea that Martian dark areas darken in spring and summer because of the growth and proliferation of a dense cover of Martian vegetation. . . .

To discover whether the Martian environment could actually support living organisms of the sort found on earth, a number of scientists have simulated the planet's conditions in the laboratory and subjected both plants and animals to them. The results have been favorable. At NASA's Ames Research Center, for instance, Richard Young and his co-workers have found that certain bacteria can live and multiply in a daily freeze-thaw cycle of $-150°$ F. to $70°$ F. In other experiments, scientists have housed turtles and other small animals in plastic-domed "Mars jars" under conditions approaching low Martian pressure and lack of oxygen, and found that they survive. Some microorganisms, indeed, have survived a combined jolt of ultraviolet light, thin atmosphere, carbon dioxide, and low water content. Moreover, with an increase of water content —such as might be expected in microenvironments during the Martian spring—some earth soil microorganisms actually multiply and thrive.

Other scientists are designing instruments to be landed on Mars' surface, where they will search for life, automatically. A typical device will send out a grapple or an arm or a sticky string to gather samples of Martian surface material; it will then retrieve this material and examine it, either by testing for metabolic activity or with a microscope. It is possible that Mars only supports microorganisms, but even if there are larger forms of life there still must be microorganisms, so the designers have deliberately concentrated on detecting the smallest forms of life.

The design of such instruments is not easy. Various assumptions about the nature of life on Mars have to be built into every Martian life detector. For example, a device to examine metabolic activity must assume that Martian organisms will like the food that we send along, or that they will give off some particular gas such as carbon dioxide in a process of photosynthesis similar to that on earth. It is conceivable that Martian organisms could be so different from terrestrial ones that all the early instruments would give negative results even though the planet teemed with life. But we have to look for the type of life we know first.

For this detection to succeed, however, the greatest care must be taken not to contaminate the planet with terrestrial microorganisms. For this reason both the United States and the Soviet Union have publicly committed themselves to sterilizing all vehicles intended for landing on Mars—or on any other planet.

In the next decades we shall be sending our instruments to Mars, and they are sure to yield us much exciting new information. A thorough reconnaissance will obviously be followed by the landing of automatic devices—but there is a debate as to whether man himself ought to plan to attempt a landing on Mars because of the contamination problem. It can only be resolved after detailed unmanned studies of this intriguing planet.

Mercury, the closest planet to the sun, is also the smallest and swiftest of the nine planets. In diameter it is only half again as big as the earth's moon, and it whirls around the sun at a speed of thirty miles per second. When nearest to the sun, a distance of some twenty-eight million miles, Mercury receives more than ten times as much radiation as the earth and the moon. The maximum temperature has been measured at Mount Wilson and Mount Palomar as about 800°F. At its farthest distance from the sun, some forty-three million miles, Mercury is still hot enough (545°F.) to melt tin and lead. Fred Whipple has suggested that Mercury, rather than Pluto, should have been named for the god of the underworld.

Little is known about the planet Mercury, not only because of its small size, but because its nearness to the sun makes it hard to study. However, recent radar studies have brought us new knowledge of the planet and upset one important, long-standing assumption. Until a few years ago, astronomers firmly believed that Mercury spins on its axis once a year, always turning the same face to the sun as the moon does to the earth. Put another way, its rotation period was thought to be synchronous with its orbital period, both being 88 earth days. Astronomers' maps of the surface of Mercury seemed to confirm this hypothesis, and there were theoretical reasons for believing that the planet, so close to the sun, would be "locked" to it as the moon is to the earth.

In line with this belief, one side of Mercury was thought to be blazing hot, and the other side perpetually cold—in fact, not too far from absolute zero. But in 1964, radio astronomers using a 250-foot dish antenna in Australia found the temperature on the dark side to be about 62°F. Perhaps, it was thought, heat moves from the bright side of the planet to the dark side by conduction through the planet or by convective air currents. Neither explanation was convincing. Conduction through the planet would be a slow and inefficient process, and as for convection, the atmosphere seemed much too thin to be capable of conveying the amount of heat needed to warm up the dark side.

The puzzle of the warm temperature on the dark side was solved in 1965 when Gordon Pettengill and Rolf B. Dyce, using the powerful radio telescope at Arecibo, found that the earlier supposition of the rotation period of Mercury was wrong—that in fact it was 59 days (plus or minus five), almost exactly two thirds of its orbital period of 88 days. Thus all parts of the planet receive sunlight, though Mercury's night is quite long—about 176 earth days.

This startling discovery prompted other astronomers to recheck the drawings that had been made of the surface of Mercury during the previous forty years. They found that three other values for the rotation period would explain the features observed just as well as the 88-day period, and one of them was the 59-day period observed by Pettengill and Dyce.

This illustrates a point about the limitations of scientific evidence. The early drawings of Mercury were consistent with an 88-day period, but that is not the same thing as *proving* an 88-day period. It has been shown time and again that a scientific theory which satisfactorily explains all the information known about a subject may have to be completely revised on the finding of one new fact.

From what little is known about Mercury, the planet presents a stark picture, somewhat like the moon in pitted, desolate appearance. Only on the dark side would temperatures be bearable. If there is an atmosphere at all, it probably would not contain oxygen, a gas too light to be held by the small planet with its weak gravitational pull. If a visitor beheld the sun, it would look enormous, and he would be risking lethal radiation. Astronomers agree that Mercury would be one of the last places to look for signs of life. Because of its lack of atmosphere, however, the planet should be much easier to explore than the shrouded and mysterious Venus, and because of its closeness to the sun, Mercury may be an ideal site for an instrumented research station, which could study our nearest star. . . .

THE JOVIAN PLANETS AND BEYOND

A spaceship leaving Mars would have to cross a great gulf of space— more than four times the earth-sun distance—to reach Jupiter, the next major planet outward from the sun and the largest of a very different planetary group, of which Saturn, Uranus, and Neptune are the other members. These are known as the Jovian planets—giant worlds shrouded in heavy atmospheres that whirl about their axes faster than any of the terrestrial planets.

Of the four Jovians, Jupiter, the most impressive, is appropriately

named for the king of the gods. It is thirteen hundred times larger than
the earth in volume; two and a half times the mass of all the other planets combined; and attended by twelve satellites, more than any other
planet. Escape velocity from Jupiter's surface is estimated at 130,000
miles per hour, compared with a mere 25,000 to get free from the earth.

Through the telescope Jupiter shows up as a flattened disk, somewhat
like a squashed volley ball, with dark belts that parallel the planet's
equator. These belts are separated by yellowish bands, or zones. Both
belts and zones change color. By focusing on a patch of color an observer can follow its motion and determine the rate of rotation of the atmosphere (we can only guess at the rate of rotation of the planet itself).
Jupiter's atmosphere, like that of the gaseous sun, actually rotates faster
at the equator than it does at the poles; the cloud layers make one complete revolution every 9 hours and 55 minutes at typical temperate zone
latitudes, as compared with 9 hours and 50 minutes at the equator. Excluding the asteroids, Jupiter is the fastest spinning planet in the solar
system. In contrast, however, it takes its time revolving around the sun
—nearly twelve years to cover the three billion miles of its orbital path.
The inclination of Jupiter's equator to its orbit is so small that there are
no seasons on the planet.

Jupiter's moons move in such a peculiar fashion that they would confuse anyone trying to figure out a usable calendar from their movements.
The four tiny outermost moons swing through their orbits as far out as
10,000,000 miles from the planet, while the closest of the inner moons is
no farther than 112,000 miles away, about half the distance of our own
moon from earth. The outermost moons revolve around Jupiter in a retrograde direction, that is, in paths opposite that of the planet's rotation
and opposite the inner moons' orbits. . . .

One of the earliest ideas about Jupiter's composition, suggested in the
1930's by astronomer Rupert Wildt, was that the planet probably contained great amounts of hydrogen and helium. These are gases light
enough to account for its over-all low density, which is about one fourth
that of the earth; they are also the most abundant atoms in the sun and
stars. Wildt also proposed that (as in the primitive days on earth) hydrogen might have combined with nitrogen and carbon, which are also
fairly abundant cosmically, to form ammonia (NH_3) and methane
(CH_4). The presence of these gases might account for the dark absorption bands in the Jovian spectrum.

Picking up this thought, Theodore Dunham of the Mount Wilson Observatory introduced methane and ammonia separately into a long pipe,
transmitted a light beam through them, and investigated the results with
a spectroscope. He found that the spectral lines and bands in the labora-

tory specimens agreed in wavelength with the dark bands of the spectrum of Jupiter. In addition, the existence of these two gases implied that the Jovian atmosphere was filled with hydrogen, the original hydrogen, it is thought, from which the planet was formed. While earth and the other terrestrial planets lost much of their original complement of light gases because of their relatively weak gravity and high upper-atmospheric temperatures, the Jovian planets evidently retained their hydrogen and helium because of their strong gravity and lower temperature.

More recent spectroscopic observations have confirmed that hydrogen and helium comprise the bulk of Jupiter's atmosphere and that methane and ammonia are present in small amounts—around 0.1 per cent. The temperatures in and above the clouds of Jupiter, as measured spectroscopically and by determining the infrared radiation emitted by the planet, are less than $-95°$ F. This is not cold enough to condense out hydrogen, helium, or methane, but ammonia probably will be condensed at such temperatures, and the visible clouds of Jupiter are believed to be made up of ammonia rather than of water vapor as on earth. If hydrogen has combined with carbon and with nitrogen on Jupiter to form methane and ammonia, it seems reasonable to expect that it has also combined with oxygen to form water, since oxygen is just as abundant cosmically as carbon and nitrogen. This would have to happen, however, at much greater depths than those into which we have penetrated as yet, where the temperatures are far too low for it to condense out. Astronomers think it quite likely that water condensation clouds exist in Jupiter's lower atmosphere, below the ammonia clouds, although the presence of water has never been detected directly.

Jupiter's atmosphere is very extensive and much deeper than that of the earth. We have no idea how far down the atmosphere extends before the surface is reached. We do not even know whether there *is* a surface on Jupiter in the ordinary sense. It is quite possible that the atmosphere becomes denser and denser with increasing depth, eventually becoming quite thick and gradually fading into what we would call a solid, without there even being the clear demarcation of a surface. Pressures in the planet's interior are so great that the hydrogen there may be formed into metallic hydrogen, a material unknown on earth.

Seen through a telescope, Jupiter is a brilliant, swirling mélange of browns, yellows, oranges, blues, and reds. Yet in ordinary visible light hydrogen, helium, methane, ammonia, and water are entirely colorless; some other molecule must be present in Jupiter's clouds. In 1955 Francis Owen Rice, a chemist at Catholic University, suggested that the coloring material might be chemical fragments known as free radicals. On earth free radicals are rare and short-lived because they are rapidly destroyed by collision with other molecules. But at the low temperatures and low densities that characterize Jupiter's upper atmosphere, some free radicals could exist—and some of these molecular fragments, composed of carbon, hydrogen, nitrogen, and oxygen atoms, are indeed brightly colored.

Harold Urey had a more dramatic idea. To him it seemed more likely that the colors were due to organic molecules that were produced by interactions of methane, ammonia, and hydrogen in the Jovian atmosphere. Carl Sagan and his colleagues tested this idea in the laboratory by simulating the atmosphere of Jupiter, supplying energy to the mixture of gases, and then investigating the results with a computer. They found that simple organic molecules were produced quite readily, along with some more complex ring-shaped organic molecules that were brightly colored. More recently, instrumented rockets have taken ultraviolet spectra of Jupiter that independently suggest that quite complex organic molecules exist both in and above the planet's clouds. This is not altogether surprising, since the Jovian atmosphere today is thought to be very similar to the primitive earthly atmosphere in which the organic molecules arose that may have been the precursors of life on earth, and Jupiter's atmospheric temperatures are not too extreme, in the lower reaches, to support the more complex organic molecules of life itself.

The Jovian atmosphere's most striking feature is a huge, apparently permanent oval shape known as the great Red Spot. Twenty-five thousand miles long and eight thousand miles wide, the spot lies in a bright band called the south tropical zone. Its color varies from brick red to pink to faint gray; in the 1880's the spot grew so pale that astronomers concluded it was disappearing. But the spot is still there, and so is its mystery. . . .

Jupiter continues to surprise astronomers. Recently, Frank Low of the University of Arizona, using a germanium bolometer, an instrument highly sensitive to infrared radiation, determined that Jupiter actually gives off more energy than it receives from the sun. Jupiter is probably too cold a body to be generating heat from thermonuclear reactions in its interior, as stars do. More likely, the planet is undergoing a process of slow contraction, generating heat as its materials are squeezed toward the center, a process common in the early history of stars, before their interior temperatures are hot enough to initiate thermonuclear reactions.

Another recent Jovian surprise occurred when radio astronomers found that Jupiter was emitting radio waves of great intensity. Planets were originally regarded as unlikely sources of radio emission, but in 1955 Kenneth L. Franklin and Bernard F. Burke, using a radio telescope near Washington, D.C., noticed tremendous bursts of radio waves coming from Jupiter. The bursts, some as powerful as ten million kilowatts at their source, sometimes occurred in rapid succession, the product of "storms" lasting as long as two hours.

In addition to these rapid bursts of activity, astronomers discovered a continuous, unchanging emission at long radio wavelengths. For this to be coming from some hot region of the Jovian atmosphere would mean that temperatures there were in excess of 180,000° F., far too hot to be reasonable for the upper atmosphere. This represented a puzzle until 1959, when Frank Drake of Cornell suggested the alternative explanation that the emissions are caused not by planetary heat but by electrons being impelled at very high velocities by the planet's strong magnetic field—a Jovian analogue of the Van Allen radiation belt around the earth.

More recently, scientists at the California Institute of Technology, using a microwave interferometer, have confirmed that this continuous emission from Jupiter does indeed come from a gigantic radiation belt extending to a distance several times the planet's radius.

The shorter radio bursts are probably also connected with this belt, perhaps caused by the sporadic dumping into the atmosphere of charged particles that ordinarily would be trapped by the Jovian magnetic field.

For all its vast distance from the earth, Jupiter is the closest and most accessible of the Jovian planets. These are in a much earlier stage of evolution than are the terrestrial planets, and when, in a few years, we start to fly spacecraft toward Jupiter to investigate the mysteries of its atmosphere, we may be granted an unrivaled opportunity to realize what earth was like in its primordial state.

Half a billion miles of blackness separate Jupiter from Saturn, the second of the Jovian planets and by far the most striking. The two planets are similar in several ways: Saturn is only slightly smaller, its day is only half an hour longer, it has almost as many moons (ten), and its composition is probably very nearly identical. Saturn has bands of color parallel to its equator, like those of Jupiter, though they are not as distinct. But Saturn's density is only a fraction of Jupiter's—in fact, Saturn is so light it would float in water—and hence its gravitational pull is milder.

The most distinctive features of Saturn, of course, are the thin, flat rings that gird the planet at its equator. Ancient Babylonian and Egyptian astronomers knew of Saturn's existence and the Romans made the planet a symbol of the harvest, but identification of the rings had to await the invention of the telescope.

Even then, the rings of Saturn eluded discovery. Their rotational axis is tilted to Saturn's orbit, and there are periods during the planet's thirty-year trip around the sun when the rings, as seen from earth, are too thin to be observed when viewed edgewise. Galileo saw the rings approximately face-on in 1610, and drew ear-like appendages that he described as being "like two servants supporting an old gentleman." Two years later they seemed to be gone. Disputes about these elusive appendages went on for half a century, until the eminent Dutch scientist Christian Huygens established their existence and explained why they could not always be clearly seen.

Huygens actually thought Saturn had only one ring, but in 1675 Cassini detected a black line or gap that divided the ring into two rings. This gap is now known as Cassini's division. The third and innermost ring—dark and very faint and known as the crape ring—went undetected until 1850, when George Bond discovered it while observing Saturn through a telescope at Harvard.

From Huygens' time onward, the rings invited speculation about their origin and composition. Like most of his predecessors, William Herschel, the great eighteenth-century astronomer, thought the rings were solids, and for most of his life he believed that the Cassini division was not a true gap at all but a dark band in the continuous solid. In 1791 he abandoned this notion, having convinced himself that the blackness of the gap consistently matched that of the background sky.

The idea that the rings were solid remained alive until 1859, when James Clerk Maxwell proved that they could not be solid (or liquid either, for that matter), but would have to be made up of billions of small particles. Only in that form, with each small particle constituting a kind of independent moon, could the rings be stable. Otherwise, the gravitational forces of nearby Saturn would tear the rings apart. In 1917, two British astronomers, Maurice Ainslie and John Knight, observed Saturn's outer ring passing in front of a distant star. The star did not disappear entirely, proving that the ring is transluscent and suggesting that it is indeed made up of particles. . . .

Gerard Kuiper's spectrographic studies have led him to conclude that the particles in the rings, if not actually hoarfrost, are coated with ice. . . .

Saturn hides its surface beneath a thick, opaque atmosphere, very much like Jupiter's. Infrared and radio measurements indicate that Saturn's cloud temperatures are colder than those of the Jovian clouds, but it is probable that the temperature rises closer to the Saturnian surface, like the temperature in the earth's own atmosphere. Saturn, like Jupiter, emits more thermal energy than it receives from the sun, but there is no sign of a strong radio emission, and it is just possible that the planet's rings sweep up the charged particles that would otherwise constitute a Saturnian Van Allen belt.

Beyond Saturn lie the three planets that are too dim and too slow in their orbits to have been known to ancient and medieval astronomers. Two of them, Uranus and Neptune, are similar in several ways. Both are three to four times larger in diameter, though far less dense than the earth; Uranus, the bigger of the two, is only two thirds as dense as Neptune. Both appear greenish in the telescope, probably because of very strong absorption of red light by the methane that is known to be prominent in their atmospheres. Uranus, the closer of the two, takes eighty-four years to orbit the sun, while Neptune takes nearly twice as long.

Uranus has one unusual characteristic: it rotates at nearly a right-angle plane of its revolution about the sun, like a ball rolling across the floor, and if one takes the upward tilting pole of the planet to be its north pole, the spin turns out to be retrograde—an intriguing characteristic Uranus shares with Venus. . . .

Pluto is so small, its orbit so highly elliptical, and its rotational period so slow (six and a half days), that some astronomers believe it may once have been a moon of Neptune that somehow, in a cataclysmic event, escaped Neptune's pull and became a planet in its own right. However, it has recently been shown that Pluto and Neptune are held by gravitational forces to a repeating orbital cycle of twenty thousand years; this precludes Pluto from going anywhere near Neptune, and makes the idea that Pluto is an escaped satellite less tenable. . . .

REACHING FOR THE MOON

A quarter of a million miles from earth there orbits in stately splendor our planet's only natural satellite, the moon. Worshiped in ancient times for the tides it brings and the regularity of its cycle, appreciated by lovers in all epochs of human history, the moon is at last on the verge of giving up its secrets. . . . We are at the exciting point in history when the moon is being explored directly for the first time, and although many ancient problems will soon be solved, it is almost certain that many new ones, not even dimly glimpsed at present, will arise to claim the attention of fresh generations of lunar scientists.

This exploration of the moon should provide man with a new and unique scientific perspective on the origins of his planet, obtained by learning the order in which lunar geological events occurred. On earth the layers of rock in the crust are a great calendar of past geological eras. The peaceful-looking strata actually chronicle the violent story of vast mobile sheets of ice, of cataclysmic convulsion, and of slow erosion, and they reveal in the midst of these long-gone upheavals the life and death of huge populations of strange creatures that once inhabited the globe. Slowly, as scientists have pieced together the clues in the rocks, we have come to know something of the earth's history—but not enough to answer the perplexing questions of the origin of the earth and of the solar system.

The moon may close this gap. On the moon, where there is apparently very little erosion because earth's major eroding elements, water and air, are absent, lunar geologists (or selenologists) happily see a preserved record in rock of events that took place eons ago. . . .

Since the moon lacks an atmosphere, it has no weather, and the vacuum is greater than anything we can establish on the earth. Though the moon receives about the same amount of radiation from the sun as the earth does, the absence of air and surface water causes temperature fluctuations from well above the boiling point of water to about 300 degrees below freezing on the Fahrenheit scale in the period between full and new moon. An interesting computation shows that if we could put an atmosphere on the moon like that of earth, the moon's weak gravity—one sixth that of earth—would permit almost all the atmosphere to leak away in about ten thousand years.

This is the broad portrait of the moon, but for scientists the details are being filled in rapidly. . . .

How did the craters and maria get there? Until the late nineteenth century, most astronomers thought, largely on the basis of analogy with the earth, that the craters were the product of volcanic activity. Then, in 1893, Grove Karl Gilbert, senior geologist of the U.S. Geological Survey, cast doubt on the volcanic hypothesis by pointing out that volcanoes on earth differ in appearance from many lunar craters and that none have craters as large as the moon's biggest ones. Gilbert argued persuasively that the lunar craters were caused by the impact of meteorites, and in time many astronomers came to agree with him. Today both hypotheses have their supporters, and while advocates in each camp usually are willing to agree that the moon has been the subject of both volcanic and meteoritic activity, they dispute which has caused most of the moon's surface features.

But there seems no doubt that the maria were formed by the impacts of enormous objects, perhaps during the final stages of the formation of the moon. The impact basins were subsequently flooded, possibly by lava, producing their present smooth, dark surfaces. . . .

Lunar Orbiter 5 data lend additional weight to the impact hypothesis of maria origin. Whenever the spacecraft passed over one of the five circular maria of the moon it sped up slightly, indicating it was responding to a small additional gravitational pull. According to one version of the impact hypothesis, underneath the circular maria should be either the remnants of the enormous impacting objects that produced them or denser rock produced during or immediately after the impact. Brian T. O'Leary, formerly a scientist-astronaut and now working at Cornell University, and his colleagues have suggested that previously enigmatic features of the moon's shape and motion may be due to these mass concentrations [or *mascons*] beneath the circular maria.

Although the weight of astronomical opinion is on the side of the impact hypothesis for crater origin, its adversaries account for most of the smaller lunar surface features just as readily by volcanic activity. The spokes around some of the craters, they say, are material that oozed from fissures on the lunar surface. The fact that some lunar craters are larger than any seen on earth is attributed to the moon's low gravity. . . .

Controversy has also been sparked by the long lunar fissures called rills that wind sinuously across the moonscape for distances as great as one hundred miles. One group of astronomers suggests that they must have arisen out of some lunar activity such as a moonquake; others agree with Harvard's William Pickering, who in 1903 suggested the fissures might indeed be what they resemble: dried-out stream beds. Until recently Pickering's idea was generally shunned by serious investigators, but now some eminent scientists, including Harold C. Urey and Donald Menzel, have concluded that these clefts may well be the remains of streams. Whatever atmosphere the moon has is so diffuse that a pool of liquid water would evaporate almost instantly. But frozen water could exist in the cold lunar subsurface as hoarfrost. Liquid water might be trapped below the hoarfrost layer. A meteorite hitting the moon could have ruptured the hoarfrost layer, three U.C.L.A. scientists have suggested, releasing enough water to flood the crater and flow down the outside slopes as rivers. In the intense cold of the long lunar nights, the surface of the rivers would quickly turn to ice, giving a protective blanket to the liquid water carving the rills below.

If these ideas of water on the moon should turn out to be true, then the belief in the past existence of life on our satellite—now regarded as highly unlikely—might be revived. . . .

The unanswered questions about the moon's surface all have a bearing on the most intriguing question of all: What is the moon's origin? There are three general theories of lunar origin that are currently debated among lunar astronomers. In the first, the moon is pictured as having once been a part of the earth, which was torn loose by some violent catastrophe. In the second view, the moon and the earth were formed together out of the same materials and by similar processes as a kind of double planet. In the third, the moon and the earth were formed separately and in different parts of the solar system, but the moon, in traveling past the earth, was gravitationally "captured," and locked into a satellite orbit.

One of the earliest versions of the first theory was proposed in 1880 by George H. Darwin, son of the famous evolutionist. Darwin, an eminent scientist in his own right, suggested that in its early history—billions of years ago, we would now say—when the earth was young and perhaps molten, the sun's gravitational pull raised tremendous tides in the elastic earth. At that time the earth was rotating very fast: a day lasted only four hours. The sun's pull and the earth's centrifugal force worked in partnership to cause increasingly great oscillations in the earth. Eventually, these oscillations became so great that a piece of the earth broke off and receded from the planet earth to become its moon. One piece of evidence in favor of this theory is that the moon's density, while far less than that of the earth, is similar to that of the rocks in the earth's crust. In 1881, Osmond Fisher, an English geologist, went so far as to suggest that the Pacific Ocean now covers the area torn from the earth.

Subsequent mathematical analysis, however, seemed to show that it was impossible to achieve the conditions necessary to keep the ejected mass locked in orbit around the earth.

As a result of this attack, the theory slipped into oblivion. However, the notion of the earth giving birth to the moon was revived a few years ago by three astronomers—Thomas S. Lovering, Donald U. Wise, and John A. O'Keefe. These men, independently of one another, showed that the rotational instability of the earth would increase under the conditions assumed to exist when it was formed—that is, when heavier materials were sinking toward the earth's center to form its core. Still, the energy that would have been required for the breakaway is many times greater than that apparently available at the time.

The second theory of lunar origin—that the earth and moon were formed together by accretion from the available solar materials—also suffers defects. For one thing, the theory does not explain why, if the earth and the moon came out of the same mass of dust and gas, the moon ended up with a density only 60 per cent as great as the earth's. For another, it is difficult to see how the earth and the moon kept the critical distance from each other—never close enough to collide and never far enough for the moon to escape—all the while they were growing and solidifying and thus changing their gravitational relationship.

The third theory—the capture theory . . . would eliminate the problem of the difference in average density between the earth and the moon. But it presents other puzzles, notably the need to explain how the capture occurred, and where the moon, with its relatively great bulk, came from in the first place. One fascinating variation of this theory is that the moon, larger than it is now, approached close to the earth and was torn apart because the tidal forces due to the earth's gravitation exceeded the lunar gravitational forces that were helping to hold the moon together. Some of the mass landed on earth and became the continents, the remainder staying in orbit. These little moons eventually coalesced into a single moon.

The samples of lunar material returned by Apollo astronauts in the Apollo 11 and 12 missions is just beginning to be analyzed, digested and crosscorrelated by scientists. There are indications of past melting and evidence of very great age of the lunar material. Important information on the origin and early history of the moon is surely available in such samples, but there has not yet been time for it to emerge.

Harold Urey, whose own version of the capture theory postulates that the moon and the earth almost collided and actually traded some material (including water and biological material), sums up the present state of affairs in lunar astronomy: "All explanations for the origin of the moon are improbable."

Selections from the Summary of Apollo 11 Lunar Science Conference * (1970)

On 24 July 1969 the first samples from our sister planet, the moon, were returned to earth for direct scientific investigation. Prior to this, our understanding of the extraterrestrial universe derived from study of electromagnetic radiation from stars and planets, from study of cosmic rays, and from analysis of meteorites. Meteorites were, until the return of Apollo 11, the only extraterrestrial objects we could actually hold in our hands and scrutinize in the laboratory. Unlike meteorites, the lunar samples come from a good sized planetary object whose location is well known. . . .

The variety and sophistication of the techniques used in the study of these samples draws heavily on the experience gained in the last 25 years in the study of meteorites and terrestrial petrology and mineral-

* From the summary prepared by the Lunar Sample Analysis Planning Team, *Science*, Vol. 167, pp. 449–451, January 30, 1970. Copyright 1970 by the American Association for the Advancement of Science. Reprinted by permission.

ogy. The scope and quality of the analyses applied to the returned samples far exceed anything we could hope to achieve with remotely controlled devices even with the most advanced methods of modern technology. . . .

It is not surprising that at this early stage in a new era of study of a planet that there should be diversity in interpretation of observations and even some differences between the observations themselves. We do not propose to resolve these differences in this introductory summary. Our objective is to furnish a guide to the results and conclusions that are still in a state of ferment.

The samples from Tranquillity Base consist of basaltic igneous rocks, microbreccias, which are a mechanical mixture of soil and small rock fragments compacted into a coherent rock, and lunar soil. The soil is a diverse mixture of crystalline fragments and glassy fragments with a variety of most interesting shapes; it also contains small fragments of iron meteorites. Most of the rock fragments are similar to the larger igneous rocks and apparently were derived from them; the rocks in turn were probably once part of the underlying bedrock. A small number of the crystalline fragments are totally different from any of the igneous rocks of the Tranquillity site. There is a strong possibility that these fragments represent samples from the nearby Highlands.

Many of the rock surfaces and individual fragments in the soil show evidence of surface erosion by hypervelocity impacts. Examination of the surfaces of glassy objects which are themselves formed by impact processes shows that they contain beautifully preserved microscopic pits as small as 10 microns in diameter, which are the result of impacts by tiny high velocity particles. There is also evidence that the impact processes are accompanied by local melting, splashing, evaporation, and condensation. . . .

The ages of the basaltic crystalline rocks were determined by ^{87}Rb-^{87}SR and ^{40}K-^{40}Ar methods and show that they formed at 3.7×10^9 years ago. This age is supported by U-Th-Pb data. These results show that igneous rocks of the Sea of Tranquillity were melted (possibly extruded) and crystallized about 10^9 years after the formation of the moon. A single exotic rock fragment yielded an age of 4.4×10^9 years and indicates variability in age between different areas on the lunar surface. The relatively young age of the basalts shows that the moon has not been a completely dead planet from its formation but has undergone significant differentiation, at least locally, in a thin lunar crust. The time period prior to the 3.7×10^9 year event almost certainly is recorded in older Highland areas; it represents an interval where the earth's record has been obliterated. Therefore, much of the surface of the moon is of great importance in understanding the early evolution and differentiation of planets.

Some samples of soil and breccia give concordant Pb-U-Th and Sr-Rb ages of 4.6×10^9 years. These samples, which are aggregates of highly varied rocks, minerals, and glasses of younger age, appear to provide an effective average of the lunar crust. This crust, originally of material 4.6×10^9 years old, underwent differentiation at later times but without the injection into the lunar crust of younger differentiated material which is enriched in rubidium, uranium, and thorium.

All of these observations show unequivocally that the materials on the surface of the moon are the product of considerable geochemical and petrological evolution. Thus, the ancient planetary surface we have sampled gives us a picture of the early evolutionary processes of a terrestrial planet which have hitherto been obscured from view. . . .

The existence of complex carbon compounds in extraterrestrial materials would have great significance for the origin of life both on earth and other planets. The search for important protobiological compounds, such as purines, pyrimidines, amino acids, and porphyrins, was carried on with some of the most sophisticated and sensitive analytical techniques ever devised. Nevertheless, no unambiguous identification of indigenous compounds was made at extremely low levels of detection (usually less than 10 parts per billion). Contaminants from the exhaust from the lunar module and from terrestrial handling of the samples have been detected at these levels. The bulk carbon content of the soils, breccias, and igneous rocks ranges from 50 to 250 parts per million. The highest concentrations occur in the fine-grained portion of the soil; the lowest concentrations occur in the igneous rocks. Some methane and carbon monoxide are released by acid dissolution, but the majority of the carbon is not released until samples are heated above $800°C$ in a vacuum, when it comes off as carbon monoxide and carbon dioxide. The concentrations of carbon observed in the soils and breccias are approximately those expected from solar wind deposition. Graphite has been observed, but the precise nature of most of the lunar carbon remains to be defined.

Micropaleontological examination of the lunar sample by optical microscopy and by electron and scanning electron microscopy produced uniformly negative results. An intensive search for viable organisms employing a multitude of environmental and media combinations produced negative results, as did the one quarantine study.

The Apollo 11 samples were collected from a very tiny fraction of the moon's surface. Nevertheless, they have given a vast new insight into the processes that have shaped this surface and have established some significant limits on the rates and mechanisms by which it evolved. The results reported do not resolve the problem of the origin of the moon. However, the number of constraints that must be met by any theory have been greatly increased. For example, if the moon formed from the earth, it can now be stated with some confidence that this separation took place prior to 4.3×10⁹ years ago. Furthermore, such a hypothesis must now take account of certain definite differences in chemical composition. There is clear evidence in the chemical and petrologic observations that the surface of the moon is variable in both composition and age. It is therefore of great scientific importance to obtain materials from a variety of terrains and sites. Samples from the Highlands and from deeper in the lunar crust will probably be the next important milestones in the advance of lunar science.

Editor's Note on Radio Pulses from Our Galaxy (1970)

Since astronomers are agreed that life is probably not a rare phenomenon in the universe and that there is even a good possibility of intelligent life on other planets beyond our own solar system, it may seem odd that we have had no visitors from other worlds. However, we must remember that the distances to other solar systems are so great that, traveling by the fastest means known to man, it would take hundreds of thousands of years to reach the nearest star. Our advanced civilization has existed for only a few centuries. For other intelligent beings to reach us would require that their evolution had started thousands or millions of years before our own. If so, perhaps they visited the earth long before we were intelligent enough to know.

Light and radio signals, however, make the trip to Tau Ceti in 10.8 years. The possibility of communicating by radio is intriguing enough that over the past decade the radio astronomers have been watching for signals that might be interpreted as messages. It was suggested that these would be most likely to take the form of regular radio pulses, perhaps a sequence like the cardinal numbers: 1, 2, 3, 4, etc.

Then in 1967, the radio astronomers at the University of Cambridge searching the skies for quasars (see p. 438) with a very sensitive antenna system covering 4½ acres, stumbled upon a strange phenomenon. A weak but very regular radio signal appeared to be coming from a body no larger than a planet and situated relatively near us in our own galaxy. The exciting idea was immediately suggested that these might be messages from another civilization. The L. G. M. or "Little Green Men" theory seemed possible in the absence of any scientific explanation for a

Figure 9-20. *One-mile radio telescope at the Mullard Radio Astronomy Observatory, Cambridge, is being used to observe fainter radio sources than had previously been detected. Two of the three 60 foot paraboloids which make up the telescope are shown above. The reflector in the background can be moved on a rail track which is accurate to ⅛th inch over half a mile. Observations are made simultaneously on wavelengths of 74 and 20.7 cm and, after amplification by low noise parametric amplifiers, the signals from the three aerials are combined in receivers in the central laboratory. The outputs of the receivers are then punched on paper tape for processing by computer. Once the region to be observed has been selected, the operation of the radio telescope then becomes entirely automatic.* (Cambridge University)

pulsed radio signal of such rapidity (1.33730113 seconds) and of such amazing regularity (now known to one part in 100 million).

Soon, however, other pulsing sources were discovered in different locations in the Milky Way, each one having a very rapid pulse with its own characteristic frequency. As more "pulsars" were discovered it seemed less likely that they could all represent messages from advanced civilizations. A study of the energies involved showed that the objects must be much more massive than planets. Furthermore, if the sources were traveling in orbits around a nearby star, this motion should show up as a Doppler effect and careful analysis of the pulses showed no evidence of this. Astronomers soon came to the conclusion that these pulses were due to a natural phenomenon.

Many explanations of pulsars have been proposed in the last few years. The most favored theory is that they belong to a special class of "neutron stars," representing an extremely dense state of matter, millions of times more dense than the matter in white dwarfs. A thimbleful of matter from a neutron star would weigh hundreds of millions of tons. These stars are believed to have magnetic fields higher than any other known class of objects and the pulses are thought to be caused by very rapid rotation. Forty-six pulsars have now been found and research is progressing at such a rate that more information—and possibly new theoretical interpretations—are certain to be forthcoming by the time this book is in print.

The Human Response

Editor's Note. The first two selections below were written in response to the theory that the planets were formed by the accidental collision of our sun with another star. The last two selections were written in response to the modern theory. The difference in the human implications is striking.

The Mysterious Universe *
by Sir James Jeans (1930)

Standing on our microscopic fragment of a grain of sand, we attempt to discover the nature and purpose of the universe which surrounds our home in space and time. Our first impression is something akin to terror. We find the universe terrifying because of its vast meaningless distances, terrifying because of its inconceivably long vistas of time which dwarf human history to the twinkling of an eye, terrifying because of our extreme loneliness, and because of the material insignificance of our home in space—a millionth part of a grain of sand out of all the sea-sand in the world. But above all else, we find the universe terrifying because it appears to be indifferent to life like our own; emotion, ambition and achievement, art and religion all seem equally foreign to its plan. Perhaps indeed we ought to say it appears to be actively hostile to life like our own. For the most part, empty space is so cold that all life in it would be frozen; most of the matter in space is so hot as to make life on it impossible; space is traversed, and astronomical bodies continually bombarded, by radiation of a variety of kinds, much of which is probably inimical to, or even destructive of, life.

Into such a universe we have stumbled, if not exactly by mistake, at least as the result of what may properly be described as an accident. The use of such a word need not imply any surprise that our earth exists, for accidents will happen, and if the universe goes on for long enough, every conceivable accident is likely to happen in time. It was, I

* From Sir James Jeans, *The Mysterious Universe,* Cambridge University Press, New York, 1930. Reprinted by permission.

think, Huxley who said that six monkeys, set to strum unintelligently on typewriters for millions of millions of years, would be bound in time to write all the books in the British Museum. If we examined the last page which a particular monkey had typed, and found that it had chanced, in its blind strumming, to type a Shakespeare sonnet, we should rightly regard the occurrence as a remarkable accident, but if we looked through all the millions of pages the monkeys had turned off in untold millions of years, we might be sure of finding a Shakespeare sonnet somewhere amongst them, the product of the blind play of chance. In the same way, millions of millions of stars wandering blindly through space for millions of millions of years are bound to meet with every sort of accident, and so are bound to produce a certain limited number of planetary systems in time. Yet the number of these must be very small in comparison with the total number of stars in the sky. . . .

Just for this reason it seems incredible that the universe can have been designed primarily to produce life like our own; had it been so, surely we might have expected to find a better proportion between the magnitude of the mechanism and the amount of the product. At first glance at least, life seems to be an utterly unimportant by-product; we living things are somehow off the main line.

A Free Man's Worship *
by Bertrand Russell (1903)

Such, in outline, but even more purposeless, more void of meaning, is the world which Science presents for our belief. Amid such a world, if anywhere, our ideals henceforward must find a home. That man is the product of causes which had no prevision of the end they were achieving; that his origin, his growth, his hopes and fears, his loves and his beliefs, are but the outcome of accidental collocations of atoms; that no fire, no heroism, no intensity of thought and feeling, can preserve an individual life beyond the grave; that all the labours of the ages, all the devotion, all the inspirations, all the noonday brightness of human genius, are destined to extinction in the vast death of the solar system, and that the whole temple of Man's achievement must inevitably be buried beneath the debris of a universe in ruins—all these things, if not quite

° From Bertrand Russell, *Mysticism and Logic,* George Allen & Unwin, Ltd., London, 1903. Reprinted by permission.

beyond dispute, are yet so nearly certain, that no philosophy which rejects them can hope to stand. Only within the scaffolding of these truths, only on the firm foundation of unyielding despair, can the soul's habitation henceforth be safely built. . . .

Brief and powerless is Man's life; on him and all his race the slow, sure doom falls pitiless and dark. Blind to good and evil, reckless of destruction, omnipotent matter rolls on its relentless way; for Man, condemned to-day to lose his dearest, to-morrow himself to pass through the gate of darkness, it remains only to cherish, ere yet the blow falls, the lofty thoughts that ennoble his little day; disdaining the coward terrors of the slave of Fate, to worship at the shrine that his own hands have built; undismayed by the empire of chance, to preserve a mind free from the wanton tyranny that rules his outward life; proudly defiant of the irresistible forces that tolerate, for a moment, his knowledge and his condemnation, to sustain alone, a weary but unyielding Atlas, the world that his own ideals have fashioned despite the trampling march of unconscious power.

Of Stars and Men *
by Harlow Shapley (1958)

You may say that these are but speculations, insecurely founded, and that you choose to believe and reason and worship otherwise. And I must reply that you should follow your inclination. But you are invited to think seriously of the cosmic facts. Let us hope that an easement is not sought in comforting tradition alone, or in resort to crude irrationality. The new knowledge from many sources—from the test tube, from the extended radiation spectrum, the electron microscope, experimental agriculture, and the radio telescope; from mathematical equations and the cosmotrons—the revelations from all these make obsolete many of the earlier world views. The new discoveries and developments contribute to the unfolding of a magnificent universe; to be a participant is in itself a glory. With our confreres on distant planets; with our fellow animals and plants of land, air, and sea; with the rocks and waters of all planetary crusts, and the photons and atoms that make up the stars—with all these we are associated in an existence and an evolution that inspires respect and deep reverence. We cannot escape humility. And as groping philosophers and scientists we are thankful for the mysteries that still lie beyond our grasp.

There are those who would call this attitude their philosophy, their religion. They would be loath, I hope, to retreat from the galaxies to the earth; unwilling to come out of the cosmic depths and durations to concern themselves only with one organic form on the crust of one small planet, near a commonplace star, at the edge of one of the galaxies. They would hesitate to retreat to that one isolated spot in their search for the Ultimate. May their kind increase and prosper!

A Personal View *
by Fred Hoyle (1960)

With the clear understanding that what I am now going to say has no agreed basis among scientists but represents my own personal views, I shall try to sum up the general philosophic issues that seem to me to come out of our survey of the Universe.

It is my view that man's unguided imagination could never have chanced on such a structure as I have put before you. No literary genius could have invented a story one-hundredth part as fantastic as the sober facts that have been unearthed by astronomical science. You need only compare our inquiry into the nature of the Universe with the tales of such acknowledged masters as Jules Verne and H. G. Wells to see that fact outweighs fiction by an enormous margin. One is naturally led to wonder what the impact of the New Cosmology would have been on a man like Newton, who would have been able to take it in, details and all, in one clean sweep. I think that Newton would have been quite unprepared for any such revelation, and that it would have had a shattering effect on him. . . .

Next we come to a question that everyone, scientist and nonscientist alike, must have asked at some time. What is man's place in the Universe? I should like to make a start on this momentous issue by considering the view of the out-and-out materialists. The appeal of their argument is based on simplicity. The Universe is here, they say, so let us take it for granted. Then the Earth and other planets must arise in the way we have already discussed. On a suitably favored planet like the Earth, life would be very likely to arise, and once it had started, so the argument goes on, only the biological processes of mutation and natural selection are needed to produce living creatures as we know them. Such creatures are no more than ingenious machines that have evolved as strange by-products in an odd corner of the Universe. No important connection exists, so the argument concludes, between these machines and the Universe as a whole, and this explains why all attempts by the machines themselves to find such a connection have failed.

Most people object to this argument for the not very good reason that they do not like to think of themselves as machines. But taking the argument at its face value, I see no point that can actually be disproved, except the claim of simplicity. The outlook of the materialists is not simple; it is really very complicated. The apparent simplicity is only achieved by taking the existence of the Universe for granted. For myself there is a great deal more about the Universe that I should like to know. Why is the Universe as it is and not something else? Why is the Universe here at all? Is it true that at present we have no clue to the answers to questions such as these, and it may be that the materialists are right in saying that no meaning can be attached to them. But throughout the history of science, people have been asserting that such and such an issue is inherently beyond the scope of reasoned inquiry, and time after time they have been proved wrong. Two thousand years ago it would have been thought quite impossible to investigate the nature of the Universe to the extent I have been describing it to you . . . And I dare say that you yourself would have said, not so very long ago, that it was impossible to learn anything about the way the Universe is created. All experience teaches us that no one has yet asked too much. . . .

If there is one important result that comes out of our inquiry into the nature of the Universe it is this: when by patient inquiry we learn the answer to any problem, we always find, both as a whole and in detail, that the answer thus revealed is finer in concept and design than anything we could ever have arrived at by a random guess. And this, I believe, will be the same for the deeper issues we have just been discussing. I think that all our present guesses are likely to prove but a very pale shadow of the real thing; and it is on this note that I must now finish. Perhaps the most majestic feature of our whole existence is that while our intelligences are powerful enough to penetrate deeply into the evolution of this quite incredible Universe, we still have not the smallest clue to our own fate.

Suggestions for Further Reading

° Asimov, Isaac: *The Universe, From Flat Earth to Quasar,* Walker and Company, New York, 1966. Chapters 1–3 and 7–11 deal with the solar system and theories of stellar evolution.

———: *Is Anyone There?,* Doubleday & Company, Inc., Garden City, New York, 1967. A collection of speculative essays dealing with science, conjecture, and science fiction.

Clarke, Arthur C.: *The Promise of Space,* Harper & Row, New York, Evanston, and London, 1968. Chapters 22–26 discuss the planets and possibilities of life within our own solar system; Chapters 26–30 discuss the possibility of contacting intelligent life in other planetary systems. Arthur Clarke writes in a clear and interesting style for the layman.

Holmes, Captain David C.: *The Search for Life on Other Worlds,* Bantam Books Inc., New York, 1967. Describes new space age techniques for detecting life on other planets.

° Jastrow, Robert, *Red Giants and White Dwarfs,* Signet Book, New American Library, New York, 1969. This book describes the evolution of stars, the elementary facts of the evolution of life on earth, and the chances of finding it on the other planets in our solar system. An easy, non-technical presentation with excellent illustrations.

° Ohring, George: *Weather on the Planets,* Doubleday & Company, Inc., Garden City, New York, 1966. This Science Series outlines the facts known about the atmospheres of the planets in 1966. Since satellites and space probes are constantly bringing in new facts, no book on this subject can be entirely up-to-date.

° Schatzman, E. L.: *The Structure of the Universe,* translated from the French by Patrick Moore, World University Library, McGraw-Hill Book Company, New York, 1968. Present scientific ideas on the evolution of stars.

° Shapley, Harlow: *Of Stars and Men,* The Beacon Press, Boston, 1958, 1964. We have used several selections from this stimulating and well-written book and strongly recommend it in its entirety.

Shklovskii, I. S., and Carl Sagan: *Intelligent Life in the Universe,* Holden-Day, Inc., San Francisco, London, and Amsterdam, 1966. Contains fascinating speculations and further information on all the questions raised in this series of readings.

° Sullivan, Walter: *We Are Not Alone,* Signet Book, New American Library, New York, 1964. This interesting book, written for the layman, gives the historical background and development of present scientific theories concerning other life in the universe.

° Paperback edition

10

Why Explore Space?

The space age has dramatically brought to our attention the fact that science has become the concern of every citizen today. These readings describe the development of the concepts and the technology that made it possible to put a man on the moon. They also examine the question of how we responded to the challenge of the space age, both as a nation and as a part of "all mankind."

Why Explore Space?

Introduction

TODAY WE HAVE TAKEN the first step toward the realization of
one of the oldest dreams of man. Since ancient times men have been
challenged by thoughts of outer space; indeed it is the same challenge
that has led to the development of knowledge about the universe. An
understanding of the relative positions of the heavenly bodies, the nature of the earth's atmosphere, and the laws that govern the motion of
all bodies: these concepts were necessary foundations for man's venture
into space. In this sense the scientists we have been studying—
Copernicus, Torricelli, Galileo, Kepler, and Newton—were the real pioneers of the space age. The desire to know, coupled with the urge to
adventure and discovery, has led inevitably to the exploration of the

Figure 10-1. "Voyage to the Moon," a mid-nineteenth—century illustration
by Paul Gustave Doré. (The Bettmann Archive, Inc.)

525

Figure 10-2. In Jules Verne's "From the Earth to the Moon! in 97 hours and 20 minutes," written in 1877, this gun was designed to shoot a rocket to the moon. In Verne's story the launching-gun is the project of a Baltimore Gun Club, despairing because there is no war in prospect.

To convince the Gun Club that they should attempt to launch a rocket to the moon, the Gun Club President Barbicane "calmly continued his harangue:

"'There is no one among you, my brave colleagues, who has not seen the moon, or at least heard speak of it. Don't be surprised if I am about to discourse to you regarding this Queen of the Night. It is perhaps reserved for us to become the Columbuses of this unknown world.'

"'Three cheers for the moon,' roared the Gun Club, with one voice." (The Bettmann Archive, Inc.)

universe. Space science is simply the most recent phase in this exploration.

Through the centuries of scientific advance men's imaginations have leaped out beyond actual knowledge to speculations and dreams about journeys to other worlds. There are mentions of space travel in early Egyptian and Persian literature, and, of course, we all recall the Greek myth of Icarus' ill-fated flight. Bruno believed in the existence of a plurality of worlds, and in 1634 Kepler wrote one of the first science fiction tales of a journey to the moon (*Somnium: Or the Astronomy of the Moon*). Although Kepler transported his hero to the moon by supernatural means he was the first writer to base his description of the moon on the new knowledge gained by the telescope. Four years later, an English author, Bishop Godwin, published *Man in the Moone*. Again in this story the means of transport was supernatural. However, when Godwin's hero stepped out on the moon, he immediately noticed the smaller pull of gravity and could jump to great heights. This concept is extraordinary when we remember that Godwin was writing a half-century before Newton formulated the law of gravitation.

During the next two hundred years several notable contributions were made to the fiction of space travel. Bishop Wilkin's *A Discourse Concerning a New World* proposed that colonies might be planted on the moon (British colonies, of course!). To Cyrano de Bergerac goes the credit for first suggesting rocket propulsion. Jules Verne (*From the Earth to the Moon*, 1865) used the same mechanical principle, projecting his men into space with a spacegun and altering the course of their capsule by firing rockets in flight (see Figure 10-2). Verne's story is especially remarkable in the close parallels that exist between this imaginary account and the actual occurrence a hundred years later.

Artificial satellites were described for the first time by Edward Everett Hale in *The Brick Moon*, published in 1869. Then, in the first year of the twentieth century, appeared a most romantic and imaginative space story by H. G. Wells—*The First Men in the Moon*. Scientifically, however, Wells's account represented a regression from Verne's since the space ship was propelled by "cavorite," a substance that was said to act as a gravity insulator. Wells never explained how such a substance could have been found on earth. With such properties it should have departed into space long ago.

The major breakthrough that turned these dreams into practical reality was the development of rocket propulsion. Several pioneers working in different parts of the world were responsible for bringing the possibilities of rocket propulsion before the scientific world. The first man was a Russian, Konstantin Tsiolkovski. In 1898 he wrote an article on the principles of rocket flight through space. This article did not immediately receive much attention from other scientists. Then in 1919 the Smithsonian Institution published a small pamphlet written by Robert Goddard, a professor of physics at Clark University in Massachusetts. The pamphlet was entitled *A Method for Reaching Extreme Altitudes* and brought Goddard a grant to study the possibilities of rocket development. This study, like Tsiolkovski's, created very little interest. Goddard, however, worked on alone for many years, supported by grants from the Guggenheim and Carnegie Foundations, and in 1935 one of his rockets reached an altitude of 7500 feet. In the meantime a German professor, Hermann Oberth, had published a book on space flight (1923). This book did stimulate professional interest and started activity on an international front. The Russians rediscovered their prophet, Tsiolkovski. Hermann Oberth and Willy Ley were instrumental in founding the German Rocket Society, which helped to create popular interest in the subject.

However, it was largely the impetus of the military applications and the advent of the Second World War that brought about rapid development in rocket design. The German research institute at Peenemünde, under the direction of Walter Dornberger and Wernher von Braun, produced the V-2 rocket, which was the largest and most spectacular product of this war effort. The Russians were also proceeding with the development of a large rocket, and the United States had one project for a small, high-altitude rocket, which culminated, after the war, in the rocket called the WAC Corporal. At the end of the war the research and planning staff from Peenemünde surrendered to the American forces and was brought to America. The Russians captured the German production engineers of the V-2 rockets.

Throughout its short history, the challenge of man's venture into space has been closely intertwined with national interest. The interdependence of the two aspects was dramatically apparent in the space race between the United States and the Soviet Union. In the public mind, the pure motivations aising from man's curiosity and the desire to explore were overshadowed by the drive for national prestige and the fear of losing the cold war with Russia. Certainly the immense outlay of money that was needed for space exploration would not have been immediately forthcoming were it not for military and prestige factors. Some responsible citizens doubted whether such a vast outlay of national resources was warranted to send a man to the moon. Other writers pointed out that both spiritually and materially science has always more than repaid the efforts that have gone into it. They said that the problem is not so much whether space should be explored—the advance of science makes this exploration inevitable—but how fast it should be attempted. President Kennedy said in his address to Congress, of May 25, 1961, "This decision demands a major national commitment. . . . Every citizen of this country as well as the members of Congress should consider the matter carefully in making their judgment . . . and there is no sense in agreeing, or desiring, that the United States take an affirmative position in outer space unless we are prepared to do the work and bear the burdens to make it successful. . . . In a very real sense, it will not be one man going to the moon—it will be an entire nation." [1]

It is very interesting to see the change of emphasis that took place between the initiation of the race to the moon and its realization in 1969 with the landing of Apollo 11. When man took his first step on another world, this "giant leap" was taken—not for one nation—but for all mankind. In some subtle way, as Toynbee had intimated much earlier, the social apparatus that was unpacked at the end of the voyage had undergone "a sea change into something rich and strange!" [2]

1. See pp. 576, 575–576, 574–575.
2. See p. 624.

Figure 10-3. *The earth seen rising over the grey dead surface of the moon.* (NASA)

Ah! my dear comrades, it will be rather curious to have the earth for our moon, to see it rise on the horizon, to recognize the shape of its continents, and to say to oneself, 'There is America, there is Europe'; then to follow it when it is about to lose itself in the sun's rays!

Jules Verne
Round the Moon (1873)

The Science of Space

Beginnings [*]

by Arthur C. Clarke (1968)

NOTHING TO PUSH AGAINST

There was a time, not very long ago, when any writer or lecturer on space flight had to devote a good deal of effort to convincing his audience that rockets *could* provide thrust in the vacuum of space, where, obviously, there is "nothing to push against." This infuriating phrase—so true, yet so misleading—is seldom heard now that the capabilities of rockets in space have been amply demonstrated. Nevertheless, most people would probably find it very hard to explain how a rocket does manage to function in a medium where all other forms of propulsion are useless.

It is no answer to say glibly, as do some writers, that the rocket operates by "pure reaction." *Every* form of propulsion does that; it is impossible to conceive, even by the wildest flight of imagination, of one that does not.

A *reaction* is simply an opposing thrust or force. When a man walks, the friction between his feet and the ground makes the Earth move backward, ever so slightly. If there were no reaction—if, for example, he were standing on a sheet of completely frictionless ice—there could be no movement.

So it is with automobiles, ships, and aircraft. They all react, through tires, screws, or propellers, on the medium that supports them. This fact was first clearly recognized by Sir Isaac Newton and embodied in his third law of motion—"To every action there is an equal and opposite reaction"—a statement of such deceptive simplicity that it may seem self-evident.

This equality of action and reaction is universally true, but in most of

[*] Abridged from Arthur C. Clarke, *The Promise of Space*, pp. 23–78 (including twelve line drawings). Harper & Row, Publishers, Inc., New York. Copyright © 1968 by Arthur C. Clarke. Reprinted by permission of Harper & Row, Publishers, Inc.

the cases in everyday life we are aware of only the action; the reaction is unobservable. Why this is so is obvious when one considers the case of a man jumping. He imparts an equal reaction to the Earth; but as the mass of the Earth is about 100 sextillion (100,000,000,000,000,000,000,000) times greater than his, the velocity he gives to it is smaller in exactly the same ratio. Only in rather exceptional or dramatic cases is the re-action obvious; the most familiar example is the recoil produced by the firing of a gun. But whether it is obvious or concealed, the reaction is always there, and no movement of any kind is possible without it.

We can best visualize the mode of operation of a rocket by consider-ing what Einstein used to call a "thought experiment," an experiment which no one would actually perform but which illustrates some princi-ple. Imagine a man on a light sled, which also carries a large pile of bricks, and assume that the sled is resting on a sheet of smooth, abso-lutely frictionless ice.

The man takes one of the bricks and throws it horizontally. Newton's third law (and common sense, which is not *always* wrong) tells us that the action of throwing the brick produces an equal reaction on the sled. But because the sled (plus cargo) weighs much more than the brick, it moves off at a correspondingly smaller velocity. Again, there would be an exact proportionality. If the vehicle's weight was a hundred times that of the brick, it would move at one-*hundredth* of its velocity.

However, this velocity would not be lost, since we have assumed that the ice is completely frictionless. Even if the sled's acquired speed were only a few inches a minute, it would retain this speed indefinitely. (We are also, of course, assuming that there is no air resistance.)

Now the passenger throws away another brick, at exactly the same ve-locity as before. The speed of the sled at once jumps again, but by a fractionally greater amount this time, for it is now a little lighter owing to the loss of the first brick. And as more and more bricks are thrown overboard, it will continue to gain speed, each time by a slightly greater amount as its mass diminishes.

We can learn several important lessons from this simple analogy; in fact, it teaches almost everything that is necessary to know about rocket propulsion.

First of all, it is obvious that what happens to the bricks after they have left the sled does not matter in the least; all the recoil or thrust is produced *during* the act of throwing. The bricks could sail on forever or could crash into a wall six inches away—it would make no difference to the sled. The method of propulsion is, therefore, *independent of any ex-ternal medium.*

As the bricks are used up, so the weight of the vehicle steadily dimin-

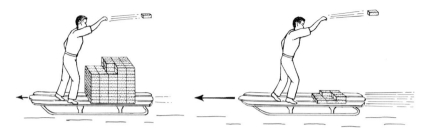

Figure 10-4. The rocket principle.

ishes. (To forestall objections from purists: the words "weight" and "mass" are used interchangeably here, as there is no need to make the distinctions that will be necessary later.) The last bricks will, therefore, produce much greater effect than the first; and the difference can be very large if the mass of the sled has been substantially reduced. If the "empty" weight of the sled is only half that of its full weight, the very last brick will produce twice the gain in speed of the first one. Consequently, not only does the sled's *velocity* increase during the experiment, its *acceleration* does so as well.

The analogy with the rocket should now be clear, the main difference between the two cases being that a rocket ejects matter continuously and not in separate lumps, so that it produces a steady thrust instead of a series of jerks. If the man on the sled were pumping water out of a nozzle, the analogy would be exact.

Thus it is possible to have a completely self-contained propulsion system that can operate in a vacuum. Yet, though the logic is impeccable, for a long time even highly qualified engineers and scientists remained unconvinced. They felt in their bones—and some readers may sympathize—that though such arguments were sound for devices that ejected solid masses, like the sled discussed above, they did not apply to a rocket that released a "mere" stream of gas into an infinite vacuum. In fact, some savants denied that any combustion was possible in these circumstances; it was for this reason that Goddard went to the trouble of firing small rockets in vacuum chambers. This did not stop the *New York Times* from printing an editorial in 1920, in which it expressed hopes that a professor at Clark College was only *pretending* to be ignorant of elementary physics, if he thought that a rocket could work in a vacuum.[1] The writer would doubtless have been surprised to know that

1. I have been mean enough to reprint this editorial, despite Mrs. Goddard's kind-hearted protests, in *The Coming of the Space Age* (Des Moines: Meredith Press, 1967).

one day the *Times* would receive the National Rocket Club's award for aerospace reporting—at the Robert H. Goddard Memorial Dinner.

The critics overlooked the fact that mass is mass, whether it be in the form of solid lumps or the most tenuous vapor. If the bricks in our thought experiment were ground into fine sand before being ejected, it would make no difference to the result. Similarly if they were volatilized, the final speed of the sled would be exactly the same as before, provided only that the ejection speed of the material remained unaltered.

So the answer to the old question, "What does a rocket push against?" should now be obvious. It pushes against its own combustion products.

A further and more subtle question is then often asked: "Just *where* inside the rocket is the thrust developed?" Essentially, a rocket motor—and this is true whether it is solid- or liquid-fueled—consists of an enclosed space (the combustion chamber) containing hot, expanding gases which can escape in only one direction, through a nozzle or orifice.

For simplicity, suppose that the combustion chamber is spherical (which is actually the case for some high-efficiency solid rockets) and that there is at first *no* orifice; the chamber is completely sealed. The combustion products are then unable to escape, and they produce the same thrust over the entire interior surface of the sphere. All forces balance out and so (assuming that the chamber does not burst) there is no movement in any direction.

Now pierce a hole in one side of the chamber. The pressure exactly opposite this hole will be unbalanced; there will therefore be a net force producing movement toward the left. The other forces will still cancel each other, and their only effect will be to exert pressure on the wall of the chamber, which must therefore be built strongly enough to withstand them.

The situation shown in Figure 10-5 (a) and (b) is that reproduced in the familiar experiment of blowing up a balloon, releasing the neck and letting it jet around the room. This demonstration, though perfectly

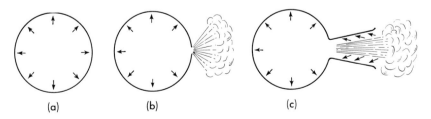

(a)   (b)   (c)

Figure 10-5. *The forces in the rocket engine.*

sound, is not really convincing; a skeptic could always argue that the balloon's gyrations were produced by reaction against the air.

A simple hole like that shown in (b) would result in a highly ineffi-cient performance; most of the escaping gas would expand sideways and do no useful work. Matters can be much improved by the addition of a nozzle (c); when the released gases expand, they press against it as shown, and so provide additional thrust.

Anyone who has followed this argument should now understand that all the thrust of a rocket is generated inside the combustion chamber and nozzle and that any surrounding medium plays no essential part in the process. This is not to say, however, that it has absolutely no effect on the performance of a rocket motor. When any rocket flies inside the atmosphere, the surrounding air actually *hinders* the expansion of the exhaust gases. For this reason the thrust of a rocket increases by 10 per cent or more as it leaves the atmosphere and enters the vacuum of space —the only environment where it can function with full efficiency.

It also follows—and this is perhaps even harder to accept—that a rocket gets no additional thrust at takeoff if the jet impinges on some fixed object, such as the ground or the launching pad. Indeed, this must be avoided at all costs, since the reflected stream of hot gases can cause great damage to the vehicle.

Almost all rockets that have been built so far have obtained their thrust from chemical reactions; burning substances have generated hot gases that escape from a nozzle. However, there are endless ways of producing the same effect: *any* power source may be used, from a nu-clear reactor to an electric battery. And any material may be used to provide the jet: solids, liquids, gases, electrons, ions, subatomic particles. As long as they have mass and can be aimed in a definite direction, they will give thrust.

Perhaps in the far future there may be spacecraft propelled by the swiftest "jet" that can exist—beams of pure light of unimaginable inten-sity, created by generators brighter than a billion suns. But they will still be rockets, in the direct line of descent from the crude vehicles which, in our time, first broke through the barrier of the atmosphere.

POWER FOR SPACE

By the early 1930's there was plenty of rocket theory but very little practice. Though a few small test vehicles had flown, their performance had not been impressive, especially when set against talk of travel to the Moon and planets. Most people who heard that Goddard's rockets had ascended a few thousand feet probably reacted in the same way as the shortsighted but typical newspaper editor who said of the Wright brothers' first hop off the ground: "57 seconds? If it had been 57 minutes, that *might* have been news."

What the skeptics failed to realize was that, even when the basic theory is completely sound, it requires millions of man-years and billions of dollars to develop a new technology. A man like Goddard could design an entire rocket vehicle, and it might seem to the layman that all that then had to be done was to send the drawings to the workshop. But it is not as easy as that; even the simplest liquid-propellant rocket contains dozens of components, all of which have to function perfectly, and most of which have to be specially built. Apparently straightforward devices such as valves to control the flow of propellants, gyroscopes for steering, pumps for feeding fuel into the combustion chambers, and reliable parachute-ejection mechanisms may demand months of development and dozens of tests. When one considers the mishaps that have plagued programs with virtually unlimited funds and manpower, it seems a miracle that any of the pioneering experimenters ever got their rockets off the ground.

As is well known, rockets fall into two distinct categories, one based on solid, the other on liquid propellants. "Solid" rockets, of which the ordinary back-garden, or Fourth of July, fireworks are the most familiar example, have been in existence for many centuries. Precisely how long is still a matter of debate, but they were certainly recorded in Chinese literature around A.D. 1200.

Although it has been said that the Chinese invented gunpowder and proved their culture by using it only for fireworks, the rocket refutes this, for they employed it with great effect against the Mongols at the siege of K'ai-fung-fu, north of the Yellow River, in 1232. News of the invention reached Europe very quickly, and the rocket was soon in common use both as a firework and as an impressive but usually unreliable weapon. It was not until the end of the eighteenth century, however, that its military application was taken very seriously in the Western world. Then, once again, the demonstration came from the Orient, when the Indian prince Tipu Sahib of Mysore used it against the British at the Battle of Seringapatam (1792). Although Tipu lost (his opponent, Charles Cornwallis, was rather luckier than at Yorktown eleven years earlier), the havoc wrought by his rocket artillery created a great impression. It came to the notice of Colonel William Congreve (not to be confused with the dramatist of the same name) who developed large war rockets with ranges of more than a mile and weights of up to 42 pounds. For a while it seemed that the rocket might replace the gun, as indeed Congreve believed that it would, but the great improvements in artillery soon made it obsolete, except for campaigns against ill-equipped natives.

In the meantime, however, it had found another use, as a launcher of rescue lines to ships stranded offshore. Between 1850 and 1940 the rocket was used almost exclusively for pyrotechnics and lifesaving; if the totals could be added up it might yet turn out that the rocket has saved more lives than it has taken.

No simpler propulsive device than a solid-fueled or powder rocket can be imagined. Even today its simplicity enables it to hold its own in many applications, such as the Polaris and Minuteman missiles, al-though these involve a fantastic degree of chemical, engineering, aero-dynamic, and electronic sophistication. By trial and error, over a period of centuries, the classical design shown in Figure 10-6 was evolved.

The propellant was a slow-burning form of gunpowder known as black powder, the composition of which is roughly 60 per cent saltpeter (potassium nitrate), 25 per cent charcoal, and 15 per cent sulfur; note the crude nozzle and the internal-combustion chamber formed by the conical space in the hard-packed powder charge. Stability in flight was maintained—if at all—by the trailing stick.

Cheapness and ease of manufacture were the two merits of this de-sign; in almost every other respect it was deplorable, and as a means of carrying substantial payloads any distance it was useless. One obvious improvement was to remove the dead weight of the stick and to obtain stability by small fins or canted exhaust nozzles, which made the rocket spin in flight like a rifle bullet; but graver defects were not so easily

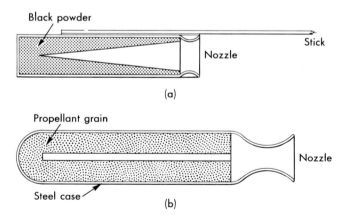

Figure 10-6. *Solid-propellant rockets: (a) firework, (b) modern solid.*

remedied. The most serious—for a device which is to be used as a propulsive engine, not as a missile—is lack of controllability. Once started, the powder will burn until it is all used up. The acceleration produced is also extremely high, so that the rocket reaches maximum speed very quickly and then wastes all its energy against air resistance.

But a much more fundamental objection to the classical powder rocket is that its propellants are really very feeble. It was a long time before this was appreciated, because the spectacular performance of an ascending rocket gives an impression of great energy.

And yet, pound for pound, such mixtures as ordinary gasoline or kerosene, with the correct proportions of oxygen, give *several times* the energy of gunpowder. True, they do not burn so rapidly, but that is also an advantage, for the last thing wanted is an explosion. On the contrary, what is desired is a controlled release of energy, and preferably a reaction that can be stopped or started at will. Liquid mixtures, which can be pumped and metered, are ideal for providing this. Some typical examples, with their performances, are listed in Table 3, page 558.

One seldom-mentioned advantage of liquid propellants is that some of the most powerful combinations (e.g., kerosene and liquid oxygen—"lox") are also very cheap, costing only a few cents per pound. Since large rockets contain hundreds or even thousands of tons of fuel, this is no trivial matter.

On all counts, therefore—performance, control (including ability to stop and restart), and economy—liquid-propellant mixtures appeared much superior to solid or powder ones. For this reason almost all space-flight discussions from the time of Tsiolkovsky onward were based on liquid propellants—usually alcohol, or a hydrocarbon, or hydrogen, burning with oxygen.

Yet technological progress has a curious way of doubling back on itself. During World War II, after the liquid-propellant rocket had been fully proved, new types of solid propellant were discovered which greatly narrowed the "energy gap." Much more surprising, ways were found of controlling, stopping, and even restarting solid-propellant motors, which have now been built in sizes (20 feet in *diameter!*) beyond all reasonable expectations of a few years ago. Such giant motors are valuable as strap-on boosters to provide additional thrust at takeoff; but it seems unlikely that they will displace the liquid-propellant systems that have dominated space exploration since the 1940's and which will do so at least for several decades to come.

Reduced to their simplest elements, these systems have to comprise the following items:

1. Fuel tank, 2. Oxidizer tank, 3. Fuel pump, 4. Oxidizer pump, 5. Pump motor(s), 6. Combustion chamber, 7. Guidance system, 8. Structure, 9. Payload. . . .

The first time all the items worked successfully in a really large rocket was on October 3, 1942, when a V-2 missile rose from its launching pad at Peenemünde and plunged into the Baltic 120 miles away. That evening General Walter Dornberger, who directed all German Army rocket development, told his colleagues, "Today the spaceship was born!" . . .

Luckily for the Allies, Hitler . . . did not believe in rockets. He had dreamed that the V-2 would never cross the English Channel, so did not give it the support that might have changed the progress of the war. For this, both victors and vanquished may well be thankful. The first nuclear chain reaction was achieved in the same month as the first V-2 flight. If southern England had been evacuated as a result of rocket bombardment, the invasion of Europe might never have taken place—and, almost certainly, the atomic bomb would have been used first against Germany, not Japan.

After the collapse of Germany in the spring of 1945, Dr. von Braun and more than a hundred of his top men surrendered to the United States Army—a few jumps ahead of the advancing Russians. Stalin, to his loudly expressed annoyance, secured only a handful of the senior scientists, but several hundred production engineers and technicians were "persuaded" to work in the U.S.S.R for some years. Perhaps even more

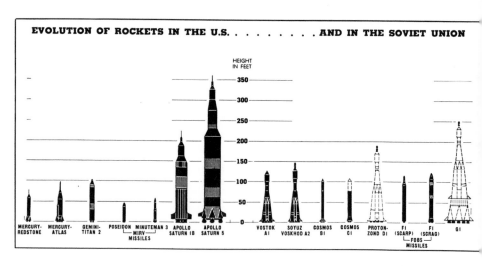

Figure 10-7. (© 1969 by The New York Times Company. Figure added by editor.)

valuable to the Russians was the capture of the vast underground V-2 plant at Nordhausen, intact apart from one hundred missiles surreptitiously whisked away by the United States Army a few hours earlier. This operation was quite illegal, since Nordhausen and all its equipment were in the zone already assigned to the U.S.S.R. Many Americans have since wished that the Army had compounded its felony by blowing up the whole plant, instead of leaving it in full working order for its grateful and incredulous allies.

However, since brains are always more valuable than hardware, the United States had much the better bargain; it merely failed to exploit it, for more than five years. By then it was the old story of too little and too late, for the U.S.S.R. had achieved a head start that would take more than a decade to overcome.

Apart from the tremendous increase in size, from the 14 tons of the V-2 to the awesome 3,000 tons of the Saturn 5, the improvements in rocket design during the 1950's and 1960's lay more in increased sophistication and reliability than in major changes of concept. Engines became more efficient as their combustion-chamber pressures were raised from less than 300 to about 1,000 pounds per square inch, with even higher values now in sight. Fuels of greater energy content than alcohol —e.g., kerosene and, finally, liquid hydrogen, the ultimate chemical fuel —were developed. The airshiplike fins of the V-2 shrank and, frequently, vanished altogether from such missiles as Atlas and Titan, which make few concessions to aerodynamics and are virtually flying cylindrical storage tanks. Steering is now effected by pivoting the engine, or engines—just like the outboard motor on a small boat. This is a more complex but more efficient arrangement than the Goddard V-2 solution of rudders in the jet exhaust.

Not so obvious merely by looking at the designs are improvements in materials and constructional techniques, which have steadily reduced the dead, or empty, weight of rocket structures. In the V-2 the propellant tanks were enclosed inside an outer skin; in later rockets the tanks themselves form the main body of the vehicle. The material of which they are built is sometimes so thin that they may have to be pressurized to prevent them from collapsing; they are, in effect, metal balloons, capable of holding a dozen or more times their own weight in fuel and oxidizer.

As a result of all these improvements, the range of single-stage rockets of the V-2 type rose from 200 miles in 1942 to well over 1,000 miles by the 1950's. Though such performances would have permitted scientific, and even manned, flights some hundreds of miles into space, they were not adequate for the task of escaping completely from the Earth. To understand why this was the case, let us now look at the obstacles on the road to the planets.

ESCAPE FROM EARTH

Any project for leaving the planet Earth has to take account of two dominant factors—the presence of the atmosphere and the force of gravity. They are not independent; if gravity were weaker, the atmosphere would be less dense, and would also extend farther out into space, since it would not be held so tightly to the surface. Such a situation prevails on Mars, where the gravity is one-third and the air pressure only one-hundredth of ours. On the Moon conditions are even more extreme; the gravity is one-sixth of Earth's, and the atmosphere has leaked away completely.

To the would-be space traveler, our atmosphere is both a help and a hindrance. On the way out, its resistance cuts back the speed attained by an ascending spacecraft, and extra fuel has to be carried to overcome this loss. The penalty decreases with the increasing size of the vehicle and is relatively unimportant for very large rockets, though it dominates the performance of very small ones.

This is a consequence of the well-known square-cube law. The mass of a body increases as the cube of its dimensions; its surface area, only as the square. Thus larger bodies have proportionately less area, and therefore less air resistance, than smaller ones. If a cannonball and a marble are thrown at the same speed, the cannonball will go much farther. It was this effect which, for more than a thousand years, obscured the true laws of motion.

Fortunately the very mode of operation of the rocket minimizes the influence of air resistance. Where the atmosphere is densest, at ground level, the rocket is moving at its lowest speed. When it has gained appreciable velocity, the atmosphere is already thinning rapidly. And by the time the rocket has reached its maximum speed, it is in frictionless space.

On the return from space, the atmosphere is almost wholly beneficial; it acts as a 100-mile-deep cushion, absorbing the enormous velocity of re-entry. None of our past feats and future plans for manned space travel would be possible if we had to do all our braking by rocket power alone. Thanks to heat shield and parachute, the final landing on Earth can be achieved without expenditure of energy.

The question "Where does the atmosphere end?" is one that cannot be answered simply—which is rather unfortunate, since it is now a matter of great legal importance. According to international law, most countries claim jurisdiction over vehicles traveling through their "airspace," whatever that may be. Attempts to define it now run to a good many million words.

The atmosphere, in reality if not in law, has no definite end; it slowly thins out into the near (but not perfect) vacuum of interplanetary space. For every 3 miles of altitude, the air density is approximately halved. Men can live and work without artificial aids at heights of 3 to 4 miles if they are given time to adapt themselves. But 5 miles marks the limit of human endurance for sustained periods; Mount Everest (6 miles high) is already beyond that limit. A man can exist there for some time without breathing gear, but he cannot exert himself.

Signposts are always helpful on any road, and the list below is an attempt to establish a few on the road to space. Their positions are only approximate and in some cases debatable. As far as an unprotected man is concerned, even 10 miles up is already "space"; at the other extreme, 100 miles is not high enough if one wishes to establish a satellite in a permanent orbit.

Table 1. The Earth's Atmosphere

Height, Miles	Temperature, Degrees F.	Pressure, Atmospheres	Characteristics
0	−100 to +100	1	Sea level
5	0 to −100	4/10	Limit of unaided human life
8	−70	2/10	Limit with oxygen mask
20	−40	1/100	Limit for aircraft
25	0	1/1,000	Limit for balloons
60	−100	1/1,000,000	Ionosphere
70	+100	1/1,000,000,000	Meteors burn up
100	500 to 1,500	1/1,000,000,000,000	Satellites re-enter; "space" begins legally?
600	1,000 to 3,000	10^{-15}	Upper limit of aurora

NOTE: *For very high altitudes, the figures shown are merely representative; they can vary widely. Thus, at 600 miles, the daily temperature variation can be more than 1,000 degrees!*

The height at which a satellite, moving at orbital speed of 18,000 mph, encounters catastrophically increasing atmospheric drag is almost exactly 100 miles. For many practical purposes, therefore, this may be regarded as the beginning of space. Below this altitude, no pure spacecraft is capable of prolonged free orbital flight.

There are a number of ways, not yet widely exploited, in which the atmosphere may be used to assist departure from the Earth. Balloons have been used as platforms to carry small rockets to great altitudes before ignition, and there have been many studies of schemes for using atmospheric oxygen for the early stages of departure. All these involve great engineering complications and in most cases appear to be more trouble than they are worth; but the time may well come when spacecraft receive a considerable part of their initial boost by jet- or ramjet-propelled lower stages, capable of flying back to their launching sites for re-use after each mission.

One of the first facts that scientists discovered when they started making balloon ascents in the eighteenth century was that it becomes rapidly colder as one goes upward. Indeed, this is obvious to anyone who has ever done any mountaineering. At great heights, though the Sun may be shining in the clear sky, it is always extremely cold, and it is possible to get sunburn and frostbite simultaneously.

It is cold at great altitudes, despite the increased strength of the unhindered sunlight, because the air is too thin to absorb much heat; it can thus no longer act as a thermal bath, warming all bodies immersed in it. The temperature recorded by a thermometer (shielded from the Sun, of course) reaches a minimum of −60 degrees F. at an altitude of about 10 miles; then, surprisingly, it starts to climb again. At 30 miles' altitude it has risen to the freezing point; this zone of relative warmth coincides with the existence of a layer of ozone, very tenuous but vital for the protection of life on Earth, as it blocks the Sun's dangerous ultraviolet rays.

Thereafter the temperature falls again to a second low, and about 60 miles up it reaches a new minimum of −100 degrees F. But now we are approaching the ionosphere, where incoming solar radiation produces intense electrical activity. So the temperature starts to rise again, very rapidly. Soon it is hotter than at sea level; 100 miles up (the frontier of space) the temperature reaches the boiling point and continues to rise rapidly to 1,000 degrees or beyond.

These facts, which were discovered in the 1930's, were gleefully seized upon by some critics of space flight to prove that any vehicle would melt as soon as it left the atmosphere. I well remember one newspaper article that had the sensational title "We Are Prisoners of Fire." Others suggested that shooting rockets into the inferno overhead would cause it to leak downward and burn up the world.

The explanation of this paradox is that at such extreme altitudes, the atmosphere is so thin that the word "temperature" no longer has its conventional meaning. At ground level the molecules of nitrogen and oxygen which compose the air are so tightly jammed together that they travel, on the average, only a few millionths of an inch before they collide with their neighbors. The air thus behaves as a continuous fluid, and a thermometer immersed in it will give a definite reading, just as it would in a bath of water.

One hundred and fifty miles up, however, the situation is entirely different. The molecules have to travel, not millionths of an inch, but something like *one mile* before they encounter each other. Although their individual velocities may be those that correspond to temperatures of thousands of degrees, they are so few and far between that the amount of heat they actually contain is negligible. A thermometer immersed in them would give no meaningful reading at all.

A good analogy of this situation is provided by the common "sparkler" firework. This gives off showers of incandescent sparks so bright that their temperatures appear to be several thousands of degrees. But when they fall on the hand, they produce no sensation whatsoever; they contain so little matter that their heat capacity is negligible. So it is with the air of the ionosphere and beyond.

As far as *out*going spacecraft are concerned, therefore, the atmosphere is not very important; it is little more than the scenery along the road. But that road, of course, winds uphill all the way, because of the inescapable influence of gravity. That is the force which has bound us so long to our native planet and which even now taxes our skill and resources to the utmost when we attempt to leave it.

Gravity may be one of those fundamental, irreducible entities which has no "explanation"; it simply *is*. Despite immense efforts, scientists have made little progress in understanding it, and none in modifying or controlling it. In the seventeenth century Sir Isaac Newton discovered the law of gravitation, which makes apples fall and keeps stars in their courses, but his great law was a description, not an explanation. Though it could predict, with amazing accuracy, the movements of bodies under the influence of gravity, it said nothing about the mechanism, if any, of this universal force.

Almost three hundred years later, Einstein's General Theory of Relativity introduced some subtle modifications to the Newtonian picture. It replaced the idea of a force acting between two bodies with that of curved space—a concept that only mathematicians can grasp and which so far has had not the slightest practical application. For the purposes of space travel, it is as if the General Theory had never been formulated; astronauts will always base their calculations on Newton's law. Any de-

viations from it are so tiny that they will cause about as much concern
as does the curvature of the Earth to an architect when he is planning a
house. . . .

It is a fact of everyday experience that moving upward against the
force of gravity involves work, and therefore a source of energy must be
available. A climbing man obtains this energy from the food he has
eaten; he would be doing well if he climbed one mile on one meal. An
ascending rocket must get the energy it needs for its mission from the
fuel and oxidizer it carries in its tanks. Mountaineer and rocket thus
face similar problems, but they solve them in very different ways.

A climbing man expends energy at a more or less constant rate and
moves at a fairly uniform speed. He can also stop at any point and rest,
without falling back and losing any of the altitude that he has gained.
But a vertically rising rocket cannot do this, since it has no support;
during every second of flight, gravity is inexorably deducting 32 feet per
second from its speed. For this reason spaceships and mountaineers
have to use entirely different strategies to attain their objectives.

Yet the spaceship has one advantage: the force that it is fighting diminishes with increasing altitude, according to the inverse-square law
first enunciated by Newton. For the nonmathematically minded, this
simply means that if you double your distance from the Earth's center,
gravity is reduced to a quarter; increase the distance three times, it is
reduced to one-ninth; ten times, to one-hundredth; and so on. Thus for
small distances the weakening of gravity is very slight, but at great distances it fades away rapidly. Though it never becomes zero, for almost
all practical purposes it may be ignored after a few million miles.

Some actual figures may help to make the picture clearer. One
hundred miles up—where the closest satellites orbit, just before they
re-enter the atmosphere—gravity still has 99 per cent of its sea-level
value. The average man, standing on bathroom scales at the summit of a

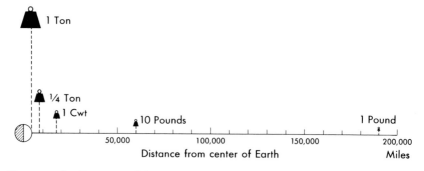

Figure 10-8. *Gravity and distance.*

100-mile-high mountain, would observe that he had lost about two pounds of weight, and would not be aware of the difference. (The astronaut whizzing past him at 18,000 mph at exactly the same altitude, of course, feels no weight at all. The reason for this will be discussed in detail later; for the moment it is necessary only to note that the mountaineer is supported, while the astronaut is in free fall. Anyone who has ever had a chair suddenly jerked away from underneath him will appreciate the distinction.)

At an altitude of 1,000 miles—a quarter of the Earth's radius—gravity is cut to 64 per cent of its sea-level value. To reduce it to one-half, it is necessary to climb to 1,700 miles. These heights are of course very modest in terms of rocket performances, but they show why gravity is a constant, invariant factor in our everyday lives.

Figure 10-8 expresses the same facts in graphical form, with the Earth drawn to scale. Note that at the Moon's distance of 240,000 miles the force of gravity is almost too small to be indicated; nevertheless, it is still powerful enough to keep the enormous mass of our solitary natural satellite firmly chained in its orbit.

The steady falling off of gravity with distance from the Earth gives us a mental picture, or model, from which a great deal can be learned. Climbing out of the Earth's gravity field is rather like ascending a slope which is at first very steep but which slowly flattens out until at last it becomes almost horizontal. Thus the early stages of the ascent are extremely difficult and require the expenditure of a great deal of energy, whereas the final ones require practically no energy at all.

We can set a precise numerical value to the height of this imaginary gravitational hill, up which we must climb in order to escape from the Earth. When we calculate the *total* amount of work which has to be done to leave our planet completely, we obtain a surprisingly simple and mathematically beautiful result. It turns out that lifting a body right away from the Earth, and out into the depths of space, requires exactly the same amount of work as lifting it *one* Earth radius—4,000 miles— against a constant gravity pull *equal to that at sea level*. So if you desire a mental picture of the energy needed to escape from the Earth, imagine climbing a mountain 4,000 miles high, assuming that there is no falling off in gravity on the way to the top. Alternatively, imagine climbing a 4-mile-high mountain 1,000 times: it comes to exactly the same thing.

To make this mental image more realistic, and more useful, it is best to turn the mountain upside down—to convert it into a pit or crater. From the gravitational point of view, we dwellers of the Earth's surface are in the position of people at the bottom of a gigantic funnel, 4,000 miles deep. To escape, we have somehow to climb the steeply sloping walls; near the bottom they are almost vertical, but eventually they flatten out into an endless plain. This horizontal plain represents gravitationless space, across which we can travel for immense distances with very little expenditure of energy—once we have reached it.

The above calculation is one case of a completely general law, applying to every body in the universe. To escape from any star, planet, or moon demands as much work as moving vertically through the radius of that body, against a gravity field equal to that at its surface. Let us anticipate a little and see what this implies in the case of our nearest neighbor, the Moon.

We have already said that its surface gravity is one-sixth of Earth's; its radius is very nearly one-fourth of our planet's. Hence the work required to escape from it is only $\frac{1}{6} \times \frac{1}{4}$, or $\frac{1}{24}$th, of that needed to leave Earth. Any lunarians are thus at the bottom of a gravitational crater a mere 170 miles, not 4,000 miles, deep. This shows how very much easier it is to escape from the Moon than from the Earth.

Figure 10-9 shows these two imaginary gravitational craters (the Moon's is so small that it has been necessary to exaggerate it). Remember that this picture is only a mathematical model, having no more physical reality than the isobars familiar to millions on the television weather charts or the contour lines on a map. But a study of it can give a very clear impression of the energy requirements for the Earth-Moon journey, and we shall return to it again when we study this in more de-

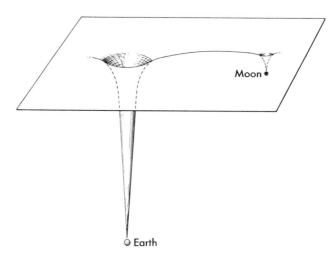

Figure 10-9. *The gravitational fields of earth and moon.*

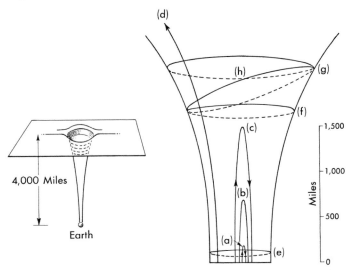

Figure 10-10. *Orbits in the earth's gravitational crater.*

tail. For implicit in this simple model is the entire theory of flight from world to world; we can use it to visualize the orbits of spaceships and space probes not only around the Earth but also, as we shall see later, around the Sun itself.

OTHER ORBITS, OTHER MOONS

During the first decade of the Space Age, the Earth acquired almost a thousand new moons—some very small and temporary, others weighing many tons and with lifetimes comparable to that of our natural Moon. Virtually overnight, what had been one of the most impractical and theoretical fields of mathematics—celestial mechanics—became a branch of engineering affecting the prestige and destiny of nations.

Celestial mechanics is horribly complex; many pure mathematicians would also consider it horribly ugly, despite the grandeur of its subject matter. Even some of its simplest problems have remained unsolved after centuries of effort, though answers can always be obtained to any desired degree of accuracy by brute-force methods using giant electronic computers. These machines have revolutionized the subject; it is not too much to say that space research would be as impossible without them as without the rocket itself.

Luckily, all the important basic ideas in this area can be grasped without any knowledge of mathematics at all, by the use of the model described in the last chapter—the 4,000-mile-deep crater which is the analogue of the Earth's gravitational field. The lower slopes of this model are shown in Figure 10-10; if we imagine that it is made of some smooth, hard material such as perfectly frictionless glass, we can use it to demonstrate the movements of space vehicles in the neighborhood of the Earth. All that we have to do to reproduce them for any initial conditions of velocity and direction is to see what happens to an object when it is projected along the slope, like a marble flicked inside a wineglass.

The case of a vertical launching is the simplest; obviously, the greater the initial speed, the higher the object will rise before gravity reduces its velocity to zero. Then it will fall back, gaining speed, until it returns to the starting point—at its initial velocity.

Although this example may seem rather trivial, we can learn several very important points from it. The first is that velocity can never be lost in space; it can be exchanged for altitude, but it can always be exchanged back again. In general, no matter what path or orbit a body takes around a planet, when it comes back to the same altitude (or distance) it will always be moving at the same speed (though not necessarily in the same direction; in this case the speed has been reversed, but the *energy* is the same).

It is also clear that as the velocity of projection increases, the altitude reached increases *even more rapidly* because of the steadily flattening slope. Figure 10-10 shows, to scale, the heights reached by bodies launched away from the Earth at speeds of (a) 5,000, (b) 10,000, and (c) 15,000 mph.

It is obvious that as the speed of projection increases still further, there will be a certain critical velocity at which the body will never fall back. Though it will lose most of its initial speed during the ascent, it will still have some velocity left when it reaches the "rim" of the crater, and so it will continue to move on outward forever. The speed that is just sufficient to climb out of a gravitational field is known as the velocity of escape; for the Earth its value is about 25,000 mph, or 7 miles per second.

If a body starts off with more than this critical speed, it will still have something in hand when it leaves the gravitational pit. However, the excess velocity cannot be obtained by simple subtraction, because we are really dealing with *energies*, not merely velocities. A body starting from the bottom of the crater at a speed of 7 miles per second reaches the top at zero velocity; but one beginning at 8 mps emerges at considerably

more than the one mps that might be expected. The rather curious results that follow are shown in this table:

Initial Velocity, MPS	Final Velocity, MPS
7	0
8	4
9	5½
10	7

This matter is not very important as far as the present argument is concerned, but in the case of actual missions it has a great effect on fuel requirements and flight times.

Now consider the case of a body projected not vertically up the
gravitational slope, but horizontally—at right angles to it. If its speed is
adjusted properly, it can remain orbiting at a constant altitude, like a
motorcyclist in the "Wall of Death" popular in circus sideshows.
(Though this analogy is a good one, it is not quite accurate, because
friction operates here, and the rider has to keep his engine running to
counteract it.)

This is the now-familiar case of a satellite in a circular orbit, at a constant distance from the Earth. It is obvious that the higher the satellite,
the more slowly it needs to move to preserve its position. A very close
satellite requires an orbital speed of 18,000 mph (5 mps), whereas a
distant one like the Moon need move at only about 3,600 mph (1 mps).

It also follows that as altitude or distance increases, the time to complete one orbit increases at an even greater rate, for not only is there
more distance to be covered, but the speed in orbit is less. This fact is
expressed in Kepler's famous third law—"the square of the time for one
revolution increases as the cube of the radius"—which was the clue that
led Newton to his law of gravitation.

We have by no means exhausted the possibilities of our model, for let
us now consider the case of an object projected along its surface at some
arbitrary speed or inclination. If it does not have enough velocity to
maintain itself at the height where it enters into the system, it will drop
down the slope, but as it does so it will gain speed, like any falling
body. When it has reached its lowest point—and its greatest velocity—
it will start to rise again, continually retracing the same inclined curve.
This, of course, is the analogue of a satellite in an elliptical orbit.

In Figure 10-11 the elliptical orbit has been drawn so that it touches
two circular ones; it can lie anywhere, but this has been done to illustrate another idea—transfer from one orbit to another. It will be real-

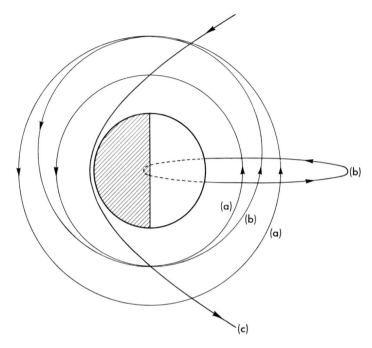

Figure 10-11. *Orbits around the earth:* (*a*) *circular,* (*b*) *elliptical,* (*c*) *parabolic.*

ized that at the upper point of contact, an object in the elliptical orbit is moving more slowly than one in the circular orbit here; this is why it falls back to the lower level again. If it is to remain in the higher orbit —make a rendezvous—it must be given an additional impulse.

Conversely, at the lower point of contact the body in the transfer orbit is moving too fast for a rendezvous; it must therefore receive an impulse to slow it down if its orbit is to be "circularized."

Before we leave, for the moment, this highly instructive model, there is just one more case to be considered, that of a body entering the gravitational field from a great distance and at a considerable speed.

This is the exact reverse of the "escape-velocity" case; the object will gain speed as it slides down the wall of the crater. It will gain so much, in fact, that it will whip completely around the bottom of the crater and rise out of it, to disappear once more toward infinity—eventually regaining all its original speed, though it will be heading in some different direction.

There are fun-fair operators, doubtless unversed in celestial mechanics, who have learned to profit by this example. They display attractive vases, which can be won by anybody who can toss a ball into them. It looks easy, because the openings of the vessels are quite wide; but the trick is almost impossible, because any ball that does get into the smooth interior promptly emerges again with barely diminished speed. The astronomical analogue, of which there are several examples every year, is the comet which enters the solar system from the depths of space, does a hairpin turn around the Sun, and then heads out once more toward the stars.

A careful study of the perhaps rather unlikely model in Figure 10-11 will, therefore, give a very good idea of all the possible trajectories and orbits of a space probe or satellite moving in the gravitational field of the Earth. More than that, it may be generalized for *all* celestial bodies —the Moon, the Sun, or any of the planets. Only the numerical values are different; thus, for the Moon, escape velocity is only 1½ miles per second (compared with 7 mps for Earth). For the Sun, however, it is an enormous 400 miles per second; some idea of the forces raging there can be gathered from the fact that solar eruptions frequently exceed this speed, so that matter is continually escaping from the Sun. It must be remembered that Figure 10-11 is a *model,* not a map; it shows the characteristics of possible orbits, not their actual shapes in space.

It will be seen that there are two possible classes of orbits, closed and open. The closed ones are circles and ellipses; they repeat themselves indefinitely. All real orbits are in fact ellipses; the circle is the theoretical, limiting case of the ellipse with zero eccentricity—a state of perfection which does not exist in nature, though it has been approached by some artificial satellites. Venus has the most perfectly circular orbit, with an eccentricity of 0.0068. A synchronous satellite like Early Bird has an orbital eccentricity of only 0.0005—ten times better.

The open orbits, which never repeat themselves but lead off to infinity, are hyperbolas or parabolas. Like the circle, the parabola is a limiting case which exists only in theory. It is the orbit of a body which has exactly *enough* velocity to escape—not one micron per millennium more or less. So in practice, all escape orbits are really hyperbolas.

❊   ❊   ❊

The critical speed of 5 miles a second [is] necessary to establish an orbit, and is quite independent of the size or mass of the satellite concerned; it applies equally to the 100-ton-payload of Saturn 5 and the barely visible wire hairs launched by the millions in the notorious "Needles" experiment. If Earth had a natural moon just outside the atmosphere, it would have to orbit at this speed; and since at 18,000 mph it takes one and a half hours to circumnavigate the globe, this would be the duration of the "month." A month that was shorter than the day sounds an odd phenomenon; but stranger things happen on other planets.

At greater distances from the Earth, less speed is necessary to counter the weakening gravitational pull, so the period of revolution steadily increases; 1,075 miles up, it is exactly two hours, or one-twelfth of a day. For many purposes, such as regular tracking from ground stations, it is convenient to have satellites "geared" to exact ratios of the Earth's rotation; the twelve-hour orbit is particularly useful. But the most valuable of all is the twenty-four-hour, or synchronous, orbit, which permits a satellite to hover apparently motionless over one spot on the globe. The idea of a body hanging fixed in the sky seems more than a little uncanny, but of course such a "geostationary" satellite is not really motionless. It is merely turning through space at the same rate as the Earth; and in order to do that at a distance of 26,000 miles from the center of

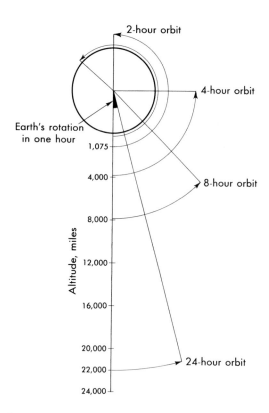

Figure 10-12. *Angular speeds of satellites.*

rotation, it has to move along its orbit at almost 7,000 mph—very far from standing still. Figure 10-12, which is drawn to scale, shows the altitudes at which these various phenomena occur.

Although it is simplest to talk about circular orbits, all real ones are elliptical. In some cases, the ellipticity—or eccentricity, to use the correct term—can be very high, a satellite coming to within a few hundred miles of the Earth and then rising tens of thousands of miles out into space at its far point. Oddly enough, the eccentricity does not affect the period of a satellite; that is determined only by the length of its longest axis. Elliptical orbits are sometimes referred to as "egg-shaped," but this is incorrect, since eggs are—usually—more pointed at one end than the other. An ellipse is perfectly symmetrical about both axes.

The dimensions of an orbit and its eccentricity are its two most important characteristics; but they do not define it completely, for it can be tilted at any angle to the Earth's axis. From the practical point of view, it is easiest to launch a satellite in the equatorial plane, because when it takes off it can get the maximum boost from the 1,000 mph of spin available there. Unfortunately, equatorial sites tend to be politically unstable, though in 1967 the Italians neatly avoided this in their San Marco project by launching from an oil-drilling rig in the Indian Ocean. (For *very* large spacecraft, ocean launches may one day be mandatory.)

At the other extreme from the equatorial orbit is the polar one, which is slightly more difficult to achieve, since it cannot take advantage of the Earth's spin. This handicap is more than offset by the fact that a polar satellite, in the course of a few revolutions, can observe the entire surface of the globe, whereas an equatorial one is limited to low latitudes.

It is also perfectly possible to launch a satellite *against* the direction of the Earth's spin, so that it has what is known as a retrograde orbit. If it were exactly retrograde—moving from east to west, as the natural celestial bodies appear to do—it would require an extra 2,000 miles an hour of launching speed to overcome the effects of the Earth's rotation, and there would be little point in such a fuel-wasting procedure. But satellites that are very slightly retrograde—a few degrees "backward" from the orbit over the poles—do have some valuable properties (see p. 556).

The ground track of a satellite—the path it traces over the surface of the Earth—may be almost as important as its orbit. It determines the regions from which the satellite can be observed by ground stations, either electronically or optically. It also determines the areas of Earth which the satellite itself can observe and the frequency with which it can view them, obviously a deciding factor for meteorological or reconnaissance satellites.

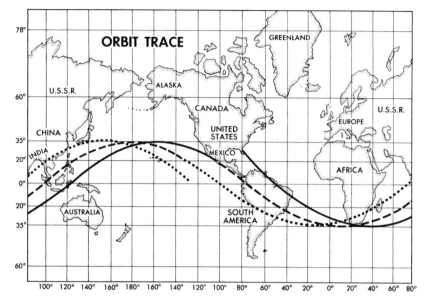

Figure 10-13. *Ground track of low-inclination orbit.*

The very simplest case is that of the equatorial satellite; it remains always above the equator and retraces the same path forever. It will go over the same spot once in every orbit, at regular intervals determined by its period of rotation. The only exception is the stationary satellite, which always stays over one spot and may thus be regarded as having an infinite period.

A satellite in a polar—90-degree—orbit, on the other hand, will weave a pattern embracing the entire Earth as it shuttles from north to south, and the planet turns beneath it. A similar kind of basketwork pattern, but spanning a narrower band of latitude, is traced out by satellites launched in inclined orbits. The typical Cape Kennedy ground track, made familiar to millions by the Mercury and Gemini flights, is that shown in Figure 10-13.

Satellites with unusual periods and angles of inclination can produce the most extraordinary ground tracks, forming loops, apparently going backward, and so on. All these cases can be worked out with the aid of a globe, as long as one remembers that the plane of the orbit stays fixed in space, while the Earth revolves at a uniform speed inside it. Highly instructive models can be purchased from educational stores to illustrate all these cases, so no attempt will be made to describe them here.

The stability of an orbit is obviously a matter of the greatest practical importance, especially for expensive satellites, which must operate for many years to pay for themselves—either scientifically or commercially. However, there are two kinds of stability to be considered. The first involves a satellite's lifetime—how long it will remain aloft before it re-enters the atmosphere and is burned up (or recovered). The second and more subtle point is: how long will the satellite remain in its *original* orbit?

As far as lifetime is concerned, the simple answer is that a satellite will stay up forever, as long as no part of its orbit ever enters the atmosphere. But if the lowest point of the orbit (the perigee) is close enough to Earth for there to be any appreciable air resistance, the satellite is bound to come down sooner or later. Every time it slices into the atmosphere, it loses a little of its energy, so it does not rise quite so far out into space the next time around. The far point (apogee) therefore steadily descends, coming closer and closer with each revolution—though perigee remains almost unaffected. To put it expressively though not quite accurately, the satellite spirals in toward Earth.

At last, the orbit becomes perfectly circular; apogee and perigee are identical. The orbit is now wholly inside the atmosphere; resistance is acting over the entire path, and the satellite has only a few hours of life remaining to it. Unless it is a heavily protected re-entry vehicle, designed to survive the thousands of degrees of heat produced by boring through 5 miles of gas in every second, it will burn up in a spectacular display of artificial meteors.

The period of a satellite in the closest possible orbit, just before the final catastrophe, is almost precisely 90 minutes (the very last orbit is completed in about 84 minutes). This means that, by pure coincidence, during the last few days of its life every satellite makes *exactly* 18 revolutions in every 24 hours, and so retraces the same path over the surface of the turning Earth, day after day.

So, in the spring of 1958, millions of people throughout Europe were able to watch, at almost the same time every night, and moving along the same track through the constellations, the brilliant star of Sputnik 2 carrying the corpse of the dog Laika. This nightly death-watch for a dying satellite may well be repeated, with far deeper emotions, in the years ahead, as some future space mission terminates in a celestial funeral pyre.

This fate can never befall a satellite whose perigee is more than 1,000 miles from Earth—still very close indeed, as cosmic distances go. But though an orbit may be stable, it is not necessarily permanent; there are forces at work which may slowly change it.

Among these are the gravitational attractions of the Sun, Moon, and planets, engaged in an endless tug-of-war. Their influence, however, is very small, at least for satellites close to the Earth. But one "perturbation" which is not small is that due to the Earth itself.

If our planet were a perfect sphere, with a uniform distribution of matter inside it, a satellite would always repeat the same orbit. But in the real case, the Earth has a pronounced equatorial bulge, as well as other less conspicuous dents and bumps. The polar flattening produces a most important effect known as precession, which is well demonstrated in the case of a spinning top.

When a top loses its speed and begins to fall over, the downward pull of gravity has a paradoxical effect on its behavior. The axis of the top, which until now has been fixed vertically in space, starts to trace out a conical path. Anyone who has ever played with a toy gyroscope and has noticed how it appears to move at right angles to the direction in which a force is applied has observed the phenomenon of precession in its clearest form.

A satellite whirling around the Earth is in effect an enormous gyroscope, several thousand miles in radius, and the plane of its orbit tends to remain fixed in space. This indeed happens, when the orbit is directly over the equator, and its axis coincides with the Earth's. But when the orbital plane is tilted, the attraction of the Earth's equatorial bulge can then come into play, and the orbital plane begins to twist, so that after a few thousand, or a few million, revolutions it may have precessed around a complete circle.

By selecting the right inclination, one can choose any rate of precession desired. This has been used to advantage, in the case of some meteorological and reconnaissance satellites, to produce an orbit whose plane makes one revolution every year. Such an orbit is called "sun synchronous"; it exactly cancels out the Earth's annual rotation around the Sun, and a satellite moving in it passes over the same spot on the Earth at the same time every day. The United States Air Force's Samos reconnaissance satellites have this useful characteristic, so that they can rephotograph the same areas under identical illumination. To do this, they have to be launched into a slightly retrograde orbit, tilted about six degrees backward from the axis of the Earth. Who would have thought, even a decade ago, that the intricacies of celestial mechanics would one day be of military importance?

Magnetic and electrical effects in space can also produce minor effects upon satellite orbits; more surprisingly, so can the pressure of sunlight . . . Feeble though it is, as it acts continuously it can produce large effects on satellites of low density, like the Echo balloons. These huge but flimsy structures have been "blown" hundreds of miles out of their original orbits by the pressure of solar radiation.

To sum up, then, it is possible to establish satellites in orbit around the Earth at almost any distance, eccentricity, and angle of inclination. . . .

THE PRICE OF SPEED

We have seen earlier that the speed required for even the simplest and easiest space mission—orbiting the Earth—is about 5 miles a second, or 18,000 miles an hour. To escape completely requires 7 miles a second (25,000 mph), but once this critical speed is attained, a whole range of possibilities opens up, as shown by the table below.

Ignoring for the present the rather odd fact that it is twice as difficult (in terms of velocity) to reach the Sun as the nearest star, this shows that a very slight extra speed over the mimimal escape velocity brings the closer planets within range, as has already been demonstrated by the various Mars and Venus probes.

Tsiolkovsky used the phrase "first cosmic speed" for the orbital velocity of 5 mps, and "second cosmic speed" for the escape velocity of 7 mps; these expressions are still employed in contemporary Russian space writings. Even the lower velocity seemed so wildly beyond hope of attainment at a time when airplanes could barely reach 100 mph that one can hardly blame the early critics of astronautics for their skepticism.

Table 2. Mission Launch Speeds

	Launch Speed	
Mission	*MPH*	*Miles / Sec.*
Close Earth orbit	18,000	5
Escape from Earth	25,000	7
Voyage to Mars or Venus	26,000	7
Voyage to Jupiter	32,000	9
Voyage to Pluto	35,000	10
Voyage to nearest star	37,000	10
Voyage to Sun	70,000	20

And yet Tsiolkovsky and his successors had shown in complete theoretical detail how such speeds might be attained by means of rockets, provided that certain engineering problems could be overcome. It was all a question of getting a sufficiently high exhaust velocity and an efficient enough structure. These were the two vital factors; everything else was secondary.

Let us go back to the brick-carrying sled used to demonstrate the rocket principle, and consider how its performance—that is, its final velocity after all its "propellant" has been expended—depends on these two parameters.

Common sense, without any mathematical aids, tells us that the final velocity will be directly proportional to the speed of ejection of the bricks. If the bricks are thrown out at 20 mph and the sled reaches a final speed of 1 mph, then it will reach 2 mph if the ejection speed is increased to 40 mph, assuming exactly the same number of bricks are thrown out as before.

Thus the exhaust speed of a rocket is its most important characteristic, for its final speed, at "all burnt," is directionally proportional to this. Table 3 gives some values for a few representative propellants.

Table 3. Rocket Propellants

Propellant	Exhaust Velocity, MPH
Black powder (firework)	700
Modern solid propellant	3,000 to 5,500
Alcohol-oxygen (V-2)	6,200°
Kerosene-oxygen (Atlas, Saturn)	6,500°
Hydrogen-oxygen (Centaur, Saturn)	8,500°
Hydrogen-fluorine	9,000°
Hydrogen (nuclear rocket)	20,000 and up
Ions (electric rocket)	40,000 and up

° *Sea-level values. Exhaust velocities in vacuum could be 10 to 15 per cent higher.*

When these figures are compared with the mission requirements shown in Table 2, it will be seen at once that the old-style powder rocket is pitiably inadequate. But even the modern, liquid-propellant rockets have exhaust speeds which are only a fraction of the "first cosmic velocity." To perform any space mission, therefore, we must build rockets capable of traveling *several times as fast as their exhaust speed.*

At first sight, it may seem impossible for a rocket to attain a speed greater than that of the jet that propels it; I have known able mathematicians who intuitively dismissed the idea. But it must be remembered that the jet exhaust always leaves the rocket at the same speed, *whatever* the velocity of the rocket itself may be relative to some arbitrary external point. As long as there is any fuel aboard, the jet will continue to give the same thrust, and the rocket will continue to accelerate.

To return to the analogy of the sled and its load of bricks, each brick gives the same impetus to the vehicle, whether the sled is standing still or moving at 100 miles an hour across the ice. Since the fuel is carried along with the vehicle and shares its speed at any time, the sled's velocity cannot affect its performance.

The fact that rocket exhaust speeds are considerably less than those needed for space missions does not, therefore, make them impossible; it merely makes them difficult. We can see how difficult if we look again at the man on the sled and ask ourselves what amount of propellant he would have to throw off in order for his vehicle to reach "exhaust speed," that is, the speed with which he is throwing out the bricks.

It is easy to see what the *minimum* weight must be. If all the propellant could be ejected simultaneously, in one explosive effort, and if its weight equaled that of the empty sled, then the velocities would also be matched. After the Big Bang, we would have two equal masses moving in opposite directions, with equal speeds (Figure 10-14a).

In this case, the initial mass of the system (vehicle plus propellant) would be twice that of the empty, or final, mass. It would be said to have a "mass ratio" of 2. Such a ratio presents no engineering difficulties, though it is a good deal higher than usual for surface vehicles; the average automobile has a mass ratio of about 1.03, since only about 3 per cent of its weight is fuel. It is attained by some aircraft, which can carry their own empty weight in fuel.

However, the explosive, or "instant-burning," case we have described is not applicable to the rocket, where combustion takes place over a period of time which may last for several minutes. And certainly the man

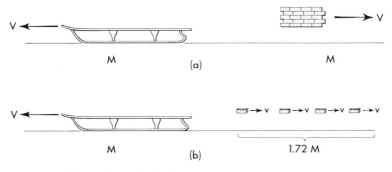

Figure 10-14. *The rocket-velocity law.*

on the sled would require a considerable time to throw out a mass of bricks more than equal to his own weight!

This alters the situation a good deal, reducing the efficiency of the system. Because all the propellant is *not* ejected at once, work has to be done to accelerate the unused material, up to the moment until it is finally discarded. This means that more propellant has to be carried—and more propellant has to be carried to accelerate *that*, in an infinite but fortunately diminishing series. The very last brick on the sled has to be carried to the bitter end; when it is finally jettisoned, it has almost reached the velocity of the payload; yet all the work done to accelerate it to that point is a complete waste, though an unavoidable one.

It is straightforward, though tedious, to calculate the additional mass of bricks now needed to bring the vehicle up to "exhaust speed" without using any higher mathematics. (Anyone who likes to try may assume that the propellant mass is split into first 1, then 2, then 4, 8, 16, etc., bricks. As the individual units get smaller and smaller, he will see that the answer converges to a limiting value.) In the case of a real rocket, where there is a continuous flow of material, the calculus has to be used, and it can be easily shown that in order for the vehicle to reach the speed of its exhaust, the mass ratio must be increased from 2 to the somewhat higher value of 2.72. Thus the vehicle has to eject 1.72 times its empty weight of propellant (Figure 10-14b). The 0.72 is the penalty we have to pay for carrying along part of the fuel until it is needed; it might be much worse.

The now-primitive V-2, it is interesting to note, had a mass ratio considerably higher than 2.72. Its loaded weight was 28,000 pounds, its empty weight 8,500 pounds, and the ratio of these two figures is 3.3. In theory, therefore, a V-2 could travel faster than its exhaust (5,000 mph); that it actually achieved only 3,600 mph was due to air resistance and gravity losses. It could have attained its theoretical performance in the vacuum of space.

Now let us be more ambitious. What load of bricks has the sled to carry, if its final speed is to equal *twice* that of its "exhaust"?

It turns out that we have to square the mass ratio, thus increasing it from 2.72 to 2.72^2, or 7.4. In other words, the sled has to carry 6.4 times its own empty weight in propellant.

Similarly, for three times the exhaust speed, the mass ratio has to be cubed, giving a value of almost exactly twenty, and so on. There is no theoretical limit to the process, but clearly the practical difficulties are increasing very rapidly. Is it possible to construct a vehicle strong enough to stand the accelerations of flight whose empty weight—including payload!—is only one-twentieth of its weight when loaded

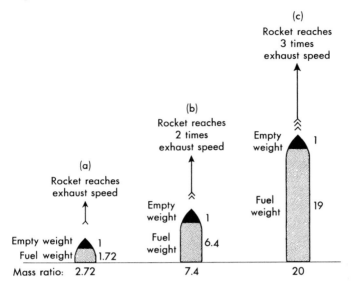

Figure 10-15. *Rocket mass ratios.*

with propellants? But this is what has to be done if we are to build a rocket which can fly three times as fast as its own exhaust.

These results are shown in diagrammatic form in Figure 10-15. In each case the empty weight of the rocket is assumed to be the same—say, one ton. That one ton, remember, must cover the weight of the propellant tanks, the rocket engine, the control system, the payload—*everything*.

The old V-2, as we have seen, is slightly better than case (a). Today's best liquid-propelled rockets can surpass (b), and there are some solid-propellant rockets that can even match case (c). The makers claim that their mass ratio of twenty beats that of nature's most efficient container, the egg. However, this figure applies to the rocket motor only, and the complete vehicle would bring us back to something poorer than case (b) again. We can conclude, therefore, that it is not practical to build a rocket with a final speed more than twice that of its exhaust. (There may be exceptions for vehicles built to operate exclusively in space, where very light structural materials and novel techniques can be employed.)

The greatest exhaust speed for conventional propellants, listed in Table 3, is 8,500 mph. Since orbital velocity is 18,000 mph, it appears impossible to build a rocket, using hydrogen and oxygen, to become a satellite of Earth.

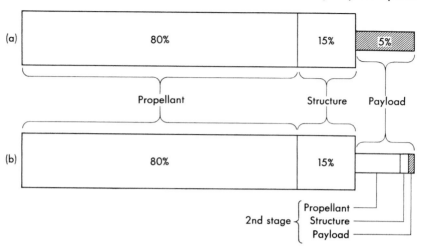

Figure 10-16. *Weight breakdowns of one- and two-stage rockets.*

The way out of the dilemma is the simple, effective, but expensive device of the step, or multistage, rocket. All the calculations given above refer to *single*-stage rockets, where the structure, dead weight, etc., which begin a mission are used right through to the end. But it is obvious that if we make the payload of our rocket *another* complete rocket, which starts to operate only when the first stage has exhausted its fuel and has been dropped off, we can achieve a much higher final speed. In fact, if the two rockets have the same propellants and the same mass ratio and are identical except in size, the final speed will be doubled.

How the step rocket works (and why it is expensive) is shown in Figure 10-16.

Let us suppose that one can design a rocket with a weight breakdown of 80 per cent propellant, 15 per cent structure, and 5 per cent payload, as shown in Figure 10-16a. (It is possible to do a good deal better than this, but these values have been chosen to give round numbers.) This being a single-stage rocket, its 5 per cent payload might be able to attain a theoretical final speed of 10,000 mph. In practice, air resistance and gravity could reduce this to about 8,000 mph. Even if the *whole* of the payload were replaced by propellant, the mass ratio would show only a slight improvement—in this case, from 5 to 6.7. The resulting increase in speed—at the cost of zero payload!—would only be 2,000 mph.

If, however, we take the payload and make it a second rocket, with the same weight breakdown as the first (Figure 10-16b), then this stage will have the same performance. But it would start where the first had left-off, both in velocity and in altitude. It could add another 8,000 mph to the speed it already possessed, giving a grand total of 16,000 mph.

In actual practice it would do a good deal better than this. Since the second stage would not begin firing until it was scores of miles above the ground, it would lose very little velocity through air resistance, and its engine would operate at maximum efficiency. Moreover, it would no longer be ascending vertically; it would have started to curve over toward the horizontal. This means that gravity drag (which acts in the downward direction) would be less effective in reducing its speed.

For all these reasons, the second stage would approach its theoretical performance of 10,000 mph, so that its payload would achieve 18,000 mph, or orbital velocity.

The reason why a step rocket can travel so much faster than a single-stage rocket, whatever its size, is not hard to see. In a single-stage rocket, as the propellants become used up we have the situation where a now unnecessarily powerful, and therefore excessively heavy, engine is trying to accelerate a lot of useless dead weight. The big propellant tanks, for example, still have to be carried along even when they are virtually empty, so that in a sense the overall efficiency of a rocket steadily decreases as its propellants are consumed. In the last seconds of firing time the engine is wasting most of its effort imparting velocity to structural mass which is no longer serving any useful purpose. The only way to improve the situation is to throw away the empty rocket and to start again with a new, scaled-down one.

There is no limit to the number of stages that may be employed in a step rocket or, therefore, to the speed which may be attained by the final stage when the earlier ones have been discarded. The practical disadvantage of the step principle, of course, is that after two or three stages the ultimate payload is an extremely small fraction of the initial takeoff weight. In Figure 10-16 the payload of the single-stage rocket is 5 per cent, but that of the two-stage rocket is only 5 per cent of 5 per cent, or a mere 1/4 per cent. For multistage vehicles, such as the Saturn 5 designed for the lunar mission, the percentage of payload is even less; but that is the penalty that has to be paid for achieving high velocities.

It is sometimes asked, "Why do we need such high speeds for space flight? Could not a rocket leave the Earth at a fairly low, steady velocity, running its engines at some modest thrust level, rather than going all-out to reach the escape velocity of 25,000 mph as quickly as possible?"

Yes, it could—if it had a virtually infinite source of energy. The slower the rate of climb in the Earth's gravitational field, the more wasteful is the expenditure of fuel. This will be obvious if we look at two extreme cases.

Suppose the rocket burns all its fuel instantly, so that it acquires escape speed while it is still at ground level. (We know, of course, that air resistance, as well as engineering factors, would make this impossible in practice. But there are some anti-ballistic-missile missiles—ABM's—that do have incredible accelerations even in the lower atmosphere and will serve to illustrate the principle.)

In this theoretical case, *none* of the fuel has to be lifted against the earth's gravitational field; it therefore imparts all its energy to the rocket and wastes none lifting itself. It all stays near the ground—but the rocket escapes from the Earth. We are, virtually, dealing with the case of a gun-launched vehicle, and the propellant is used at maximum efficiency. To look at it in another way, because the whole process of reaching escape speed takes zero time, gravity—which normally reduces the speed of any vertically climbing missile by 20 mph every second—has no time to act.

Now consider the other extreme, the case of a rocket which takes off so slowly that it merely hovers at a fixed altitude. It burns its entire load of propellant merely balancing itself against gravity, and gets nowhere.[2]

Clearly, the nearer we can get to the first case, the more efficient the operation and the smaller the total amount of propellant needed. Escaping from Earth is difficult enough as it is, without aggravating the problem by making unnecessary concessions to gravity. We have to climb out of the gravitational crater as quickly as possible; the more time we spend lingering on its lower slopes, the more we shall slip back toward the bottom.

The above argument also throws light on a fallacy often advanced by critics of space flight in the early days; sometimes trained mathematicians, who should have known better, fell headlong into the trap. Here is a splendid specimen from a speech by one Professor Bickerton, delivered to the British Association for the Advancement of Science in 1926:

This foolish idea of shooting at the moon is an example of the absurd length to which vicious specialisation will carry scientists working in thought-tight compartments. Let us critically examine the proposal. For a projectile entirely to escape the gravitation of the earth, it needs a velocity of 7 miles a second. The thermal energy of a gram at this speed is 15,180 calories. . . . The energy of our most violent explosive—nitro-glycerine—is less than 1,500 calories per gram. Consequently, even had the explosive nothing to carry, it has only one-

2. General Dornberger has given a vivid eyewitness account of this happening at Peenemünde; see *The Coming of the Space Age* (Des Moines: Meredith Press, 1967).

tenth of the energy necessary to escape the earth. . . . Hence the proposition appears basically impossible. . . .

In this relatively short passage Professor Bickerton managed to compress two major errors. One would have thought it obvious that an *explosive* was the last substance suitable for a rocket propellant; in any event, nitroglycerin contains considerably less energy than equal weights of typical propellants like kerosene and liquid oxygen. This fact of elementary chemistry had been carefully pointed out years before by Tsiolkovsky and Goddard.

Bickerton's second error is the "energy fallacy" in its purest form. What does it matter if the nitroglycerin (or other propellant) contains only a fraction of the energy necessary to lift *itself* away from the Earth? It never has to do so.

What it has to do is to impart that energy to a suitable payload, and (if it were not for air resistance) that could all be done at ground level. Thus even Bickerton's own argument merely proves that at least ten pounds of nitroglycerin would be required to send one pound to the Moon. For actual space vehicles, most of the propellant is burned within a hundred miles of Earth, and so lifts itself only a fraction of the way out of the Earth's gravitational field. When Luna 2 impacted on the Moon just thirty-three years after Professor Bickerton proved it was impossible, its several hundred tons of kerosene and liquid oxygen never got very far from the Soviet Union—but the half-ton payload reached the Mare Imbrium.

The passage I have quoted is also worth studying for another reason. It demonstrates how men who should be scientifically trained can let prejudices and preconceived beliefs distort their logic, so that they commit almost childish errors when attempting to prove their points. Space flight, and aviation before it, attracted a lot of this nonsense; and even today . . . there are circles in which it is a popular pastime to "prove" that though interplanetary flight is perfectly feasible, we shall never, *never* be able to reach the stars.

MOONRISE IN THE WEST

The use of rockets for high-altitude research—the dream which Goddard had pursued but had never realized in his own lifetime—began immediately after World War II. Many of the V-2's captured by the United States Army were launched from the White Sands Proving Ground, New Mexico, with their warheads replaced by instruments which would radio their observations back to ground stations. In addition, smaller rockets, such as the Aerobee and the Viking, were developed purely for scientific purposes.

This work was very modestly funded and would probably have led to the achievement of manned space travel around the middle of the twenty-first century. Fortunately or unfortunately, depending upon one's point of view, the main impetus for rocket development was being provided not by man's quest for knowledge but by his instinct for survival. As in the case of the V-2, the military were quietly providing the real money.

Although the United States possessed, in the Peenemünde team, the most experienced rocket designers in the world, it showed no inclination to use them. Its monopoly of the atom bomb (confidently expected to last for many years) and its fleet of B-29 Superfortresses made such futuristic weapons as long-range ballistic missiles appear unnecessary. Moreover, theoretical studies showed that it would require rockets weighing several hundred tons to deliver nuclear warheads over useful distances. No competent engineer doubted that such vehicles could eventually be developed, but the cost would be enormous and there seemed no justification for a high-priority program.

The Soviet Union—and specifically Joseph Stalin—thought otherwise. From their point of view, intercontinental missiles made excellent sense. They also had a long tradition of interest in rockets, going back to Tsiolkovsky, and until the mid-1930's their engineers had probably led the world in this field—though this fact was not generally known or believed outside the Soviet Union.

In addition, they had now acquired the priceless background of German wartime research, a great deal of hardware (including Peenemünde and a complete V-2 factory), the very few top-ranking scientists and engineers who had not thrown in their lot with Dr. von Braun's team, and more than a thousand technicians. This was a prize of no small value, but to imagine that it was responsible for establishing the Soviet lead in space is absurd. The United States had the better bargain, as Stalin was quick to point out. According to one eyewitness (Tokaty), he berated General Serov as follows: "This is absolutely intolerable. . . . We occupied Peenemünde, but the Americans got the rocket engineers. . . . How and why was this allowed to happen?"

Although a carrier for their first, heavy atom bombs would have to weigh several hundred tons, the Soviet Union decided to go ahead and develop it. When perfected in the late 1950's, it was also large enough to launch heavy satellites—and to carry the first man into space. Whether this had been planned from the beginning, or whether it was a lucky bonus, is perhaps one of those questions which even the Soviets themselves cannot now answer.

When it was finally revealed to the outside world at the Paris Air Show in the spring of 1967 (ten years after its first historic flights), the giant Sputnik carrier proved to be of a highly original design, bearing scarcely any resemblance to the V-2 formula. . . .

The American long-range rocket program remained in limbo until about 1950, though well over one billion dollars was spent on the development of jet-propelled guided missiles such as Navaho and Snark, which were no more than robot airplanes, capable of cruising only at relatively low speeds inside the atmosphere. All were abandoned after the ICBM breakthrough.

The two events which suddenly revived American interest in long-range rockets were the outbreak of the Korean War in 1950 and the realization that thermonuclear (or fusion) weapons could soon be built. These would not only be hundreds of times more powerful than the first fission bombs, but also lighter; they could be delivered by vehicles weighing about a hundred tons, instead of several hundred. And so, after several false starts, were initiated the programs which led first to ballistic missiles of intermediate range such as Redstone, Jupiter, and Thor, and later to the true intercontinental missiles Atlas and Titan.

All this was going on during the period 1950–55, and meanwhile the scientists were also getting involved. Many of the younger ones had been using sounding rockets to explore the upper atmosphere and had pioneered new fields of research in meteorology, astronomy, and geophysics. But this work was as frustrating as it was exciting, for sounding rockets could spend only a few minutes reporting from space before they fell back into the atmosphere. The obvious answer was the artificial satellite, which could stay aloft indefinitely, behaving, as one wit put it, like a Long-playing Rocket.

There was much discussion, therefore, of scientific satellites around 1950; as early as 1951, the British Interplanetary Society sponsored a congress in London on "The Artificial Satellite." By an accident of history, all these studies (some of which went into great technical detail) appeared just at the time when the scientific community was planning the greatest global research effort ever conceived—the International Geophysical Year (1957–1958).

The United States committee of the IGY, under the chairmanship of Dr. Joseph Kaplan, recommended that a satellite be launched as part of the nation's program; the suggestion was approved by the government, and the White House made the announcement to a somewhat startled world on July 29, 1955. Little notice had been paid to the statement on April 15, 1955, in the Soviet press that an Interdepartmental Commission on Interplanetary Communications had been set up to develop satellites for meteorological purposes, so when the U.S.S.R. repeated this item the day after the American announcement, it was received with amused skepticism.

. . . On October 4, 1957, when to the utter astonishment of the world the U.S.S.R. did exactly what it had said it was going to do, and orbited the first artificial satellite of Earth, it was at once obvious that a full-sized ICBM vehicle—not a small, purely scientific rocket—had been used as a launcher. Radar and optical observations showed that though the announced payload was 184 pounds, the empty final stage also circling the Earth must weigh several tons; when the even larger Sputnik 2 went into orbit carrying the dog Laika on November 3, 1957, photographs obtained by powerful tracking cameras proved that the combined structure of payload and final stage was more than 60 feet long.

[Fortunately for the United States, the Army's Redstone rocket, being developed by the Von Braun team, used hardware that already existed, and so could be ready in a very short time.] Dr. von Braun had made repeated attempts to get authorization for his project, and though they were all turned down, he continued the struggle. . . .

Not until the Navy's first Vanguard test vehicle had exploded spectacularly on the pad (December 6, 1957) was the Army given permission to go ahead. On January 31, 1958, the United States had its first satellite in orbit and could obtain some consolation from the fact that it had made the most important discovery of the IGY. For it was Explorer 1 that detected the wide-ranging, invisible halo of the Van Allen radiation belt.

So, in circumstances more dramatic than any novelist could have contrived or any scientist would have desired, Earth's new moons came into being. Millions of people were to see Sputnik 2, still catching the sunlight on the edge of space, gliding slowly from west to east across the ancient constellations. Few could have remained unmoved by the knowledge that they belonged to the first generation to set its sign among the stars.

Response to the Soviet Challenge

Selections from the Report of the Select Committee on Astronautics and Space Exploration

House Report No. 2710, 85th Congress, 2nd Session (1959)

III. A SPACE PROGRAM FOR AMERICA

The circumstances under which this committee was created relate to events which have recently changed the course of world history and the development of civilization for all time to come.

The creative forces of modern science and technology, through which man is beginning to realize some of his finest powers of reason and constructive ability, are now moving like the breakup of an Arctic river in the spring.

What might be called the "intellectual freezeup" of our objective understanding of the universe, together with our relative inability to use its forces, has been showing tentative cracks and shifts for some time. Now the changes have accelerated, so that today we seem about to witness a major loosening of the ice jam of our intellectual past. We may, in fact, be ready to move swiftly into a scientific world as different from our early technical endeavors as a warm summer is from the frigid season which precedes it.

This is not a change which can be forestalled, for the growth of these forces is more powerful than any political government on earth.

The loud report of the first nuclear explosion at Alamogordo, N. Mex., in 1945 merely signaled an attainment which had become inevitable with the calculations of Leibnitz, Planck, and Einstein.

The appearance of manmade satellites in the skies, traveling millions of miles at speeds which would bring incandescence in the atmosphere, has been an even more vivid sign that a new age is dawning. If we are unprepared, we may be swept away. At best, our situation is dangerous. It finds us vulnerable not only to the threat of nuclear, chemical, and biological war, with space employed as a medium of attack, but to the more subtle and equally threatening danger which lies in moral, psychological, and political fields.

It is possible, with foresight and skill, to master these tremendous changes. But nothing short of a national effort by all those armed with power, ability, and devotion can guarantee safe navigation of the swift currents of change which are now coming into view. Success is almost certain to mean unparalleled progress for tomorrow. Failure, as the history of the complacent, wealthy, and unresponsive nations of the past attests, very probably points to a new Dark Age.

Philosophic observations such as these are doubtless unusual in a formal report to the Congress. Nevertheless, the members of this committee have been so impressed by the nature of their work that they feel it necessary to abandon the prosaic approach in order to convey some sense of the enormity of the problems to be faced.

As this report and its supporting staff studies will show, every phase of modern life is somehow bound up with the new space technology. This fact poses two questions:

(1) Will the United States move steadily to marshal the resources of the free world and insure that no threat of force from space will be able to control or destroy us?

(2) Will the United States provide the leadership to unite the world in dedicating its space capabilities to peaceful purposes?

Both the threat and the high potential of the peaceful use of space have emerged in unmistakable outline from the studies of the select committee during the past year. We can think of no other report to the Congress of equal gravity in its implications for our future. . . .

C. POLICY IMPLICATIONS FOR THE SPACE PROGRAM

1. *Inexorable changes in society and political power will follow the development of space capabilities; failure to take account of them would virtually be to choose the path of national extinction.* Space developments already have changed the course of world history as inevitably as the discovery of America changed the history of Europe. The space challenge must be met by the adaptability, the energy, and the vision of the American people. Informed citizens, progressive business managements, and the organs of Government, including the Congress, must understand these facts, create public awareness, and act decisively in their respective fields to put this scientific knowledge and space capability to practical use. This must be done with sound and understanding cooperation—particularly, in the field of Government, between the legislative and executive branches.

2. *What program the United States could achieve and what it will in fact achieve may be two very different things.* . . . We must warn that a

policy of drift or neglect, or the frequent changing of goals would inevitably lead to our falling far short of the benefits and the capabilities that can be achieved in the next decade.

3. *Budget pressures in the short run should not be the primary basis for decisions on space programs which are inherently long range, and which involve the very survival of the Nation.* Budget problems of the Nation are very real. The budget represents an attempt to assign national resources with the greatest efficiency for the meeting of overall national goals. Stability of prices and the lowest practical level of taxes and Federal debt are worthy objectives in the interest of economic growth and flexibility. However, it is inevitable that certain expenditures on space must be made within the foreseeable future, and that delay will raise costs.

4. *This Nation should not make inadequate short-run expenditures on its space program at substantial risk to its survival a few years later.* The American people prefer economic and fiscal stability and soundness, but must view them in long-run terms.

5. *The best advice obtainable by the committee supports the view that within a decade peaceful applications of space development to weather prediction and long-range communication alone will more than pay back to the economy all the funds previously required to achieve these capabilities.* Six satellites could handle the entire overseas mail volume of the World, as the Atlas satellite has already demonstrated in rudimentary form. Better weather predictions are expected to save the United States economy about $4 billion a year.

6. *The greatest benefits of space development and exploration in all probability cannot even be predicted today.* This has been true of countless past discoveries. The search which discovered America was intended to find a direct route to the Orient; the prize which was found was much more important. More recently, the byproducts of nuclear reactors seemed at first an unwanted nuisance. But already the economic benefits of radioactive isotopes bring a return great enough to pay the capital charges on our entire peaceful and military investment in the atomic field. What may come in resources, in understanding of the forces of nature, in solving many of the present mysteries of life may be beyond all immediate human understanding. This is not the kind of challenge man can safely ignore.

7. *Although engineering secrets related to national defense deserve the utmost protection, the greater part of the space program will progress more rapidly without the shackles of an undue security control.* No country has a monopoly on scientific knowledge, and history has shown that inventions and discoveries are frequently made independently at

much the same time in different countries of the world. Because the space effort is so complex, almost every branch of human study has made some contribution to its success. Compartmentalization of research findings, within a narrow, need-to-know framework, whether done in the name of national security or to protect proprietary interests, can only slow up our progress. Free exchange of information among scientists and engineers is particularly important for a country which is making a late start in developing its space capabilities.

8. *Full scientific and technical cooperation among the nations of the free world is essential to their joint survival and to the fastest growth of the American space program.* The recent report on international cooperation issued by this committee traced in detail the compelling reasons for international cooperation in space. Partly they are psychological. Partly they arise from the need for worldwide tracking and control facilities. Partly they are necessary in order to tap a rich pool of scientific talent in other countries which now is underused and not marshaled. Finally, the space program will soon reach such size that it will require the combined effort of many nations. It is our hope that the exploration of space will in time absorb the friendly common efforts of all nations. In the meantime the free world cannot afford to delay its own space development on a united plan while waiting for improved relations with the Communist bloc.

9. *Scientific education in the United States stands in need of critical review.* No phase of the soul-searching produced by the launching of the sputniks excited a more sympathetic response from an alarmed people or from educators and legislators than the realization that American education had failed to measure up to the needs of the hour. Public education fell under intense scrutiny, from teachers' salaries to the neglect of mathematics and science studies in the secondary schools and the institutions of higher learning. The performance of the Soviet schools in some of these respects presented a challenging contrast.

The Soviet Union appears to put a greater total emphasis on education. The difference would be hard to measure, even in terms of money and manpower, partly because the term "education" in Soviet use includes some activities which are not regarded by Americans as education at all. There is also good evidence that schooling is generally more intense in the Soviet Union than in the United States. In addition, the Soviet curriculum includes more study of mathematics and the physical sciences.

While the Soviet schools are also believed to have many shortcomings, such as rote learning and excessively strict "discipline," it behooves the United States to recognize those aspects of Soviet education which are better suited than its own to the demands of the future. At the same time, any changes which are made must be consistent with American objectives, ideals, and conditions. It is the view of this committee that prudent and imaginative steps must still be taken to give to mathematics and science the needed emphasis in school curricula. These steps should be designed not only to meet Soviet educational competition, but should, irrespective of Soviet standards, raise American standards to cope with the Space Age. The committee intends its use of the word "prudent" to imply an emphasis that does not neglect the humanities in favor of mathematics or related subjects.

EPILOGUE

As this committee concluded the meeting at which this report was adopted, the news of the Soviet lunar probe [1] became public. This committee has been consistent during the year of its life in believing that the Soviet Union has a distinct lead in astronautics over the United States, although it also believes that with time and toil the United States can surpass the Soviet achievements. Many prominent Americans, however, have minimized the Soviet advantage at various times during the year. The latest Soviet achievement proves the soundness of the view of this committee that our scientific race, not alone in space but in the broader realm of science, is serious and urgent and demands the utmost effort by this Nation.

Editor's Note. On April 12, 1961, the Russians launched the first manned space flight. Yuri Gagarin made one orbit of the earth in Vostok 1 at an altitude of 188 miles. On May 5th Alan Shepard made one suborbit (altitude 117 miles) in Freedom 7.

Selection from President Kennedy's Address to Congress (May 25, 1961)

If we are to win the battle for men's minds, the dramatic achievements in space which occurred in recent weeks should have made clear to us all the impact of this new frontier of human adventure. Since early in my term, our efforts in space have been under review.

1. Lunik II, the first satellite to hit the moon, Sept. 12, 1959, weighed 858 pounds. It contained scientific instruments and a Soviet coat of arms. Ed.

With the advice of the Vice President we have examined where we are strong and where we are not, where we may succeed and where we may not. Now it is time to take longer strides—time for a great new American enterprise—time for this nation to take a clearly leading role in space achievement.

I believe we possess all the resources and all the talents necessary. But the facts of the matter are that we have never made the national decisions or marshaled the national resources required for such leadership. We have never specified long-range goals on an urgent time schedule, or managed our resources and our time so as to insure their fulfilment.

Recognizing the head start obtained by the Soviets with their large rocket engines, which gives them many months of lead time, and recognizing the likelihood that they will exploit this lead for some time to come in still more impressive successes, we nevertheless are required to make new efforts.

For while we cannot guarantee that we shall one day be first, we can guarantee that any failure to share this effort will make us last. We take an additional risk by making it in full view of the world—but as shown by the feat of Astronaut Shepard, this very risk enhances our stature when we are successful.

But this is not merely a race. Space is open to us now; and our eagerness to share its meaning is not governed by the efforts of others. We go into space because whatever mankind must undertake, free man must fully share.

I therefore ask the Congress—above and beyond the increases I have earlier requested for space activities—to provide the funds which are needed to meet the following national goals:

First, I believe that this nation should commit itself to achieving the goal, before this decade is out, of landing a man on the moon and returning him safely to earth. No single space project in this period will be more exciting, or more impressive, or more important for the long-range exploration of space; and none will be so difficult or expensive to accomplish.

Including necessary supporting research, this objective will require an additional 531 million dollars this year and still higher sums in the future. We propose to develop alternate liquid and solid fuel boosters much larger than any now being developed, until certain which is superior.

We propose additional funds for other engine development and for unmanned explorations—explorations which are particularly important for one purpose which this nation will never overlook—the survival of the man who first makes this daring flight. But in a very real sense, it will not be one man going to the moon—it will be an entire nation. For all of us must work to put him there.

Second, an additional 23 million dollars, together with 7 million dollars already available, to accelerate development of the Rover nuclear rocket.

This is a technological enterprise in which we are well on the way to striking progress, and which gives promise of some day providing a means for even more exciting and ambitious exploration of space—perhaps beyond the moon, perhaps to the very ends of the solar system itself.

Third, an additional 50 million dollars to make the most of our present leadership by accelerating the use of space satellites for world-wide communications. When we have put into space a system that will enable people in remote areas of the earth to exchange messages, hold conversations, and eventually see television programs, we will have achieved a success as beneficial as it will be striking.

Fourth, an additional 75 million dollars—of which 53 million is for the weather bureau—to give us at the earliest possible time a satellite system for world-wide weather observation. Such a system will be of inestimable commercial and scientific value; and the information it provides will be made freely available to all the nations of the world.

Let it be clear—and this is a judgment which the members of the Congress must finally make—let it be clear that I am asking the Congress and the country to accept a firm commitment to a new course of action—a course which will last for many years and carry very heavy costs, 531 million dollars in the fiscal year 1962 and an estimated 7 to 9 billion dollars additional over the next five years.

If we are to go only halfway, or reduce our sights in the face of difficulty, in my judgment it would be better not to go at all.

This is a choice which this country must make, and I am confident that under the leadership of the space committees of the Congress and the appropriations committees you will consider the matter carefully. It is a most important decision that we make as a nation; but all of you have lived thru the last four years and have seen the significance of space and the adventures in space, and no one can predict with certainty what the ultimate meaning will be of the mastery of space.

I believe we should go to the moon. But I think every citizen of this country as well as the members of Congress should consider the matter carefully in making their judgment to which we have given attention over many weeks and months, as it is a heavy burden; and there is no sense in agreeing, or desiring, that the United States take an affirmative position in outer space unless we are prepared to do the work and bear the burdens to make it successful. If we are not, we should decide today. Let me stress also that more money alone will not do the job. This decision demands a major national commitment of scientific and technical manpower, material and facilities, and the possibility of their diversion from other important activities where they are already thinly spread.

It means a degree of dedication, organization, and discipline which have not always characterized our research and development efforts. It means we cannot afford undue work stoppages, inflated costs of material or talent, wasteful interagency rivalries, or a high turnover of key personnel.

New objectives and new money cannot solve these problems. They could, in fact, aggravate them further—unless every scientist, every engineer, every service man, every technician, contractor, and civil servant involved gives his personal pledge that this nation will move forward, with the full speed of freedom, in the exciting adventure of space.

_____

Editor's Note. In response to President Kennedy's strong appeal Congress voted the funds for the greatest technological endeavor in the history of mankind. Although the die had been cast the arguments for and against the program continued to build up and reached a crescendo in the Spring of 1963. On June 10th and 11th the Aeronautical and Space Sciences Committee held a hearing to consider the protests which were being raised, especially among scientists, against the outlay of such an enormous effort to put a man on the moon before 1970.

Hearing of the Aeronautical and Space Sciences Committee [*]
(1963)

While agreeing that a manned lunar landing should eventually be carried out, Abelson argued that "most of the important scientific questions concerning the moon and other planets could be studied soon at relatively low cost employing unmanned vehicles." Furthermore, he said, while it is claimed that "vast frontiers of knowledge" will be opened by putting a man in space, "no one has delineated any impressive body of questions which are to be studied."

"Making man a part of the scientific exploration of space has two im-

[*] From Lloyd V. Berkner *et al.*, "Scientists on Space: Senate Group Hears Criticism and Support for Manned Lunar Landing Program," *Science*, Vol. 140, pp. 1195–96, June 14, 1963. Copyright 1963 by the American Association for the Advancement of Science. Reprinted by permission.

portant drawbacks," he continued. "It increases costs and it will probably slow down, at least for some years, the pace of getting valuable
results. . . . Our recent Mariner II probe to Venus cost a few tens of
millions. To send man on a comparable mission might cost a hundred
billion dollars and could not be done for years."

Abelson also raised questions about the emphasis that NASA is placing on scientific research. "One of the most puzzling aspects of the
NASA Program," he said, "is the continued failure to land electronic
equipment on the moon. After a trajectory of more than 100 million
miles, Mariner II scored a fine success in exploring Venus. Why can't we
hit the moon, which is comparatively in our own backyard? Why was
there insufficient backup of the five Ranger vehicles which failed? I
have the feeling," he added, "that scientific exploration of the moon has
been accorded a low priority, that the Apollo program is distorting scientific priorities and at least indirectly slowing progress."

Kusch, too, said that manned space exploration is a legitimate eventual goal, but he questioned the high priority that has been assigned to
landing men on the moon and returning them in this decade. And he
suggested that earthly needs, "the preservation and repair of our continent," might be a more appropriate goal for a national effort on the
scale of the moon landing.

"I do not think that the present space exploration effort can be justified on the grounds that it will have a visible effect on the lives of people other than through the pride they may feel in its achievement or the
vicarious sense of adventure that they may experience."

"It is my belief," he continued, "that the present space program attempts too much too fast. There is not enough time for profound
thought, for imagination to play over the demanding problems that
occur. . . . I find it difficult to believe that the exploration of space is a
more compelling goal than the exploration of the planet that we inhabit.
My very real sympathy for the space program does not extend to a belief that it should be the overriding national effort at this time."

Statement of Eight Scientists Endorsing the Lunar Program *
(1963)

Some members of the scientific community have criticized the Apollo
project, which is aimed at the achievement of the manned lunar landing

* From M. L. Ewing, "Space Controversy: Senate Committee To Hear Scientists
on Moon Program," *Science,* Vol. 140, p. 1078, June 7, 1963. Copyright 1963 by the
American Association for the Advancement of Science.

in this decade. The critics assert that the scientific benefits of space research can be gained by heavier reliance on robot instruments, with the manned flight program carried out at a slower and less expensive pace.

This criticism raises important issues regarding the motives which underlie the United States space effort. In 1961 the Congress responded to the call by President Kennedy for a vigorous space program, including a commitment to the manned lunar landing within the decade, by voting overwhelmingly in favor of the funds requested. The support was reaffirmed in 1962.

Was this support tendered for scientific reasons primarily, or was it motivated by a broader concern with national interests and national goals?

We believe that the support given to the enlarged space program by the people and the Congress was not based primarily on scientific grounds. We believe it was based on a conviction that this program will, for many reasons, make an important contribution to the future welfare and security of the United States.

On this basis we take issue with those of our scientific colleagues who criticize the Apollo program by contending that it does not have scientific value. We regard the criticism as invalid for two reasons.

First, man-in-space makes an essential contribution to the scientific objectives of lunar exploration. The exploration of space will pose an immense variety of challenges, unexpected opportunities and unforeseen obstacles. In the early stages of experimentation, automatic apparatus is effective. In later stages, when important questions have to be answered by difficult experiments, very complicated instruments must be developed to attempt a crude imitation of human judgment and flexibility. Robot instruments will always play an important role in the exploration program, but situations are bound to arise in which the human performance is indispensable for achievement of the scientific objectives. A sound approach requires both the development of automatic instrumentation and a vigorous program to achieve an early capability for manned exploration.

Second, science plays an important role in lunar exploration but is not the sole objective of that project. The momentum and significance of the lunar program are derived from its place in long range United States plans for exploration of the solar system. The heart of those plans is man-in-space. Although it is the responsibility of the scientist to see that research is a strong element within the framework of the program, nevertheless, the impetus of the program is not derived from scientific research alone. Therefore, the pace of the program cannot be set only by the steady flow of scientific developments. It is essential that it be influenced also by the urgencies of the response to the national challenge. In making these remarks we wish to stress that the space effort is a national program which warrants the interest, criticism and active participation of the entire scientific community.

Maurice L. Ewing, Robert Jastrow, Joshua Lederberg, Willard F. Libby, Gordon J. F. MacDonald, Lyman Spitzer, Harold C. Urey, and James A. Van Allen.

Pros and Cons °

by Eugene Rabinowitch (1963)

The two [following] articles were written by scientists who have rendered great service as advisers and organizers of American scientific efforts. The difference between their views shows how controversial the space program emphasizing manned moon flight is, even among scientists. Arguments on both sides are strong and carried by deep conviction.

One view both authors have in common: the American space effort should not be influenced by desire for sensational success, yearning for "victory" in a "race" with the Soviet Union. But as for the rest, the conflict between their opinions is sharp and deep. This editor—probably in common with many other scientists—has done much soul-searching without arriving at a decision between the two attitudes.

He would like only to contribute two remarks to the debate. One is that listing worthwhile aims in science, education, and welfare, which could be achieved by saving the $30 billion now earmarked for a manned expedition to the moon, would be fully convincing only if the likelihood existed that these billions actually would be used so constructively—for aid to underdeveloped countries, for increased educational and research facilities at home, for medical and welfare services for ourselves and others. Unfortunately, this is not likely. Space exploration is the only large-scale national program—in addition to military preparedness—that commands broad support in Congress and has caught the imagination of the public at large. Billions not spent on it are more likely to be used to purchase more luxury articles than on building hospitals or increasing foreign aid.

A second point not mentioned by Weaver and Berkner is the present and, even more, the potential role of the space program in international cooperation, bridging the moat between the East and West.

In the last two years, cooperation in space has been the one field in which positive agreement has been achieved between the Soviet Union and the United States. In facing cosmic space, the quarrels and struggles between different factions of humanity appear petty and irrelevant; the dimensions and costs of exploring the solar system—not to speak of venturing beyond it—are so enormous, and so obviously call for common effort, that in the midst of the Cold War the need for cooperation impresses itself on nations and their leaders, despite their bitter rivalry and conflict. Common effort could help to create bonds and foster the trust which comes from participating in a common enterprise.

If space exploration could help bring together the two alienated parts of humanity and reduce, even slightly, the danger of all-destroying nuclear war, that alone would make worthwhile investing in it many billions of dollars.

The Compelling Horizon *

by Lloyd V. Berkner (1963)

The development of our space capabilities during the past decade has opened altogether new vistas to science and to new and powerful technologies. Space vehicles operating above the atmosphere have provided opportunities to view the Earth and its associated physical phenomena in completely new perspectives; to peer into the universe over the whole spectrum of radiant energies; to explore interplanetary space to which the Earth is coupled as it courses in its orbit; and to visualize rather clearly the solution to technological problems of visiting the moon and the nearby planets. Significant scientific questions about our environment in the universe, which were in the realm of fantasy only a few short years past, can now be asked quite sharply. (For the enormous range of scientific experimentation opened by space exploration, see *Science in Space* by L. V. Berkner and Hugh Odishaw. McGraw-Hill, 1961.)

The case for the conquest of space can be classified under three major headings in order of their ultimate contribution to the growth of man's welfare and to the rise of his stature as a civilized being: science in space, technology in space for advancement of man's civil pursuits, and military technologies.

* From *Bulletin of the Atomic Scientists*, May, 1963.

Science in space will, over the coming decades, undoubtedly exercise an immense influence over man's development. The new techniques of scientific observation and synthesis growing out of our space capabilities are altogether revolutionary to many sciences. Certainly astronomy and astrophysics will make giant strides during the years ahead, as their major base of observation moves into space. Likewise, outer-atmospheric physics and meteorology are making dramatic advances through the space observations already achieved. Already our concepts of the interplanetary medium, the magnetosphere of the Earth, the solar wind, basic knowledge of solar physics, and solar-atmospheric effects are evolving rapidly as space experiments are completed. The immediate prospect of lunar and planetary explorations must expand enormously our horizons of the physical knowledge of the universe and, perhaps, of our comprehension of the origins and evolution of life itself.

These new vistas have inspired men everywhere to a new level of aspiration. Having broken the confining chains that have kept men bound to a two-dimensional existence, the vision of free movement truly into a three-dimensional universe has given man a new challenge to his spirit. Out of this new level of aspiration, the political implications of space exploration principally arise. In a world involved in a political struggle between contesting ideologies, man instinctively recognizes that the rewards and acclaim will necessarily go to the side that can best satisfy these new aspirations.

There are those who protest that space exploration is too expensive—that diversion of resources to this or that cause would serve man's purpose. But they miss the point that without the effort to meet the challenge of these aspirations, neither space exploration nor their favorite causes will be served. Human society—man in a group—rises out of its lethargy to new levels of productivity only under the stimulus of deeply inspiring and commonly appreciated goals. A lethargic world serves no cause well; a spirited world working diligently toward earnestly desired goals provides the means and the strength toward which many ends can be satisfied. If the challenges growing out of man's vision of space achievement increase his productivity by a hundredth, then the whole bill for space exploration is paid; any additional effort becomes a margin for advancement toward other goals.

Space exploration promises, of course, much more than the spiritual rewards of man's enlarged comprehension of his status in the universe. Already the technology of communications via space relay is far advanced and our physical capability to provide long-distance channel capacity for conveyance of information will increase by many orders of magnitude. Thus, we can look to a world more closely bound together by an adequate system of communications; a world revolutionized in its organization through the impact of ready communications to stimulate its commerce and unify its cultural and esthetic patterns.

Likewise, advancing enormously is our knowledge of the weather— our most controlling environmental influence. New data are being acquired in the Tiros and Nimbus programs on storm systems, their dynamics, and their distribution. Future applications of satellites to meteorology offer even more powerful potentialities. Already studies are proceeding of satellite communication systems ("Reapers") for quick collection and assembly of meteorological data from a worldwide system of free balloons at tropospheric heights and meteorological buoys floating widely over the oceans. So powerful is the promise of these satellite meteorological systems that we can confidently foresee great advances during the next few years in our ability to forecast and predict weather conditions all over the world, and perhaps ultimately to control them to some extent.

Again, satellite systems of immense benefit for navigation by ships at sea or planes flying the oceans are close to operational reality.

The economic benefits from these systems alone—communications, meteorology, navigation—may well repay our nation fully for the costs of our investment in the space effort.

Aside from the variety of new technologies that emerge from our direct employment of space technique, there is the "fallout" of technical ideas that occurs from pressing any technology to its limits. In the conquest of space, men, ideas, and materials are pushed beyond previous limits and capabilities. The seemingly impossible is captured and brought within the range of daily employment of our whole technology; and this occurs usually only under the intense pressure toward some highly desired goal. Perhaps this is why war is usually so productive in the advance of technology, and perhaps the intense aspiration toward more peaceful goals, such as space exploration, are justifiable substitutes for war in their mass motivation toward the most advanced goals. At any rate, the technological dexterity derived from programs such as "space" are of immense value to a country that aspires to leadership.

As long as military forces remain a necessity for world stability, the military potentialities of space cannot be ignored. Space vehicles can doubtless play a significant role in surveillance, reconnaissance, intelligence, and early warning. These applications should not be denigrated because of their military character and we, as a nation, would be seriously remiss not to exploit fully their potentialities. Actually, they may be important contributors to world peace and stability, since they may remove otherwise serious national doubts whose pressure could generate great danger in a highly aroused and frustrated world. Certainly, space vehicles can be used as bomb carriers conceivably useful to certain strategies. In this role, however, such carriers are merely an extension of existing missile strategies and appear generally inferior to them.

In viewing the space effort, one must be seriously disturbed by the voices that attempt to reduce the programs merely to the spectacle of an athletic contest to "land a man on the moon." This attitude may be suitable at the track, in the stadium, or at a prizefight. However, in terms of our space goals, these opinions are diversionary and achievement of only this objective would have very limited and even derogatory consequences. They demonstrate a very fundamental lack of appreciation for man's basic aspirations and motivations in the program. For man's aspiration to conquer space has an intellectual quality through his desire to comprehend more clearly and definably his place in the universe.

The world is quick to recognize a phony. So the side that puts the space race on the level of the mere athletic contest is certain to lose ignominiously. Man demands that the conquest of space be undertaken at the scientific level, where the answers advance his comprehension of the nature of the universe, and add to his stature and dignity by broadening understanding of his place in it. Consequently, if the real objectives of our space program are to be achieved, the preparation and support for an adequate program of science in space is imperative. The long lead time required to think out the objectives of a significant and sophisticated program and to prepare the scientific apparatus does not permit us to relegate the scientific preparation to second place; if we do, the program of science will be entirely superficial. Therefore, the engineering of vehicles and preparation of scientific programs must go hand in hand in a carefully planned and balanced undertaking. Man, or his instruments, goes into space to see and measure and understand. Thus, exploration of space must be a *scientific* exploration so that the consequent feats will satisfy man's aspirations. Anything less endangers the purpose of the whole space program.

But beyond the safeguards of purpose rendered by science are the essential safeguards that science can yield in the very engineering underlying the venture of man himself into space and in the development of his vehicles and spacecraft. In undertaking these efforts at the very limits of technology, it is rarely possible to foresee all of the major factors that may interdict success in future space operations under hostile and strange environmental conditions. Inevitably, as an engineering program at the limits of technology comes up against some dead end, a rapidly moving program is delayed with immense expense and great national frustration. If such programs are accompanied in reasonable balance by scientific research that finds and elaborates unexpected questions, great monetary savings can be effected and frustrations avoided.

Hence, the short-range savings which might be achieved by elimination of a balanced program of research in association with the necessary engineering will, without question, be vastly exceeded by subsequent costs and national losses as unexpected dead ends appear. A policy of "penny wise and pound foolish" in funding space research should be avoided at all costs or shortsightedness will most certainly catch up with us.

Fortunately the leaders of the NASA and of the Space Council have recognized the major importance of an adequate scientific program in support of space exploration goals. Moreover, these goals have been under constant scrutiny by the Space Science Board of the National Academy of Sciences. Through a variety of carefully considered reports, and through constant discussion, there is general agreement among scientific leaders and government administrators on the major features of the required scientific program. Nevertheless, this agreement of goals must be understood and supported by the public at large, if we are to avoid degeneration of the program to unproductive effort. The scientific cream of the program involves less than 10 per cent of the total costs, yet it embraces essentially all of those elements from which the ultimate success of the space program will be ultimately measured.

In our efforts toward conquest of space, we should ever remember that we are motivated by the spirit that forced the explorers of the past to accept the dangers of extreme and hostile environments. That spirit is expressed vividly in the words of the great scientific explorer Fridtjof Nansen: "The history of the human race is a continuous struggle from darkness toward light. It is therefore of no purpose to discuss the use of knowledge; man wants to know and when he ceases to do so he is no longer man."

. . . The widespread and growing belief that the first visit to the Moon will bring back some kind of a scientific Holy Grail is probably the biggest popular delusion of all time . . .

Geminus, *The New Scientist,* November 8, 1962

Dreams and Responsibilities °

by Warren Weaver (1963)

I have the great privilege of living in a democracy where differences of opinion are permissible and proper. I want to discuss an aspect of the U.S. space program—and particularly the "moon shot"—on which my views differ from the views of some others. But it must be understood that I am in no sense a space scientist; and that I am very much concerned that I do not make any pretense of being such.

I also want to emphasize that I do not in the least assume that I am "speaking for science" or for any group of scientists. Although I have of course been influenced by the conversations I have had on these matters with many other scientists, I am speaking only for myself. I am an interested and concerned citizen who was trained first as an engineer, then as a mathematician, and who has had the privilege of a long and wide experience with research in many fields of science both in the U.S. and in other parts of the world.

Two entirely different sets of considerations bear on the question of how much money and effort (manpower, equipment, physical facilities, etc.) the U.S. should devote to its space program. The first set of considerations includes defense, national prestige, and international relations. The second set includes matters related only to scientific research —to obtaining new knowledge.

Now as to the first of these two sets of considerations, I wish to make it clear that I do not consider myself competent in those areas. I most strongly deplore scientists making pronouncements outside their areas of competence, especially when, implicitly at least, they invoke the power and prestige of science to advance their purely personal and perhaps poorly supported opinions.

I am moreover deeply concerned to be, and to act as, a patriotic citizen of my country. I would find it most distressing to have to take any position opposing a considered and firm policy of my country.

Therefore, if I could be assured that in the judgement of competent, experienced, and highly placed general government authorities, the expenditures proposed in our space program are necessary and justified because of their relation to defense, to national prestige, and to our world position, then I would accept that decision.

But—and we now come to the heart of the matter—these programs

° From *Bulletin of the Atomic Scientists*, May, 1963.

are often brought forward to the public as *justified on scientific consid-erations.* The great power and prestige of science are invoked to per-suade the public, Congress, and perhaps even the top government offi-cials, of their propriety and necessity.

It is at this point only that I disagree. I do not think that scientific considerations justify the proposed magnitude of the program, and even more emphatically I do not believe that scientific considerations justify its frantic, costly, and disastrous pace.

Although the money costs are simply astronomical, I think this the least serious side of the matter. I think the manpower and facilities re-quirements are more serious. If we put the suggested effort into the space program we will simply have to curtail certain other scientific ac-tivities. Such is the glamour of this field that we will, moreover, have a generation of young men largely headed for the moon, so to speak, rather than headed for a wide range of other scientific careers. I espe-cially deplore the major emphasis upon the physical sciences, to the ex-tent that this must unhappily involve a diminution of effort on the bio-logical and medical sciences.

It is of course intellectually important that we gain more knowledge about the universe in which we live. But the fact is that a very large fraction of the scientific data which we need to collect in space and bring back down to the earth can be collected by purely electrical and mechanical packages of equipment and can, through the marvels of modern telemetry and radio, be returned to listening and recording de-vices on the earth.

The total cost of this kind of experimentation, although substantial, is small indeed as compared with the costs of attempting to put men on the moon and, eventually, even on the more distant planets. It is this frantic and to me rather cheaply dramatic aspect of the space program that seems primarily objectionable.

The sum of $30 billion, which is undoubtedly an underestimate of the total cost of "putting a man on the moon," is a sum so large that the or-dinary human being can simply not grasp its magnitude. It was for that reason that I wrote a short article (in *Saturday Review,* August 4, 1962) to point out what alternative things could be done with $30 billion.

With that sum one could give a 10 per cent raise in salary, over a ten year period, to every teacher in the U.S. from kindergarten through uni-versities (about $9.8 billion required); could give $10 million each to 200 of the better smaller colleges ($2 billion required); could finance seven-year fellowships (freshman through PhD) at $4,000 per person per year for 50,000 new scientists and engineers ($1.4 billion required); could contribute $200 million each toward the creation of ten new medical schools ($2 billion required); could build and largely endow complete universities with liberal arts, medical, engineering, and agricultural faculties for all fifty-three of the nations which have been added to the United Nations since its original founding ($13.2 billion required); could create three more permanent Rockefeller Foundations ($1.5 billion required); and one would still have left $100 million for a program of informing the public about science.

I want to emphasize that I am sure that the individuals who are energetically sponsoring the United States space efforts are competent and sincere and patriotic individuals. But I do not think that the magnitude and character of this effort is justified on scientific grounds alone, and I have in fact never discussed this matter with any responsible and experienced scientist in the United States who has not agreed with the general views which I have here expressed.

Milestones on the Road to Space

Opening Skies [*]
by Arthur C. Clarke (1968)

MAN IN ORBIT

The first serious students of astronautics had taken it for granted that men would be the most important payloads that rockets would carry into space. Tsiolkovsky and Oberth had written of little else, and though the cautious (as well as more practical-minded) Goddard had confined his few public statements to instrument-carrying vehicles, his private notebooks leave no doubt as to where his interest lay. In these, he went so far as to speculate about refueling bases on the planets—a far cry indeed from "A Method of Attaining Extreme Altitudes."

One reason for this attitude was that these pioneers saw, much more clearly than many who came later, that space travel was the next stage in man's exploration of his environment. Today's controversies between the protagonists of manned and unmanned spacecraft would have seemed to them as pointless as the theological disputes of the Middle Ages.

The conditions which men would encounter in flight beyond the atmosphere were well understood long before any rockets had entered space, and the more conscientious science-fiction writers had quite accurately described ways of coping with them. The rise of aviation medicine from the 1920's onward put these speculations on a more scientific basis, and when the Space Age dawned in 1957 there was only one serious unknown. Every condition that could be encountered in space was reproducible in the laboratory, with the single exception of weightlessness. Rocket flights with animals (especially dogs and monkeys) in the decade up to 1957 had shown that the apparent absence of weight could be endured for several minutes, so at least there was no danger, as some had feared, of the heart promptly running amok when the gravitational

[*] Abridged from Arthur C. Clarke, *The Promise of Space*, pp. 80–122 (including twelve line drawings). Harper & Row, Publishers, Inc., New York. Copyright © 1968 by Arthur C. Clarke. Reprinted by permission of Harper & Row, Publishers, Inc.

load was removed from it. But the effects of really prolonged weightlessness were still unknown, and there were plenty of dire predictions from the pessimists. I can well recall crossing swords at an international conference, as late as 1963, with a distinguished biologist who stated categorically that lack of weight could not be tolerated for more than a week.

The reason why an astronaut is normally weightless is perhaps more misunderstood than anything else in the whole business of space travel. It is nothing to do with being "beyond the pull of the Earth's gravity" —not that such a thing is literally possible in any case. As we saw [earlier] most manned orbital flight takes place in regions where gravity has diminished by only a few per cent from its sea-level value. Yet, despite this, astronauts weigh exactly nothing once their rockets have ceased to thrust.

This confusing paradox is largely due to poor semantics. On Earth we tend to use the words "weight" and "gravity" interchangeably, because in almost all terrestrial situations the two phenomena occur together. But they are really quite separate entities and can exist independently of each other; in space they normally do so. On occasion, however, they can be separated even on Earth, as will now be demonstrated. (This is another Einsteinian "thought experiment"; carry it out at your own risk.)

If you place a set of bathroom scales on a trapdoor and stand on the platform, the pointer will indicate your weight. But if the trapdoor is opened, the reading on the scales will drop instantly to zero. Nothing has happened to the Earth's gravity—but your weight has vanished!

For weight is a *force*, normally produced by gravity, and you cannot feel a force if it has "nothing to push against," to use a familiar phrase. You do not feel any force when you push against a swinging door; you cannot feel any weight when you have no support and are falling freely. And an astronaut, except when he is firing his rockets or re-entering atmosphere, is *always* falling freely. The "fall" may be upward or downward or sideways; the direction does not matter, as long as it is free and unrestrained.

This is why one can feel weightless even in the presence of gravity. A man in free orbit around one of those dense dwarf stars whose gravitational field is a million times as great as Earth's would still feel completely weightless.

And, conversely, one can feel "weight" even if there is no gravity; acceleration can produce an identical effect. If you were standing on those bathroom scales out in deep space, billions of miles from any celestial body, they would again register zero weight. But attach a rocket motor and start it firing—then weight would return. At an acceleration of 32 feet per second per second the scales would read correctly: you would be under "1 g," and unless you had some other means of discovering the truth, there would be no way of telling that you were not standing on the surface of the Earth. The absolute equivalence of weight due to gravity and that due to accelerations is one of the cornerstones of Einstein's General Theory of Relativity. As we shall see in the next section, it also provides us with a means of generating weight artificially, should that be desired.

When Sputnik 2 launched the dog Laika into orbit, the experiment proved two things. It demonstrated that higher animals (presumably including man) could endure long periods of weightlessness, and it showed that the Soviet Union was intensely interested in space *travel*, not only space science. Later experiments, in which dogs were safely recovered after several days in orbit, gave the Soviet Union further valuable information, but it was generally believed that the first manned flights would be fairly brief suborbital or ballistic shots, like those with which the United States did in fact open its own program (Shepard, Grissom, 1961). Hence it was a great surprise to most people when the Soviets went straight for orbit on April 12, 1961, and Yuri Gagarin circled the world in 89 minutes aboard Vostok 1. There were men still alive who had been born when Jules Verne had dared to suggest that this feat might be accomplished in eighty *days*, and millions who could well recall when it was first done in as many hours.[1]

From the opening of the Space Age to the first man in orbit was only 3½ years—largely because the Soviets had made their very first ventures into this new ocean with boosters already large enough to carry a man, and merely had to develop the life-support systems. The United States not only had to do this but also had to stretch its existing rockets to perform a task for which they had never been designed. By brilliant improvisation, the Atlas ICBM was upgraded and "man-rated," so that less than a year after Gagarin, John Glenn was able to perform three orbits of the Earth in the Project Mercury capsule Friendship 7 (February 20, 1962).

The four Mercury flights were followed by the still more successful Gemini launches, ten in all, each involving two astronauts. The Vostok was succeeded by Voskhod, carrying as many as three men. During the first five years of manned space flight (1961–66) a truly extraordinary record of achievement and safety was established, neither the Americans nor the Soviets (contrary to numerous reports) suffering any casualties.

---

1. In *Profiles of the Future* I have given reasons for doubting if this will ever be done in eighty seconds, though it may ultimately be achieved in about one-eighth of a second. That "ultimately" is quite a long way off.

By a sad coincidence, the first fatalities in both space programs occurred within a few weeks of each other during tests of the *third* generation of manned vehicles (Apollo and Soyuz). And in each case, the disasters occurred virtually at ground level, not in space—which remains the safest medium for transportation yet discovered.

A list of the outstanding events in this first half decade of manned space flight gives a good idea of the remarkable rate of progress. The nationalities involved have been carefully omitted; if any reader finds it hard to remember who did what, it may occur to him that, perhaps, it may not be as important as he had imagined.

Since a lunar round trip takes less than two weeks, this series of flights proved that there were no outstanding physiological barriers between Earth and Moon. Although numerous minor problems and difficulties had been encountered, these had all been overcome, and it seemed that man could do anything that he wished in space. Adaptation to weightlessness had been astonishingly easy; although it caused housekeeping and, above all, sanitary [2] annoyances, most astronauts found it a delightful experience, and some were afraid that they might become addicted to it. Myriads of skindivers had known this for years.

Date	*Achievement*
April 12, 1961	First man orbits Earth
August 6–7, 1961	First man spends full day in orbit
August 11–15, 1962	First launch of two spacecraft; first near-rendezvous
October 12, 1964	First multimanned (3) spacecraft; first "shirt-sleeve" (no spacesuits) environment; first non-astronaut passenger
March 18, 1965	First man to leave spacecraft in orbit
March 23, 1965	First manned orbital maneuver
June 3–7, 1965	First manned propulsion outside spacecraft
August 21–29, 1965	First men spend week in orbit
December 4–18, 1965	First men spend two weeks in orbit; first sustained space rendezvous
March 16, 1966	First docking of two spacecraft

However, it is important to realize that although *weight* is nonexistent in orbit, *mass* or inertia remains quite unaffected. It is just as difficult to

2. To the invariable question "How *do* they manage?" the answer is "Not very well." At least one American astronaut had a rather damp flight.

set a given object in motion aboard a spaceship as it is on Earth, and it requires just as much effort to stop it again. The inherent laziness of matter—its tendency to keep on doing the same thing—is independent of gravity. Although this fact can be used to advantage, it can also cause problems.

Astronauts engaged on the early extravehicular activities found it very difficult to control their movements, because these continued even after the initial force was applied. Hampered as they were by their clumsy pressure suits, they found even the simplest tasks exhausting; they needed both hands merely to keep themselves in position. However, by the time the Gemini series of flights had terminated, suitable constraints, tethers, and handrails had been tested, and the astronauts were able to carry out all their assigned tasks without difficulty.

What is perhaps surprising is that a man can step out of a vehicle hundreds of miles above the Earth and drift along beside it for hours without any sense of vertigo or disorientation. It is true that all those who have experienced this have been highly selected, trained, and motivated; it may well be doubted that the average person would enjoy it. But once again the adaptability of the human organism may astonish us; in view of the universal fear of heights, who would have believed a century ago that flight would be possible for almost everybody?

There was never any doubt that the other problems of maintaining life in space could be solved by straightforward engineering techniques. What made the building of "Life Support Systems" very difficult in practice were the contradictory requirements of extreme reliability and minimum weight; the Mercury capsule, in particular, was a tour de force of expensive engineering. Everything, including a heavy heat shield, had to be included in the 3,000-pound payload which was the maximum that the Atlas could inject into orbit. The much more powerful Titan booster used for the Gemini flights could orbit 8,000 pounds, but even this was little enough to keep two men comfortable in space for fourteen days. By contrast, Gagarin's Vostok weighed 10,000 pounds, and the three-manned Voshkod more than 12,000, so from the very beginning the Soviet Union was operating under much less severe weight restrictions.

This reflected itself in many details of design. For example, the Soviet space capsules had sufficient braking ability to touch down softly on land, with their crews inside (though the astronauts landed separately on the earlier flights). The American spacecraft had to splash down at sea, with all the resulting complications, expense, operational restraints, and possible dangers of a mid-ocean recovery. And perhaps even more important, the cabin atmosphere in the Soviet spacecraft was normal air, whereas to save weight and reduce complexity, the American designers elected to use pure oxygen. There was nothing wrong with this decision per se, but by a series of disastrous errors and oversights it led to the loss of three lives, tens of millions of dollars, numerous reputations, and perhaps a year of time on the journey to the Moon.

The air we breathe is normally under a pressure of slightly less than 15 pounds per square inch, and one-fifth of it consists of oxygen; the remaining four-fifths is nitrogen (with a trace of other gases) and plays no part in respiration.

From the physiological point of view, therefore, a pure-oxygen atmosphere at three-pounds-per-square-inch pressure is just as good as air (one-fifth oxygen) at five times that pressure. From the engineering point of view it has several advantages: the risk of leaks is smaller, the pressure cabin need not be so strong, extravehicular activities are easier, and the whole air-purification system is simplified. The fact that in the 1967 Apollo disaster the capsule under test contained (a) pure oxygen at *full sea-level pressure* (and a little more); (b) inflammable substances that had accumulated unnoticed, and (c) electrical equipment that may have been faulty, does not mean that there was anything basically wrong with the design.

Under normal conditions a man uses about two pounds of oxygen per day—a surprisingly small quantity—which presents no storage problems for flights of a few days or even a few weeks. After the "combustion" of the food which provides the human machine with energy, the gaseous exhaust products are carbon dioxide and water vapor; these must be continuously removed, otherwise the atmosphere will quickly become unbreathable. Various types of CO_2 absorber and water separator have been available for decades, largely as a result of submarine technology, though they have had to be carefully redesigned to work in the weightless condition. In principle, oxygen can be regenerated from the carbon-dioxide absorber, but the additional complications are not worth it except for the very long-duration missions involved in planetary flights or permanent space stations.

A man requires even less food than oxygen—about 1½ pounds per day for a 3,000-calorie diet. But this is the *dry* weight, assuming that it is completely dehydrated, as is the case with the freeze-dried foods used on the Gemini and Apollo missions. An astronaut also needs 5 to 6 pounds of water for drinking and for reconstituting the food.

However, the water problem is much simpler than the oxygen one, for it is easily extracted from the atmosphere, purified, and reused. In addition, the electrical generating system of fuel cells used on both Gemini and Apollo actually produces water as the reaction proceeds, so ample supplies are available for both consumption and toilet purposes.

The temperature of a spacecraft has to be regulated very accurately, for though men can survive for limited times over an extraordinary range (from above boiling point in very dry air, down to far below freezing), for optimum working conditions the cabin temperature should not stray outside 70–80 degrees Fahrenheit. In order to keep within these limits, a manned spacecraft usually has to be *cooled.*

This will come as a great surprise to those who have heard about the "intense cold" of outer space. But temperature, like color, is a property of matter; as space is a vacuum, it can be neither hot nor cold. Only an object in space can have any temperature, and the value of this will depend in a rather complicated way on the heat falling upon it from an outside source (usually the Sun, but possibly a nearby planet), its own rate of radiation into space, and any internal sources of heat (electrical, metabolic) it may possess.

A large manned spacecraft may generate many kilowatts of heat from its equipment and the bodies of its crew. (One kilowatt is the power of the average portable electric heater.) If this were all trapped by an efficient insulating system, neither machines nor men could survive for more than a very few hours. The excess heat has therefore to be radiated away into space by suitable cooling fins or surfaces. At the same time, it is just as essential to see that too much is not radiated away. The empty universe can absorb unlimited amounts of heat, as anyone who has stood under a clear sky on a still winter's night can testify. If the spacecraft's radiating system is *too* good, the temperature inside will start heading for absolute zero (−460 degrees Fahrenheit). It won't get there, of course; whether it levels off at minus 300 or only minus 100 depends partly on the vehicle's location.

If the spacecraft is in full sunlight, every square yard facing the Sun will receive almost 1½ kilowatts of solar heat, and it may easily get too warm. If it is in shadow—on the night side of Earth, for example—it will be shielded from this intense source of heat and will tend to get too cold. It must therefore have some way of adjusting its radiation to varying conditions, and this can be done by opening and closing reflecting screens. The fact that this problem has been already solved for the worst possible case—a close satellite that swings from midnight to midday every forty-five minutes—proves that this matter is fully under control.

A much more difficult heating problem is that encountered in re-entering the atmosphere; for a long time it was not even certain if this could be solved. The energy of a body moving at orbital speed is enormous; anyone who has picked up a rifle bullet immediately after it has hit a target will know that it is uncomfortably hot, and an object traveling at 18,000 miles an hour has at least thirty times as much energy. There is, in fact, no substance which would not be completely vaporized if all its orbital energy were converted into heat. This problem had to be solved as part of the ICBM program; if missile nose cones could not be brought safely back into the atmosphere, there seemed little hope that fragile human cargoes could survive the same treatment.

The answer was found in 1952 by H. J. Allen, chief of the high-speed research division of the Ames Aeronautical Laboratory. For almost half a century aircraft had been growing slimmer and more streamlined, and it seemed logical to assume that this process would continue as even greater speeds were attained. But this was a case where intuition, and even advanced mathematics, was completely misleading. At the hypersonic velocities of re-entry, where temperatures of up to 12,000 degrees Fahrenheit were encountered, all the needle-nosed models melted down within seconds.

Allen realized that the opposite approach was needed. By using a blunt body—the very reverse of streamlined—a powerful shock wave would be produced ahead of the missile, and most of the frictional heat would be carried off in a sheath of incandescent air; only a small fraction would leak back into the capsule itself. So evolved the inelegant, approximately conical shape first made famous by the Mercury vehicles; the Soviets, presumably using similar arguments, chose a completely spherical design, so that Gagarin flew around the world in a giant cannonball that might have come straight out of Jules Verne's novels.

Curiously enough, nature had given a hint that flattened, rounded shapes would best survive re-entry. The small, glassy meteorites known as tektites often assume this form, as they are fused and molded by their passage through the upper atmosphere. The importance of this minor astronomical curiosity, which might have saved the United States a few hundred million dollars had it been realized earlier, was first noticed by the meteorite expert H. H. Nininger, at the very moment that Allen was circulating his highly secret findings to the skeptical missile makers.[3]

Even with the blunt-nosed configuration, enough heat would reach the forward part of the spacecraft to produce temperatures of several thousand degrees, and it was therefore necessary to provide additional protection. After experiments with various alternative systems, a saucer-shaped plastic and fiberglass "ablation shield" was developed, which slowly burned or charred away during re-entry. Millions of Americans will remember the alarm felt during John Glenn's three-orbit flight, when a faulty indicator lamp suggested that his heat shield had come

---

3. The problems of the needle-nosed re-entry vehicle have now been solved, and it is now used in certain applications, especially for warheads designed to frustrate antimissile missiles.

adrift. If this had really been the case, nothing could have saved him from a meteoric fate.

In the decades before man went into space, there was much concern over the human body's ability to withstand the accelerations involved. However, space travel does not *necessarily* demand high acceleration; the time will come, though perhaps not in this century, when it will cause no more physical stress than the takeoff of a jet airliner. But for the reasons given earlier fuel economy requires that today's rocket vehicles perform their task as quickly as structural considerations allow,[4] and this involves peak accelerations (at the moment before engine cutoff, when the propellant tanks are almost empty) of up to eight gravities. Even higher accelerations are encountered during the return through the atmosphere, when up to 12 g may be experienced briefly. (At 12 g, a 170-pound man weighs one ton.) But thanks to form-fitting couches and prior training, the astronauts felt little more than momentary discomfort; they were even able to continue talking while their apparent weight increased almost tenfold and their blood became as dense as molten metal.

Anyone who has ever witnessed the takeoff of a large launch vehicle will have marveled that a human being can survive even within several hundred yards of such a continuous concussion; the unimaginable volume of sheer sound produced by a multimillion-horsepower rocket engine is not even faintly conveyed by radio or TV. But this, too, has proved to be little problem to the astronauts in their double-walled, insulated capsules. Rockets, like jets, leave most of their noise behind them. Though they may disturb whole countries, they do not inconvenience their passengers.

Apart from weightlessness, the greatest unknown hazards of space in the days before men entered it were meteoroids and radiation. There had been many conflicting estimates of their possible dangers, and much of the early experimental work with space probes was devoted to resolving these.

Meteoroids come in all sizes, from barely visible specks of dust to giant boulders, or even small mountains like the object which produced the famous Arizona meteor crater. They move at velocities of anything from 7 to 40 miles per second with respect to the Earth, depending upon whether they are overtaking it or meeting it head-on. Needless to say,

4. Ideally, a rocket should take off from the launch pad at the highest possible acceleration. The huge propellant tanks, however, would become impracticably massive and heavy if they had to withstand accelerations of several gravities when full. It turns out, after all the calculations ("optimization studies") are made, that for a large liquid-propellant rocket the best compromise involves a lift-off at the surprisingly low acceleration of about one-fourth of one gravity.

there is no hope of providing protection against the larger varieties; the only safety lies in statistics. Fortunately, those statistics are quite reassuring.

A meteoroid weighing as much as one ounce is exceedingly rare: 100 square feet of spaceship would experience an impact with such a giant about once every million days, or three thousand years. The numbers rise rapidly as the size of the particles decrease; for a 1/100-ounce meteoroid, the waiting period between impacts would be only about three years, but an object as small as this presents little danger to the hull of a spacecraft. The best proof of the relative harmlessness of meteoroids is the millions of hours of successful operation that robot probes—some with extremely fragile structures—have now accumulated beyond the atmosphere.

Meteoroids may be a nuisance—not a danger—to windows and optical elements (especially telescope mirrors) in space; they may eventually produce a kind of sand-blasting effect which could degrade performance. Additional protection for critical areas may be needed on very long voyages, especially through the asteroid belt between Mars and Jupiter, where there seems to be a great deal of space junk. And they may be a slight hazard to men wearing spacesuits, which naturally cannot provide as great a safety margin as the metal hull of a spaceship.

Even on those rare occasions when one of these cosmic bullets does penetrate the wall of a spacecraft, the damage is not likely to be serious; in most cases the small hole produced would merely add to the existing inevitable air leakage and could be easily sealed. (Self-sealing materials, like those used in aircraft fuel tanks, could be employed if necessary.) Even if the hull damage were quite serious—but short of catastrophic—there would normally be ample time for the crew to put on spacesuits and then set about repair and repressurization. It takes many minutes for all the atmosphere of a space cabin to escape, even through quite a large hole.

This seems a good point to deal with one of the most persistent myths of the Space Age—the almost universally held idea that exposure to vacuum would not only be instantly fatal, but could result in the victim exploding because of internal pressure. There was never any reason for believing this, and it has now been disproved experimentally. Dogs and chimpanzees have survived vacuum for astonishing lengths of time—up to 3 minutes—with no permanent ill effects, though they normally lose consciousness after about 15 seconds. A man who was psychologically and physiologically prepared for the experience (which does not even seem painful, though it is probably uncomfortable) would have at least a quarter of a minute of useful consciousness—a great deal of time in an emergency. And even after he had lost consciousness, he would recover if he could be repressurized within one or two minutes.

The human body is a tough piece of engineering, and, as every skin-diver knows, all its internal airspaces open into the surrounding medium so that pressure quickly equalizes. Swimming upward 10 feet, which takes only a few seconds, produces the same drop in pressure as opening an Apollo capsule to the vacuum of space.

The remaining hazard—radiation—has also turned out to be not so serious as was once feared, although the discovery of the Van Allen belt caused a momentary flurry of alarm. Beneath these zones of trapped radiation, astronauts can orbit for months without risk, and all long-range space missions will pass through the belts so swiftly that they can be ignored.

The real radiation danger may be the Sun—especially around its 11-year peaks of maximum activity, one of which coincides with the first Apollo flights. Several times a year, tremendous eruptions known as "flares" occur on the surface of the Sun, and these spray high-speed, charged particles (mainly the nuclei of hydrogen atoms) throughout the Solar System. It is possible to provide some shielding against these, at least for short journeys like the flight to the Moon. On longer missions, such as voyages to Venus or Mars, the situation may be more serious, and spaceships may have to be provided with a special "storm cellar" for protection against the occasional but possible lethal solar outbursts. Even this is not certain, and judging by the way in which all the other space bogeys have evaporated, no one will be surprised if solar flares also turn out to be more spectacular than dangerous.

Dr. Charles Stark Draper, president of the International Academy of Astronautics and head of the M.I.T. Instrumentation Laboratory, which developed the Apollo guidance system, once made the rather startling remark, "Space is a *benign* environment." It is beginning to appear that this is indeed the case. Certainly it is not so implacably, relentlessly hostile as the Antarctic or the ocean depths. It presents problems, as does all new territory, and we have to proceed with caution as we move into it. But it also presents tremendous opportunities which we now have the skill to exploit.

SCIENTIFIC SATELLITES

It is rather amusing, in the second decade of the Space Age, to look back on the hopes and predictions of those who first proposed the launching of artificial satellites—and to see how modest they were in the light of later achievement. In 1954, for example, the Space Flight

Committee of the American Rocket Society prepared a report for the National Science Foundation on "The Utility of an Artificial Unmanned Earth Satellite." Some of the points it rather diffidently made were:

ASTRONOMY AND ASTROPHYSICS. A satellite could overcome some of the limitations on observations made through the atmosphere. GEODESY (INCLUDING NAVIGATION AND MAPPING). The size and shape of the earth, the intensity of its gravitational field, and other geodetic constants might be determined more accurately. Practical benefits to navigation at sea and mapping over large distances would ensue. GEOPHYSICS (INCLUDING METEOROLOGY). The study of incoming radiation and its effects upon the earth's atmosphere might lead eventually to better methods of long-range weather prediction.

Within five years of the committee's report, all these forecasts had been amply demonstrated; within ten, the sciences mentioned were undergoing something like a revolution. And this was entirely due to the fact that, for the first time, it had been possible to establish observing stations above the atmosphere.

In the first few years of the Space Age scores of "scientific" satellites were launched, carrying instruments for hundreds of experiments. Some were extremely simple; others, like the Soviet Union's giant "Proton" satellites, nothing less than orbiting physics labs. Even to list them would take pages, and to describe the results they have obtained has already required many volumes. At the Goddard Space Flight Center, Maryland, where the instrument readings are stored for later analysis after they have been relayed to earth, hundreds of thousands of reels of magnetic tape are stacked in endless rows while the scientists try to cope with the flood of new knowledge pouring down from the stars.

Only a few typical or unusually interesting satellites will be described here, together with a sampling of the results they have obtained. But which of those results are the most important we may not know for generations. Only time will tell what secrets are now hidden away in the vaults at Goddard, waiting to demolish long-held theories or to establish new ones.

For simplicity, it would be hard to beat the "balloon" satellites, of which Echo 1 was the first and most famous. On June 24, 1966, NASA launched a singularly perfect specimen, the 100-foot-diameter Pageos, which looks exactly like a giant, highly polished ball bearing. Made of mylar film 0.0005 inch thick, Pageos weighed only 120 pounds and when inflated in orbit was half a million times larger than the canister into which it had been skillfully packed. Moving in a polar orbit at an altitude of 2,600 miles (period 181 minutes), it is easily visible to the naked eye (Figure 10-17).

Its purpose is geodesy—the mapping of the Earth to a degree of pre-

Figure 10-17. *The Pageos satellite.* (NASA)

cision never before possible. Surveyors thousands of miles apart can observe it simultaneously and photograph it as it moves across the stars;
when analyzed, these photographs will allow points on the Earth to be
fixed to within about ten yards. Similar results have also been obtained
by using satellites carrying flashing strobe lights (Anna 1-B, 1962) or
mirrors reflecting back laser beams (Geos 1, 1965), but Pageos requires
the minimum of ground equipment.

High-precision studies of orbits made possible by satellites of this type
have already revised our knowledge of the Earth's shape. It is not a simple flattened sphere (ovoid); there are bumps and bulges which will tell
us much about its evolution and the distribution of matter in its interior.
Of course, these deviations from the ideal shape are very small—utterly
invisible to the eye of the astronaut looking back at his home from a few
thousand miles away. But they are of great importance scientifically,

and it is a curious thought that knowledge of the Earth's interior can best be obtained by going far out into space.

However, most of the instrumentation aboard satellites has been designed to study the environment through which they pass, and undoubtedly the greatest discovery yet made is that due to the very first United States satellite, Explorer 1 (1958). This revealed, as someone put it expressively, that "space is radioactive," and for a while there was considerable alarm about the effects of this discovery upon orbiting astronauts.

We now know, thanks to Explorer 1 and its much more elaborate successors, that our planet is surrounded by a huge radiation belt, roughly doughnut-shaped, with the Earth in the hole. The inner part of the belt —named after its discoverer, Dr. James Van Allen—consists mostly of positively charged protons (hydrogen nuclei) and reaches its maximum intensity at a height of about 500 miles. In the outer zone, negatively charged electrons predominate, with their maximum intensity at 10,000 miles. At one time it was believed that there were two separate belts, but it is now known that they merge into each other, though they are separated by a region (about 8,000 miles high) where the intensity of radiation is a minimum. There is also very little radiation over the poles; it all lies above the equatorial and temperate zones.

The great radiation belt is produced by streams of electrons and protons from the Sun, which have become trapped in the Earth's magnetic field. As a result there is an extremely complicated interrelation between solar and terrestrial magnetic activity, both of which vary with time. For tens of thousands of miles around the Earth there is an invisible cloud of electronic and protonic "weather," with its storms and winds and calms, never suspected until our generation.

The great radiation belt is not symmetrical; the gale of charged particles "blowing" from the Sun compresses it on the daylight side of Earth and makes it trail out on the night side. The doughnut is therefore badly distorted—three or four times thicker on one side than on the other. At its outer fringes it merges imperceptibly into the (very weak) general background of radiation between the planets.

Almost every scientific satellite launched from the Earth—as well as many space probes on greater journeys—has carried instruments to measure the ever-changing phenomena in the great radiation belt. A good example is the Orbiting Geophysical Laboratory (OGO), of which the third, and first fully successful one, was launched on June 6, 1966.

OGO 3 has an orbit specifically designed to sample an enormous volume of the near-Earth environment. Its perigee is only 170 miles up, but its apogee is 75,800 miles from Earth, or one-third of the way to the Moon. Completing this very elongated path once every two days, it reports on the energy and concentration of the protons and electrons in the radiation belt, fluctuations in the Earth's magnetic field, radio propagation characteristics, cosmic rays, interplanetary dust, radio noise—to mention only some of the twenty-one separate experiments it conducts.

One of the most important—and most uncertain—characteristics of the space environment before artificial satellites became available was the frequency of meteoroids; some pessimists believed that any spacecraft would be riddled by cosmic machine-gun fire as soon as it left the protective blanket of the atmosphere. In fact, meteoroids have turned out to be so rare that it is quite difficult to accumulate reliable statistics about them, and rather heroic efforts were needed to do so. Perhaps the most impressive of these were the launchings of the three huge Pegasus satellites (February 16, May 27, July 30, 1965) by the last three of the Saturn 1 rockets. In orbit, the Pegasus satellites extended vast "wings," almost a hundred feet across, which consisted of thin aluminum panels, varying in thickness between 1.5 and 16 thousandths of an inch. These panels were connected to electrical circuits which reported any meteoroid penetrations, and signaled them back to Earth. After many months of successful operation, the three Pegasus satellites showed conclusively that, at least for flights of short duration, meteoroids were not a serious danger.[5]

The ionosphere—that electrified layer in the upper atmosphere which reflects radio signals back to Earth, and so makes long-distance communication possible around the curve of the globe—is one piece of near-space that has been of tremendous scientific, commercial, and military interest for half a century. It is not surprising, therefore, that dozens of satellites have been launched to investigate it. One of the first and most successful was the Canadian Alouette (lark), sent into a 600-mile-high polar orbit on September 29, 1962. It carried a radio transmitter which swept continuously across the VHF (very-high-frequency) band; after the signals from this "topside sounder" had passed through the ionosphere, they were picked up by ground stations and their intensity gave a measure of the electron density in this region. Ionospheric probing was also the main purpose of the first United Kingdom satellite, Ariel 1 (April 26, 1962). Its instruments measured and sampled the charged particles between 242 and 754 miles above the Earth.

From the scientific point of view, perhaps the most exciting satellites are those which, although they may be within a few hundred miles of

---

5. Note on terminology: a *meteoroid* (usually micrometeoroid) is a small solid object moving through space; a *meteor* is the streak of light it produces when it enters the atmosphere; a *meteorite* is what remains in the rare cases when it reaches the ground. The words are often used interchangeably.

the Earth, are looking outward to deep space. For by lifting our instruments through a distance which is quite trivial, even by terrestrial standards, we have been able to obtain a completely new view of the universe.

Until our age, astronomers had to make all their observations from the bottom of the atmosphere. As a result, they were rather like color-blind men straining their eyes through a fog—or, to use a well-known and not inaccurate analogy, like fish peering upward from the bottom of a muddy pool.

On a clear, moonless night, when we look up at the stars, it seems that there is nothing to obscure our view. But this is an illusion—the result of evolutionary necessity. Our eyes have, naturally enough, adapted themselves to use the light which passes through the atmosphere, and that is only a small fraction of the radiations that fall upon the Earth from space. Most of them, luckily for us, are completely blocked by the 100-mile-thick gaseous shield above our heads. . . .

The first Orbiting Astronomical Observatory was launched on April 8, 1966. It contained a battery of telescopes (the largest having an aperture

Figure 10-18. *Unmanned orbital telescope.* (American Institute of Aeronautics and Astronautics, Inc.)

of 16 inches), and with its 440,000 separate parts and 30 miles of wiring
was, at that time, one of the most complex satellites ever developed. It
was injected into a perfect orbit, but within a few hours something went
wrong with its power supply and its signals slowly faded out. In its first
attempt to see the unknown universe of ultraviolet stars and nebulae,
the United States had gambled enough to build half a dozen Mount Palomar telescopes—and had lost. . . .

The heartbreaking demise of the first multimillion-dollar Orbiting Astronomical Observatory—which might have been saved had there been
a man on the spot with a screwdriver—strongly suggests that the very
large and complex scientific satellites of the future will be designed for
easy servicing, even if they are not permanently manned. As a step in
this direction, a large number of experiments were carried out by the
Mercury and Gemini astronauts.

. . . the robot probes which have been leaving the atmosphere in such
numbers have already started a revolution, comparable to that which
began three and a half centuries ago when Galileo pointed his first
crude "optic tube" at the heavens. Every breakthrough in instruments
produces a corresponding breakthrough in knowledge; the satellites
have given us new eyes and ears, and for a long time to come we will be
dazzled and deafened by the information they bring us. But, later, we
will begin to understand.

FIRST HARVEST

In the last section we glanced briefly at the new knowledge now
being obtained by artificial satellites; in this one we shall look at their
practical uses. It is essential to realize, however, that this is a very arbitrary distinction, and the dividing line is constantly on the move. All
really great advances in technology, as opposed to mere gadgetry, arise
from scientific discoveries which at the time seem to be of no relevance
to everyday life. Electric power and light were made possible because
men like Faraday played with magnets and coils of wire, trying to understand the workings of nature and sometimes even taking a perverse
pride in the invariably mistaken belief that their discoveries would
never be of use to anybody. Yet always, a generation or two later (a
decade or two later, nowadays), the Edisons come along and turn their
"pure" science into billion-dollar industries.

As yet, the magnetic fields in space, the great radiation belt, the har-
monics of the Earth's gravitational field, the solar wind, and similar exotic phenomena have little value in the marketplace. But their time will
come.

Meanwhile, there are types of satellite which have obvious and immediate practical uses, which everyone can appreciate and many millions can indeed share. These are the so-called applications satellites, which do not gather scientific facts, but work for a living. (Many do both, for applications satellites usually carry instrumentation of various kinds.) . . .

A series of Transit satellites [aids to navigation] was launched between 1960 and 1964; they included the first satellites ever to be powered by nuclear energy, and the system became operational in July, 1964. Although at first its main customers were the Polaris submarines, it is now available for use by all ships which fit the fairly simple receivers and computers required; oceanographic survey vessels have particularly benefited by it.

More advanced systems are now being developed which will involve satellites fixed in the 24-hour stationary orbit above the equator; this arrangement will be much simpler and will permit a far wider range of users. The time will come, in the not-too-distant future, when a wristwatch-sized computer turned to the Navsat network will tell a man exactly where he is anywhere on the surface of the globe, and no one need ever again be lost, even in the remotest corner of the world.

Now that millions of TV viewers are accustomed to seeing, in their regular weather forecasts, photographs of cloud cover over whole continents, it may seem surprising that anyone ever doubted the utility of meteorological satellites. Yet their value was not at first obvious even to the experts, as I can testify from personal knowledge.

When the American Museum of Natural History's Hayden Planetarium asked me to arrange its 1954 symposium on space flight, I wrote to Dr. Harry Wexler, chief of research of the U.S. Weather Bureau, suggesting that he should present a paper on the meteorological uses of satellites. I was somewhat taken aback when he replied that they would be of very little value. After brooding awhile I wrote again, challenging him to demonstrate this—if only to stop us space cadets from wasting the valuable time of the meteorological authorities. To his credit, Dr. Wexler accepted the challenge; by the time he had written his paper, he had converted himself completely. Afterward he became the United States' chief protagonist for this new research instrument and played a major role in the development of meteorological satellites until his death in 1962. Perhaps I should add that Dr. Wexler's attitude was precisely correct and demonstrates all the stages (skepticism, inquiry, enthusiasm) a scientist *should* pass through when confronted with some novel and (in this case literally) far-out idea.

The first meteorological satellite, Tiros (Television and Infrared Ob-

servation Satellite), was launched on April 1, 1960, into an almost circular orbit a little more than 400 miles above the Earth. Because its orbit
was inclined to the equator at an angle of 48 degrees, during the course
of every few revolutions it ranged over half the surface of the globe. As
the first of its 22,500 photographs was received by the ground stations,
the meteorologists realized that they had, as one of them put it, "gone
from rags to riches overnight."

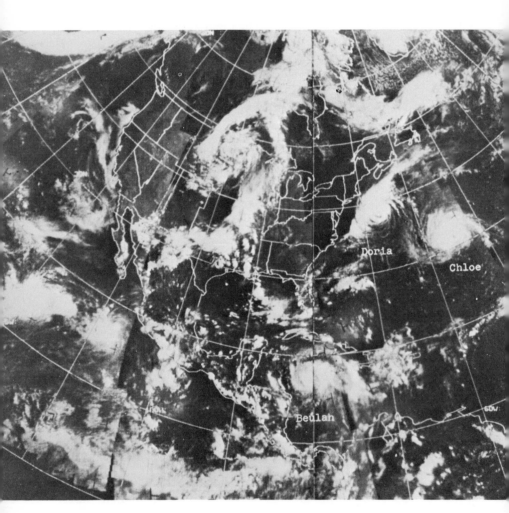

*Figure 10-19. Three hurricanes visible on a map-photo compiled from pic-
tures taken by the ESSA #5 weather satellite in September, 1967, and transmitted via ATS 1.* (NASA)

Seven further satellites were launched in the Tiros series between April, 1960, and December, 1963, all into similar orbits with periods of 100 minutes. Most of them equaled or exceeded their designed life, and between them they sent back to earth several hundred thousand photographs. In addition, they carried instruments that could measure the flow of heat from our planet back into space—information vital to the meteorologist but previously unobtainable.

The last of the Tiros series—9 and 10—were even more successful; they were launched into high-inclination (81–82-degree) orbits so that they passed almost over the poles and gave virtually global coverage. Similar orbits were used by the still more advanced Nimbus and ESSA (Environmental Science Services Administration) satellites, which have continued and extended the work begun by Tiros, establishing the world's first operational weather-satellite system. Today any electronic enthusiast with a few hundred dollars and a certain amount of ingenuity can build a simple ground station that can interrogate the ESSA satellites as they pass overhead, and can read off, on a cathode-ray tube or commercial facsimile receiver, the weather picture for a thousand miles around him. This Automatic Picture Transmission (APT) system was introduced with Tiros 8 in December, 1963, and has made the multimillion-dollar satellites freely available to any country or any individual who cares to use them.

The first high-definition photos of an entire hemisphere, made by a satellite sufficiently far from Earth to show it as a planet, were obtained in December, 1966, from the first of the Applications Technology Satellites. Stationed over mid-Pacific, ATS 1's special electronic camera produced superb studies of changing cloud patterns over almost half the Earth; these were later combined to give speeded-up movies so that meteorologists could watch the circulation of the atmosphere with their own eyes, thus learning in a few minutes facts which might never have been revealed in years of ground-based observations.

It is probable that the various "metsats" have already paid for themselves many times over. They have detected hurricanes far out at sea, hours before their existence could have been discovered in any other way. They have improved the quality of weather forecasting, with all that this implies to human wealth, productivity, safety, and happiness. It has been claimed that really accurate prediction of rain, snow, monsoons, and other meteorological phenomena will eventually be possible, thanks to the new knowledge from this source—with savings that have been estimated at *tens* of billions of dollars a year. Looking even further ahead, if weather control or modification ever becomes possible (or desirable, which is not at all the same thing), it can hardly be attempted without the complete understanding of atmospheric processes that only satellites can provide.

The ATS satellites, built by the Hughes Aircraft Company and exploiting the technology its engineers developed for Syncom and Early Bird, may be regarded as a series of space buses for testing various practical applications of satellites. In addition to obtaining meteorological information by their own onboard cameras, they can serve as communications links in an elaborate data-collecting system, relaying information from dozens of ground stations. This information includes readings from rainfall gauges, oceanographic buoys, met balloons—and possibly beacons attached to large land and marine animals for zoological research.

Probably the most interesting, and doubtless the most advanced, of the applications satellites are those about which nothing has been published; in some cases, even their names are classified. I refer, of course, to the military reconnaissance satellites.

In his 1929 book Hermann Oberth had already pointed out that a manned space station could be used to watch the movements of warships. With the development of TV techniques and camera-carrying capsules that could be recovered from orbit, satellites became of intense interest to the military, particularly after the 1960 U-2 debacle had proved the vulnerability of reconnaissance aircraft in the missile age. It is true that a satellite can also be destroyed, perhaps more easily than it can be launched. But there is a very important distinction; even the people who use reconnaissance aircraft admit that they are illegal and apologize when they are caught, for they are operating in another country's airspace. Satellites, however, fly only in the no-man's-territory beyond the atmosphere and are at liberty to take as many pictures as they please. It is true that for a while the Soviet Union considered that they were provocative and unfriendly, but since it too started using them in large numbers, very little has been heard of this objection.

The United States Air Force has, naturally, been most active in this field; it orbited its first Samos reconnaissance satellite on January 31, 1961, and since then has launched dozens of anonymous payloads, mostly into polar orbits at fairly low altitudes, so that they can thoroughly scrutinize the whole earth. The quality of the resulting photographs may perhaps be judged by some of those sent back from the Moon by the Orbiter vehicles. Where there is physical recovery of the capsules (as happens with the Discoverer satellites), the definition may well be much higher. Although haze and cloudiness set operational limits to the system, the photographs brought back by the Gemini astronauts show the astonishing amount of detail that can be observed from space when the atmosphere is clear.

There are some military satellites which do not depend on light waves and so are less affected by weather conditions. The Midas satellites were designed to spot ICBM launchings by detecting the immense amounts of infrared radiation produced by rocket exhausts.

Other space vehicles listen in to radar and communications networks; yet others are involved in precision mapping and navigation. (The Transit program, mentioned previously, was classified for some years because of its military applications.) And particular mention should be made of the VELA, or Sentry, satellites, which swing slowly along almost circular orbits 70,000 miles above the Earth, waiting to detect clandestine nuclear explosions.

The Soviet Union has its counterpart to this program, though it talks about it even less than does the United States. As long ago as December, 1965, it reached number one hundred in its rather mysterious Cosmos series, most of which return to Earth after a few days' traveling along close, high-inclination orbits. It may be doubted if their purpose is always entirely scientific.[6]

On balance, these satellites have probably had a stabilizing effect upon international affairs; they have made a reality of President Eisenhower's imaginative "Open Skies" proposal. The advance announcement of Chinese nuclear tests by the United States proves that it is now impossible for one country to conduct military preparations without the knowledge of the two super powers; nor can these hide anything from each other. It has been stated that the United States reconnaissance satellites have already paid for the *entire* space program—for by revealing that the Soviet Union's missile deployment was not as fast as had been feared, they allowed the Department of Defense to establish more modest goals for its own ICBM program. The Samos satellites have been worth many times their weight in gold to the United States taxpayer.

A few months before Sputnik 1 opened the Space Age, the following wild-eyed prophecy appeared in print: "It may seem premature, if not ludicrous, to talk about the commercial possibilities of satellites. Yet the airplane became of commercial importance within thirty years of its birth, and there are good reasons for thinking that this time scale may be shortened in the case of the satellite, because of its immense value in the field of communications" (*The Making of a Moon,* Clarke, 1957). The first $100 million of Comsat stock went on the market seven years later (June 2, 1964) and promptly disappeared into myriad safety deposit boxes.

6. Some of these—e.g., Cosmos 57 (February, 1965)—have exploded into hundreds of fragments, to the great annoyance of the satellite-tracking networks. It has been suggested that this was to prevent them from descending on United States territory.

The idea of employing satellites as radio relays, so that all possible wavelengths—including light, if desired—could be used for communications purposes, now seems a rather obvious one, and it is somewhat surprising that it did not appear until 1945. It is true that Oberth, in his 1929 classic, *The Road to Space Travel*, mentioned that manned space stations could signal to remote parts of the Earth by flashing *heliograph* mirrors, which today seems a very primitive idea. We tend to forget that the astonishing developments in electronics, miniaturization, and communications techniques which now permit us to control robots on the surface of the Moon, or in orbit around Mars, have become possible only since World War II. Willy Ley once pointed out that when Oberth wrote his book, the only long-range radio stations in existence used antenna systems acres in extent, supported by towers hundreds of feet high. The idea that this sort of equipment might one day be squeezed into a hatbox would then have seemed slightly more fantastic than space travel itself. Even as late as 1945 I still assumed that communications satellites would be large, *manned* structures. Several years of battling with balky electronics had convinced me that it was essential to have a servicing engineer on the spot; I have modified this position only slightly.

The simplest type of communications satellite is passive, an orbiting radio mirror which reflects signals back to Earth without itself modifying them in any way. Such was the giant Echo balloon, launched on August 12, 1960, into a 1,000-mile-high orbit, and for a long time one of the most conspicuous objects in the night sky. Echo 1 (and its slightly larger successor Echo 2, launched into a near-polar orbit on January 25, 1964) was used for many test transmissions of speech, teletype, and facsimile and clearly demonstrated the potential value of satellites for communications. However, passive systems (though they are simple, have nothing to go wrong, and can provide an unlimited number of circuits) are extremely inefficient; only a tiny fraction of the power beamed at the Echo balloon actually fell upon it, and an even smaller fraction of that power was picked up by ground stations. Although some of these limitations may be overcome (for example, by replacing the spherical reflector by one so shaped that it sends a much larger signal back to Earth), passive systems appear to be largely of historic interest.

Active satellites are true relays, receiving the signal from the ground station, amplifying it, and rebroadcasting it at greatly increased power (and at a different frequency, to avoid interference). Such a system, though complex, is millions of times more efficient than a passive one; it received its first public demonstration with Telstar 1 (July 10, 1962). Though the United States Army's earlier Atlas-Score (December 18,

1958) and Courier (October 4, 1960) had provided a very limited experimental service with radio signals only, Telstar heralded the age of intercontinental television when it inaugurated the first live transatlantic program on July 23, 1962. . . .

The United States moved swiftly to set up an operational communications-satellite system; the result was the remarkable semipublic, semiprivate, national-international Communications Satellite Corporation, established by Act of Congress in 1962. Comsat's first child, Early Bird, was launched on April 6, 1965 [7] and placed in service on June 28; it could provide either one TV channel *or* 240 voice (telephone) circuits, but not both. Because of this limitation, the initial rates were high, and before long Comsat was receiving squawks of protest from indignant customers. Exactly the same thing had happened 99 years before, when the first successful Atlantic telegraph went into operation. It had taken just under a century to progress from cable to satellite.

In the spring of 1967 Early Bird was joined by two larger brothers—Intelsat 2 over the Pacific and Intelsat 3 over the Atlantic. With these three satellites [in synchronous orbit, 22,000 miles above the equator so that they appear fixed in the sky], all the world's TV networks could be linked together, and the first global telecast was broadcast on June 27, 1967.

Meanwhile, the Soviet Union had not been idle and had launched Molniya (lightning) communications satellites of its own, into unusual, highly elliptical orbits, with a perigee only 300 miles up and an apogee 24,000 miles high. At first it was thought that Molniya 1 had failed to go into the synchronous orbit, but it was soon realized that its high inclination to the equator (65 degrees) and period (almost exactly 12 hours) permitted it to arc slowly high over Russia at the same time each day. For a country in northern latitudes, such an orbit had some advantages over the synchronous, equatorial one.

The first generation Comsats were all low-powered devices, so that their signals could be picked up only by sensitive receivers coupled to large antenna systems; the ground stations using them cost $1 million or more and were linked to the various national television or telephone networks. However, many experts believe that the *real* communications revolution will start when Comsats are large and powerful enough to

7. I was present at Comsat headquarters on that memorable occasion and watched the launch on closed-circuit TV. The three-stage thrust-augmented Thor Delta booster was still on the way up when Vice-President Humphrey started to give us one of his little speeches. The circuit to Cape Kennedy was switched off, and it occurred to me that if anything went wrong now, everyone in the United States would know it *except* the staff of Comsat. They were all listening to the Vice-President.

broadcast directly into the home, bypassing the ground stations completely. Only in this way will it be possible to open up the undeveloped countries—Africa, South America, much of Asia—which have never had, and now may never require, surface communications networks.

Direct *radio* (voice) broadcasting from Comsats to simple ground receivers is already technically possible and could have immense social, political, and educational consequences. TV broadcasting, which requires much more power, will take a little longer, but even here the problems are not so much technical as they are economic and political —especially the latter, for direct broadcasting obliterates national and linguistic boundaries and means, among many other things, the end of censorship. It is not surprising that some countries are very worried about it.

Whole volumes and innumerable international conferences have been devoted to the social impact of space communications.[8] Within a lifetime, they may change our world out of recognition and alter the patterns of business and society at least as much as the telephone has done. They may give us instant "newspapers," with updated hourly editions flashed on to portable receivers no bigger than this book; they may make all telephone calls local ones, so that it will be just as quick and cheap to call a friend at the antipodes as in the next apartment; they may result in the swift establishment of English (or Russian, or Mandarin . . .) as a global language; they may result in the disintegration of the cities and a great reduction in travel, as telecommunications plus telecontrol will allow most men at the executive grade to live wherever they please. And there may be even more dramatic changes, for good or bad, that no one can foresee today—any more than Samuel Morse or Thomas Edison could have imagined that one day a quarter of the human race would watch the same pictures and hear the same sounds.

Whether we like it or not, the world of the communications satellites will be one world. In the long run, the Comsat will be mightier than the ICBM. It will put the clock back to the moment before the building of the Tower of Babel—when, according to *Genesis* 11: "The Lord said: Behold, they are one people, and they have all one language, and this is only the beginning of what they will do; and nothing that they propose to do will now be impossible for them."

8. I have combined my essays on the subject in *Voices from the Sky*, which also contains a "Short Pre-history of Comsats, or: How I Lost a Billion Dollars in my Spare Time." For the propaganda uses of Comsats, see the short story "I Remember Babylon," in *Tales of Ten Worlds*. Though this is required reading by Comsat staff, I do not wish to raise any false hopes.

A Fateful Step into a Vast Unknown °
by Walter Sullivan (July 20, 1969)

If the Lunar Module carrying two American astronauts lands safely on the moon this afternoon, it will indisputably be a landmark in human history. But it will also be unique in that such a large proportion of mankind will bear it witness.

In the past, as a rule, only a few were aware of the great events that altered the course of man's destiny. Often centuries of perspective were needed before their significance became evident.

There were scientific discoveries that set in motion new epochs of human thought, such as those of Archimedes, Pythagoras, Copernicus, Newton, Darwin and Einstein. There were those who ultimately shaped the religious beliefs of millions—Moses, Jesus, Buddha and Mohammed.

There were great battles that turned the course of history, such as Marathon and Waterloo. And there were the great voyages, such as those that opened the new world to the old.

But the world was in no real sense a witness to these events, and but dimly conscious—if at all—that it would never be the same again after them.

The drama now coming to its climax as three American astronauts circle the moon, prior to a landing by two of them, represents a special kind of milestone in that it is the first voyage of human beings to another celestial body—and the entire world is watching.

The question that tantalizes is whether this is comparable merely to the ascent of Everest—achievement of the ultimate goal of conquest of the earth's geography and, to some extent, the end of an era. Or whether it is, in fact, the start of an era—a first step toward the other planets and perhaps, even, toward the stars.

In some respects the Apollo 11 mission is reminiscent of early efforts to reach the North and South Poles. Then, as now, brave men set forth to penetrate an environment utterly hostile to man. Without their special clothing and supporting parties, they would have been lost.

Yet Apollo differs from the brave feats of such past explorers in a basic way. Those who set forth across unknown seas and continents, from the Vikings to Columbus, the explorers of the North American hinterland and the polar regions, did so on their own.

Figure 10-20. *July 16, 1969, Apollo 11 blast-off for the moon.* (NASA)

Anxious friends and supporters bade them farewell and waited months or years, aware that they might already have perished through some mishap that would never come to light.

A TRIP FOR ALL

Perhaps the last of these lonely adventurers was Charles A. Lindbergh, who, on his 1927 flight across the Atlantic, carried no radio. Only through his shouted greetings to fishermen off the Irish Coast and the dramatic appearance of his plane over Paris did the waiting world learn of his success.

Now we ride all the way with the Apollo astronauts. Through an extraordinarily elaborate complex of monitoring systems, specialists here via radio telemetry keep track of hundreds of functions not only inside the spacecraft but inside the astronauts themselves.

And finally, through television [relayed by satellite] almost the whole world watches. It was estimated that more than half a billion people witnessed the memorable start of the journey from Cape Kennedy

Wednesday morning. The audience probably will be a good deal larger when astronauts Neil A. Armstrong and Edwin E. Aldrin Jr. step out onto the moon.

Next Steps in Space °

Thomas O. Paine (1969)

Mankind entered a new era at Tranquility Base—an era in which travel will be reckoned not in thousands of miles but in millions and bil-

Figure 10-21. *Footprint of the first man to walk on the moon.* (NASA)

° Condensed from *National Geographic,* December 1969. Internatic nal copyright by the National Geographic Society, Washington, D.C. Reprinted by permission.

Figure 10-22. Edwin Aldrin joins Neil Armstrong 19 minutes later. The life-support system on his back contains oxygen, water for cooling, electric-power supply, and radio equipment. (NASA)

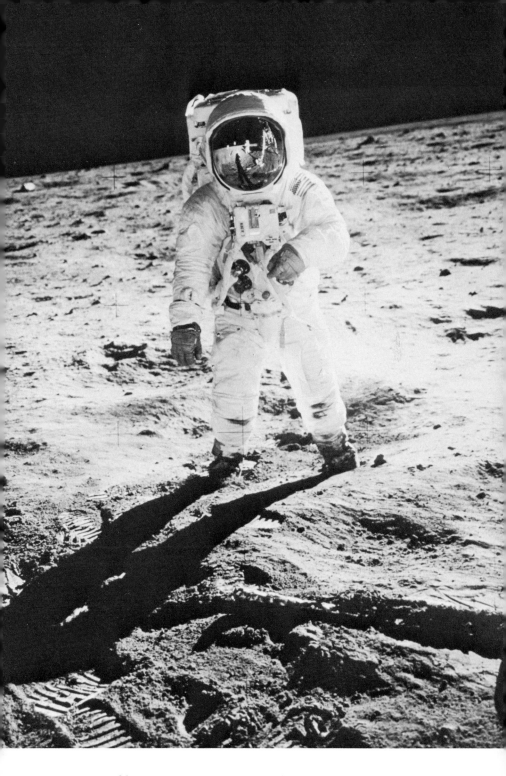

Figure 10-23. *Aldrin's visor reflects his own long black shadow, the lunar module, and Neil Armstrong as he took this picture.* (NASA)

lions. Space is an endless frontier for our children, and for all future generations.

I believe that men will drive onward in the years ahead to Mars, to the moons of Jupiter, and to other new worlds in our vast solar system. Some of these destinations are attainable in this century, some even within the next two decades. If we give full rein to our growing space capabilities, we can rapidly establish a bridgehead in the heavens in the next dozen years.

Figure 10-24. Scientific experiments being set up on the moon. The seismic unit (four seismometers in one) is powered by solar panels and nuclear heaters to help protect it against the severe cold of the lunar night. This device measures tremors and lunar quakes. Behind the seismic unit is the laser reflector. Consisting of a hundred prisms, this super-mirror reflects powerful pulses of laser light beamed from the earth in order to measure very precisely the distance from the earth to the moon. (NASA)

In the mid-1970's, for example, we could begin to assemble in earth orbit a permanent manned station. Gradually enlarged, it would become the work site of perhaps a hundred scientists.

In the late 1970's we could establish on the moon a base camp that could be occupied for months or even years.

In the 1980's we could send men to Mars—a voyage that would test our technology and equipment for travel to Venus and other planets later on.

In addition to these manned ventures, we will learn more about our solar system from unmanned probes. Several already are scheduled for the 1970's. These include flights to orbit Mars, others to land there, fly-bys of Jupiter, and the first multiple-planet flight, for which the targets will be first Venus, then Mercury.

All these are exciting prospects. But they raise the most fundamental of questions: To what goals in space should we now commit ourselves as a Nation?

My own belief is that we should press forward vigorously with a balanced program—scientific and technological development as well as exploration. Of course our goals, and the pace at which we strive to attain them, must reflect our national will, and there are well-informed and reasonable men who feel we should proceed more slowly.

REWARDS OF PROGRAM ALREADY GREAT

It has been said that we should concentrate all our resources on problems here at home. But I believe it would be a tragedy to foreclose our future in space. I believe our Nation can and must do these things simultaneously—not just one at a time.

Space exploration already has made life better on earth. Satellites, to mention just one development, have been of enormous benefit. They provide more accurate data to weather forecasters, aid mariners and aircraft pilots in fixing their positions, and give map makers hitherto unobtainable details of the earth's surface. In the years ahead, they will find undiscovered mineral deposits and sources of fresh water; make global agricultural surveys and detect diseased crops; and even help in the fight against pollution of air and water.[1]

And the conquest of space is everywhere lifting men's horizons and

1. See "Remote Sensing: New Eyes to See the World," by Kenneth F. Weaver, *National Geographic*, January 1969.

HERE MEN FROM THE PLANET EARTH
FIRST SET FOOT UPON THE MOON
JULY 1969, A. D.
WE CAME IN PEACE FOR ALL MANKIND

NEIL A. ARMSTRONG
ASTRONAUT

MICHAEL COLLINS
ASTRONAUT

EDWIN E. ALDRIN, JR.
ASTRONAUT

RICHARD NIXON
PRESIDENT, UNITED STATES OF AMERICA

Figure 10-25. Mementos left on the moon: the plaque on the abandoned landing stage (signed by the astronauts and President Nixon), and a shoulder patch of the Apollo 1 mission honoring Astronauts Gus Grissom, Edward White, and Roger Chaffee, who died in launch simulation fire. Also left but not photographed here were medals commemorating Yuri Gagarin and Vladimir Komarov who lost their lives in the Soviet space effort, a silicon disk etched with goodwill messages from leaders of 73 countries, and an olive branch symbolizing peace. (NASA)

spirits. Not only have global satellite communications brought nations closer, but—as Col. Frank Borman's warm reception in the Soviet Union showed—space achievements are crossing the barriers that divide men on earth.

Although other targets will come within reach, the moon will occupy man for many years. . . .

We have other immediate tasks: to make space travel simpler, more reliable, and much cheaper. How can we achieve these goals?

First, we must develop re-usable rocket planes, able to shuttle hundreds of times between earth and earth orbit. Even a Volkswagen would be prohibitively expensive if we threw it away after each drive— and each Saturn V rocket costs $150,000,000.

Second, we should harness the great potential of nuclear power for deep-space flight—that is, beyond earth orbit. Our most powerful chemical rockets cannot deliver to far-off destinations the heavy payloads manned flight demands. Nuclear rockets can.

Third, a permanent station in earth orbit would enable us to conduct needed research in many fields and would serve as an operations base for deep-space ventures.

Designers already can envision the re-usable craft we will need to shuttle between earth and the orbiting space base. These large rocket planes would take off vertically from earth, fly to orbit, discharge their cargo, return to earth, and land horizontally, using wings, like conventional aircraft. They could carry a dozen passengers—physicists and astronomers, perhaps—into space. They could haul 10 tons of supplies and deploy and recover unmanned satellites. Similarly, re-usable nuclear shuttles would link the space base to a base in lunar orbit, and to other stations.

Research would be a major task in the earth-orbiting space station, with flight operations increasing as more men traveled outward from earth. Scientists would investigate the effects of zero gravity on men, animals, and plants; study the heavens without the interference of earth's atmosphere; and develop new uses for earth-scanning satellites.

NUCLEAR ROCKETS TO PROBE DEEP SPACE

. . . The nuclear rocket promises to be the work horse of deep-space flight. At the NASA-Atomic Energy Commission test center in the Nevada desert, a prototype already has achieved twice the thrust per pound of fuel of our most powerful chemical rockets.

We call the prototype NERVA—nuclear engine for rocket vehicle application. Perhaps its most astonishing characteristic is its diminutive size; the reactor is no larger than a household refrigerator. Yet it generates more horsepower than Hoover Dam.

The nuclear rocket develops its great thrust by transferring the extreme heat of uranium fission—3,640 ° F. in our prototype—to hydrogen propellant. The superheated hydrogen then exhausts at great velocity through the rocket's nozzle. The rocket probably will be ready in the 1980's for a manned trip to Mars and return—an odyssey that will span a year and eight months.

Although Mariners 6 and 7 provided us with a wealth of new data as they flew by Mars last summer, we still do not know if there is, or ever has been, life there. We expect to learn more from two additional Mariner spacecraft scheduled to orbit Mars in 1971, and from two unmanned Vikings which will attempt to soft-land instruments in 1973.

Every two years Mars swings into a favorable position for travel from earth. In the 1980's, an excellent Mars launching date, or "window," will open on October 3, 1983.

TWO CRAFT WOULD MAKE MARS VOYAGE

Although NASA has no plans now for a manned voyage to Mars, the general procedure for a trip beginning on that date is clear.

We would propel from earth orbit (to avoid radioactive contamination of the earth) two spacecraft about 250 feet long, each fitted with three nuclear rockets. Each would carry a crew of six—and some 1,800 meals per man. Two of the three rockets would be shed after launch and "parked" in space for reloading and future use.

On the 251-day journey to Mars, the two spacecraft would be joined nose to nose; thus one craft would be evacuated in case of trouble. A slow spinning motion would create centrifugal force to relieve the effects of weightlessness on the crew.

Separating, each ship would briefly retrofire its unused rocket, braking to enter Mars orbit, on June 9, 1984. The ships would remain for 80 days, each sending down a three-man laboratory for a month's exploration on the surface.

Firing their rockets again to leave Mars, the two spacecraft would swing around Venus on the way home, letting the pull of that planet's gravity hasten their homeward trip. On May 25, 1985, they would reach earth orbit, 601 days after leaving it. The crews would then catch the shuttle to earth—rather like men catching a bus home from work. The deep-space ships would remain in earth orbit: resupplied, they would be ready for a new voyage with a new crew.

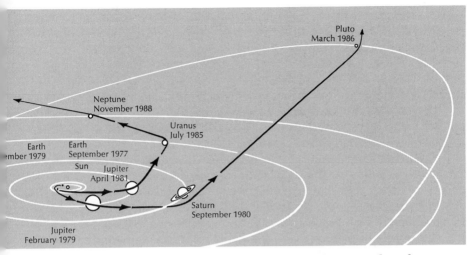

Figure 10-26. *"Grand Tours": Two unmanned spacecraft may explore the remote outer planets in the late 1980's. Gravitational pulls will bend their courses and reduce flight times from decades to only 8½ years to Pluto, and 9 years to Neptune on a launch 26 months later.* (Reproduction by special permission of the National Geographic Society, © 1970. Figure added by editor.)

What surprises would the astronauts bring home from Mars? No one can say—just as no one can say what explorers eventually may find on the moons of Jupiter, or on Pluto.

Such uncertainty inevitably attends the conquest of new horizons; explorers since the beginning of time have been unable to envision the full impact of their achievements.

Often, like Columbus, they made confident assessments which time proved wrong. It usually remained for those who followed to find the real significance of the explorer's effort, and to reap benefits far greater than were anticipated. There is little doubt in my mind that the benefits of space travel will emerge in the same way.

Challenge and Response *

by Arnold J. Toynbee (1946)

Growth is achieved when an individual or a minority or a whole society replies to a challenge by a response which not only answers that
challenge but also exposes the respondent to a fresh challenge which demands a further response on his part. . . .

We find that "virgin soil" produces more vigorous responses than land
which has already been broken in and thus rendered "easier" by previous occupants. Thus, if we take each of the affiliated civilizations, we
find that it has produced its most striking early manifestations in places
outside the area occupied by the parent civilization. The superiority of
the response evoked by new ground is most strikingly illustrated when
the new ground has to be reached by a sea passage. . . . These overseas
migrations have in common one and the same simple fact: in transma-
rine migration the social apparatus of the migrants has to be packed on
board ship before it can leave the shores of the old country, and then be
unpacked again at the end of the voyage. All kinds of apparatus—
persons and property, techniques and institutions and ideas—are subject to this law. Anything that cannot stand the sea voyage at all has to
be left behind, and many things—not only material objects—which the
migrants do take with them, have to be taken to pieces, never perhaps
to be reassembled in their original form. When unpacked, they are
found to have suffered "a sea change into something rich and strange."

Suggestions for Further Reading

Clarke, Arthur: *The Promise of Space*, Harper & Row, New York, Evanston,
London, 1968. In addition to the selections we have used here, the reader will
find in this book an excellent survey of the scientific knowledge and technology of space exploration.

Ley, Willy (ed.): *Harnessing Space*, The Macmillan Company, New York,
1963. Based upon reports by the Committee on Aeronautical and Space Sciences, U.S. Senate, the Committee on Science and Astronautics, and other
government agencies.

° From Arnold J. Toynbee, *A Study of History*, Oxford University Press, New York
and London, 1947. Reprinted by permission.

Ley, Willy: *Rockets, Missiles, and Men in Space,* The Viking Press, New York, 1968. A revision of a classic of space science (*Rockets, Missiles, and Space Travel*) written by one of the great pioneers in this field.

Shelton, William Roy: *American Space Exploration, The First Decade,* Little, Brown and Company, Boston, Toronto, 1967. A survey of the events from Sputnik to the first lunar orbiter.

Von Braun, Wernher, and Frederick I. Ordway III: *History of Rocketry and Space Travel,* Thomas Y. Crowell Company, New York. A comprehensive and authoritative history of astronautics.

11

What Are the Values and Limitations of Science?

Does science create "a gigantic robot driving toward material progress," or does it create and foster the very values that we hold most dear? Now, having seen scientists at work and having watched the evolution of some important scientific ideas, we are in a position to evaluate these opposing points of view.

What Are the Values and Limitations of Science?

Introduction

"WHAT HOPES AND FEARS does the scientific method imply for mankind? I do not think that this is the right way to put that question. Whatever this tool in the hand of man will produce depends entirely on the nature of the goals alive in this mankind. Once these goals exist, the scientific method furnishes means to realize them. Yet it cannot furnish the very goals. The scientific method itself would not have led anywhere, it would not even have been born without a passionate striving for clear understanding." These words from *Out of My Later Years* by Albert Einstein were written in 1950 and are even more pertinent today as the need becomes more urgent for a decision about the role that science should play in our society. Einstein says that science is a tool and to use any tool properly we must understand what it cannot do as well as what it can do. A tool that is excellent for one purpose may be crude and destructive when used for another. Scientific understanding has given us increasing power over nature. Now it is becoming more and more important that we understand science itself so that we may use it effectively to achieve the goals that we desire.

In the preceding ten parts we have been following the story of man's striving for understanding of the universe. We have seen how some of the great scientists worked out their contributions, have watched ideas grow and change, and have seen how new techniques led through greater accuracy to a new level of understanding. This has been a story of the power of reason, of man's rational approach to his environment. It has been brilliantly successful. But today men are beginning to wonder whether, like most success stories, it will have a happy ending. Many voices are being raised, decrying the effect that science has had on our way of life. They point to the emphasis on materialistic values, to the vast power that science has placed in the hands of people who may misuse it, to the amoral and pragmatic assumption that success is the only criterion. While others insist that these evils derive from a distortion of the aims and the spirit of science, that ". . . we are trying to employ the body of science without its spirit." [1] They say that the rational approach, which is embodied in scientific method, has fostered and encouraged the very values that we consider most important in Western civilization: independence of thought, freedom of speech, respect for truth, cooperation, justice, and human dignity.

1. See p. 669.

One of the reasons for disagreement about the effect of science on society is the failure to distinguish between pure science and technology. The distinctions between the two are explored in these readings and it is shown that pure science and technology differ in their aims as well as in the ways they can be exploited or controlled by society. Scientific principles can be applied to achieve specific practical goals by an intensive application of money and effort but a basic scientific discovery cannot be pressure-cooked in the same manner. We have seen, for instance, how the technology of space travel was developed in a very short time by teams of scientists working in enormous research centers and the expenditure of billions of dollars. But we should not forget that the really important scientific breakthrough that made space travel possible was made by men working alone: Goddard, Oberth, and Tsiolkovski. And the fundemental scientific principles on which the space age was built were conceived by Galileo, Kepler, and Newton several hundred years ago.

There is no doubt that technology has vastly increased our material well-being. Some people fear that it has outgrown all control and has become "a gigantic robot driving toward material progress."[2] The suggestion is made in these readings that society can control technology more successfully than it can control basic science. In technology the results can be more clearly predicted and the benefits or hazards more accurately foreseen. It is then a matter for society to furnish the goals and establish the priorities.

The explosive growth of technology has been sparked by a steadily increasing demand for more things, more power, more speed, and more efficient methods of production. Thus technology can be said to be the result of our materialistic and pragmatic society. Is it also the cause?

Some critics contend that the materialistic philosophy derived directly from the Newtonian picture of the world as a giant mechanism. But this world model—as Sullivan reminds us in these readings—is undergoing a drastic revision in the light of the discoveries of modern physics. If this change in fundamental scientific theory were understood by a large portion of our population, would it have a significant effect on our social philosophy? Throughout the history of science there have been a number of instances where an important revision in scientific theory has brought about a really fundamental change in the way people think about themselves.

Among the authors in this part there is striking disagreement concerning the nature of scientific truth and the role it plays in fostering a regard for truth in society. The pragmatic philosophy defines truth as util-

2. See p. 661.

ity. This idea was suggested by the way in which scientific theories change and are replaced by new ideas. In choosing between two scientific theories, the one that works is right. From this concept it is only a short step in thinking to say that whatever can be made to work, by force if necessary, is also right; and with enough power, man can even make truth.

Sharply contrasted with this interpretation is Bronowski's contention that an almost fanatical regard for truth is the motivating force and necessary condition for the practice of science. Although scientific truth is always provisional and changes from time to time, it changes in response to new and more accurate knowledge. Thus it represents a closer and closer approximation to a true understanding of the universe.

The opposition between these different attitudes is a central one in deciding whether science is suited to our democratic goals or whether it leads inevitably to a mechanistic and inhuman type of society. Having watched the evolution of some of the great scientific ideas and studied the role of facts and the nature of scientific truth, we are now in a better position to evaluate these opposing views and to take a backward look at the scientific process as a whole in order to appreciate its strengths and recognize its limitations. The values which derive directly from the scientific method itself have remained relatively constant over the years, and have had a cumulative influence on our ideas and modes of thought. It is these values, as Whitehead expresses it, that have molded "the very springs of action of mankind." [3]

3. See p. 671.

Limitations and Dangers of Science

Materialism *

by Alfred North Whitehead (1925)

[Scientific materialism is] the fixed scientific cosmology which presupposes the ultimate fact of an irreducible bruto matter, or material, spread throughout space in a flux of configurations. In itself such a material is senseless, valueless, purposeless. It just does what it does do, following a fixed routine imposed by external relations which do not spring from the nature of its being. It is this assumption that I call 'scientific materialism.' Also it is an assumption which I . . . challenge as being entirely unsuited to the scientific situation at which we have now arrived. It is not wrong, if properly construed. If we confine ourselves to certain types of facts, abstracted from the complete circumstances in which they occur, the materialistic assumption expresses these facts to perfection. But when we pass beyond the abstraction, either by more subtle employment of our senses, or by the request for meanings and for coherence of thoughts, the scheme breaks down at once. The narrow efficiency of the scheme was the very cause of its supreme methodological success. For it directed attention to just those groups of facts which, in the state of knowledge then existing, required investigation.

The Limitations of Science †

by J. W. N. Sullivan (1933)

Scientific method, as we see from the work of its founders, Copernicus, Kepler, Galileo, began by quite consciously and deliberately selecting

* From Alfred North Whitehead, *Science and the Modern World*, The Macmillan Company, New York. Copyright 1925 by The Macmillan Company; copyright 1953 by Evelyn Whitehead. Reprinted by permission.

† From J. W. N. Sullivan, *The Limitations of Science*. The Viking Press, Inc., New York. Copyright 1933; copyright © renewed 1961 by The Viking Press, Inc. Reprinted by permission.

and abstracting from the total elements of our experience. From the total wealth of impressions received from nature these men fastened upon some only as being suitable for scientific formulation. These were those elements that possess *quantitative* aspects. Between these elements mathematical relations exist, and these men were convinced that mathematics is the key to the universe. It is interesting, in view of its immense importance, to know how they came by this belief, for it was by no means a general persuasion. It was not part of the dominant Aristotelianism of their time. It seems to have been in part due to an innate prejudice, very proper to born mathematicians, and in part to the neo-Platonic philosophy that was prevalent, at that time, in South Europe. This philosophy contained important Pythagorean elements, and gave to the mathematical aspects of the universe a much more exalted position than they occupied in the current Aristotelian outlook. Copernicus became acquainted with this philosophy during his stay in Italy, and also with the fact that some of the ancient Greek philosophers had put forward the hypothesis that the earth was in motion. Being led, in this way, to take the sun as his centre of reference, Copernicus found, as we have already said, that a great harmony was bestowed on the motions of the heavenly bodies. With the incorrect ideas of dynamics prevalent at the time, Copernicus's theory was open to grave objections. As a physical explanation of phenomena it was certainly no better than the Ptolemaic theory. Nevertheless, Copernicus was confident that its superior aesthetic charm would be sufficient to commend it to mathematicians.

This expectation was justified . . . Kepler . . . shows how greatly the aesthetic charm of the new theory appealed to him. To the mind of Kepler, however, the claims of the theory were greatly reinforced by the dignified position it gave to the sun, for Kepler, in a vague and mystical fashion, was a sun-worshipper. . . .

Kepler had a preconceived idea as to the sort of thing the universe is. He did not approach the facts with the docility and lack of prejudice that is proper to the ideal scientific investigator. His deepest conviction was that nature is essentially mathematical, and all his scientific life was an endeavour to discover nature's hidden mathematical harmonies. Galileo, also, had no doubt that mathematics is the one true key to natural phenomena. It was this persuasion that gave these men their criterion for selection amongst the total elements of their experience. They confined their attention to those elements amongst which mathematical relations exist. Bodies, for instance, have for their measurable aspects size, shape, weight, motion. Such other characteristics as they possess were regarded as belonging to a lower order of reality. The real world is the world of mathematical characteristics. In fact, our minds are so constructed, Kepler said, that they can know nothing perfectly except quantities.

With Galileo this separation of the mathematical from the other qualities became a perfectly clear and definite doctrine. Kepler had supposed that the non-mathematical qualities actually did belong to bodies, but that they were somehow less real. Galileo went further than this, and stated that the non-mathematical properties are all entirely subjective. They have no existence at all apart from our senses. Thus colours, sounds, odours and so on exist, as such, wholly in our minds. They are, in reality, motions of some kind or another in the external world, and these motions, impinging on our senses, give rise to these sensations of colour, sound, and so on. It is the mind that peoples the world with the songs of birds, the colours of the sunset, etc. In the absence of mind the universe would be a collection of masses of various sizes, shapes, and weights, drifting, without colour, sound, or odour, through space and time. . . .

We have seen that, so far, the developing scientific outlook owed its main features to the predilections of the mathematician. And the main assumption of the philosophy accompanying this scientific achievement is that the real may be identified with the quantitative. Compared with the fully developed modern scientific outlook, we see that these early men of science were too prone to legislate for the universe on the basis of certain *a priori* assumptions—assumptions which were really expressions of their dominating mathematical predilections. The modern outlook differs from theirs by the more tentative character of its assumptions. Although mathematics has hitherto proved itself by far the most powerful tool for the scientific investigation of nature, it is no longer an article of belief that nature is necessarily mathematical. We now realize that mathematical deductions, however rigorous, must always be checked by experiment. Even Galileo, who strikes us as the most modern of all these early workers, tells us that often he did not consider experimental checks on his mathematical reasoning to be necessary, and undertook the experiments solely in order to convince his opponents. The first perfect fusion of the mathematical with the empirical outlook was accomplished by Isaac Newton.

Newton was not the first, of course, to realize the need of experiment. All his predecessors had realized this to some extent, and such men as Gilbert, Harvey, Boyle, were complete empiricists. But Newton was the first who ideally combined, in his own method, the two ways of approach. This combination was so unusual that none of his contemporaries understood it. It is still unusual in practice, although recognized as the ideal scientific method in theory. Mathematicians are still prone to have what the experimentalists consider an excessive confidence in their mathematical deductions, and the experimentalist sometimes seems to the mathematician to display an unnecessary degree of caution. Newton not only combined the attitude of both types; he also combined their powers. He was not only a supreme mathematician; he was also a great experimentalist. . . .

Although Newton did not share the mathematical *a priorism* of his predecessors, he shared their other assumptions. Like them, he dispensed with final causes, and found the cause of a phenomenon in its immediately preceding physical conditions. Also, he proceeded as if science formed a self-enclosed system, that is to say, as if a complete account of phenomena could be given in the terms mass, velocity, force, etc., which science had isolated, and without bringing in such concepts as beauty, purpose, etc., which did not form part of the scientific outlook. It is not possible to say that Newton held these opinions dogmatically, since he himself said that science, in the mathematical form it had assumed, was an adventure, which might have to be replaced by a truer method. But for practical working purposes he certainly made these assumptions, and his immense success caused these assumptions to be unquestioningly accepted by the whole scientific world. . . . This was, in essentials, the outlook of science for nearly two hundred years, and this assumption still profoundly colours scientific thinking.

What is called the modern "revolution" in science consists in the fact that the Newtonian outlook, which dominated the scientific world for nearly two hundred years, has been found insufficient. It is in process of being replaced by a different outlook, and, although the reconstruction is by no means complete, it is already apparent that the philosophical implications of the new outlook are very different from those of the old one.

Science has become self-conscious and comparatively humble. We are no longer taught that the scientific method of approach is the only valid method of acquiring knowledge about reality. Eminent men of science are insisting, with what seems a strange enthusiasm, on the fact that science gives us but a partial knowledge of reality, and we are no longer required to regard as illusory everything that science finds itself able to ignore. But the enthusiasm with which some men of science preach that science has limitations is not really surprising. For the universe of science, if accepted as the final reality, made of man an entirely accidental by-product of a huge, mindless, purposeless, mathematical machine. And there are men of science sufficiently human to find such a conclusion disconcerting. Even the sturdy Victorians who preached the doctrine betray at times a despairing wish that things were not so. We need not be surprised, therefore, to find that the discovery that science no longer compels us to believe in our own essential futility is greeted with acclamation, even by some scientific men.

This change in the scientific outlook seems to have taken place suddenly. It is not yet sixty years since Tyndall, in his Belfast Address, claimed that science alone was competent to deal with all man's major problems, and it is not yet twenty years since Bertrand Russell, contemplating the scientific answers, said that "only on the firm foundation of unyielding despair can the soul's habitation henceforth be safely built." But, in truth, so far as these remarks sprang from the conviction that the sole reality is "matter and motion," their foundations had already been undermined. The attempt to represent nature in terms of matter and motion was already breaking down. That attempt was at its most triumphant by the end of the eighteenth century, when Laplace was emboldened to affirm that a sufficiently great mathematician, given the distribution of the particles in the primitive nebula, could predict the whole future history of the world. The fundamental concepts isolated by Newton had proved themselves so adequate in the applications that had been made of them that they were regarded as the key to—everything.

The first indication that the Newtonian concepts were not all-sufficient came when men tried to fashion a mechanical theory of light. This endeavour led to the creation of the ether, the most unsatisfactory and wasteful product of human ingenuity that science has to show. For generations this monster was elaborated. Miracles of mathematical ingenuity were performed in the attempt to account for the properties of light in terms of the Newtonian concepts. The difficulties became ever more heartbreaking until, after the publication of Maxwell's demonstration that light is an electromagnetic phenomenon, they seemed to become insuperable. But the ether, by this time, had become too complicated to be credible. It was not only complicated; it was ugly, and ugliness in scientific theories is a thing no scientific man will tolerate if he can possibly help it. Copernicus was a true judge of the scientific temperament when he showed himself confident that the aesthetic charm of his theory would suffice to enable it to make its way against the insufferably complicated theory of Ptolemy. The construction of ethers became a decaying industry, and largely because there was so little demand for the product. For it had dawned on men of science that there was, after all, nothing sacrosanct about the Newtonian entities. It might be that his list of ultimates, mass, force, and so on, was not exhaustive. Instead of reducing electricity to these terms, it might be better to add it to the list. This was done. After a certain amount of hesitation, and a few last desperate efforts to make electricity mechanical, electricity was added to the list of irreducible elements.

This may seem to have been a simple and obvious step to take, but it was, in reality, of profound significance. For the Newtonian concepts were all of a kind that one seemed to understand intimately. Thus the mass of a body was the quantity of matter in it. Inertia was that familiar property of matter which makes it offer resistance to a push. Force was a notion derived from our experience of muscular effort. Of course, all these concepts, in order to be of use to science, had to be given quantitative expression. They entered our calculations as mathematical symbols. Nevertheless, we supposed that we knew the nature of what we were talking about but in the case of electricity its nature is precisely what we did not know. Attempts to represent it in familiar terms—as a condition of strain in the ether, or what not—had been given up. All that we knew about electricity was the way it affected our measuring instruments. The precise description of this behaviour gave us the mathematical specification of electricity and this, in truth, was all we knew about it.

It is only now, in retrospect, that we can see how very significant a step this was. An entity had been admitted into physics of which we knew nothing but its mathematical structure. Since then other entities have been admitted on the same terms, and it is found that they play precisely the same role in the formation of scientific theories as do the old entities. It has become evident that, so far as the science of physics is concerned, we do not require to know the nature of the entities we discuss, but only their mathematical structure. And, in truth, that is all we do know. It is now realized that this is all the scientific knowledge we have even of the familiar Newtonian entities. Our persuasion that we knew them in some exceptionally intimate manner was an illusion. So far as the science of physics is concerned, the old entities and the new are on the same footing—the only aspects of them with which we are concerned are their mathematical aspects.

With this realization it is no long step to Eddington's position that a knowledge of mathematical structure is the only knowledge that the science of physics can give us. Of all the philosophical speculations which have been hung on to the new physics, this seems to be the most illuminating and the best-founded. It seems to be true that "exact" science is a knowledge of what Eddington calls "pointer-reading"—the readings on an instrument of some kind. We assume, of course, that these readings refer to various qualities of the external world, but all we actually know about these qualities, for the purposes of exact science, is the way they affect our measuring instruments. As Eddington [1] says:

Leaving out all aesthetic, ethical, or spiritual aspects of our environment, we are faced with qualities such as massiveness, substantiality, extension, duration,

1. "The Domain of Physical Science," essay in *Science, Religion and Reality*.

which are supposed to belong to the domain of physics. In a sense they do belong; but physics is not in a position to handle them directly. The essence of their nature is inscrutable; we may use mental pictures to aid calculations, but no image in the mind can be a replica of that which is not in the mind. And so in its actual procedure physics studies not these inscrutable qualities, but pointer-readings which we can observe. The readings, it is true, reflect the fluctuations of the world-qualities; but our exact knowledge is of the readings, not of the qualities. The former have as much resemblance to the latter as a telephone number has to a subscriber.

. . . Science . . . would be confined to remarking, of the Big Four at Versailles, that they numbered four, to use one of Eddington's illustrations.

The fact that science is confined to a knowledge of structure is obviously of great "humanistic" importance. For it means that the problem of the nature of reality is not prejudged. We are no longer required to believe that our response to beauty, or the mystic's sense of communion with God, have no objective counterpart. It is perfectly possible that they are, what they have so often been taken to be, clues to the nature of reality. Thus our various experiences are put on a more equal footing, as it were. Our religious aspirations, our perceptions of beauty, may not be the essentially illusory phenomena they were supposed to be. In this new scientific universe even mystics have a right to exist.

The outlook just described may fairly be said to be a result of the new scientific selfconsciousness. It is more than a mere speculation. . . .

The humanistic importance of this outlook, in the minds of its authors, seems to be that it leaves us more free to attach the traditional significance to our aesthetic, religious or, compendiously, mystic experiences. It does not actively reinforce any particular religious interpretation of the universe, but it cuts the ground from under those arguments which were held to prove that any such interpretation is necessarily illusory. This it does by showing that science deals with but a partial aspect of reality, and that there is no faintest reason for supposing that everything science ignores is less real than what it accepts. . . . The statements of science are true within the limited region it claims for itself—the region of mathematical structure. If you are curious about the mathematical structure of the material universe, then science can give you information on that point.

In pursuit of its aim science has shown itself to be accommodating and flexible. The science of physics has shown itself willing, at different times, to adopt quite different principles and concepts provided they promised to be helpful in its one great aim of giving a mathematical description of nature. Where principles and concepts have persisted past their time of usefulness, this has been due to the inertia of mental habit.

No scientific principles are sacrosanct; no scientific theory is held with religious conviction. Nevertheless, most scientific men believe that there is a final scientific truth about the universe to which successive scientific theories ever more closely approximate. But this is an article of faith. Science is still an adventure, and all its "truths" are provisional.

This atmosphere of provisional hypothesis and practically verifiable statements constitutes what has been called the "homely air" of science, and is one of its great charms. Science has adopted the pragmatic criterion of truth, namely, success, and as a result science has been successful. Indeed, it would not be too much to say that, judged by the criterion of general assent and practical efficacy, it is the most successful activity that man has yet hit upon. As Professor Levy [2] says:

Any such criterion of truth may not be one that commends itself to the professional philosopher. It may be that science ignores subtleties that appear vital to academic philosophy, that it skims easily over the surface of reality. The very principle that scientists must leave no differences behind may narrow the range to superficial agreement, and restrict the nature and number of the isolates it may form. Whether or not this be so, science can at any rate look upon itself as a united movement that has left in its wake a body of tested knowledge, while philosophy is still broken up into disunited schools of thought.

The criticism has, indeed, been made that science pays for its success by its superficiality. All the deepest problems of mankind, it has been pointed out, lie outside science. If neither philosophy nor religion can present any such "body of tested knowledge," it is because they have not been content with such cheap victories. There is doubtless some truth in this criticism, and it is probably true that the problems with which science deals are intrinsically inferior in human interest to those dealt with by either philosophy or religion.

Nevertheless, the actual atmosphere of science, the manner in which it goes about its work, is quite exceptionally agreeable. It is in the scientific attitude, as much as in the scientific results, that the true value of science is to be found. If the man of science has not aimed high, according to the philosopher, he has at least aimed with a single heart, with a docility in face of the facts, with an impersonal purpose to serve which is not always found amongst our prophets and philosophers, and which it is almost impossible to find elsewhere.

There is probably no other period in history, since modern science began, when the particular values it incorporates were so rarely to be encountered in other human activities. The human tendencies to prize certitude and fear knowledge, to indulge emotion at the expense of rea-

2. *The Universe of Science.*

son, were probably always as strong as they are today, but the circumstances of the time did not show them up in so pitiless a light. The mass of men were probably always as impatient of that careful, honest verification that is the very essence of science as they are today, but there was no huge popular press to bear daily witness to the fact.

If we are to judge from what seems the overwhelming evidence provided by such activities as politics, business, finance, we must conclude that the attention and respect accorded to science is directed wholly to its results, and that its spirit is the most unpopular thing in the modern world. Yet it could very reasonably be claimed that it is in its spirit that the chief value of science resides. This can be asserted without abating anything of our claim for the values of its results. Knowledge for the sake of knowledge, as the history of science proves, is an aim with an irresistible fascination for mankind, and which needs no defence. The mere fact that science does, to a great extent, gratify our intellectual curiosity, is a sufficient reason for its existence.

But it must be pointed out that this value is inextricably intertwined with another value—the aesthetic value of science. If scientific knowledge consisted of a mere inventory of facts, it might still be interesting and even useful, but it would not be one of the major activities of the mind. It would not be pursued with passion. It could, at best, only exert a somewhat more intensified form of the attraction we feel for a timetable. Indeed, the greatest single testimonial to the fact-gathering power of science, resting, as it does, on centuries of labour and ingenuity, is probably the *Nautical Almanac*—useful, but not entrancing, reading.

For science to have inspired such ardour and devotion in men it is obvious that it must meet one of the deepest needs of human nature. This need manifests itself as the desire for beauty. It is in its aesthetic aspect that the chief charm of science resides. This is true, be it noted, for scientific men themselves. To the majority of laymen, science is valuable chiefly for its practical application. But to all the greatest men of science practical applications have emerged incidentally, as a sort of by-product.

Science and Technology in the Affluent Society *

by Jerome B. Wiesner (1963–64)

Technology in the mid-twentieth century is in a curious state. There is no doubt that it is a most dominant force in modern life. It is the engine

* From "Technology and Society," a lecture delivered at the Johns Hopkins University, Baltimore, 1964; and "Science in the Affluent Society," a lecture presented

that propels modern society, and for this it is widely acclaimed, but it has also become the object of much critical examination by experts and laymen alike. Technological activities—research and development—are among this nation's fastest growing enterprises, and public funds provide much of the support for them. Widespread understanding of the importance for such support is, therefore, very desirable.

As essential background, I would like to clear up a continuing source of misunderstanding that appears in discussions of technological development; the ambiguity that exists between technology and scientific research, their objectives and their methods. This confusion is by no means surprising since modern technology has become highly dependent upon basic scientific knowledge for much of its progress. In turn, scientific research in many fields is only possible because of the elaborate and sensitive tools that technology has provided. The large and powerful particle accelerators, the electron microscope with which to explore the world of cells and viruses, and the electronic computer to calculate problems which only a few years ago were beyond the scope of human comprehension, are only three of a large number of scientific tools which have extended enormously our ability to measure, observe and understand the world around us. This close alliance between science and technology, though relatively new, is so complete that the average person, and indeed many scientists and engineers as well, fail to distinguish any difference between them.

In the beginning, technology did not depend upon science. The inventions that provided the basis for the industrial revolution were invented by practical men and based upon art, observation and common sense. In the first stage of industrialization, man was exercising his ingenuity in the exploitation of the things he found around him. The factory, with its power machines, its use of unskilled or semi-skilled labor doing simple repetitive operations, the utilization of raw materials like iron, coal, copper, etc., improvements in transportation growing from the development of the railroad and the steamboat, are all samples of this inventiveness. Most important, of course, was that with the introduction of machines man had begun a continuing process of extending human capabilities, first by augmenting muscle power through harnessing the almost limitless energy sources found in nature, later by speeding communications by electrical means, and most recently augmenting mental activities by the introduction of computing machines to replace human effort in

at the Fourth Scientific Session of the Centennial Celebration of the National Academy of Sciences, Washington, D.C., 1963, in *Where Science and Politics Meet,* McGraw-Hill Book Company, New York. Copyright © 1961, 1963, 1964, 1965 by Jerome B. Wiesner. Reprinted by permission.

menial, repetitive activities and to assist him in performing difficult or lengthy calculations.

The fact that scientific research had little or no effect on early technology does not mean that scientists did not exist or were not working. They did and were, and during the period of the industrial revolution the foundations were laid for modern physics, chemistry and biology. However, it was not until the middle of the nineteenth century that extensive practical use was made of the accumulating scientific knowledge. Only then did men begin to discover that they could exploit the available knowledge of chemistry and electricity for useful purposes. This discovery had to wait until an extensive body of scientific knowledge existed and was widely available. Chemists learned to synthesize organic materials and set up research laboratories for obtaining the new knowledge required to meet their applied objectives. It was in the field of chemistry that research methods were first used in a systematic manner to develop new products. The application of electricity was somewhat more haphazard in the beginning. The scientific observations of Gilbert, Henry and Maxwell were seized upon by the inventors of the electric motor, the electric generator, the telegraph, telephone and other devices. Not until the end of the century were research methods applied to the exploitation of electrical phenomena, first by Thomas Edison who, in reality, was more of an inventor than a scientist, and later by many technologists in the laboratories of such industries as the General Electric Company and the predecessors of the American Telephone and Telegraph Company. Thus, it was in these fields—chemistry and electricity —that the merger of scientific inquiry and technology first occurred, that the power of the scientific method was applied to solving useful problems, and that the great value of the thorough understanding of physical phenomena was demonstrated. In exploiting electrical phenomena, technologists deal with fields and electrons and waves which can only be observed indirectly and understood through scientific research. It is not surprising, therefore, that the electronics industry should depend very heavily upon basic research and be the sponsor of much fundamental work.

Modern technology still requires invention. The vacuum tube, the transistor, memory devices for computers and new materials tailored with specific properties are all inventions. But they are inventions made by men with special knowledge; inventors who have an understanding of a scientific field and who base their creations on an intimate familiarity with that field, just as the early inventor called upon his first-hand experience of the world that he could see and feel to provide the working substance of his ingenuity.

This is the nature of modern, scientifically-based technology. The first requirement is the existence of a body of scientific knowledge. Then the technologist must have a thorough understanding of the underlying science to use it as the basis for an invention in the solution of a specific problem. Also, more likely than not, he will find that the scientists who first explored the field that he is exploiting left large areas of ignorance which must be filled before his task can be completed. This can only be done by further fundamental research. Because it has as its specific goal the acquiring of knowledge to solve a specific problem, such additional work is often called "applied" or "directed research" (though it is obvious that in another context it would be regarded as fundamental or basic research).

Who does this applied work? It depends upon the field. Generally, the exploitation of new areas of science, such as the recent efforts in solid state physics or nuclear physics, is initiated by scientists who are very well acquainted with the fundamentals of the subject. These men usually create a reservoir of technological information and train the technologists, applied scientists and engineers. Right here one can see the cause of the confusion that exists in differentiating between scientific research and technology.

I have gone into this detail to show the deep dependence of modern technology upon fundamental scientific knowledge and the interplay that must exist between scientist and technologist as new scientific information is employed for useful purposes. In planning scientific programs, it is important to understand the essential difference between fundamental research done to achieve a deeper understanding of physical phenomena and technological efforts based upon such knowledge, but undertaken to meet a specific need or to solve a specific problem. Interesting and potentially useful basic research is often difficult to defend before non-scientists, because it does not have a ready application.

The research scientist is primarily motivated by an urge to explore and understand, but society supports him primarily because experience has demonstrated how essential such work is for continuing progress in technology. Halt the flow of new research and the possible scope of technical developments will soon be limited and ultimately reduced to nothing. Incidentally, scientific knowledge need not be exploited immediately once it exists. It is available for all to use forever.

Thus, acquiring scientific knowledge is a form of capital investment. Unlike most other capital investments, it does not become obsolete nor can it be used up. Technological developments are also a form of capital investment, though somewhat less enduring. To be sure, a more efficient process will also yield its benefits endlessly, but usually technological developments tend to become obsolete as better methods, devices and processes emerge.

The past two decades have seen unprecedented expansion in research and development activities. United States' expenditures for research and development have doubled approximately every five years during this period, and currently (May 1964) amount to about 19 billion dollars per year, with nearly 15 billion dollars provided by the federal government. Approximately 70 percent of the federal funds are spent for defense and space activities; the remainder for a broad range of health, natural resource, basic research, transportation and educational activities. Approximately 10 percent of the total research and development funds have been used to support basic research.

As a consequence of this expansion, there have been major advances in our understanding of the physical world and dramatic increases in the capabilities of our technology. We have such scientific achievements as the comprehensive understanding of atomic structure, a theory of the chemical bond, an understanding of atomic nuclei, the major advances in fundamental biological phenomena, particularly at a cellular and sub-cellular level, and the development of information theory and feedback and control theory. While in most fields understanding is by no means complete—in fact, there is a long way to go in many of them—there is enough knowledge to permit almost miraculous technological achievements. The familiar ones include jet aircraft, television, atomic energy and radar, and a vast array of new materials, including plastics and metals, missiles and space craft, and the electronic computer.

Of all recent developments none is as far-reaching as the high-speed electronic computer. Created initially to facilitate routine numerical calculations, computers have now reached the point where they can perform logical operations as well. So powerful, versatile and pervasive is the electronic computer that its impact has been likened to the effect of the introduction of machines during the industrial revolution.

✧   ✧   ✧

The general interest in science is greater now than ever before. It is evidenced in the newspapers and other publications and in Congress too. There are at the moment several Congressional committees examining the purposes and methods of government-financed scientific and technological activities. While such interest is not new, the point of view seems to be. Until recently, most discussions of science consisted of uncritical praise. Now the situation is changed. Serious questions are being asked, and many of them reflect deep-seated concern about the character and purposes of the nation's scientific and technological undertakings, reflecting a clear desire to become more familiar with these processes.

Why has the mood changed so? Has science changed, or was too much expected from science in the past, or have the nation's needs changed? I don't believe that the answers really lie in this direction. What has changed, I believe, is the scope of the motivation for supporting research through the federal government. There has been a broadening of emphasis from a primary need to support military development to a wider purpose encompassing the entire spectrum of social needs. The military objective is still of great significance, to be sure, but now this is only one of many reasons for the new research and development activities. . . .

When the increases in the budget were primarily for improved military security, they were easier to understand and therefore easier to justify than they are today. In retrospect, the country was very fortunate —I hope this is not misunderstood—that this incentive for research support did exist following World War II. I doubt if there was at that time a sufficient appreciation of the general importance of scientific research to have made possible the creation of our present large and very competent scientific establishment on the basis of needs for the general welfare.

In reviewing the political scene, President Kennedy once observed that there are cycles in the affairs of men. And in science, too, we can observe this phenomenon. We have experienced a heady period of growth, and it is time to review objectives, assess accomplishments and adjust stressed institutional arrangements to present needs. . . .

To the extent that federally supported research and development is justified for social purposes other than national security, it will be judged by different standards, less well-defined, and more controversial as well.

There is concern about waste of funds and imbalances in the federal programs. There is concern about distortion of our federal activities and our universities. There is worry that the unanticipated but inevitable side effects of new technology are causing more and more difficult problems which then require further, expensive remedial actions. Agricultural surpluses, air and water pollution are good examples. The unpredictable consequences of the widespread effects of the use of pesticides is another. Technological unemployment is yet another. There is some feeling that federal research and development expenditures are responsible for unbalanced economic development between geographic areas of the country and between different industries. There is fear also that too large a fraction of the ablest youngsters are being attracted into science and engineering. Underlying it all is the belief that the whole activity is beyond the comprehension of the individuals in the government who are responsible for it, be they in the Executive Branch or in the Congress.

Some of these worries stem from real problems, involve the need for policy actions and should be considered and deliberated. Others stem from misunderstandings which we should strive to eliminate. First of all, research activities and development are frequently not distinguished. They are mixed together and called science. Even most engineers and scientists fail to make a clear enough distinction between activities carried out solely for the purpose of adding to existing scientific knowledge and work that is performed because it may be able to satisfy some practical need. And this confusion is at the root of much misunderstanding among non-scientists who are called upon to make decisions or pass judgments about technical matters. It isn't that most people fail to recognize that there is a spectrum of activities with pure research at one end and hardware development at the other. The problem is rather a failure to understand that the methods and motivations of each are different, and that often research, new knowledge, is necessary in order to achieve a practical goal.

Not only are research activities quite different from development work, but so are their costs. We have estimated that the fiscal year 1963 obligations for basic research by the federal government are approximately 1.4 billion dollars out of the approximately 15 billion dollars devoted to research and development, and of this 1.4 billion, one-half billion are funds for space science and include the cost of boosters and launching operations. I quoted this figure during a Congressional hearing and was later told by Congressman Puchinski that he hadn't realized that basic research was so small a portion of the total. "Maybe," he said, "we were focusing on the wrong problem when we focused our inquiry on basic research expenditures rather than on those for development."

In my view, development activities, the creation of useful new devices, should only be undertaken if there is a clear-cut requirement for a new product after it has been developed. This is reasonable, for it is ordinarily possible to make satisfactory predictions about the probable cost and performance of a proposed new device, whether it is an aircraft, a computer, a chemical processing apparatus, or a nuclear power plant. It also possible to make a decision about the desirability of a given development. Furthermore, because development efforts are generally much more costly than research, one should apply rigorous tests of need before starting new efforts.

In the case of exploratory development and applied research, there is reason to be more venturesome. Here the search is to see if practical applications of new knowledge are possible. For example, there is an effort today to explore possible new uses of the laser, the source of coherent light developed recently, and elements of this work should be carried to the point of demonstrating the feasibility of underlying concepts. Here, too, work beyond that point should be permitted only if the ultimate capability is needed. These criteria are applicable to all but the basic research segment of the research and development effort. In other words, a basis exists by which administrators and legislators could establish the need for most items of the more than 90 percent of the federal expenditures for research and development. I am not saying that they could or should pass on the technical validity of proposals or judge between competing ways of accomplishing a given objective. This must still be the function of experts, but then the experts will be passing on means, not ends. Corporation executives, government budget officers and department heads and members of Congress have traditionally made such decisions with confidence and with access to scientific and engineering advice; there is no reason why they cannot continue to successfully.

But the choices in the field of basic research must be left to the scientists. This is why I place so much emphasis on the matter of distinctions. Even here, others will need to make decisions regarding the overall level of effort, and if that level is less than is required to support all of the worthwhile research that scientists want to do at any given time, it will be necessary for the scientists to make decisions regarding the support for the different disciplines. Not that this would be easy either. I could write an entertaining book about our efforts to make decisions in the field of high energy accelerators. . . .

From the point of view of the academic institutions, while federal support has made research programs stronger, other problems have been created. First of all, research activities are larger than they would be if the universities still depended primarily upon private philanthropy for support as they did before World War II. Bigness, however, is not completely an advantage, and many scientists deplore the fact that modern research often requires large, expensive equipment with the concomitant need for team work and a high degree of organization to make effective use of such facilities. Laymen echo these laments without recognizing that many scientific problems today cannot be studied without such tools. Money hasn't created the problem in this case. In other research fields, the size of activities is related to the level of support, and in them it is true that government support has set that level. As we have already indicated, the motivation for doing this was to insure that rapid progress be made in a field regarded as highly important.

The government turned to the universities for increased basic research because they had traditionally been a source of new knowledge in the sciences. Such support was, of course, welcomed by the universities, particularly by the scientists, though many administrators were troubled by the long-term implications of this course. With this financial help, scientific research in the universities prospered and so did the scientific departments involved. Thus, this activity undertaken for the general public welfare proved very valuable to the academic institutions, too. There was a not wholly anticipated but vital by-product of these actions: the increased level of research made opportunities for more graduate students.

I am very frequently asked whether there are no limits to the growth of research activities, and I find it impossible to give a simple answer to the question. There is no foreseeable limit to the amount of productive scientific research that can be done. Clearly, a lack of interesting and important work is not in sight. Many more people have the potential to do effective, creative research than are currently engaged in such work or are studying to become scientists. It appears then that the amount of research will be set by the willingness of the country to pay for it.

Science is the soul of the prosperity of nations and the living source of all progress. Undoubtedly the tiring discussions of politics seem to be our guide— empty appearances! What really leads us forward is a few scientific discoveries and their application.

Louis Pasteur (1822–95)

Astronomy and the State °
by A. C. B. Lovell (1959)

The expenditure of large sums of money by the State on fundamental scientific research, although inevitable, is full of long-term dangers. In [another] lecture I gave an example of the restriction of free astronomical inquiry in Russia today. In Great Britain the dangers are enhanced by the difficult and peculiar situation which exists in the Universities. The great scientists of the past were able to achieve their tremendous

° Abridged from A. C. B. Lovell, *The Individual and the Universe,* Harper & Row, Publishers, Inc., New York. Copyright © 1958 by A. C. B. Lovell. Reprinted by permission of Harper & Row, Publishers, Inc.

successes with so little expense that the University organizations have grown up without difficulty on the basis of reasonable equality of expenditure as between arts and science. The problem of absorbing the vast new instruments of science in conventional University departments is exceedingly difficult, and in fact the desirability of doing so is in question because of the fear of destroying the traditional balance of activities. It seems to me that those charged with the administration of the Universities in Great Britain are today faced with a delicate and perilous situation which is without precedent. A failure to absorb these great new scientific projects into a framework where the traditional University freedom of inquiry can flourish will be fraught with grave dangers to scholarship and scientific education.

In America the dangers at present are of a different kind. The financing of the radio astronomical projects is already heavily biased by the investments of the armed services, and . . . it seems likely that these investments by the military authorities will dominate the American radio astronomical scene. At present the demands of the services for specialized work from these groups of scientists is small. There is, indeed, no reason for it to be otherwise, since the lines of work likely to be pursued with these telescopes by the free volition of the scientists is of great fundamental interest to the services. However, this may not always be so. I am unable to discover any safeguards which would enable this vast fabric of fundamental astronomical research to survive a change of emphasis or opinion in the armed services concerned. It is indeed fortunate that so far in Great Britain these new developments have been contained within a University framework where freedom is a prized and jealously guarded possession.

The main concern of the astronomer is with highly abstract and remote topics. Some of this work can still be pursued by the astronomers working in isolation from the daily turmoil of existence. But we are moving into a new epoch in which even the study of the remote parts of the universe demands a close partnership between the astronomer and the State. The instruments which are destined to solve the problems of the origin of the universe may well rise from the rocket range. . . . In these new fields of work we are dealing with an expenditure on an unprecedented scale, in which an experimental test of only a few minutes may cost more than the largest telescope yet built.

Those concerned with the astronomical sciences are therefore faced with an entirely new situation. It is understandable that some astronomers, brought up in the tradition of the peaceful isolation of the observatory dome under the starlit sky, do not receive with enthusiasm these new developments, in which their instruments are launched from the rocket range under the glare of publicity. Others are happy to join in the initiation of this new era of observation which would be impossible but for the political and military divisions of the world which have forced the governments to an expenditure which would never be borne as a budget for fundamental scientific work alone. . . . The pursuit of the good and the evil are now linked in astronomy as in almost all science.

In so far as Great Britain is concerned I tried to show that the recovery of our heritage in the astronomical sciences is in acute danger, and that the peril lies deep in our national well-being. The pessimists say that we cannot compete and that within the next decade the scientific and technological superiority of the U.S.S.R. over the West will be complete. I do not believe that this will necessarily be the case, because I think that the restraints on freedom may reduce the effectiveness of Russian science, and may counteract to some extent their enormous superiority in scientific manpower and finance. On the other hand, one thing seems beyond contention. Fundamental research in astronomy or any other subject is an essential component in the welfare of modern civilization. Unless the West overcomes its present parsimonious attitude to science and technology, then the relative quality of our civilization will decline, and our influence will pass to other peoples. I have referred to the great sums now being spent, particularly in the U.S.A., but in so far as they support research, these represent a quite negligible proportion of the national budget. Neither in Great Britain nor the U.S.A. have we yet provided the facilities to saturate our existing scientific manpower. Therein lie my grounds for personal pessimism. Moreover, our danger rests, not in our limited potential, but in those amongst us who think of science and astronomy in terms of the sacrifice of a television set or motor car today so that our grandchildren can get to the moon. Alas, the issues at stake are of a different order of gravity. The fate of human civilization will depend on whether the rockets of the future carry the astronomer's telescope or a hydrogen bomb.

Selection from President Eisenhower's Farewell Address to the Nation (January 17, 1961)

In the councils of Government, we must guard against the acquisition of unwarranted influence, whether sought or unsought, by the military-industrial complex. The potential for the disastrous rise of misplaced power exists and will persist. We must never let the weight of this combination endanger our liberties or democratic processes. We should take nothing for granted. Only an alert and knowledgeable citizenry can compel the proper meshing of the huge industrial and military machinery of defense with our peaceful methods and goals, so that security and liberty may prosper together.

Akin to, and largely responsible for the sweeping changes in our industrial-military posture, has been the technological revolution during recent decades. In this revolution, research has become central; it also becomes more formalized, complex, and costly. A steadily increasing share is conducted for, by, or at the direction of the Federal Government. Today, the solitary inventor, tinkering in his shop, has been overshadowed by task forces of scientists in laboratories and testing fields. In the same fashion, the free university, historically the fountainhead of free ideas and scientific discovery, has experienced a revolution in the conduct of research.

Partly because of the huge costs involved, a Government contract becomes virtually a substitute for intellectual curiosity. For every old blackboard there are now hundreds of new electronic computers.

The prospect of domination of the nation's scholars by Federal employment, project allocations, and the power of money is ever present—and it is gravely to be regarded.

Yet, in holding scientific research and discovery in respect, as we should, we must also be alert to the equal and opposite danger that public policy could itself become the captive of a scientific-technological elite.

It is the task of statesmanship to mold, to balance and to integrate these and other forces, new and old, within the principles of our democratic system—ever aiming toward the supreme goals of our free society.

Science, Liberty and Peace *

by Aldous Huxley (1946)

"If the arrangement of society is bad (as ours is), and a small number of people have power over the majority and oppress it, every victory over Nature will inevitably serve only to increase that power and that oppression. This is what is actually happening."

It is nearly half a century since Tolstoy wrote these words, and what

* From Aldous Huxley, *Science, Liberty and Peace*. Harper & Row, Publishers, Inc., New York, 1946. Reprinted by permission of Harper & Row, Publishers, Inc.

was happening then has gone on happening ever since. Science and technology have made notable advances in the intervening years—and so has the centralization of political and economic power, so have oligarchy and despotism. It need hardly be added that science is not the only causative factor involved in this process. No social evil can possibly have only one cause. Hence the difficulty, in any given case, of finding a complete cure. All that is being maintained here is that progressive science is one of the causative factors involved in the progressive decline of liberty and the progressive centralization of power, which have occurred during the twentieth century. . . .

On many fronts nature had been conquered; but as Tolstoy foresaw, man and his liberties have sustained a succession of defeats. . . .

The Impact of Science on Society [*]
by Bertrand Russell (1951)

Man has existed for about a million years. He has possessed writing for about 6,000 years, agriculture somewhat longer, but perhaps not much longer. Science, as a dominant factor in determining the beliefs of educated men, has existed for about 300 years; as a source of economic technique, for about 150 years. In this brief period it has proved itself an incredibly powerful revolutionary force. When we consider how recently it has risen to power, we find ourselves forced to believe that we are at the very beginning of its work in transforming human life. What its future effects will be is a matter of conjecture, but possibly a study of its effects hitherto may make the conjecture a little less hazardous.

The effects of science are of various very different kinds. There are direct intellectual effects: the dispelling of many traditional beliefs, and the adoption of others suggested by the success of scientific method. Then there are effects on technique in industry and war. Then, chiefly as a consequence of new techniques, there are profound changes in social organization which are gradually bringing about corresponding political changes. Finally, as a result of the new control over the environment which scientific knowledge has conferred, a new philosophy is growing up, involving a changed conception of man's place in the universe. . . . This philosophy, if unchecked, may inspire a form of unwisdom from which disastrous consequences may result. . . .

It is hardly till the time of Lister and Pasteur that medicine can be said to have become scientific. The diminution of human suffering owing to the advances in medicine is beyond all calculation.

Out of the work of the great men of the seventeenth century a new outlook on the world was developed, and it was this outlook, not specific arguments, which brought about the decay of the belief in portents, witchcraft, demoniacal possession, and so forth. I think, there were three ingredients in the scientific outlook of the eighteenth century that were specially important:

(1) Statements of fact should be based on observation, not on unsupported authority.

(2) The inanimate world is a self-acting, self-perpetuating system, in which all changes conform to natural laws.

(3) The earth is not the center of the universe, and probably Man is not its purpose (if any); moreover, "purpose" is a concept which is scientifically useless.

These items make up what is called the "mechanistic outlook," which clergymen denounce. It led to the cessation of persecution and to a generally humane attitude. It is now less accepted than it was, and persecution has revived. To those who regard its effects as morally pernicious, I commend attention to these facts.

Observation versus Authority. To modern educated people, it seems obvious that matters of fact are to be ascertained by observation, not by consulting ancient authorities. But this is an entirely modern conception, which hardly existed before the seventeenth century. Aristotle maintained that women have fewer teeth than men; although he was twice married, it never occurred to him to verify this statement by examining his wives' mouths. He said also that children will be healthier if conceived when the wind is in the north. One gathers that the two Mrs. Aristotles both had to run out and look at the weathercock every evening before going to bed. He states that a man bitten by a mad dog will not go mad, but any other animal will (*Hist. An.* 704a); that the bite of the shrewmouse is dangerous to horses, especially if the mouse is pregnant (ibid., 604b); that elephants suffering from insomnia can be cured by rubbing their shoulders with salt, olive oil, and warm water (ibid., 605a); and so on and so on. Nevertheless, classical dons, who have never observed any animal except the cat and the dog, continue to praise Aristotle for his fidelity to observation.

The conquest of the East by Alexander caused an immense influx of superstition into the Hellenistic world. This was particularly notable as regards astrology, which almost all later pagans believed in. The Church condemned it, not on scientific grounds, but because it implied subjection to Fate. There is, however, in St. Augustine, a scientific argument against astrology quoted from one of the rare pagan skeptics. The argument is that twins often have very different careers, which they ought not to have if astrology were true.

At the time of the Renaissance, belief in astrology became a mark of the free thinker: it must be true, he thought, because the Church condemned it. Free thinkers were not yet any more scientific than their opponents in the matter of appeal to observable facts.

Most of us still believe many things that in fact have no basis except in the assertions of the ancients. I was always told that ostriches eat nails, and, though I wondered how they found them in the Bush, it did not occur to me to doubt the story. At last I discovered that it comes from Pliny, and has no truth whatever. . . .

When Galileo's telescope revealed Jupiter's moons, the orthodox refused to look through it, because they knew there could not be such bodies, and therefore the telescope must be deceptive.

Respect for observation as opposed to tradition is difficult and (one might almost say) contrary to human nature. Science insists upon it, and this insistence was the source of the most desperate battles between science and authority. There are still a great many repects in which the lesson has not been learned. Few people can be convinced that an obnoxious habit—e.g. exhibitionism—cannot be cured by punishment. It is pleasant to punish those who shock us, and we do not like to admit that indulgence in this pleasure is not always socially desirable. . . .

The Dethronement of "Purpose." Aristotle maintained that causes are of four kinds; modern science admits only one of the four. Two of Aristotle's four need not concern us; the two that do concern us are the "efficient" and the "final" cause. The "efficient" cause is what we should call simply "the cause"; the "final" cause is the purpose. In human affairs this distinction has validity. Suppose you found a restaurant at the top of a mountain. The "efficient" cause is the carrying up of the materials and the arranging of them in the pattern of a house. The "final" cause is to satisfy the hunger and thirst of tourists. In human affairs, the question "why?" is more naturally answered, as a rule, by assigning the final cause than by setting out the efficient cause. If you ask "why is there a restaurant here?" the natural answer is "because many hungry and thirsty people come this way." But the answer by final cause is only appropriate where human volitions are involved. If you ask "why do many people die of cancer?" you will get no clear answer, but the answer you want is one assigning the efficient cause.

This ambiguity in the word "why" led Aristotle to his distinction of efficient and final causes. He thought—and many people still think—that both kinds are to be found everywhere: whatever exists may be explained, on the one hand, by the antecedent events that have produced it, and, on the other hand, by the purpose that it serves. But although it is still open to the philosopher or theologian to hold that everything has a "purpose," it has been found that "purpose" is not a useful concept when we are in search of scientific laws. We are told in the Bible that the moon was made to give light by night. But men of science, however pious, do not regard this as a scientific explanation of the origin of the moon. Or, to revert to the question about cancer, a man of science may believe, in his private capacity, that cancer is sent as a punishment for our sins, but *qua* man of science he must ignore this point of view. We know of "purpose" in human affairs, and we may suppose that there are cosmic purposes, but in science it is the past that determines the future, not the future the past. "Final" causes, therefore, do not occur in the scientific account of the world. . . .

Man's Place in the Universe. The effect of science upon our view of man's place in the universe has been of two opposite kinds; it has at once degraded and exalted him. It has degraded him from the standpoint of contemplation, and exalted him from that of action. The latter effect has gradually come to outweigh the former, but both have been important. I will begin with the contemplative effect.

To get this effect with its full impact, you should read simultaneously Dante's *Divine Comedy* and Hubble on the *Realm of the Nebulae*—in each case with active imagination and with full receptiveness to the cosmos that they portray. In Dante, the earth is the center of the universe; there are ten concentric spheres, all revolving about the earth; the wicked, after death, are punished at the center of the earth; the comparatively virtuous are purged on the Mount of Purgatory at the antipodes of Jerusalem; the good, when purged, enjoy eternal bliss in one or other of the spheres, according to the degree of their merit. The universe is tidy and small: Dante visits all the spheres in the course of twenty-four hours. Everything is contrived in relation to man: to punish sin and reward virtue. There are no mysteries, no abysses, no secrets; the whole thing is like a child's doll's house, with people as the dolls. But although the people were dolls they were important because they interested the Owner of the doll's house.

The modern universe is a very different sort of place. Since the victory of the Copernican system we have known that the earth is not the center of the universe. For a time the sun replaced it, but then it turned out that the sun is by no means a monarch among stars, in fact, is scarcely even middle class. There is an incredible amount of empty space in the universe. The distance from the sun to the nearest star is about 4.2 light years, or 25×10^{12} miles. This is in spite of the fact that we live in an exceptionally crowded part of the universe, namely the Milky Way, which is an assemblage of about 300,000 million stars. This assemblage is one of an immense number of similar assemblages; about 30 million are known, but presumably better telescopes would show more.[1] The average distance from one assemblage to the next is about 2 million light years. But apparently they still feel they haven't elbow room, for they are all hurrying away from each other; some are moving away from us at the rate of 14,000 miles a second or more. . . . And as to mass: the sun weighs about 2×10^{27} tons, the Milky Way about 160,000 million times as much as the sun, and is one of a collection of galaxies of which about 30 million are known. It is not easy to maintain a belief in one's own cosmic importance in view of such overwhelming statistics.

So much for the contemplative aspect of man's place in a scientific cosmos. I come now to the practical aspect.

To the practical man, the nebulae are a matter of indifference. He can understand astronomers' thinking about them, because they are paid to, but there is no reason why *he* should worry about anything so unimportant. What matters to him about the world is what he can make of it. And scientific man can make vastly more of the world than unscientific man could.

In the pre-scientific world, power was God's. There was not much that man could do even in the most favorable circumstances, and the circumstances were liable to become unfavorable if men incurred the divine displeasure. This showed itself in earthquakes, pestilences, famines, and defeats in war. Since such events are frequent, it was obviously very easy to incur divine displeasure. Judging by the analogy of earthly monarchs, men decided that the thing most displeasing to the Deity is a lack of humility. If you wished to slip through life without disaster, you must be meek; you must be aware of your defenselessness, and constantly ready to confess it. But the God before whom you humbled yourself was conceived in the likeness of man, so that the universe seemed human and warm and cozy, like home if you are the youngest of a large family, painful at times, but never alien and incomprehensible.

In the scientific world, all this is different. It is not by prayer and humility that you cause things to go as you wish, but by acquiring a knowledge of natural laws. The power you acquire in this way is much greater and much more reliable than that formerly supposed to be acquired by prayer, because you never could tell whether your prayer would be favorably heard in heaven. The power of prayer, moreover,

---

1. Current estimates of total number observable with our best instruments are as high as 100,000,000,000. Ed. (1970)

had recognized limits; it would have been impious to ask too much. But the power of science has no known limits. We were told that faith could remove mountains, but no one believed it; we are now told that the atomic bomb can remove mountains, and everyone believes it.

It is true that if we ever did stop to think about the cosmos we might find it uncomfortable. The sun may grow cold or blow up; the earth may lose its atmosphere and become uninhabitable. Life is a brief, small, and transitory phenomenon in an obscure corner, not at all the sort of thing that one would make a fuss about if one were not personally concerned. But it is monkish and futile—so scientific man will say—to dwell on such cold and unpractical thoughts. Let us get on with the job of fertilizing the desert, melting Arctic ice, and killing each other with perpetually improving technique. Some of our activities will do good, some harm, but all alike will show our power. And so, in this godless universe, we shall become gods. . . .

Science, ever since the time of the Arabs, has had two functions: (1) to enable us to *know* things, and (2) to enable us to *do* things. The Greeks, with the exception of Archimedes, were only interested in the first of these. They had much curiosity about the world, but, since civilized people lived comfortably on slave labor, they had no interest in technique. Interest in the practical uses of science came first through superstition and magic. The Arabs wished to discover the philosopher's stone, the elixir of life, and how to transmute base metals into gold. In pursuing investigations having these purposes, they discovered many facts in chemistry, but they did not arrive at any valid and important general laws, and their technique remained elementary. . . .

Science used to be valued as a means of getting to *know* the world; now, owing to the triumph of technique, it is conceived as showing how to *change* the world. The new point of view, which is adopted in practice throughout America and Russia, and in theory by many modern philosophers, was first proclaimed by Marx in 1845, in his *Theses on Feuerbach*. He says:

The question whether objective truth belongs to human thinking is not a question of theory, but a practical question. The truth, i.e. the reality and power, of thought must be demonstrated in practice. The contest as to the reality or nonreality of a thought which is isolated from practice, is a purely scholastic question. . . . Philosophers have only interpreted the world in various ways, but the real task is to alter it.

From the point of view of technical philosophy, this theory has been best developed by John Dewey, who is universally acknowledged as America's most eminent philosopher.

This philosophy has two aspects, one theoretical and the other ethical. On the theoretical side, it analyzes away the concept "truth," for which it substitutes "utility." It used to be thought that, if you believed Caesar crossed the Rubicon, you believed truly, because Caesar did cross the Rubicon. Not so, say the philosophers we are considering: to say that your belief is "true" is another way of saying that you will find it more profitable than the opposite belief. I might object that there have been cases of historical beliefs which, after being generally accepted for a long time, have in the end been admitted to be mistaken. In the case of such beliefs, every examinee would find the accepted falsehood of his time more profitable than the as yet unacknowledged truth. But this kind of objection is swept aside by the contention that a belief may be "true" at one time and "false" at another. In 1920 it was "true" that Trotsky had a great part in the Russian Revolution; in 1930 it was "false." The results of this view have been admirably worked out in George Orwell's "1984."

This philosophy derives its inspiration from science in several different ways. Take first its best aspect, as developed by Dewey. He points out that scientific theories change from time to time, and that what recommends a theory is that it "works." When new phenomena are discovered, for which it no longer "works," it is discarded. A theory—so Dewey concludes—is a tool like another; it enables us to manipulate raw material. Like any other tool, it is judged good or bad by its efficiency in this manipulation, and like any other tool, it is good at one time and bad at another. While it is good it may be called "true," but this word must not be allowed its usual connotations. Dewey prefers the phrase "warranted assertibility" to the word "truth."

The second source of the theory is technique. What do we want to know about electricity? Only how to make it work for us. To want to know more is to plunge into useless metaphysics. Science is to be admired because it gives us power over nature, and the power comes wholly from technique. Therefore an interpretation which reduces science to technique keeps all the useful part, and dismisses only a dead weight of medieval lumber. If technique is all that interests you, you are likely to find this argument very convincing.

The third attraction of pragmatism—which cannot be wholly separated from the second—is love of power. Most men's desires are of various kinds. There are the pleasures of sense; there are aesthetic pleasures and pleasures of contemplation; there are private affections; and there is power. In an individual, any one of these may acquire predominance over the others. If love of power dominates, you arrive at Marx's view that what is important is not to understand the world, but to change it. Traditional theories of knowledge were invented by men who loved contemplation—a monkish taste, according to modern devotees of mechanism. Mechanism augments human power to an enormous degree. It is therefore this aspect of science that attracts the lovers of power. And if power is all you want from science, the pragmatist theory gives you just what you want, without accretions that to you seem irrelevant. It gives you even more than you could have expected, for if you control the police it gives you the godlike power of *making truth*. You cannot make the sun cold, but you can confer pragmatic "truth" on the proposition "the sun is cold" if you can ensure that everyone who denies it is liquidated. I doubt whether Zeus could do more.

This engineer's philosophy, as it may be called, is distinguished from common sense and from most other philosophies by its rejection of "fact" as a fundamental concept in defining "truth." If you say, for example, "the South Pole is cold," you say something which, according to traditional views, is "true" in virtue of a "fact," namely that the South Pole is cold. And this is a fact, not because people believe it, or because it pays to believe it; it just *is* a fact. Facts, when they are not about human beings and their doings, represent the limitations of human power. We find ourselves in a universe of a certain sort, and we find out what sort of universe it is by observation, not by self-assertion. It is true that we can make changes on or near the surface of the earth, but not elsewhere. Practical men have no wish to make changes elsewhere, and can therefore accept a philosophy which treats the surface of the earth as if it were the whole universe. But even on the surface of the earth our power is limited. To forget that we are hemmed in by facts which are for the most part independent of our desires is a form of insane megalomania. This kind of insanity has grown up as a result of the triumph of scientific technique. Its latest manifestation is Stalin's refusal to believe that heredity can have the temerity to ignore Soviet decrees, which is like Xerxes whipping the Hellespont to teach Poseidon a lesson.

"The pragmatic theory of truth [I wrote in 1907] is inherently connected with the appeal to force. If there is a non-human truth, which one man may know while another does not, there is a standard outside the disputants, to which, we may urge, the dispute ought to be submitted; hence a pacific and judicial settlement of disputes is at least theoretically possible. If, on the contrary, the only way of discovering which of the disputants is in the right is to wait and see which of them is successful, there is no longer any principle except force by which the issue can be decided. . . . In international matters, owing to the fact that the disputants are often strong enough to be independent of outside control, these considerations become more important. The hopes of international peace, like the achievement of internal peace, depend upon the creation of an effective force of public opinion formed upon an estimate of the rights and wrongs of disputes. Thus it would be misleading to say that the dispute is decided by force, without adding that force is dependent upon justice. But the possibility of such a public opinion depends upon the possibility of a standard of justice which is a cause, not an effect, of the wishes of the community; and such a standard of justice seems incompatible with the pragmatist philosophy. This philosophy, therefore, although it begins with liberty and toleration, develops, by inherent necessity, into the appeal to force and the arbitrament of the big battalions. By this development it becomes equally adapted to democracy at home and to imperialism abroad. Thus here again it is more delicately adjusted to the requirements of the time than any other philosophy which has hitherto been invented.

"To sum up: Pragmatism appeals to the temper of mind which finds on the surface of this planet the whole of its imaginative material; which feels confident of progress, and unaware of non-human limitations to human power; which loves battle, with all the attendant risks, because it has no real doubt that it will achieve victory; which desires religion, as it desires railways and electric light, as a comfort and a help in the affairs of this world, not as providing non-human objects to satisfy the hunger for perfection. But for those who feel that life on this planet would be a life in prison if it were not for the windows into a greater world beyond; for those to whom a belief in man's omnipotence seems arrogant; who desire rather the stoic freedom that comes of mastery over the passions than the Napoleonic domination that sees the kingdoms of this world at its feet—in a word, to men who do not find man an adequate object of their worship, the pragmatist's world will seem narrow and petty, robbing life of all that gives it value, and making man himself smaller by depriving the universe which he contemplates of all its splendor."

In Western Culture science is a pragmatic pursuit; it is the discovery of useful facts, whatever they may be. Its virtue lies in the honesty and accuracy with which these facts are gathered and in the completeness of the pattern that, as part of formulated knowledge, they finally compose. This factualness of science makes it blind to the differences between the trivial and the significant, the odious and the exquisite, the good and the bad; indeed the identification of science with the realm of discoverable fact has largely removed it from most basic human concerns and made it into a gigantic robot driving toward material progress.

Henry Margenau
"Perspectives of Science" (1959) °

° From Henry Margenau, "Perspectives of Science," *The Key Reporter*, Autumn, 1959. Copyright United Chapters of Phi Beta Kappa, Washington, D.C. Reprinted by permission.

The Spirit of Science

Science and Human Values °

by Jacob Bronowski (1956)

1

My theme is that the values which we accept today as permanent and often as self-evident have grown out of the Renaissance and the Scientific Revolution. The arts and the sciences have changed the values of the Middle Ages; and this change has been an enrichment, moving towards what makes us more deeply human. . . .

2

The concepts of value are profound and difficult exactly because they do two things at once: they join men into societies, and yet they preserve for them a freedom which makes them single men. . . .

The society which I will examine is that formed by scientists themselves: it is the body of scientists.

It may seem strange to call this a society, and yet it is an obvious choice; for having said so much about the working of science, I should be shirking all our unspoken questions if I did not ask how scientists work together. The dizzy progress of science, theoretical and practical, has depended on the existence of a fellowship of scientists which is free, uninhibited and communicative. It is not an upstart society, for it derives its traditions, both of scholarship and of service, from roots which reach through the Renaissance into the monastic communities and the first universities. The men and women who practice the sciences make a company of scholars which has been more lasting than any modern state, yet which has changed and evolved as no church has. . . .

3

The values of science derive neither from the virtues of its members, nor from the finger-wagging codes of conduct by which every profession reminds itself to be good. They have grown out of the practice of science, because they are the inescapable conditions for its practice.

Science is the creation of concepts and their exploration in the facts. It has no other test of the concept than its empirical truth to fact. Truth is the drive at the center of science; it must have the habit of truth, not as a dogma but as a process. Consider then, step by step, what kind of society scientists have been compelled to form in this single pursuit. If truth is to be found, not given, and if therefore it is to be tested in action, what other conditions (and with them, what other values) grow of themselves from this?

First, of course, comes independence, in observation and thence in thought. I once told an audience of school children that the world would never change if they did not contradict their elders. I was chagrined to find next morning that this axiom outraged their parents. Yet it is the basis of the scientific method. A man must see, do and think things for himself, in the face of those who are sure that they have already been over all that ground. In science, there is no substitute for independence.

It has been a by-product of this that, by degrees, men have come to give a value to the new and the bold in all their work. It was not always so. European thought and art before the Renaissance were happy in the faith that there is nothing new under the sun. John Dryden in the seventeenth century, and Jonathan Swift as it turned into the eighteenth, were still fighting Battles of the Books to prove that no modern work could hope to rival the classics. They were not overpowered by argument or example (not even by their own examples), but by the mounting scientific tradition among their friends in the new Royal Society. Today we find it as natural to prize originality in a child's drawing and an arrangement of flowers as in an invention. Science has bred the love of originality as a mark of independence.

Independence, originality, and therefore dissent: these words show the progress, they stamp the character of our civilization as once they did that of Athens in flower. From Luther in 1517 to Spinoza grinding lenses, from Huguenot weavers and Quaker ironmasters to the Puritans founding Harvard, and from Newton's heresies to the calculated universe of Eddington, the profound movements of history have been begun by unconforming men. Dissent is the native activity of the scientist, and it has got him into a good deal of trouble in the last years. But if that is cut off, what is left will not be a scientist. And I doubt whether it will be a man. For dissent is also native in any society which is still growing. Has there ever been a society which has died of dissent? Several have died of conformity in our lifetime.

Dissent is not itself an end; it is the surface mark of a deeper value. Dissent is the mark of freedom, as originality is the mark of independence of mind. And as originality and independence are private needs for the existence of a science, so dissent and freedom are its public needs. No one can be a scientist, even in private, if he does not have independence of observation and of thought. But if in addition science is to become effective as a public practice, it must go further; it must protect independence. The safeguards which it must offer are patent: free inquiry, free thought, free speech, tolerance. These values are so familiar to us, yawning our way through political perorations, that they seem self-evident. But they are self-evident, that is, they are logical needs, only where men are committed to explore the truth: in a scientific society. These freedoms of tolerance have never been notable in a dogmatic society, even when the dogma was Christian. They have been granted only when scientific thought flourished once before, in the youth of Greece.

4

I have been developing an ethic for science which derives directly from its own activity. It might have seemed at the outset that this study could lead only to a set of technical rules: to elementary rules for using test tubes or sophisticated rules for inductive reasoning. But the inquiry turns out quite otherwise. There are, oddly, no technical rules for success in science. There are no rules even for using test tubes which the brilliant experimenter does not flout; and alas, there are no rules at all for making successful general inductions. This is not where the study of scientific practice leads us. Instead, the conditions for the practice of science are found to be of another and an unexpected kind. Independence and originality, dissent and freedom and tolerance: such are the first needs of science; and these are the values which, of itself, it demands and forms.

The society of scientists must be a democracy. It can keep alive and grow only by a constant tension between dissent and respect, between independence from the views of others and tolerance for them. The crux of the ethical problem is to fuse these, the private and the public needs. Tolerance alone is not enough; this is why the bland, kindly civilizations of the East, where to contradict is a personal affront, developed no strong science. And independence is not enough either: the sad history of genetics, still torn today by the quarrels of sixty years ago, shows that. Every scientist has to learn the hard lesson, to respect the views of the next man—even when the next man is tactless enough to express them.

Tolerance among scientists cannot be based on indifference, it must be based on respect. Respect as a personal value implies, in any society, the public acknowledgements of justice and of due honor. These are values which to the layman seem most remote from any abstract study. Justice, honor, the respect of man for man: What, he asks, have these human values to do with science? The question is a foolish survival of those nineteenth-century quarrels which always came back to equate ethics with the Book of Genesis. If critics in the past had ever looked practically to see how a science develops, they would not have asked such a question. Science confronts the work of one man with that of another and grafts each on each; and it cannot survive without justice and honor and respect between man and man. Only by these means can science pursue its steadfast object, to explore truth. If these values did not exist, then the society of scientists would have to invent them to make the practice of science possible. In societies where these values did not exist, science has had to create them.

Science is not a mechanism but a human progress. To the layman who is dominated by the fallacy of the comic strips, that science would all be done best by machines, all this is puzzling. But human search and research is a learning by steps of which none is final, and the mistakes of one generation are rungs in the ladder, no less than their correction by the next. This is why the values of science turn out to be recognizably the human values: because scientists must be men, must be fallible, and yet as men must be willing and as a society must be organized to correct their errors. William Blake said that "to be an Error & to be Cast out is a part of God's design." It is certainly part of the design of science.

There never was a great scientist who did not make bold guesses, and there never was a bold man whose guesses were not sometimes wild. Newton was wrong, in the setting of his time, to think that light is made up of particles. Faraday was foolish when he looked in his setting for a link between electromagnetism and gravitation. And such is the nature of science, their bad guesses may yet be brilliant by the work of our own day. We do not think any less of the profound concept of General Relativity in Einstein because the details of his formulation at this moment seem doubtful. For in science as in literature, the style of a great man is the stamp of his mind and makes even his mistakes a challenge which is part of the march of its subject. Science at last respects the scientist more than his theories, for by its nature, it must prize the search above the discovery and the thinking (and with it the thinker) above the thought. In the society of scientists each man, by the process of exploring for the truth, has earned a dignity more profound than his doctrine. A true society is sustained by the sense of human dignity.

I take this phrase from the life of the French naturalist Buffon who, like Galileo, was forced to recant his scientific findings. Yet he preserved always, says his biographer, something deeper than the fine manners of the court of Louis XV; he kept "le sentiment exquis de la dignité humaine." His biographer says that Buffon learned this during his stay in England where it was impressed on him by the scientists he met. Since Buffon seems to have spent at most three months in England, this claim has been thought extravagant. But is it? Is history really so inhuman in arithmetic? Buffon in the short winter of 1738–9 met the grave men of the Royal Society, heirs to Newton, the last of a great generation. He found them neither a court nor a rabble, but a community of scientists seeking the truth together with dignity and humanity. It was, it is, a discovery to form a man's life.

The sense of human dignity that Buffon showed in his bearing is the cement of a society of equal men, for it expresses their knowledge that respect for others must be founded in self-respect. Theory and experiment alike become meaningless unless the scientist brings to them, and his fellows can assume in him, the respect of a lucid honesty with himself. The mathematician and philosopher W. K. Clifford said this forcibly at the end of his short life, nearly a hundred years ago.

If I steal money from any person, there may be no harm done by the mere transfer of possession; he may not feel the loss, or it may even prevent him from using the money badly. But I cannot help doing this great wrong towards Man, that I make myself dishonest. What hurts society is not that it should lose its property, but that it should become a den of thieves; for then it must cease to be society. This is why we ought not to do evil that good may come; for at any rate this great evil has come, that we have done evil and are made wicked thereby.

This is the scientist's moral: that there is no distinction between ends and means. Clifford goes on to put this in terms of the scientist's practice.

In like manner, if I let myself believe anything on insufficient evidence, there may be no great harm done by the mere belief; it may be true after all, or I may never have occasion to exhibit it in outward acts. But I cannot help doing this great wrong towards Man, that I make myself credulous. The danger to society is not merely that it should believe wrong things, though that is great enough; but that it should become credulous.

And the passion in Clifford's tone shows that to him the word "credulous" had the same emotional force as a "den of thieves."

The fulcrum of Clifford's ethic here, and mine, is the phrase "it may be true after all." Others may allow this to justify their conduct; the practice of science wholly rejects it. It does not admit that the word true can have this meaning. The test of truth is the fact, and no glib expediency nor reason of state can justify the smallest self-deception in that. Our work is of a piece, in the large and in detail, so that if we silence one scruple about our means, we infect ourselves and our ends together.

The scientist derives this ethic from his method, and every creative worker reaches it for himself. This is how Blake reached it from his practice as a poet and a painter.

He who would do good to another must do it in Minute Particulars: General Good is the plea of the scoundrel, hypocrite & flatterer, For Art & Science cannot exist but in minutely organized Particulars.

The Minute Particulars of art and the fine structure of science alike make the grain of conscience.

5

Usually when scientists claim that their work has liberated men, they do so on more practical grounds. In these four hundred years, they say, we have mastered sea and sky, we have drawn information from the electron and power from the nucleus, we have doubled the span of life and halved the working day, and we have enriched the leisure we have created with universal education and high-fidelity recordings and electric lights and the lipstick. We have carried out the tasks which men set for us because they were most urgent. To a world population at least five times larger than in Kepler's day, there begins to be offered a life above the animal, a sense of personality, and a potential of human fulfilment, which make both the glory and the explosive problem of our age.

These claims are not confined to food and bodily comfort. Their larger force is that the physical benefits of science have opened a door and will give all men the chance to use mind and spirit. The technical man here neatly takes his model from evolution, in which the enlargement of the human brain followed the development of the hand.

I take a different view of science as a method; to me, it enters the human spirit more directly. Therefore I have studied quite another achievement: that of making a human society work. As a set of discoveries and devices, science has mastered nature; but it has been able to do so only because its values, which derive from its method, have formed those who practice it into a living, stable and incorruptible society. Here is a community where everyone has been free to enter, to speak his mind, to be heard and contradicted; and it has outlasted the empires of Louis XIV and the Kaiser. Napoleon was angry when the Institute he had founded awarded his first scientific prize to Humphry Davy, for this was in 1807, when France was at war with England. Science survived then and since because it is less brittle than the rage of tyrants.

This is a stability which no dogmatic society can have. There is today almost no scientific theory which was held when, say, the Industrial Revolution began about 1760. Most often today's theories flatly contradict those of 1760; many contradict those of 1900. In cosmology, in quantum mechanics, in genetics, in the social sciences, who now holds the beliefs that seemed firm fifty years ago? Yet the society of scientists has survived these changes without a revolution and honors the men whose beliefs it no longer shares. No one has been shot or exiled or convicted of perjury; no one has recanted abjectly at a trial before his colleagues. The whole structure of science has been changed, and no one has been either disgraced or deposed. Through all the changes of science, the society of scientists is flexible and single-minded together and evolves and rights itself. In the language of science, it is a stable society.

The society of scientists is simple because it has a directing purpose: to explore the truth. Nevertheless, it has to solve the problem of every society, which is to find a compromise between man and men. It must encourage the single scientist to be independent, and the body of scientists to be tolerant. From these basic conditions, which form the prime values, there follows step by step the spectrum of values: dissent, freedom of thought and speech, justice, honor, human dignity and self-respect.

Our values since the Renaissance have evolved by just such steps. There are of course casuists who, when they are not busy belittling these values, derive them from the Middle Ages. But that servile and bloody world upheld neither independence nor tolerance, and it is from these, as I have shown, that the human values are rationally derived. Those who crusade against the rational and receive their values by mystic inspiration have no claim to these values of the mind. I cannot put this better than in the words of Albert Schweitzer in which he, a religious man, protests that mysticism in religion is not enough.

> Rationalism is more than a movement of thought which realized itself at the end of the eighteenth and the beginning of the nineteenth centuries. It is a necessary phenomenon in all normal spiritual life. All real progress in the world is in the last analysis produced by rationalism. The principle, which was then established, of basing our views of the universe on thought and thought alone is valid for all time.

So proud men have thought, in all walks of life, since Giordano Bruno was condemned to be burnt for his cosmology, about 1600. They have gone about their work simply enough. The scientists among them did not set out to be moralists or revolutionaries. William Harvey and Huygens, Euler and Avogadro, Darwin and Willard Gibbs and Marie Curie, Planck and Pavlov practiced their crafts modestly and steadfastly. Yet the values they seldom spoke of shone out of their work and entered their ages and slowly re-made the minds of men. Slavery ceased to be a matter of course. The princelings of Europe fled from the gaming table. The empires of the Bourbons and the Hapsburgs crumbled. Men asked for the rights of man and for government by consent. By the beginning of the nineteenth century, Napoleon did not find a scientist to elevate tyranny into a system; that was done by the philosopher Hegel. Hegel had written his university dissertation to prove philosophically that there could be no more than the seven planets he knew. It was unfortunate, and characteristic, that even as he wrote, on 1st January 1801, a working astronomer observed the eighth planet Ceres.[1]

6

. . . Has science fastened upon our society a monstrous gift of destruction which we can neither undo nor master, and which, like a clockwork automaton in a nightmare, is set to break our necks? Is science an automaton, and if so has it lamed our sense of values?

These questions are not answered by holding a Sunday symposium of moralists. They are not even answered by the painstaking neutralism of the textbooks on scientific method. We must indeed begin from a study of what scientists do, when they are neither posed for photographs on the steps of spaceships nor bumbling professorially in the cartoons. But we must get to the heart of what they do. We must lay bare the conditions which make it possible for them to work at all.

When we do so we find, leaf by leaf, the organic values which I have been unfolding. And we find that they are not at odds with the values by which alone mankind can survive. On the contrary, like the other creative activities which grew from the Renaissance, science has humanized our values. Men have asked for freedom, justice and respect precisely as the scientific spirit has spread among them. The dilemma of today is not that the human values cannot control a mechanical science. It is the other way about: the scientific spirit is more human than the machinery of governments. We have not let either the tolerance or the empiricism of science enter the parochial rules by which we still try to prescribe the behavior of nations. Our conduct as states clings to a code of self-interest which science, like humanity, has long left behind.

The body of technical science burdens and threatens us because we are trying to employ the body without the spirit; we are trying to buy

1. Ceres is the largest asteroid in the belt between Mars and Jupiter. Ed.

the corpse of science. We are hag-ridden by the power of nature which we should command because we think its command needs less devotion and understanding than its discovery. And because we know how gunpowder works, we sigh for the days before atomic bombs. But massacre is not prevented by sticking to gunpowder; the Thirty Years' War is proof of that. Massacre is prevented by the scientist's ethic, and the poet's, and every creator's: that the end for which we work exists and is judged only by the means which we use to reach it. This is the human sum of the values of science. It is the basis of a society which scrupulously seeks knowledge to match and govern its power. But it is not the scientist who can govern society; his duty is to teach it the implications and the values in his work. Sir Thomas More said this in 1516, that the single-minded man must not govern but teach; and went to the scaffold for neglecting his own counsel.

7

I have analysed . . . only the activity of science. Yet I do not distinguish it from other imaginative activities; they are as much parts one of another as are the Renaissance and the Scientific Revolution. The sense of wonder in nature, of freedom within her boundaries, and of unity with her in knowledge, is shared by the painter, the poet and the mountaineer. Their values, I have no doubt, express concepts as profound as those of science and could serve as well to make a society—as they did in Florence, and in Elizabethan London and among the famous doctors of Edinburgh. Every cast of mind has its creative activity which explores the likenesses appropriate to it and derives the values by which it must live.

The exploration of the artist is no less truthful and strenuous than that of the scientist. If science seems to carry conviction and recognition more immediately, this is because here the critics are also those who work at the matter. There is not, as in the arts, a gap between the functions (and therefore between the fashions) of those who comment and those who do. Nevertheless, the great artist works as devotedly to uncover the implications of his vision as does the great scientist. They grow, they haunt his thought, and their most inspired flash is the end of a lifetime of silent exploration. . . .

Whether our work is art or science or the daily work of society, it is only the form in which we explore our experience which is different; the need to explore remains the same. This is why, at bottom, the society of scientists is more important than their discoveries. What science has to teach us here is not its techniques but its spirit: the irresistible need to explore. Perhaps the techniques of science may be practiced for a time without its spirit, in secret establishments, as the Egyptians practiced their priestcraft. But the inspiration of science for four hundred years has been opposite to this. It has created the values of our intellectual life and, with the arts, has taught them to our civilization. Science has nothing to be ashamed of even in the ruins of Nagasaki. The shame is theirs who appeal to other values than the human imaginative values which science has evolved. The shame is ours, if we do not make science part of our world, intellectually as much as physically, so that we may at last hold these halves of the world together by the same values. For this is the lesson of science, that the concept is more profound than its laws and the act of judging more critical than the judgment. In a book that I wrote about poetry I said:

Poetry does not move us to be just or unjust, in itself. It moves us to thoughts in whose light justice and injustice are seen in fearful sharpness of outline.

What is true of poetry is true of all creative thought. And what I said then of one value is true of all human values. The values by which we are to survive are not rules for just and unjust conduct, but are those deeper illuminations in whose light justice and injustice, good and evil, means and ends are seen in fearful sharpness of outline.

The Power of Reason °
by Alfred North Whitehead (1925)

[We have been following] a record of a great adventure in the region of thought. It was shared in by all the races of western Europe. It developed with the slowness of a mass movement. Half a century is its unit of time. The tale is the epic of an episode in the manifestation of reason. It tells how a particular direction of reason emerges in a race by the long preparation of antecedent epochs, how after its birth its subject-matter gradually unfolds itself, how it attains its triumphs, how its influence moulds the very springs of action of mankind, and finally how at its moment of supreme success its limitations disclose themselves and call for a renewed exercise of the creative imagination. The moral of the tale is the power of reason, its decisive influence on the life of humanity. The great conquerors, from Alexander to Caesar, and from Caesar to Napo-

° From Alfred North Whitehead, *Science and the Modern World,* The Macmillan Company, New York. Copyright 1925 by The Macmillan Company; copyright 1953 by Evelyn Whitehead. Reprinted by permission.

leon, influenced profoundly the lives of subsequent generations. But the total effect of this influence shrinks to insignificance, if compared to the entire transformation of human habits and human mentality produced by the long line of men of thought from Thales to the present day, men individually powerless, but ultimately the rulers of the world.

Suggestions for Further Reading

° Bronowski, Jacob: *The Common Sense of Science*, Vintage Books, Random House, Inc., New York, 1953. Readers who have enjoyed the Bronowski selection in this part will be interested in chaps. 8 and 9: "Truth and Value" and "Science, the Destroyer or Creator."

° Conant, James: *Modern Science and Modern Man*, Anchor Books, Doubleday & Company, Inc., Garden City, N.Y., 1952. A collection of a series of lectures delivered in 1952, discussing the role of modern science and its relation to society.

Huxley, Aldous, ° *Brave New World*, Bantam Books, Inc., New York, 1963; ° *Brave New World Revisited*, Bantam Books, Inc., New York, 1960.

———: *Science, Liberty and Peace*, Harper & Row, Publishers, Inc., New York, 1946. These three books present in vivid terms the dangers of a technological society. Readers who would like to understand more thoroughly the point of view presented by Huxley in the selection in this part are urged to read all of his little book *Science, Liberty and Peace*.

° Oppenheimer, J. Robert: *The Open Mind*, Simon and Schuster, Inc., New York, 1955. A collection of lectures on atomic weapons and the significance of the atomic age. The last four lectures deal with the relationship between science as an intellectual activity and the wider culture of our times.

———Russell, Bertrand: *The Impact of Science on Society*, Simon and Schuster, Inc., New York, 1953. A vivid penetrating analysis of the effect science has had on our society.

° Snow, C. P.: *The Two Cultures and a Second Look*, Cambridge University Press, New York, 1959. A discussion of the lack of communication and understanding between science and society. Well worth reading.

° Paperback edition

BIOGRAPHIES

Aristotle (384–322 B.C.) was born in Stagira, a Greek colonial town on the northwestern shore of the Aegean. His father was a doctor who had also been court physician to the father of Philip of Macedon. Some years later, Philip's son, Alexander, became Aristotle's pupil; perhaps, his most famous one. From the time he was seventeen until he was thirty-seven Aristotle studied with Plato at the Academy. And although the characters of the minds of the two men are very different (Plato, the mathematician, the abstractor; Aristotle, the observer of the facts of nature, the humanitarian), it must have been the most profound fact in Aristotle's life to have worked with the old teacher.

Aristotle was a psychologist, logician, moralist, biologist, political thinker, and the founder of literary criticism as well as a philosopher. At the death of Alexander the Great, Aristotle fled to his mother's home in Calcis, where he died at the age of sixty-two.

Isaac Asimov was born in Russia in 1920. He was educated at Columbia University (Ph.D. in 1948) and is an associate professor of biochemistry at Boston University School of Medicine. To date he has an impressive record of more than a hundred successful books and is considered one of today's most imaginative interpreters of scientific subjects. He lives in West Newton, Massachusetts.

Francis Bacon (1561–1626) was an English philosopher and essayist whose career led him into a most difficult position in the famous dispute between Elizabeth I and Essex. Although Essex was his patron and although Elizabeth had repeatedly overlooked Bacon for political office, he served as the Queen's counsel in the trial of the rebels of 1601 and was instrumental in securing Essex's conviction for treason.

His efforts in the case brought him ill feeling on the part of the public and did not advance him in favor with the Queen. It was not until fifteen years after the coronation of James I that Bacon was appointed Lord Chancellor of England.

A year after he had completed his *Novum Organum* he was tried for taking bribes and accused of corrupt dealings in chancery suits. He was convicted, fined, and imprisoned. A general pardon was granted by the King, and Bacon was released after four days in the Tower. Although he admitted taking bribes, he denied they had ever influenced his decisions.

He did the major portion of his most important work after retirement from public life, including the *History of Henry VII* and the *History of Life and Death*.

Roger Bacon (1214–1294) was a British scholar. He studied at Oxford and possibly took holy orders. He lectured as regent master at the University of Paris (1241–47). Then, returning to Oxford about 1251, he assumed the habit of the Franciscan order. He experimented in optics, invented the magnifying glass, and formulated rough approximations of the laws of reflection and refraction. He suggested a revised calendar and a lighter-than-air machine and was one of the first to propose that medicine should make use of the discoveries of chemistry.

In 1266 he was commissioned by Pope Clement IV to write a general treatise on the sciences. He spoke out firmly in favor of experimentation and mathematics and rejected the principle of truth by authority. His works were condemned by the Franciscan order for "suspected novelties," and he was imprisoned from 1277 until just before his death at Oxford about 1294.

Lincoln Barnett is one of our most prolific and accomplished writers in science. He was born in New York City, received his B.A. from Princeton University in 1929 and his M.A. from Columbia University in 1932. During his university years he was also college correspondent for the *New York Herald Tribune.* When *Life* magazine began publication in 1937, Barnett became a staff correspondent and worked there in various departments until 1946. It was shortly afterwards that *Life* asked him to write an article on Einstein. Barnett was then working for himself and took the assignment on a free-lance basis.

The result of the research he did was spectacular. A book, *The Universe and Dr. Einstein,* evolved and was published in 1950. It became a best seller, was widely praised, won the National Book Award Special Citation, and turned Lincoln Barnett into a science writer.

Lloyd Viel Berkner was born in 1905. He received degrees in electrical engineering and physics from the University of Minnesota and George Washington University. He served as engineer on the first Byrd Antarctic Expedition (1928–30). From 1933 to 1941 he was physicist in the department of terrestrial magnetism at Carnegie Institute. After holding a number of government posts, he returned to Carnegie to head the section exploring the geophysics of the atmosphere. In 1960 he became president of the Graduate Research Center S.W. He has received many honors and has been active since 1956 on the President's Science Advisory Committee.

Hermann Bondi was born in Vienna in 1919 of American parents. In 1937 he went to Trinity College, Cambridge, to continue his study of mathematics and was interned there in 1940 because he was an Austrian subject. He was awarded his B.A. degree at about the time he was released. He returned to Cambridge in 1945 to lecture on mathematics, but because of his close contact with Fred Hoyle during the war his interest had changed from classical mathematics to theoretical astronomy. He came to the United States in 1951 as a research associate at Cornell and in 1953 gave a series of Lowell Lectures at Harvard. Since 1954 he has been professor of applied mathematics at King's College, University of London.

Besides his interest in cosmology—he proposed the steady-state theory of the expanding universe, together with Thomas Gold, in 1948—Bondi is interested in travel, other than space, and in the design of children's toys.

William Henry Bragg (1862–1942) was born and educated in England. Then for twenty-three years he was professor of physics and mathematics at Adelaide University in Australia. Returning to England in 1909, he was awarded the Nobel prize for physics and the Barnard Gold Medal (Columbia) in 1915. He shared both distinctions with his son, W. L. Bragg, also a physicist interested in X-ray defraction. Together they developed the X-ray spectrometer, which made possible the elucidation of the arrangements of atoms and crystals. In 1920 William Henry Bragg was knighted. *The Universe of Light* was the last of Sir William's major works. It was preceded by *The World of Sound* and *Concerning the Nature of Things*. He died in 1942, at the age of 80.

Percy Williams Bridgman (1882–1961) took both his undergraduate and graduate degrees from Harvard and joined the Harvard faculty in 1910. From 1926 to 1950 he was Hollis Professor of Mathematics and Natural Philosophy. He was the recipient of many honors including the Nobel prize for physics in 1946. His work in high-pressure phenomena, proving that viscosity increases tremendously with pressure (except for water), led to the production of synthetic diamonds.

Jacob Bronowski is well known for his work in two fields often thought incompatible: mathematics and literature. He was born in Poland in 1908 and educated as a mathematician, earning his Ph.D. from the University of Cambridge in 1933. He was senior lecturer in mathematics at University College until 1942, when he left that post to do wartime research. In 1953 he came to the United States as a visiting professor at the Massachusetts Institute of Technology. Besides this scientific career, he has made a considerable reputation as a radio dramatist and won the Italia prize for the best dramatic work broadcast in Europe during the 1950–51 season. His combination of scientific and literary ability has made him important to the modern movement of scientific humanism in England.

Herbert Butterfield was born on October 7, 1900. He was educated in England and came to the United States in 1924 as a visiting fellow at Princeton University. He edited the *Cambridge Historical Journal* from 1938 until 1952 and was vice-chancellor of the University of Cambridge from 1959 to 1961. He held the position of Regius Professor of History (1963–68) and Professor of Modern History (1963–68). Butterfield has been interested in statecraft for many years, and his books include *The Peace Tactics of Napoleon,* a study of the historical novel, and a study of Christianity and history.

John Christianson was born in Mankato, Minnesota, in 1934. He graduated from Mankato State College, after which he spent a year of study at the University of Copenhagen. While there, his interest in Danish history increased, and this and his interest in the sixteenth century and the history of science converge in the life of the Danish nobleman and astronomer Tycho Brahe.

Returning to the United States, Christianson did his graduate work at the University of Minnesota.

Arthur C. Clarke was born in England in 1917. He received a B.Sc. degree from King's College in London. He served for five years with the R.A.F., becoming technical officer on the first G.C.A. radar system in 1943. From 1946 to 1947 and again from 1950 to 1953 he was chairman of the British Interplanetary Society. Since 1954 he has been engaged in underwater exploration on the Great Barrier Reef of Australia and the coast of Ceylon. In addition, he has written many well-known science and science fiction books, lectured extensively, and appeared on radio and television.

I. Bernard Cohen was born in Far Rockaway, New York, in 1914. He was educated at Harvard University and was awarded his Bachelor of Science degree *cum laude* in 1937. He did his graduate work at the same university in physics, astronomy, and the history of science and received his Ph.D. in 1947.

Cohen has been on the Harvard faculty since his graduate-school days. He has also been a special lecturer at University College, London; the Sorbonne; Oxford University; as well as the University of Florence, and Cambridge University.

Cohen's avocations are traveling, tower climbing, photographing castles and fishing boats, and writing about science, which he does frequently and extremely well.

James Bryant Conant was awarded his Ph.D. from Harvard University in 1916. During World War I, he spent a year in the research division of the chemical warfare service and then returned to Harvard to teach. He became widely recognized as a brilliant organic chemist, was advanced to the rank of professor, and in 1933 was made president of Harvard.

During World War II he was instrumental in organizing American scientists for the war effort and participated in the development of the atomic bomb. At the end of the war he became senior adviser to the National Science Foundation and to the Atomic Energy Commission. In 1953 he became the United States High Commissioner for Western Germany and in 1955 was made our ambassador there. When he returned to the United States in 1957, he again pursued his interest in education and completed a study on the comprehensive high school, which has been of considerable value both to the general public and to educators. From 1963 to 1965 he served as educational adviser to the Ford Foundation in Germany.

Edward Uhler Condon was born in New Mexico in 1902. He worked as a newspaper reporter before entering the University of California, where he received his Ph.D. in 1926. He accepted a National Research Fellowship at Munich for the next year and returned to the United States in 1927 as a lecturer in physics at Columbia University. Condon has held many important positions both in private industry and in government. He has been the director of the National Bureau of Standards and scientific adviser to the United States Senate Special Committee on Atomic Energy. He was elected president of the American Association for the Advancement of Science in 1953.

Condon has published numerous research papers in *Physical Review* and written two important books: *Quantum Mechanics* (with P. Morse) and *Theory of Atomic Spectra* (with G. H. Shortley). He was professor of physics and chairman of that department at Washington University 1956–63, and has been at the University of Colorado at Boulder since 1963.

René Descartes (1596–1650), perhaps best known for his attempted geometrization of all nature and his contributions to pure mathematics, also made astonishingly brilliant contributions to theoretical physics, methodology, and metaphysics. He was the third child of a provincial French politician and was educated at the Jesuit college at La Flèche and at the University of Poitiers and graduated in law. When he was twenty-three a "marvellous science" was revealed to him in a dream showing him that all sciences should be connected "as by a chain," an idea which guided almost all his future work.

Descartes's analysis of reason as it is used in mathematics and application of this method to other sciences represent the end of the medieval approach to science, in which it was thought that different subject matter must be dealt with by different means.

His great, original invention, coordinate geometry, is one of man's most ingenious accomplishments.

Lee Edson is a New Yorker who has been writing about science in one way or another since the age of ten when he won a chemistry set as first prize in an essay contest. After obtaining a B.S. degree in physics and English from the City College of New York, he did graduate work in physics at Brooklyn Polytechnic Institute. Since then he has been constantly involved in teaching, writing, and consulting on scientific subjects.

Albert Einstein (1879–1955) was born in Ulm, Germany, in 1879. He was educated in Germany and in Switzerland. After receiving his Ph.D. from the University of Zurich, he tried, in vain, for a position with a university and finally took a job with the patent office in Berne. His job left him with much time for his own work, and in 1905 he published four important papers. It was the areas of physics dealt with in these papers in which he was to do most of his important work: the special theory of relativity, the establishment of the mass energy equivalence, the theory of Brownian motion, and the photon theory of light.

In 1921 he was awarded a Nobel prize for his photoelectric law and his work in theoretical physics. He emigrated to the United States in 1933 and joined the Institute for Advanced Study at Princeton. He became a United States citizen in 1940.

Einstein's death in 1955 left the science of physics vastly changed. The general theory of relativity, which demonstrates that the laws of physics are the laws of geometry in four dimensions and that these laws are determined by the distribution of matter and energy in the universe, is one of the most beautiful and profound constructions ever conceived by the mind of man.

Galileo Galilei (1564–1642) was sent at seventeen by his father to the University of Pisa to study medicine. His family was noble but impoverished, and although his father was a competent mathematician, he had carefully kept his son away from the study of mathematics because he felt it would lead to the disruption of a medical career. Galileo, however, happened to overhear a lesson in geometry, which aroused his interest, and began, with his father's reluctant permission, to study mathematics and science.

His considerable reputation rests mainly on his discoveries with the telescope. He made his first instrument in 1609 and published his first observation in 1610. Among his discoveries were the satellites of Jupiter, the phases of Venus, the configuration of the moon, and sunspots.

He was tried by the Inquisition in 1633 for lending his support to the Copernican view of the universe, which held that the earth and planets moved around the sun. He recanted under the threat of torture. He was finally allowed to return to Florence and spent the remainder of his life in seclusion working on the principles of mechanics, his observations of the moon, and the application of the pendulum to the regulation of clockwork.

George Gamow (1904–1968) was born in Odessa, Russia, in 1904. He studied nuclear physics at the University of Leningrad and received his Ph.D. there in 1928. In 1934 he came to the United States as a professor of physics at George Washington University, where he was for over twenty years. From 1956 until his death in 1968, he was professor of physics at the University of Colorado. During this time his interest moved from atomic physics to astrophysics, the theory of the expanding universe, and later to the fundamental problems of biology, including molecular genetics and the synthesis of proteins. He also published a number of popular books on science for the layman. His second wife, Barbara Perkins Gamow, wrote the humorous poem that appears on p. 436.

A. R. Hall was born in England in 1920. While a lecturer in the history of science at the University of Cambridge he published *The Scientific Revolution: 1500–1800*, which deals with all those various elements which were a part of the formation of the modern scientific attitude. Perhaps it is not strange that such a fine history should come from a lecturer at Christ College, a school whose libraries and halls were frequented by Newton and Darwin and Rutherford.

Hall spent several years in the United States as professor of philosophy, first at the University of California at Los Angeles and then at the University of Indiana. In 1963 he returned to England where he has been professor of the history of science and technology at the University of London.

Fred Hoyle was born in Yorkshire, England, in 1914. By the time he was six he had taught himself the multiplication tables up to 12 times 12. When he was thirteen, his parents bought him a 3-inch telescope and indulged him by allowing him to stay up all night using it. In 1939 he won a prize fellowship to St. John's and the next year joined an Admiralty research group. He continued to devote much of his spare time to astronomy and after the war returned to Cambridge, where he became a lecturer in mathematics at St. John's College.

Since 1956 Hoyle has been a member of the staff of both the Mt. Wilson and Palomar observatories. In 1958 he was named Plumian Professor of Astronomy and Experimental Philosophy at Cambridge. He has also written a number of books about science, including two novels and several plays.

David Hume (1711–1776), the Scottish philosopher and historian, was born in Edinburgh. He studied for a while at the University of Edinburgh and took up law. But after suffering a nervous breakdown, he gave up this subject to devote himself to philosophy. Living in France, he completed a major work, *Treatise on Human Nature*, before he was twenty-five years old. For a short time, he tutored a young lunatic nobleman. The next year (1745) he was taken as secretary to General St. Clair on a secret mission to Vienna and Turin. In

1752 he became keeper of the Advocates' Library in Edinburgh. Later he served as secretary to the ambassador in Paris and under-secretary of state for the Home Department. His *History of England* and his philosophical treatises won him wide recognition. His denial of the principle of causation, the existence of self, and the existence of God made him a very controversial figure and laid the groundwork for the empiricist and postivist schools of philosophy.

Aldous Leonard Huxley (1894–1963) was a noted English author. After attending a preparatory school, he went to Eton on a scholarship in 1908. He began to study medicine until he contracted keratitis, which left him almost blind. At eighteen he wrote a novel which he was unable to read, and when his sight was partially restored, the manuscript had been lost. He did his graduate work in literature at Oxford and in 1919 joined the staff of *Athenaeum.* His best-known works include *Chrome Yellow, Antic Hay, Point Counter Point,* and *Brave New World.*

Thomas Henry Huxley (1825–1895) was a distinguished English biologist and the head of an accomplished family. His eldest son, Leonard, became a classical scholar. One of his grandsons, Julian, became a famous biologist and philosopher; another, Aldous, became a well-known writer.

After Darwin's *Origin of the Species* was published in 1859, Huxley became an important defender of the work, and *Darwiniana* was published in 1863. The last twenty years of Huxley's life were devoted to public service more than to science, and his term on the London School Board deeply influenced the English national elementary school system. In 1890 he moved away from London because of his health and died at Eastbourne on June 29, 1895.

Christian Huygens (1629–1695), a contemporary of Newton's, was born at The Hague. When he was just twenty-six he was working with his brother on the improvement of the telescope when he discovered a new method of grinding lenses. The resulting clarification of images made possible the discovery of a satellite of Saturn and the resolution of Saturn's rings. He was elected to the Royal Society in 1663. From 1666 to 1681 he worked in France and did his major work on the pendulum. This concludes with the theorems on centrifugal force in circular motion, which were to prove so helpful to Newton. The *Treatise on Light,* from which the selection in the present work is taken, was written in Holland in 1678 and published at Leyden in 1690, where Huygens died five years later.

James Hopwood Jeans (1877–1946) was a classical mathematician and physicist widely known for his popular books on astronomy. He was born in London and educated at Cambridge and became university lecturer in applied mathematics at Cambridge and later professor of applied mathematics at Princeton University. He had marked success with applied mathematics in the fields of physics and astronomy especially. He wrote papers on many aspects of radiation, on the formation of binary stars, spiral nebulae, giant and dwarf stars, the source of stellar energy, and the evolution and radiation of gaseous stars. He was knighted in 1938 and awarded the Order of Merit in 1939.

Arthur Koestler was born in Budapest in 1905 and educated at the University of Vienna. From 1926 until 1931 he was a foreign correspondent in the Middle East, Paris, and Berlin. The next year he signed on as a member of the Graf Zeppelin Arctic Expedition. When he returned, he covered the Spanish Civil War for the *London News Chronicle* and was imprisoned for his pains by Franco. Upon his release, his interest in trouble spots led him to join the French Foreign Legion. It was after this experience that his most famous work, *Darkness at Noon*, was published. Among his other books are *Spanish Testament* and *Lotus and Robot*. His chief recreations are canoeing and chess.

Pierre Simon de Laplace (1749–1827) was born in Normandy in 1749. His genius was recognized early, as it is with many mathematicians, and he was sent to the University of Caen when he was sixteen. By the time he was twenty-four he was famous for a mathematical theorem (named after him) and had already begun to turn his attention to celestial mechanics. Between 1784 and 1786 he published a memoir which successfully explained the eccentricities of the movements of the planets in the solar system, particularly the behavior of Saturn and Jupiter, which had so perplexed Newton that he had been led to hypothesize divine intervention to set things right. His great work, the summary of the labors of three generations of mathematicians on gravitation, appeared in five volumes between 1799 and 1825.

When Napoleon became first consul, Laplace was made minister of the interior but was removed from the position because he brought "the spirit of infinitesimals into administration." After the restoration of the Bourbons, he was made a marquis. He died in Paris in 1827.

Stephen Butler Leacock (1869–1944) was born in England in 1869 and emigrated to Canada with his parents when he was six. After taking his Ph.D. at the University of Chicago he became head of the department of economics and political science at McGill University, in Montreal. Although he wrote a great deal in the fields of history and political economy, his reputation derives mainly from his humorous lectures and literary fantasies. He published some thirty books of humor, many of which are satirical. He saw in humor the power to comfort, to draw men together, to relieve what he called "the appalling inequalities of the human lot."

In respect to his concern for man and his talent for comic invention he had very few peers.

Giacomo Leopardi (1798–1837), an Italian poet, was born into a poor but noble family. When he was sixteen he could read and write fluently in Latin, Greek, French, Spanish, English, and Hebrew. A chronic invalid, he devoted his time to travel and literature, living in succession in Rome, Bologna, Florence, Milan, Pisa, and Naples. He was especially gifted as a writer of lyric poetry.

Alfred Charles Bernard Lovell was born on August 31, 1913. He was educated at the University of Bristol and in 1936 became assistant lecturer in physics at the University of Manchester. He was elected an honorary foreign member of the American Academy of Arts and Sciences in 1955. He has been director of Jodrell Bank Experimental Station since 1951. His published works include

Science and Civilization, Exploration of Space by Radio, and *The Individual and the Universe.* He has received many honors. He was decorated with the Order of the British Empire in 1946 and knighted in 1961. When not hard at work he enjoys cricket, gardening, and music.

Henry Margenau was born in Beilefeld, Germany, in 1901. He came to the United States early in his life and was educated at Midland College and Nebraska University. He received his Ph.D. from Yale in 1929. For a year after this he was a Sterling Fellow at the same university. In 1931 he became an assistant professor; in 1945, a full professor. Since 1949 he has been the Eugene Higgs Professor of physics and natural science at Yale. His main professional interest has been in high-frequency discharges and intemolecular and nuclear forces.

Sir Isaac Newton was born in 1642, the year in which Galileo Galilei died. This coincidence represents perhaps one of the most important turning points in the history of science, for although Galileo had come to rely more and more on the results of direct experiment, it was Newton who presented science with a new way of looking at the facts of observation. In Newton, the long-standing Aristotelian tradition of solving problems which dealt with natural philosophy (the study of light, motion, and sound) by speculation and meditation gave way to dealing with those problems by measurement and experiment. His celebrated work on light and optics and the laws of motion and gravity represents a transition from ancient methods to the modern manner of scientific thought.

He was knighted in 1705 at Cambridge by Queen Anne—the first time that this honor had ever been conferred for achievement in pure science. He died in 1727 and is buried in Westminster Abbey.

José Ortega y Gasset (1883–1955) was born in Madrid, an heir to a cultural and intellectual Spanish tradition which was very nearly extinguished. When he died in the same city seventy-two years later, he left behind, not only the Instituto de Humanidades, which he founded, but also a cultural and literary revival which he had deeply influenced.

His early training was classical and accomplished under the tutelage of the Jesuits. In 1898 he went to Madrid University and after this to Germany to study for four years. When he became professor of metaphysics at Madrid in 1910, he began to diverge from the neo-Kantian school of philosophy, which had so influenced him earlier. He saw man's individual and present life as his basic reality, substituted historical reason for absolute reason and the individual's perspective for absolute truth.

Because of Spanish politics he exiled himself between 1936 and 1945. He then returned to Spain, where he founded his famous institute in 1948.

Andreas Osiander (1498–1552) was the prominent ancestor of an extensive line of theologians and scholars, none of whom, however, was either as brilliant or as erratic as he. His father was a blacksmith, and Osiander was fortunate to be educated as well as he was. He became a priest in 1520, after having left the University of Ingolstadt. He involved himself in the reform of the imperial free city of Nürnberg and won over Albert von Hohenzollern to the Lutheran movement. In 1543 he drafted a preface to Copernicus's *De Revolu-*

tionibus Orbium, which was instrumental in keeping that controversial work off the Roman Catholic Index of Prohibited Books for over fifty years. He died in 1552 in the midst of a vigorous controversy over the justification of sinners, which it is generally thought he precipitated.

Thomas O. Paine was born in California in 1921. He graduated from Brown University in 1942 and then served as submarine officer and deep-sea diver for the Navy during World War II. After the war he did his graduate work at Stanford, receiving a Ph.D. in physical metallurgy. From 1949 until 1968 he worked in various capacities for the General Electric Company. At their Research Laboratories he initiated research on magnetic materials. Then he served as laboratory manager at the Meter and Instrument Department, and as manager of Engineering Applications at the Research and Development Center. In 1963 he was made manager of TEMPO, GE's Center for Advanced Studies in Santa Barbara, California. On January 31, 1968, President Johnson appointed Paine Deputy Administrator of NASA and eight months later, upon the retirement of James Webb, he became Administrator. It was during his administration that the space program reached the spectacular culmination of the Apollo 11 flight. In July 1970, Paine resigned his post at NASA.

Blaise Pascal (1623–1662) was a member of a highly gifted French family. His father was a respected mathematician, his sister, Jacqueline, was considered a literary prodigy, and Pascal himself was famous at twenty-one for the invention of an arithmetical machine he built to help his father in his calculations as well as for his *Essais pour les coniques,* which had been published when he was seventeen. Pascal was not only a mathematician but a physicist, philosopher, and superb writer as well. His *Provinciales* is considered by many to mark the beginning of modern French prose. The large part of his last years were spent at Port-Royal in religious work. By 1658 he had put together the notes which were published as *Pensées.* His friends at the monastery at Port-Royal persuaded him to help in the composing of the *Elements de géométrie.* In February of 1659 he became ill and returned to the ascetic and devotional life. He died in 1662, probably from carcinomatous meningitis, which followed the development of a malignant ulcer of the stomach.

Plato (427?–347? B.C.) as a comparatively young man became a friend and student of the great Greek philosopher Socrates, although his own primary interest at the time was politics. It was only after the death of Socrates, and perhaps as a result of Socrates' treatment at the hands of the democratic leaders of Athens, that Plato decided there was no place for a man of sincere conscience in active politics. He turned his attention to philosophy and, when he was about forty, founded the Academy as an institute for the systematic pursuit of philosophical and scientific research. It is probable that he intended his dialogues to interest the outside world in the more serious and difficult labors of this school. The Academy survived well over 800 years until it was closed by the emperor Justinian in 529.

Henri Jules Poincaré (1854–1912) was generally considered to be one of the greatest mathematicians and original thinkers of his day He was educated at the Ecole Polytechnique and the Ecole des Mines and taught in Caen and

Paris. From his first work in pure mathematics he went on to develop new mathematical techniques, and he founded the study of topological dynamics. In a paper on the dynamics of the electron published in 1906 he arrived at many of the results of the special theory of relativity independently of Einstein.

His writings in philosophy are just as interesting, original, and perhaps as important, as his work in mathematics. His mastery of French prose provided him with a wide and various audience for his work, and he produced more than thirty books and five hundred papers during his career.

Eugene Rabinowitch was born in St. Petersburg, Russia, in 1901. He was educated in Berlin and in 1926 became an assistant at the Kaiser Wilhelm Institute for Physics and Chemistry. In 1933 he went to Denmark to the Royal Academy of Science and the next year to London. He is now professor of Biophysics at the University of Illinois and also edits the *Bulletin of Atomic Scientists*. As a physicist his main interest has been in reaction kinetics; as a private person he is most interested in public affairs.

Eric M. Rogers was appointed professor of physics at Princeton University in 1957. He was born in 1902 in England, educated at Cambridge, where he worked under Lord Rutherford, and assumed his first teaching post in 1930 at Harvard University. Since 1957 he has also been on the staff of the Physical Science Study Committee and has published a most readable and rewarding book entitled *Physics for the Inquiring Mind*.

Bertrand Arthur William Russell (1871–1970) was born on May 18, 1871. He was orphaned at three and raised by his grandmother from then on. He attended Trinity College, Cambridge, and graduated with distinction.

Russell was interested in mathematics, logic (he believed that mathematics and formal logic are one), and philosophy, as well as in the problems of individual liberty. He received the English Order of Merit in 1949 and the Nobel prize for literature in 1950.

His occupation with the liberty of the individual brought him into conflict with prevailing authority on several occasions. During World War I he was fined £100 for writing a leaflet criticizing the punishment of conscientious objectors. His college deprived him of his lectureship, and he was offered one at Harvard but was refused a passport. In 1918 he was imprisoned for a pacifist article he had written and was imprisoned again in 1962 for having organized demonstrations against nuclear armament.

Giorgio Diaz de Santillana was born in Rome in 1902. He was awarded his Ph.D. in physics from the University of Rome in 1925. After this he became an instructor at the university, where he helped organize the School for History of Science. He came to the United States in 1936 and became a naturalized citizen in 1945. After lecturing at the New School for Social Research for one year and at Harvard University for three years, de Santillana joined the faculty of the Massachusetts Institute of Technology, where he became professor of the history and philosophy of science. He is the author of *Galileo's Dialogue on the Great World Systems, The Crime of Galileo, The Age of Adventure,* and *The Origins of Scientific Thought.*

Niccolò Cardinal Schoenberg was a distinguished theologian who was appointed Archbishop of Capua in 1520. He unsuccessfully urged Copernicus to publish his work on the heliocentric system but had more direct influence in prevailing upon Galileo to publish the famous *De Revolutionibus Orbium Coelestium*, which, of course, deals with the same hypothesis.

Erwin Schrödinger (1887–1961) was born in Vienna in 1887. He was educated at the university there and later became professor of physics at Stuttgart, Breslau, Zurich, and finally, in 1928, in Berlin. From there he went to Dublin to become a professor at The Institute for Advanced Studies. Most of his work was highly mathematical in character and has dealt with the physics of the atom. He extended De Broglie's theory of wave mechanics and applied it to the problem of atomic structure. In 1933 he won the Nobel prize for physics for his contribution to the new science of wave mechanics. He died in Vienna in 1961.

Harlow Shapley was born in Nashville, Missouri, in 1885. He was educated at the University of Missouri and at Princeton. After having served at the Mount Wilson Observatory, he became professor of astronomy at Harvard and from 1921 until 1952 was director of the Harvard College Observatory.

Shapley's major work consists in or is the result of extensive surveying, particularly of the Milky Way and the Magellanic Clouds. He was able to formulate the hypothesis that globular star clusters form the center of the Milky Way and that our sun is located in the outer parts of that system. He was one of the first to demonstrate that galaxies occur in clusters and that the Milky Way is a member of a local cluster of galaxies.

His works include *Star Clusters, Flights from Chaos, Galaxies, Inner Metagalaxy,* and *Of Stars and Men.*

George Bernard Shaw (1856–1950) was born in Dublin but spent most of his life in England. He was an art, music, and drama critic for London journals and established his reputation by sympathetic criticisms of the impressionist painters, Wagnerian music, and the Ibsen school of drama. He also became prominent as a Socialist. About 1892 he began writing plays and soon became known as the leading British dramatist. In addition to his many successful plays he wrote several tracts on Socialism. He was awarded a Nobel prize for literature in 1925.

J. W. N. Sullivan (1886–1937) was the son of an Irish sailor. Although he had considerable gifts as a musician and mathematician, his major work was not done in these fields themselves but in the interpretation of them to a large and interested audience. He had the ability to illumine the spirit as well as the facts of science, and perhaps his most brilliant insights are into science's aesthetic and moral values. He died in England in 1937.

Sullivan's works include the superb collection of monographs called *The Limitations of Science* and a classic work on Beethoven entitled *Beethoven: His Spiritual Development.*

Walter Seagar Sullivan was born in 1918. After graduating from Yale, he joined the staff of the *New York Times.* He was foreign correspondent in the

Far East and Germany. Since 1962 he has devoted his time chiefly to science writing and editing for the *New York Times,* as well as publishing several books on science. He has received awards for journalism and science writing.

Rudolf Thiel was born in Kaiserslautern, Germany, in 1899. His family has been interested in astronomy for generations, and he owns one of the original Fraunhofer telescopes (instruments which, because of the excellent quality of lens Fraunhofer was able to produce, were largely responsible for the renewed interest in the refracting telescope in the first quarter of the nineteenth century). Thiel began to make astronomical observations as a boy and went on to study physics at Bonn and Munich. He has published a number of books on aspects of cultural history. The writing of *And There Was Light,* from which a selection was taken for the present work, occupied him for six years.

Arnold Joseph Toynbee was born in 1889 in England. He was educated at Oxford and after several years of government work during World War I, he became professor of Byzantine and Modern Greek Language, Literature, and History at London University (1919–24). He was director of studies in the Royal Institute of International Affairs in 1925, director of the Foreign Research and Press Service (1939–43), director of the Research Department of the Foreign Office (1943–46), and Research Professor of International Affairs at the University of London. Since 1955 he has been professor emeritus. He earned a world-wide reputation for his writings on history, especially *A Study of History,* published in twelve volumes between 1934 and 1961.

Warren Weaver was born in 1894. He received his education at the University of Wisconsin and joined the faculty there in 1920. He was chairman of the department of mathematics (1928–32) and director of the division of natural sciences (1932–37). In 1932 he also became director of the division of natural sciences at Rockefeller Foundation (1932–55). He has served as trustee, officer, and adviser in many government agencies and on the board of the Sloan-Kettering Institute. In addition, he has published several books and been the recipient of many special awards and distinctions.

Alfred North Whitehead (1861–1947) was born at Ramsgate, Kent. He was educated at Cambridge, specializing in mathematics. He collaborated with Bertrand Russell in writing the *Principia Mathematica,* a monumental study of the history of mathematical and logical thought which was published in three volumes between 1910 and 1913.

In 1924 he came to Harvard University as professor of philosophy. He spent his later years trying to work out an organic philosophy which would correct our modern overemphasis on natural science and provide a guide for civilized living by correlating the data of social sciences with mathematics and the natural sciences together with data on aesthetic, moral, and religious experience. It was a prodigious task, and its worth is still debated.

The work from which the selections in the present work are taken is one of three books of a nontechnical nature which most completely express his mature philosophy. The others are *Process and Reality* and *Adventures of Ideas.* He died in 1947.

Jerome B. Wiesner was born in 1915. He received his B.S. and Ph.D. degrees from the University of Michigan. In 1942 he joined the staff of the Massachusetts Institute of Technology, serving as professor and chairman of the department of electrical engineering (1946–61), dean of the school of science (1964–66), and provost since 1966. Wiesner was special assistant to the President for science and technology and director of the Office of Science and Technology at the White House from 1956 to 1961.

William Persehouse Delisle Wightman was born in 1899. He was educated at Eastbourne College and Imperial College, London. He was for several years Head of the Science Department of Edinburgh Academy, and from 1951 was Reader in the History and Philosophy of Science at the University of Aberdeen. When he retired from this post in 1968, he and his wife moved to Oxford, where they now live. Dr. Wightman is a Fellow of the Royal Society of Edinburgh and a member of the Union internationale d'Histoires des Sciences.

Abraham Wolf (1876–1948) was born in England and educated at University College, London, and St. John's College, Cambridge. In 1915 he became an examiner in philosophy at the University of London and in 1922 took a similar position at Oxford.

Most of Wolf's academic career was spent at University College, London, where he was professor of logic and scientific method and later head of the department of the history and method of science. He has written much in both these fields and was the coeditor of the fourteenth edition of the *Encyclopedia Britannica*. He died in 1948.

GLOSSARY

The following is a glossary of terms which appear in the text and which may be unfamiliar to the reader. Definitions of these terms are included for easy reference, and in some cases supplementary information is also given. A few simple algebraic derivations have been included for the interest of those readers who prefer a mathematical explanation. Cross references in the glossary are indicated by italics in those cases in which the cross reference may throw additional light on the term being defined, for example, "metagalaxy: a system of *galaxies* or extragalactic nebulae. See page 486." The word "galaxies" is in italics to show that it is also defined in the glossary. The page number refers the reader to material in the body of the text that will provide additional help in understanding the term.

acceleration The rate of change of velocity with respect to time. $a = (v - v_0)/t$, where v_0 is the starting velocity and v the velocity at the time t. If the body starts from rest $(v_0 = 0)$, then $a = v/t$. The average velocity $(v + v_0)/2$ will be $v/2$, and the distance traveled in time t will be $(v/2)\ t$. Therefore, by simple algebra, if

$$d = \frac{v\ (t)}{2}$$

and $\quad a = \frac{v}{t}$

$$v = at$$

$$d = \frac{at\ (t)}{2} = \frac{1}{2}\, at^2$$

° We wish to acknowledge the kind permission granted to us by Penguin Books, Inc., to reprint from their *A Dictionary of Science,* by E. B. Uvarov and D. R. Chapman, the definitions in whole or in part for alpha particle, continuum, diffraction, electrons, galaxy, interference of wave motions, ion, meteor, parsec, proton, protoplasm, radioactivity, thermal neutrons, waves. We wish also to acknowledge our indebtedness to the *McGraw-Hill Encyclopedia of Science and Technology,* McGraw-Hill Book Company, Inc., New York, 1960, from which we adapted the definitions for carbon cycle, Cepheid variables, conservation, cosmic rays, diffraction grating, logarithmic scale, Fitzgerald-Lorentz contraction, proton-proton chain, refraction, relativity, spectroscope, statistics, telescope.

In case of uniform acceleration, where a is constant, distance is proportional to t^2. See page 187 and Figure 4-17. Since velocity is a *vector* quantity having both magnitude and direction, in problems involving curvilinear motion it is necessary to consider the rate of change of the different components of the velocity. In the case of uniform circular motion, there is no change in the magnitude of the velocity, but there is a constant change in the direction and therefore a constant acceleration.

acceleration of gravity Acceleration of a body falling freely in a vacuum. This varies in different localities as a result of the rotation of the earth and the distance from the center of mass of the earth. The standard accepted value at sea level is 32.2 feet per second per second. See page 187.

accuracy In every form of physical measurement there is some inaccuracy, and the degree of this inaccuracy must be known if the measurement is to have any scientific value. There are three ways of specifying a measurement that is known to be reliable to about 1 in 1,000:

$$x = 1.234 \pm 0.001$$
$$x = 1.234 \pm 0.1 \text{ per cent}$$
$$x = 1.234$$

In the third form, the last digit, the number 4, is understood to be unreliable, but we have less information about just how unreliable it is than we have in the first two forms. See the editor's note on page 224.

aerobic Pertaining to organisms that can grow only where there is free oxygen available.

Agena A 15,000-pound-thrust second stage designed for use on both Atlas and Thor vehicles. The stage features the use of "storable" liquid propellants and restart capability in space. The Thor-Agena vehicle is used in the Discoverer program, and the Atlas-Agena launched the Samos and Midas payloads.

alpha particle (α particle) Helium nucleus; i.e. a close combination of two *neutrons* and two *protons*, and therefore positively charged. Alpha particles are emitted from the nuclei of certain *radioactive* elements.

ammonia (NH_3) Pungent smelling gas, very soluble in water. Like methane it is a prime constituent of the atmospheres of the outer planets.

amplitude See *waves.*

anaerobic Pertaining to organisms that can grow in the absence of free oxygen.

angular momentum A *vector* quantity that is the product of angular *velocity* and moment of *inertia.* The law of *conservation* applies to angular as well as linear *momentum.* This law is one of the guide posts of the theoretical work on the origin of the solar system, for it tells us that the original angular momentum must still be present somewhere in the system. See page 459.

angular velocity Rate of motion through an angle about an axis, measured in degrees, radians, or revolutions per unit time.

apastron See *aphelion.*

aphelion The aphelion of an orbit is the point of farthest distance from the sun. This point is also an *apse*, being on the major axis of the ellipse. A similar term, used in connection with double stars, is *apastron.* See Figure 5-10.

approximation When we know that a number is only approximately correct, we can express this by writing it $x \approx 500$ or $x \cong 500$. Both these signs mean "equals approximately." For still rougher estimates we use the sign $y \sim 1,000$. This sign expresses the order of magnitude and means that y is nearer 1,000 than 100 or 10,000. See also *accuracy* and the editor's note on page 224.

apse or **apside** The apse of an orbit is that point at which the distance of the body from the center of the attraction is either greatest or least. The line that joins the two apse points is called the line of the apse. In an elliptical orbit this line of the apse is the major axis of the ellipse, and the two points are the points of nearest (see *perihelion*) and farthest (see *aphelion*) distance.

asteroids See *planetoids*.

astronomical unit Unit of measurement in astronomy defined as the mean distance from the earth to the sun and taken as 92,897,000 miles.

Atlas Intercontinental ballistic missile (ICBM) having a maximum speed of over 15,000 miles per hour and a range of over 6,000 miles. It uses liquid *propellant* and is designed to carry nuclear warheads. It has also been used as the first stage of several space-launch vehicles such as Atlas-Able, Atlas-Agena, and Atlas-Mercury. The Atlas-Mercury placed about 2,400 pounds in orbit.

axiom A statement accepted as a self-evident truth but incapable of formal proof. See page 22.

beta particle (β particle) Term applied to a swiftly moving *electron*, β^-, *positron*, β^+, when emitted by a *radioactive* substance.

billion The American billion is a thousand million, or 10^9; the British billion is a million million, or 10^{12}.

binary system A binary system of stars is a pair of stars located sufficiently near each other to be connected by a mutual bond of gravitational attraction that compels them to describe an orbit around a common center of mass. See page 331.

bise or **bice** A pale blue pigment prepared from blue carbonate of copper.

Boyle's law At a constant temperature the volume of a given quantity of any gas varies inversely as the pressure to which the gas is subjected; that is, for an ideal gas

$$P \times V = \text{constant}$$

where P = pressure
V = volume

Actually, the law is only approximately true, even for such gases as hydrogen and helium. Since a gas consists of a large number of small elastic particles in rapid motion and the pressure on the walls is simply the effect of bombardment, Boyle's law can be looked upon as a statistical or probability picture of the motion of the gas particles. We cannot, at any one time, give a history of any one particular particle, but we can say what the probability is that a certain number of these particles may be in some specific state of motion. See page 60.

brennschluss The German term for the instant at which the rocket fuel is shut off.

carbon cycle The carbon cycle, or, properly, the carbon-nitrogen-carbon cycle, is the name applied to a theoretically deduced chain of *thermonuclear* reactions that appears to be important as an energy-producing process in certain types of stars. The cycle has the property that C^{12} (carbon) nuclei act as *catalysts* for a chain of nuclear reactions that finally converts four *protons* into one He^4 (helium) nucleus with a release of much energy. The following series is in agreement with the experimental evidence:

$$C^{12} + H^1 \rightarrow N^{13} + \gamma$$
$$N^{13} \rightarrow C^{13} + {}_{+1}e + v$$
$$C^{13} + H^1 \rightarrow N^{14} + \gamma$$
$$N^{14} + H^1 \rightarrow O^{15} + \gamma$$
$$O^{15} \rightarrow N^{15} + {}_{+1}e + v$$
$$N^{15} + H^1 \rightarrow C^{12} + He^4$$

where ${}_{+1}e$ is a *positron*, v is a *neutrino*, and γ is a *gamma ray*. With the production of He^4 and the release of energy, a new C^{12} catalyst nucleus is created, so that the reaction can continue. See pages 468–69.

carbon dioxide (CO_2) A colorless gas with a faint smell. Formed by the oxidation of carbon and carbon compounds, it is present in our atmosphere and is used by plants. It seems to be present in the atmosphere of Venus in very large amounts.

catalyst A substance which alters the rate at which a chemical reaction occurs but which is itself unchanged at the end of the reaction.

celestial equator If the equator of the earth were projected out until it cut the apparent sphere of the sky, it would mark a line known as the celestial equator. See Figure 3-4 and page 78.

Cepheid variables A class of variable stars. Variable stars are those which have detectable changes in their intensities, often accompanied by other physical changes. The changes in brightness of variable stars may be a few thousandths of a magnitude to 20 magnitudes or more. Cepheid variables are characterized by several important relations. They are among the most luminous stars known and thus can be seen at great distances. There is a definite correlation between period and luminosity; as a result of this relationship, they are used as measuring tapes for the universe. There are two types of Cepheids, based on the classification of stars as being in type I or type II *populations*. Classical Cepheids resemble the prototype and belong to type I; they are found in large numbers in the Milky Way. Having periods ranging from 1 to 50 days, but commonly about 5 days, about 500 are recognized in our galaxy. They are quite steady in periodicity and form of light curve. Type II Cepheids are grouped with type II population stars and are found more frequently in the globular clusters. They differ from classical Cepheids in the form of the light curve and the period-luminosity relationship as well as by spatial difference.

chain reaction A reaction in which one of the agents necessary to the reaction is itself produced by the reaction so as to cause like reactions. In the neutron-fission chain reaction, a *neutron* plus a fissionable atom causes a fission resulting in a number of neutrons that in turn cause other fissions. See also *carbon cycle*.

color The term usually refers to the visible part of the electromagnetic spectrum, each color being represented by a band of wavelengths or frequencies of light. One end of this visible color range is red and the other violet, the red having a longer wavelength or lower frequency. White light consists of a mixture of the wavelengths in the visible range. A surface that reflects all of these will appear white; some surfaces, however, have the property of absorbing some of the radiations they receive and reflecting the rest. Thus, a surface that absorbs all light radiations excepting those corresponding to green will appear green by reflecting only those radiations. In the case of color seen by transmitted light, as in colored glass, the glass absorbs all the radiations except those which are transmitted. See article starting on page 261 and Figure 6-1.

comets Objects which are moving in space under the influence of the sun's gravitational field and which occasionally come close enough to the earth to be observed. As they move in toward the sun, radiation pressure causes tails to form, sometimes millions of miles in length. After passing close to the sun, they may become permanent members of the sun's family. Comets are believed to have a very low mass—less than one hundred-thousandth of the mass of the earth.

concave mirror A concave mirror curves inwards; the center of the reflecting surface is lower, or thinner, than the mirror at the edges. The curvature of a concave mirror is called negative curvature. Remembering the rule that the angle of reflection of a beam of light is equal to the angle of incidence, it is a simple matter to calculate the point at which the reflected beams will be brought to a focus. If the mirror is spherical, as in the left figure, the focal length is just one-half of the radius of curvature of the mirror. If the mirror is parabolic, as in the right figure, rays of light from a source so distant that the rays are parallel are brought to a focus at one point (called the focal point of the mirror). Conversely, if a source of light is placed at the focus, the light will be reflected in a parallel beam. Parabolic mirrors are used in reflecting telescopes to collect light from distant stars and bring it to an accurate focus.

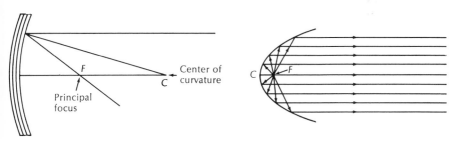

concept A broad term meaning, generally, an idea or notion but differing from perception by involving the process of abstraction. A scientific concept may be mathematical, for example, a formula; it may be denominative or simply a name given to a class of objects, for example, metal; or it may be a property of something, for example, elasticity. The term *concept* is also sometimes used to designate theoretical systems such as Newton's gravitational theory. See page 91.

conic sections When a solid circular cone is cut by a plane in various directions, the line of intersection of plane and cone defines the various figures called conic sections. A cone cut "straight across" gives a circle. If the cut is slanted, the section is an *ellipse*. With greater slant, just "parallel to the cone's edge," the section is a *parabola*. With still more slant, the section is a hyperbola. See top of page 692.

conservation laws It has been found experimentally that certain physical quantities are conserved in processes into which they enter, and that in spite of changes in form or distribution their sum total remains the same. The most commonly known of the conservation laws are those which have been generalized for *mass, energy,* and *momentum.*

    conservation of mass The total mass of a system remains constant.

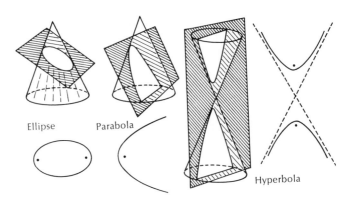

Ellipse   Parabola

Hyperbola

conservation of energy The total quantity of energy in a closed system remains constant. (This law is often called the first law of thermodynamics.) In view of the fact that energy and mass are mutually convertible, a conclusion of the special theory of relativity, the laws of conservation of mass and of energy are recognized as special cases of the conservation of mass-energy. See pages 356 and 432.

conservation of momentum: In a perfectly elastic collision, the total momentum of two bodies before impact is equal to their total momentum after impact. When velocities comparable to the speed of light are under study, the variation of mass with velocity must be considered.

continuum A continuous series of component parts passing into one another; e.g. the three space dimensions and the time dimension are considered to form a four-dimensional continuum. See page 357.

convex mirror A mirror that curves outwards, having a positive curvature. The center of curvature and the principal focus are both behind the mirror. After reflection, the rays of light leave the mirror as if they came from a point behind the mirror. This kind of image is called a virtual image.

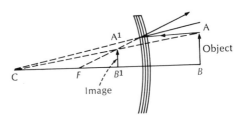

A^1

A

Object

C   F B^1 B

Image

coordinates A set of numbers that locate a point in space. The most common are the Cartesian or orthogonal coordinates in which mutually perpendicular axes are used. The values of x and y refer to distances along these axes measured from the origin o. Another commonly used system is that of polar coordinates. In general relativity the most used are gaussian coordinates, in which the surface is not a plane but of constant gaussian curvature. This gaussian curvature G is $G = 1/R_1 R_2$, where R_1 and R_2 are radii of the principal curvature.

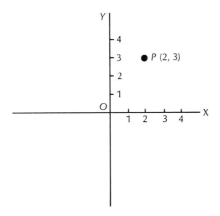

corona The outer part of the sun's atmosphere seen during an eclipse when the moon blocks out the sun's disk. See Figure 6-13. This corona extends beyond the sun's surface as much as thirty times the diameter of the sun and has a temperature of about 1000000° K.

cosmic rays The nuclei of atoms, largely hydrogen, which impinge upon the earth from all directions of space with nearly the speed of light. These nuclei with relativistic speeds are often referred to as primary cosmic rays to distinguish them from the cascade of secondary particles generated by their impact against air nuclei at the top of the terrestrial atmosphere. The secondary particles shower down through the atmosphere and are found with decreasing intensity all the way to the ground and below. Cosmic rays are detected principally by the *ionization* they produce in the matter through which they are passing. The primary cosmic-ray particles coming into the top of the terrestrial atmosphere make inelastic collisions with nuclei in the atmosphere, splitting one or both particles into a number of smaller nuclear fragments, each of which carries away some of the primary particle's energy. Individual cosmic-ray nuclei have energies as low as a few hundred million electron volts (Mev), and occasionally energies as high as 10^9 Bev. The large individual energies indicate that cosmic rays are not confined just to the solar system but are a phenomenon of at least galactic scale.

cosmical constant This constant was introduced by Einstein into the field equations of general *relativity* to make possible a static solution of these equations. This term introduced a repulsive force that increased with distance to balance the attractive force of gravity. The static solution was abandoned by Einstein and most other physicists in 1929 when observations of the *red shift* showed that the universe is not static, as astronomers had previously thought. However, by assigning different values to the cosmical constant it is possible to construct other world models like, for instance, the one proposed by Lemaître that begins expanding, then goes through a static period, and finally expands again. See page 410.

cosmogony The branch of astronomy that deals with the origin of the universe.

cosmology The branch of astronomy that deals with the general structure of the universe and the laws of space and time, as well as the study of the parts of the universe such as the origin of the solar system and the life history of the stars.

covariant When an equation transforms so that the equational relationship is undisturbed, we speak of the equation being covariant with respect to that *transformation*. Einstein in *Relativity: The Special and General Theory* gives the following example of covariance:

> Every general law of nature must be so constituted that it is transformed into a law of exactly the same form when, instead of the Space-Time variable *x*, *y*, *z*, *t* of the original co-ordinate system *k*, we introduce new Space-Time variables *x′*, *y′*, *z′*, *t′* of a co-ordinate system *k′*. In this connection the relation between the ordinary and the accented magnitudes is given by the Lorentz transformation, or, in brief, General Laws of Nature are *covariant* with respect to Lorentz *transformations*. See page 350 and footnote 1, page 351.

declination The measure of angular position of a heavenly body from the *celestial equator*. The measurement is in degrees, 0° at the equator and 90° at each pole. See page 222.

deductive reasoning The process of reasoning from a generalization that is assumed to be true to a set of specific conclusions about the system. These conclusions may then be checked by observation or experiment. See page 22.

deferent See *epicycle*.

degrees, minutes, and seconds of arc Each full circle contains 360°. Each degree contains 60 minutes and each minute 60 seconds of arc. Therefore one full circle has 21,600 minutes or 1,296,000 seconds of arc. See footnote 2, page 222, and footnote 2, page 240.

density Mass per unit volume. When expressed in grams per cubic centimeter, it is numerically equal to specific gravity.

diffraction When a beam of light passes through an aperture or past the edge of an opaque obstacle and is allowed to fall upon a screen, patterns of light and dark bands (with monochromatic light) or colored bands (with white light) are observed near the edges of the beam, and extend into the geometrical shadow. This phenomenon, which is a particular case of interference, is due to the wave nature of light, and is known as diffraction. The phenomenon is common to all wave motions. See Figures 6-8 and 6-9.

diffraction grating An optical device consisting of an assembly of narrow slits or grooves, such as *A* or *B*, that by diffracting light produces a large number of beams that can interfere with one another in such a way as to produce spectra. Since the angles at which constructive interference patterns are produced by a grating depend on the lengths of the waves being diffracted, the waves of various lengths in a beam of light *W* striking the grating will be separated into a number of spectra, produced in various orders of interference on either side of the undiffracted central image *F*. By controlling the shape and size of the diffracting grooves when producing a grating and by illuminating the grating at suitable angles *i*, a beam of light can be thrown into a single spectrum whose purity and brightness may exceed that produced by a prism. Gratings can be made with much larger apertures than prisms and in such form that they waste less light and give higher intrinsic dispersion and resolving power. A single grating can be used over a much broader range of spectrum than can any single prism, and its dispersion will vary less rapidly with wavelength. Gratings are being used increasingly in large *spectrographs* and for precise spectroscopic work, as well as in monochromators and analytical spectrographs.

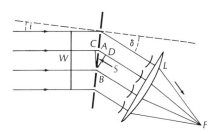

Doppler shift The amount of the effect upon the apparent frequency of a wave-train produced by relative motion of the source and the observer. The direction of the shift tells in what direction the source is moving relative to the observer: in the case of light, if the shift is toward the red or longer wavelengths, the motion is away from the observer; if the shift is toward the blue or shorter wavelengths, the motion is toward the observer. See page 291 and Figure 6-14.

dwarf stars These are stars that have a low luminosity and very high density and are considered to be the end result of stellar evolution. See Figure 9-6 and page 474.

eccentric A circle not having the same center as another contained within it. In Ptolemaic astronomy, in order to account for the observed irregularities of the planets' motions, it was necessary to elaborate the theory of *epicycles* by having the center of the deferents move in an eccentric manner.

eclipse There are two types of eclipses commonly observed on earth: the solar and the lunar. The solar eclipse occurs when the moon passes between the earth and the sun so that the disk of the moon totally or partially covers the disk of the sun. The minimum speed at which the moon's shadow moves across

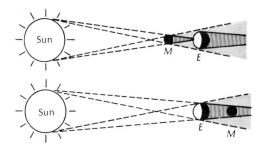

the earth's surface is about 2,300 miles per hour and the maximum speed can be as high as 5,000 miles per hour. The maximum time that a total solar eclipse can last is 7½ minutes. The lunar eclipse takes place when the earth passes between the sun and moon and the earth's shadow sweeps across the moon. A total lunar eclipse can last for 1 hour and 40 minutes.

ecliptic The apparent annual path of the sun in the sky relative to the stars; or the great circle cut out on the celestial sphere by the plane containing the orbit of the earth. See page 82.

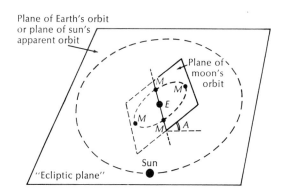

electron The electron is one of the fundamental particles of nature. It has a mass of 9.1×10^{-28} grams and a negative charge of 4.8×10^{-10} electrostatic units. See page 293. *See positron.*

ellipse The locus of a point that moves so that the sum of its distance from two foci is a constant. See also *orbits* and *conic sections.* The equation of the ellipse in rectangular coordinates is $x^2/a^2 + y^2/b^2 = 1$.

empirical A term used to describe facts or truths established by observation or experiment rather than those deduced from axioms or theory. See page 27.

energy Capacity for doing work. There are many different forms of energy: potential, kinetic, electrical, heat, chemical, atomic, and radiant. Energy is particularly important because it is a conserved quantity (see *conservation*). It cannot be created or destroyed but it may be exchanged among various bodies or converted from one form to another. The most obvious form of energy is kinetic energy, the energy of a body in motion:

$$\text{Kinetic energy} = \frac{1}{2}mv^2$$

where m is the mass of the moving body and v is the velocity.

epicycle A small circle the center of which moves around the circumference of a larger circle called the deferent. This system of motions was used in Ptolemaic astronomy to account for the observed periodic irregularities in planetary motions. See Figures 3-34 and 3-35.

equinox See *precession of equinoxes.*

escape velocity The velocity that must be reached if an object is to escape from the attractive force of a gravitating body. This velocity for the earth is approximately 7 miles per second. A speed of approximately 25 miles per second would be needed to escape from the solar system, starting from the earth's surface.

ether A hypothetical medium that was postulated to act as a medium for the transmission of the wave characteristics of light through space. This medium was assumed to be transparent, undispersive, incompressible, continuous, and without viscosity. All experimental attempts to demonstrate the existence of this medium have failed. See pages 336 and 340 to 342.

exhaust velocity See *propellant.*

exponents Numbers indicating the power of a quantity. Thus the exponent of x in x^4 is 4. The laws of exponents are methods of handling these powers and are very useful in dealing with large numbers:

Law	*Example*
$\dfrac{1}{A^a} = A^{-a}$	$\dfrac{1}{10^3} = 10^{-3}$
$A^a \times A^b = A^{a+b}$	$10^3 \times 10^5 = 10^8$
$\dfrac{A^a}{A^b} = A^{a-b}$	$\dfrac{10^6}{10^3} = 10^3$
$(\cdot A^a)^b = A^{ab}$	$(10^2)^3 = 10^6$

extrapolation The method by which data of a restricted range may be projected to establish inferences about data lying outside that range. See page 24.

Fitzgerald-Lorentz contraction The contraction of a moving body in the direction of its motion. In 1893, G. F. Fitzgerald and H. A. Lorentz proposed independently that the failure of the Michelson-Morley experiment to prove the existence of an ether and to detect an absolute motion of the earth in space arose from a physical contraction of the interferometer in the direction of the earth's motion. See page 341 and Figure 7-3. According to this hypothesis, as formulated more exactly by Albert Einstein in the special theory of *relativity* (see footnote 1, page 351), a body in motion with speed v is contracted by a factor $\sqrt{1 - v^2/c^2}$ in the direction of motion, where c is the speed of light. For the earth traveling at 18.5 miles per second in its orbit, such a contraction is about one part in 2×10^8 or about 2.5 inches along the diameter of the earth. Such an amount accounts precisely for the Michelson-Morley result, because it allows for a shrinkage in distance between the light source and the mirror when the system is moving lengthwise. See also *transformation.*

force Any agent that causes a change of *momentum* in a body. It is measured by the time rate of change of that momentum. If the speeds are low compared with that of light, a force may be defined as the product of the mass m of a body and the acceleration a that is produced by the force. Thus $F = ma$. Force is a vector quantity, requiring both a magnitude and a direction for its complete specification. See page 320.

frequency See *waves.*

galaxy Enormous cluster of stars and other heavenly bodies such as the cluster of which the solar system forms a part—Milky Way. See page 384.

  galaxies Gigantic star clusters separated by even vaster regions of space. See pages 392–93 and Figure 8-9.

gamma rays Electromagnetic waves of very short lengths, shorter than those of X rays.

geocentric A system of measurements of position in which the origin of the system (center point) is located at the center of the earth. The word is used in connection with the theory that the earth is the center of the stellar and planetary motions and that all revolve about the earth. This is called the geocentric theory. See Figure 3-23.

geodesic A curve representing the shortest distance between two fixed points. In three-dimensional Euclidian geometry, the geodesic is a straight line. If the path is constrained to the two-dimensional surface of a sphere, it is a segment of a great circle. In the non-Euclidian geometries appropriate to the general relativity theory, the geodesic is the path followed by a particle upon which no electromagnetic forces act. See Figure 7-6.

giant stars Stars that are abnormally large and very luminous. The material in these stars is extremely rarefied. The giant stars appear on the Russell diagram above the main sequence. See Figure 9-6 and page 472.

gravitation, law of The Newtonian law expressed by the equation

$$F = \frac{GMm}{R^2}$$

where M and m are the masses of two bodies, R is the distance between them, and G is the gravitational constant. See pages 319 ff.

H-H chain See *proton-proton chain*.

heliocentric A system of measurements of position in which the origin of the system (center point) is located at the center of the sun. Heliocentric is also used to denote the theory that requires the sun to be the center of the solar system and the earth to revolve around the sun. See Figure 3-23.

Hubble's constant The equation $v = R/T$ is the relation between the velocity v of a nebula and its distance R. The value $1/T$ is known as Hubble's constant.

hypothesis A tentative theory that is employed to explain certain observed facts and to serve as a guide for the investigation of other facts. See page 23.

indeterminacy principle See *uncertainty principle*.

inductive reasoning The logical method by which a quantity of data may be employed as the basis of a generalization that will be assumed valid for the explanation of subsequent data of the same kind in the same system. See pages 16 and 22.

inertia The tendency of a body to preserve its state of rest or uniform motion in a straight line. This is the property that necessitates the exertion of force upon the body to give it acceleration. The law of inertia is Newton's first law of motion, which states that material bodies not subjected to the action of forces remain at rest or move in a straight line with constant speed. See page 190.

interference of wave motions The addition or combination of waves; if the crest of one wave meets the trough of another of equal amplitude, the wave is destroyed at that point; conversely, the super-position of one crest upon another leads to an increased effect. The color effects of thin films are due to interference of light waves; beats produced by two notes of similar frequency are the result of the interference of sound waves. See also *waves, diffraction*, and Figure 6-10.

interpolation A method of inference by which previously unknown facts are obtained from known facts when the former lie within the range over which the known data are valid. See page 24.

invariant An expression involving the coefficients of an algebraic function that remain constant when a *transformation*, such as translation or rotation of *coordinate* axes, is made. Einstein and Infeld in *The Evolution of Physics* give the following example of invariance:

> Suppose we have, in our simplest case, two coordinate systems, that is two rigid rods; we draw one above the other and call them respectively the "upper" and "lower" coordinate system. We assume that the two coordinate systems move with a definite velocity relative to each other. . . .

> Take two fixed points on the upper rod and consider the distance between them. This distance is the difference in the coordinates of the two points. To find the positions of the two points relative to different coordinate systems, we have to use transformation laws. But in constructing the differences of the two positions the contributions due to the different coordinate systems cancel each other and disappear, as is evident from the drawing. We have to add and subtract the distance between the origins of two coordinate systems. The distance of two points is, therefore, *invariant*, that is, independent of the choice of the coordinate system.

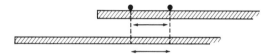

ion An electrically charged atom or group of atoms. Positively charged ions have fewer *electrons* than is necessary for the atom or group to be electrically neutral; negative ions have more. Gases can be ionized by the passage of charged particles, electric sparks, *X rays, gamma rays,* etc.

isostasy The theory of gravitational balance between relatively broad, contiguous areas of different average altitudes or topographical relief, applied particularly to the equilibrium of the major topographical features of the earth— continents and ocean basins. Thus the continents are assumed to "stand high," relative to the ocean basins, because they are lighter (or less dense) material. Whenever any large segment of the earth's crust is out of balance with its surroundings, adjustment is assumed to take place by means of solid flow in a deep subcrustal zone. Since the surfaces of the continents are constantly being worn down by erosion, if it were not for isostasy, early in the geologic history of the earth all land areas would have been reduced to ocean level.

Jupiter-C A four-stage research test vehicle comprised of the Redstone ballistic missile as the first stage and solid *propellant* upper stages. First stage thrust is 75,000 pounds. The Jupiter-C placed America's first satellite, Explorer I, in orbit on January 31, 1958.

Kepler Johannes Kepler, in the *Mysterium Cosmographicum* (1596), attempted to fit the orbits of the six known planets into the five regular solids as follows: (1) a cube inscribed within the orbit of Saturn would touch the outside of the orbit of Jupiter and contain it; (2) a tetrahedron (four sides) inscribed within the orbit of Jupiter would touch the outside of the orbit of Mars and contain it; (3) a dodecahedron (twelve sides) inscribed within the orbit of Mars would touch the outside of the orbit of earth and contain it; (4) an icosihedron (twenty sides) inscribed within the orbit of earth would touch the outside of the orbit of Venus and contain it; (5) an octahedron (eight sides) inscribed within the orbit of Venus would touch the outside of the orbit of Mercury and contain it. See Figure 1-3. That theory did not comprehend the elliptical shape of planetary orbits and was, therefore, superseded by the laws of motion twenty-two years later. Kepler's three laws of planetary motion are:

1. The orbit of every planet is an ellipse with the sun at one focus.
2. A line joining any planet to the sun sweeps over equal areas in equal times.
3. The squares of the periods of revolution of any two planets around the sun are in the same proportion as the cubes of their mean distances from the sun.

Looking more closely at these laws, we find that the second law tells us that the speed of a planet in its orbit is not constant:

Area A = Area B.

Time from A_1 to A_2 = Time from B_1 to B_2.

The planet covers more of the arc during the A^1 to A^2 passage.

Written as an algebraic statement the third, or, as it is known, the harmonic law, is as follows:

$$\frac{T_a^2}{T_b^2} = \frac{R_a^3}{R_b^3}$$

where T_a = period of planet A for one trip around the sun
          T_b = period of planet B for one trip around the sun
          R_a = mean distance of planet A from the sun
          R_b = mean distance of planet B from the sun

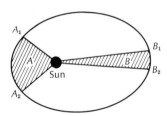

latitude The angular distance north or south from the equator of a point on the earth's surface, measured on the meridian of the point. The equator is o degrees latitude and each pole is 90 degrees north or south.

laws of motion See *Newton's laws of motion.*

laws of planetary motion See *Kepler.*

lichen Perennial plants that are in reality a combination of two plants growing together in an association so intimate that they appear as one. Such an association is called symbiosis. The components of a lichen are always an alga and a fungus.

light This term properly refers to the range of electromagnetic radiation associated with vision or the wavelength range 4×10^{-5} to 7×10^{-5} centimeter approximately. See Figure 6-1. The velocity of light, which is approximately

3×10^{10} centimeters per second or 186,000 miles per second, is considered to be the top limiting velocity in the universe.

light-year The distance a photon of light travels in one year at the velocity of 186,000 miles per second—5.88 trillion miles or 5.88×10^{12} miles.

logarithmic scale A scale so divided that the distance from any point on the scale to the origin is in proportion to the logarithm of the number used to mark that distance. The scale is often used in graphical presentation. See Figures 9-1 and 9-15.

longitude Angular distance in the east and west direction on the face of the earth, the zero point being on the meridian passing through the Greenwich Observatory.

Lorentz-Fitzgerald contraction See *Fitzgerald-Lorentz contraction.*

main sequence of stars The vast majority of stars fit within a narrow band running from the upper left-hand corner to the lower right-hand corner of the Russell diagram. This band of stars is known as the main sequence. See Figures 9-6 and 9-14.

mass There are two principal ways of describing and measuring mass. We will discuss each separately and then the relationship between them. Inertial mass is the constant m in the equation $F = ma$. It is the physical measure of the resistance of any body to a change of motion and can be measured by finding the *acceleration* produced by a known force acting on the body in question (if the speeds are small compared to the speed of light). Except at very high speeds, inertial mass is believed to be an unchangeable property of matter, having the same value anywhere in the universe. Gravitational mass is a measure of the amount of matter acted upon by a gravitational field and is proportional to the weight of the object in question. It should be borne in mind, however, that the weight (which is the *force* of attraction of the earth on a given mass) varies with location, depending on the position from the center of the earth, just as the acceleration of gravity, g, also varies in the same proportion. The relationship becomes clear if we substitute weight W for Force and g for acceleration in the equation $F = ma$, which then becomes $W = mg$. W and g both vary, depending on the position, but m remains constant. If the proper units are chosen, gravitational mass equals inertial mass and is assumed to remain constant everywhere in the universe except at speeds approaching the speed of light. See page 324.

metagalaxy A system of *galaxies* or extragalactic nebulae. See page 486.

meteoroid Solid body from outer space. A meteoroid becomes incandescent ('shooting star') on entering the earth's atmosphere owing to frictional forces set up at its surface. Consists of various materials, often metallic iron. See page 602.

methane (CH_4) A colorless and odorless gas sometimes called marsh gas. The fact that this gas remains in a gaseous state at very low temperatures permits it to be the prime constituent of the atmosphere of the outer planets.

momentum The product of the *mass* and the *velocity* of a moving body: $p = mv$. It should be borne in mind that the velocity is a *vector* quantity. For speeds approaching that of light, the variation of mass with velocity must be taken into account, and the value of m appropriate to the velocity of the body must be used in the expression for the momentum. See also *conservation of momentum* and *angular momentum* and page 327.

neutrino A fundamental particle whose existence was postulated in order to preserve the laws of *conservation* of mass and energy and conservation of momentum in certain nuclear reactions. The neutrino has no charge and probably zero rest mass.

neutron A particle possessing no electric charge and having a mass slightly greater than the *proton*. A constituent of all atomic nuclei except the normal hydrogen nucleus, which is a single proton. Owing to the absence of electric charge, the neutron can pass relatively easily through matter.

Newton's laws of motion There are three in number and form the basis of classical mechanics:

1. Every body continues in its state of rest or uniform motion in a straight line except insofar as it is compelled by external forces to change that state.
2. Rate of change of *momentum* is proportional to the applied force and takes place in the direction in which the force acts.
3. To every action there is always an equal and opposite reaction.

See pages 319 to 327.

nova and supernova A nova is a variable star that occasionally expands and becomes abnormally bright. It expands and contracts many times during its lifetime, each time losing part of its material. A supernova, on the other hand, explodes with great brilliance, destroying itself and scattering its material throughout space. Some supernova have been so bright that they were visible in daylight. Supernovae can be seen in other galaxies and give us one method of determining the distance of the galaxy. See page 475 and Figure 8-13.

opposition A planet having its orbit outside that of the earth is in opposition when the earth is in a direct line between the sun and the planet.

orbits See *Kepler* for laws of planetary motion. The derivation of elliptical orbits from Newton's law of gravitation requires either calculus or rather cumbersome geometry, with the result that we will not give it here; but the reader with some mathematical background may be interested in seeing how Kepler's third law derives from the Newtonian theory. The following diagram and proof are from Eric Rogers' *Physics for the Inquiring Mind:*

To deduce Kepler's Third Law, Newton had merely to combine his *laws of motion* with his laws of universal *gravitation*. For elliptical orbits, calculus is needed to average the radius and to deal with the planet's varying speed, but the same law then follows:

For circular orbits, suppose a planet of mass m moves with speed v in a circle of radius R around a Sun of mass M. This motion requires an inward resultant force on the planet, mv^2/R, to produce its centripetal acceleration v^2/R (see page 321). Assume that gravitational attraction between sun and planet just provides this needed force. Then

$$G\frac{Mm}{d^2} \text{ must} = \frac{mv^2}{R}$$

and distance d between m and M = orbit-radius, R. But

$$v = \frac{\text{circumference}}{\text{time of revolution}} = \frac{2\pi R}{T}$$

where T is the time of one revolution.

$$\therefore G\,\frac{Mm}{R^2} = \frac{m\,(2\pi R/T)^2}{R}$$
$$\therefore G\,\frac{Mm}{R^2} = \frac{4\pi^2\,mR^2}{T^2\,R}$$

To look for Kepler's Law III, collect all R's and T's on one side: move everything else to the other.

$$\therefore \frac{R^3}{T^2} = \frac{GM}{4\pi^2}$$

Now change to another planet, with different orbit radius R' and time of revolution T'; then the new value of $(R')^3/(T')^2$ will again be $GM/4\pi^2$; and this has the same value for all such planets since G is a universal constant and M is the mass of the sun, which is the same whatever the planet. Thus R^3/T^2 should be the same for all planets owned by the sun, in agreement with Kepler's Third Law. For another system, such as Jupiter's moons, M will be different (this time the mass of Jupiter) and R^3/T^2 will have a different value, the same for all the moons.

The planet's mass, m, cancels out. Several planets of different masses could all pursue the same orbit with the same motion. . . . Calculus predicts Law III for elliptical orbits too, where R is now the average of the planet's greatest and least distances from the sun.

These relationships make it simple to calculate the speed with which satellites will circumnavigate the earth if the average distance (R) of the satellite from the center of the earth is known, by substituting the value for G and mass of the earth in the expression $\dfrac{GM}{4\pi^2}$. The relationship is independent of the mass of the satellite and for a given average radius there is only one speed that will result in a stable orbit.

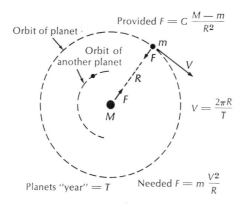

Orbit of planet

Provided $F = G\,\dfrac{M - m}{R2}$

Orbit of another planet

R

m

F

V

F

M

$V = \dfrac{2\pi R}{T}$

Planets "year" $= T$ Needed $F = m\,\dfrac{V2}{R}$

ozone (O_3) A form of oxygen, containing three atoms instead of two in a molecule. It is produced when an electric spark is passed through air. It is very active chemically and a powerful oxidizing agent.

parabola One of the families of conic sections and defined as the locus of a point that moves so as to remain equidistant from a fixed point and a fixed

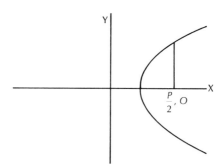

line. The standard equation in Cartesian *coordinates* is $y^2 = 2px$. See *conic sections*.

paradox A statement or proposition seemingly self-contradictory or absurd and yet explicable as expressing a truth. A famous example of a paradox is the one proposed by Zeno, concerning the race between Achilles and the tortoise. If Achilles, being a very fast runner, allows the tortoise to start first, then when Achilles starts at *A* the tortoise is at *B*. When Achilles reaches *B* the tortoise has moved on a small distance to *C* and when Achilles reaches *C*, the tortoise has progressed an even smaller distance to *D*. As this description can go on and on, apparently Achilles never catches the tortoise, but actually we know that he will.

parallax Difference in direction or shift in the apparent position of a body, due to a change in position of the observer. As the earth revolves around the sun, we see the nearby stars move against the so-called fixed stars much farther away. This periodic shift called the parallactic shift is very small in angular measure, the largest known parallactic shift being less than 1 second of arc (see *degrees*). The amount of parallactic shift is inversely proportional to the distance of the object. See footnote 4, page 102.

parsec Astronomical unit of distance, corresponding to a *parallax* of one second of arc, 19×10^{12} miles, 3.3 *light-years*.

periastron See *perihelion*.

perigee See *perihelion*.

perihelion The perihelion point of a planetary orbit is defined as the point of closest approach to the sun, helios being the word for sun. This point is also an *apse* point on the major axis of the elliptical orbit. A similar term is used in connection with work in double stars but is called periastron. The point of the moon's closest approach to the earth is called perigee. See Figure 5-10.

photon A packet or *quantum* of electromagnetic energy. The energy of a photon is hv, where h is *Planck's constant* and v is the *frequency* associated with the photon. See pages 292 to 295.

Planck's constant A very small constant that is denoted by the letter h and expressed in units of action ($h = 6.624 \times 10^{-27}$ erg sec). When multiplied by the frequency of radiation v, it gives the quantity of energy ($= hv$) contained in one *quantum*. See page 293.

plane A flat surface mathematically defined as a surface containing all the straight lines passing through a fixed point and also intersecting a straight line in space.

planetoids Known also as asteroids or minor planets, they occupy a region between Mars and Jupiter at 2.8 *astronomical units* from the sun. They vary in diameter from 500 miles to a fraction of a mile. They number in the tens of thousands, and it is believed that they may be fragments of a planet that disintegrated or one that never completely formed. See page 463.

population, stellar types I and II See pages 479 and 481.

positron The positive *electron,* an elementary particle having the same mass as the electron and a positive charge numerically equal to the negative charge of the electron.

postulate An idea assumed without proof as a basis for reasoning. See page 22.

precession of equinoxes The equinoxes are the points or nodes where the projection of the earth's orbit intersects the *ecliptic,* one point being called the vernal equinox and the other the autumnal equinox. Owing to a phenomena called precession (the same type of motion that a dying top exhibits) these two points do not remain fixed but move very slowly to the westward at such a rate that they will complete one revolution of the ecliptic in approximately 25,000 years. This westward motion of these points of intersection is called the precession of the equinoxes. See footnote 5, page 77 and page 158.

premise See *syllogism.*

pressure The *force* per unit area acting on a surface. For instance, the pressure of the earth's atmosphere at sea level is about 15 pounds per square inch. See page 45.

probability The likelihood that some particular form of event might occur. This likelihood is expressed in numerical form by the numbers between 0 and 1, a probability of 1 being certain that some event would occur and the probability 0 being certain that the event would not occur. The probability 0.5 means an equal likelihood that the event would or would not occur. This number is obtained by evaluating the ratio R/N, R being the number of elements in the set of events to take place and N being the total possibilities. For example, what is the probability of picking a spade from a deck of cards? The total possibilities are 52 of which 13 are spades, therefore the probability is $13/52$ or 0.25.

propellant A liquid or solid substance burned in a rocket for the purpose of developing a thrust. The velocity (relative to the nozzle) of the gases that exhaust through the nozzle of a rocket engine is called the exhaust velocity.

proportion A statement of equality between two ratios as $a/b = c/d$. One *variable* may be said to be proportional to another variable if they both increase at the same rate as, for instance, a is twice as large as b, or $a = 2b$. This could also be written $a/b = 2/1$, giving it the same form as the ratio above. Sometimes the notation $a \propto b$ is used, meaning a is proportional to b.

protein Class of organic compounds of very high molecular weights that compose a large part of all living matter. Protein molecules invariably contain the elements carbon, hydrogen, oxygen, and nitrogen; often also sulphur and sometimes phosphorus.

proton Positively charged particle, having a mass approximately 1840 times greater than that of the *electron* (i.e. 1.00757 atomic mass units) and charge numerically equal but opposite in sign to that of the electron. Constituent of all atomic nuclei.

proton-proton chain An energy-releasing nuclear reaction chain that is believed to be of major importance in energy production in hydrogen-rich stars. The net effect of the proton-proton, or H-H, chain is the conversion (fusion) of four protons into one He⁴ (helium) nucleus with the release of about 26 Mev of energy.

protoplasm A highly complex substance containing *protein*-like materials; essential constituent of all living cells.

pulsars Astronomical bodies that emit very regular and rapid pulsed radio signals. These bodies are believed to be small and extremely massive, possibly *neutron* stars. See page 514.

qualitative Dealing only with the nature, and not the amounts, of the substances under consideration. See page 56.

quantitative Dealing with quantities; measurements that relate the quantities of objects or phenomena under consideration.

quantum A discrete quantity of radiation (also called a *photon*). A quantum has a particle-like behavior, and an energy equal to $h\upsilon$, where h is *Planck's constant* and υ is the frequency of the radiation. See pages 292 to 295.

quasars "Quasi-stellar" sources are strong radio sources characterized by very high *red shifts*. These red shifts interpreted as distance measurements, imply that the quasars lie at great distances from us and must be extremely luminous and yet occupy a relatively small volume of space. See article starting on page 437.

radar Pulses of radio waves of very short wavelengths are transmitted from a unit, and this unit then listens for the echo that is reflected by an object intercepting the transmitted wave. The time lapse between transmission of a pulse and receipt of the echo indicates the range of the reflecting surface. This method is now being used to find the distances to planets and the sun.

radio telescope A device for receiving waves of radio frequencies emitted or reflected from astronomical sources. The wavelengths received are much longer than lightwaves, and this fact makes it difficult to design a radio telescope of high resolution. For instance, a 50-foot-diameter radio telescope for use with wavelengths around 20 centimeters has a resolving power of about 1°. See page 400, and Figure 8-12.

radioactivity The spontaneous disintegration of unstable atomic nuclei to give more stable product nuclei, usually accompanied by the emission of charged particles and electromagnetic energy, e.g. *alpha* or *beta particles* and *gamma rays*.

rainbow A bow or arc of prismatic colors that are caused by *refraction* and internal reflection of the sun's rays from small water droplets in the atmosphere. To see the rainbow the observer must have his back to the sun.

red shift See *Doppler shift*.

refraction Refraction of waves is the change in direction of propagation of any wave phenomenon that occurs when the wave velocity changes. The term is most frequently applied to visible light, but it is also applied to all other electromagnetic waves, as well as to sound and water waves. The physical nature of the effect can be visualized by considering orderly rows of automobiles driving in a column across a boundary between smooth concrete and heavy snow. If they drive perpendicular to the boundary, the rows are simply slowed up and crowded close to one another when they cross it to the snow; but if they drive at an angle, one end of each row is slowed down sooner than the opposite end, and the row swings around to a direction nearer to the perpendicular. A train of waves is affected in the same manner in passing from one medium to another. The simple mathematical relation governing refraction is known as Snell's law. Thus, if θ_1 is the angle of incidence and θ_2 the angle of refraction, Snell's law states that sin $\theta_1 = n$ sin θ_2, where n is the ratio of the indexes of refraction of the two mediums. (The index of refraction of a medium is the ratio of the speed of waves in a vacuum to their speed in the medium.) Refraction can and does occur in a single medium if its properties vary from one place to another because of changes in conditions through the portion of the medium traversed by the waves. The twinkling of stars is caused by variations in the atmosphere resulting from temperature differentials.

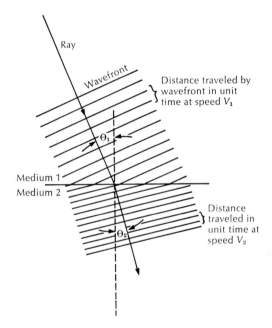

relativity A theory of the physical meaning of space and time due largely to Albert Einstein. In its present form, the theory consists of two parts: (1) the special, or restricted, theory (1905), which explains why the laws of nature appear the same to all observers moving with constant velocity relative to one another; and (2) the general theory (1916), which is the relativistic theory of gravitation and an extension of the special theory. Einstein's ideas have been of the greatest significance in the clarification of the foundations of theoretical physics, in addition to their direct contribution to experimental physics.

    special theory The special theory was based on the hypothesis that the velocity of light is constant when measured by one of a set of observers moving with constant relative velocity (see *transformation*). It resulted in two important modifications of the concepts of space and time that arose from *Newton's laws*. These modifications lead to the conclusion (1) that clocks in rapidly moving systems slow down and (2) that a rapidly moving body contracts in the direction of motion (see page 351 and *Fitzgerald-Lorentz contraction*). A third important consequence of this theory is indicated in the equation $E = mc^2$, which says that *energy* and *mass* are directly related to each other in proportion to the square of the speed of light.

general theory The general theory of relativity rests on the so-called principle of equivalence, which asserts the equivalence of gravitational forces with inertial forces, that is, the forces experienced by accelerated observers, such as persons in a rolling ship or in an automobile coming to a sudden stop. In addition, one postulates a generalized relativity principle, that the equations of mechanics have the same form for all observers, whether accelerated or not. Capitalizing on the principle of equivalence, the general theory modifies further the space and time concepts of special relativity and arrives at the notion of the curved space-time *continuum*.

retrogressions of the planets As the earth moves faster in its orbit than the outer planets, the earth "catches up" to these planets, and as it passes them they appear to move backward. This apparent backward motion is called the retrograde motion of the planets. See Figure 8-4.

right ascension The astronomical measurement of position measured eastward from the vernal equinox. The value is given in hours, minutes, seconds, which of course can be translated to degrees and may be likened to the measure of *longitude* on the earth. See *degrees, minutes, and seconds of arc.*

spectrograph An instrument by which spectra may be photographed; also, a photograph taken by such an instrument.

spectroscope An optical instrument that separates composite *light* into its components, thus producing a spectrum for visual observation to determine what wavelengths are present. Its essential elements are a slit S, a collimating lens C, for making the light from the slit parallel before entering the prism, one or more dispersing prisms P, and a telescope T, for forming images of the slit in the various wavelengths and so providing a method for viewing or photographing the spectrum. Modern spectroscopes employ the *diffraction grating* instead of the prism; the name spectroscope is also commonly applied to instruments that can separate other electromagnetic radiation than visible light into its component wavelengths.

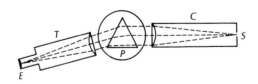

stations of the planets These stations or stationary points are the points at which a planet appears to be motionless when viewed from the earth, owing to the combined motions of the earth and the planet around the sun. See Figure 8-4.

stroboscopic Stroboscopic methods are used for viewing moving objects in such a way that they appear stationary, by using flashes of illumination of very short duration. The object being observed moves a negligible distance while the light is on and thus appears sharply defined. See Figures 4-16 and 4-18.

sun spots Huge vortexlike storms on the sun. They show very strong magnetic fields and appear dark when seen in a photograph because they are much cooler than the other areas of the sun's surface. See page 402.

supernova See *nova.*

syllogism The species of reasoning in which two statements are taken as true, making a third statement inevitably valid. The first two statements are termed premises. The third statement is called the conclusion. In the most common type of syllogism, the first premise (major) usually asserts something of a class or category, for example, "All men are mortal." The second premise (minor) proposes a certain relationship between the class and a particular entity, for example, "Socrates is a man." The conclusion asserts that "Therefore, Socrates is mortal." See pages 17 and 23.

telemeter Apparatus for recording physical events at a distance by radioing instrument readings from a rocket to a recording machine on the ground.

telescope An optical instrument that increases the visibility of distant objects. The simplest type of telescope consists of an objective lens, *O*, which forms a real image of the object, and an ocular or eyepiece *E* for magnifying and viewing this image. A telescope having an objective lens is called a refracting telescope. The objective lens can be replaced by a mirror or by a system containing refracting and reflecting elements, such as a *concave* parabolic mirror; such a telescope is a reflective telescope. The telescope used by Galileo was a refracting telescope. A converging or convex lens *O*, the objective lens, forms at its principal focus a real inverted image *R* of a distant object. Before reaching the focal plane of the objective, the converging rays are intercepted by a diverging or concave lens *E*. If the eye is placed behind *E*, an enlarged virtual image of the object is observed. Telescope builders of the seventeenth and eighteenth centuries were unable to obtain glass disks of sufficient clearness and homogeneity for the construction of large object lenses, and turned to the reflecting type of telescope. The largest reflecting telescope ever built is the 200-inch Hale Telescope at Mount Palomar, California, while the telescope at Yerkes Observatory—with a 40-inch objective lens—is the largest refractive telescope ever built.

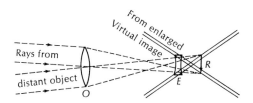

thermal neutrons *Neutrons* of very slow speed and consequently of low energy. Their energy is of the same order as the thermal energy of the atoms or molecules of the substance through which they are passing. Thermal neutrons are responsible for numerous types of nuclear reactions, including nuclear fission.

thermodynamics First law: see *conservation laws.* Second law: Heat *energy* cannot of itself pass from a body of low temperature to one of high. In an isolated system containing energy in concentrated form the transformations that occur always involve the passage to a less concentrated or more disorganized form. From this law very broad inferences have been drawn. It is contended that if this process of disorganization proceeds without hindrance it will result in a stage of uniform distribution of energy in the universe and all energy interchange will cease. See page 369.

thermonuclear A nuclear reaction that takes place only at very high temperatures.

Thor A single-stage intermediate-range ballistic missile of 150,000-pound thrust. Also used as the first stage of a series of launch vehicles for space missions such as the Thor-Able, Thor-Agena, and Thor-Delta series. Among the space vehicles launched by the Thor series are Transit, Tiros, Echo, Explorers VI and X, and the Discoverer series.

Titan Intercontinental ballistic missile (ICBM) having a maximum speed of over 15,000 miles per hour and a range over 6,000 miles. It uses liquid-*propellant* rockets in two stages (two 150,000-pound thrust engines in the first stage and one 80,000-pound thrust engine in the second stage).

trajectory The curve that a body describes in space when acted upon by two or more *forces*, as the path of a planet or a projectile. See Figure 4-17 and pages 189–90.

transformation A change of variables in an algebraic expression, as in changing from one *coordinate* system to another. Suppose that we have two coordinate systems K and K' as indicated in the accompanying figure. K' has a velocity of v to the right relative to K. Any event can be measured in space with respect to K by the three coordinates x, y, z and with regard to time by t. Relative to K', the same event would be measured by the corresponding values x', y', z', t', which, of course, are not identical to x, y, z, t. If the magnitudes x, y, z, t of an event are known, how can the magnitude of x', y', z', t' for the same event with respect to K' be found? Transformation equations establish the relationships for the solution of this problem. If the origins of the two frames are coincident at $t = 0$ and $t' = 0$, then, using the principle of the addition of velocities, the Galilei transformation sets up the relationships in the following manner:

$$x' = x - vt$$
$$y' = y$$
$$z' = z$$
$$t' = t$$

If, on the other hand, we use Einstein's condition that the speed of light in vacuo must be constant (c) as measured in both coordinate systems, the relationships below are found to apply:

$$x' = \frac{x - vt}{\sqrt{1 - v^2/c^2}}$$
$$y' = y$$
$$z' = z$$
$$t' = \frac{t - xv/c^2}{\sqrt{1 - v^2/c^2}}$$

This system of equations is known as the Lorentz transformation. These equations reduce to the Galilei transformation above when $v/c \to 0$. The Lorentz transformation equations form the basis of Einstein's special theory of relativity. See pages 351–52. For a simple algebraic derivation of the Lorentz transformations see Einstein's *Relativity: The Special and General Theory,* Appendix I.

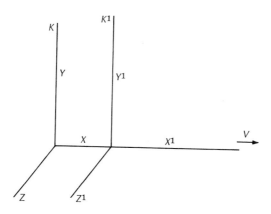

uncertainty principle Sometimes called indeterminacy principle. This principle, first stated by Heisenberg, has the mathematical formulation.

$$\Delta p \cdot \Delta q \geqq h$$

where Δq is the range of values that might be found for the coordinate q of a particle, and Δp is the range in the simultaneous determination of the corresponding component of its *momentum p*. The product of these uncertainties will always be greater than or equal to h, which is *Planck's constant,* owing to the fact that every observation is an interaction between the system and the measuring device, which will itself alter the state. See pages 298–99.

V-2 Also called A-4. A liquid-oxygen–alcohol–fuelled single-stage ballistic missile developed by the Germans at Peenemünde during World War II as a long-range artillery device. The finished missile was 47 feet long, 5½ feet in diameter, and had a gross weight at take-off of 14 tons. The motor developed 56,000 pounds of thrust and carried a 1-ton warhead 150 miles. The United States captured some 70 of these missiles, which were used for upper atmosphere research.

vacuum A perfect vacuum would be a space entirely without matter of any sort. To our knowledge there does not exist a perfect vacuum anywhere in nature. Even the space between the stars is occupied by some matter, very thin to be sure, but still not empty. The term "vacuum" is generally taken to mean a space containing air or other gas at very low pressure. See page 42.

variable A symbol to which any number of values in a given set may be assigned. See page 56.

vector Any physical quantity that requires a statement of direction as well as magnitude for its complete definition, for example, *velocity*. A vector is often indicated graphically by means of an arrow. The length of the arrow is proportional to the scalar magnitude of the vector, and the direction in which the arrow points is the direction of the vector.

velocity Velocity is the rate of change of distance with respect to time. Average velocity can be found by dividing the distance covered by the time interval: d/t. Velocity is a *vector* quantity having both magnitude and direction in which the motion takes place. See page 190.

Viking A liquid-fuelled high-altitude sounding rocket developed in 1949. The original series of vehicles was 48.6 feet long and 32 inches in diameter. It weighed 10,700 pounds and developed 20,000 pounds of thrust. Rockets of this series carried 500 pounds to an altitude of 136 miles.

violet shift See *Doppler shift.*

virus An infective agent smaller than the common microorganisms and requiring living cells for multiplication. See page 466.

Wac Corporal A liquid-*propellant* sounding rocket developed by the Army Ordnance Department capable of launching a 25-pound payload to an altitude of 19 miles. The missile was 16 feet long, and the engine delivered about 1,500 pounds of thrust for 45 seconds. It was first flight-tested in 1945.

waves The propagation of a periodic disturbance carrying *energy*. At any point along the path of a wave motion, a periodic displacement or vibration about a mean position takes place. This may take the form of a displacement of air molecules (e.g. sound waves in air), of water molecules (waves on water), a displacement of elements of a string or wire, displacement of electric and magnetic *vectors*, etc. The locus of these displacements at any instant is called the wave. Wavelength is the distance in convenient units from crest to crest or from any one place on a wave to the same place on the next wave. The amplitude is the maximum departure from the equilibrium value. Waves in which the vibration or displacement takes place in the direction of propagation of the waves are called longitudinal waves. Sound is an example of longitudinal wave motion. Waves in which the vibration or displacement takes place in a plane at right angles to the direction of propagation are called transverse waves, e.g. electromagnetic waves. See pages 278 to 283.

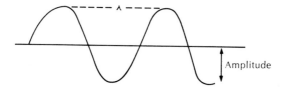

weight The *force* of attraction of the earth on a given mass is the weight of that mass. See also *mass* and page 324.

weightlessness Strictly speaking, the absence of any *force* of gravity acting on a given body. However the term is generally used in a broader sense to mean the absence of sensations of resistance to gravity. The human body does not actually feel gravity. It feels the resistance to gravity—the pressure of the feet on the floor or pavement that prevents the body from following the pull of gravity. The feeling of weightlessness, or zero-G as it is sometimes called, can occur in several different types of situations: (1) in cases in which there is actually no resultant force of gravity acting on the body, as, for example, when the body is so far from the earth that the earth's gravitational field is just balanced by the sun's field and (2) in cases in which the body is being accelerated by the force of gravity but in which, because everything around it is also being accelerated at the same rate, there is no sensation of resistance to the attraction of gravity. This second type is the situation encountered in satellites as well as in free fall. See also the discussion on pages 588–89.

X rays Electromagnetic radiation extending from the extreme ultraviolet into the gamma-ray region, that is, from 10^{-7} to 10^{-9} centimeter wavelength.

zero-G See *weightlessness* and pages 588–589.

zodiac As the earth moves around the sun, the sun appears to move through the background stars during the period of one year. The twelve different constellations through which the sun apparently moves in the course of one year are called the zodiac and indicate the position of the sun for each month of the year. It must be understood that this apparent motion of the sun is really the reflected motion of the earth in its orbit around the sun. The zodiacal constellations are Capricornus, Aquarius, Pisces, Aries, Taurus, Gemini, Cancer, Leo, Virgo, Libra, Scorpius, Sagittarius. See page 83 and Figure 3-10.

zodiacal light A faint glow extending along the zodiac, probably due to sunlight reflected from a great number of tiny particles in or near the plane of the *ecliptic*. This glow extends around the entire *zodiac*, but it shows most prominently in the neighborhood of the sun. It can best be observed in the western sky in the spring just after the sunset twilight has completely disappeared, or in the eastern sky in the fall just before the morning twilight appears. See page 459.